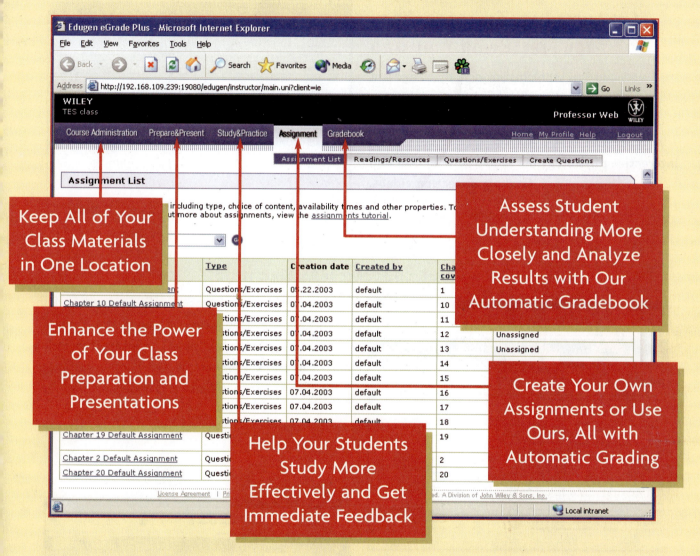

Students,
eGrade Plus Allows You to:

Study More Effectively

Get Immediate Feedback When You Practice on Your Own

eGrade Plus problems link directly to relevant sections of the **electronic book content,** so that you can review the text while you study and complete homework online. Additional resources include **interactive physiology lab simulations** and **cat and fetal pig dissection videos.**

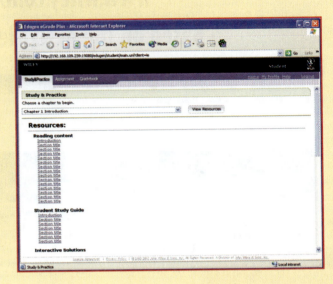

Complete Assignments / Get Context-Sensitive Help

An **Assignment** area keeps all your assigned work in one location, making it easy for you to stay "on task." You will have access to a variety of **interactive in-lab activities and lab reports,** as well as other resources for building your confidence and understanding.

Keep Track of How You're Doing

A **Personal Gradebook** allows you to view your results from past assignments at any time.

Laboratory Manual for Anatomy and Physiology

Second Edition

CONNIE ALLEN

VALERIE HARPER

Edison College

John Wiley & Sons, Inc.

SENIOR EXECUTIVE EDITOR	Bonnie Roesch
PROJECT EDITOR	Mary O'Sullivan
EXECUTIVE MARKETING MANAGER	Clay Stone
PRODUCTION MANAGER	Pamela Kennedy
PRODUCTION EDITOR	Sarah Wolfman-Robichaud
SENIOR DESIGNER	Kevin Murphy
SENIOR ILLUSTRATION EDITOR	Anna Melhorn
PHOTO MANAGER	Hilary Newman
PHOTO EDITOR	Elle Wagner
SENIOR MEDIA EDITOR	Linda Muriello
MEDIA PROJECT EDITOR	Steve Chasey
BOOK DESIGN	Nancy B. Field
COVER DESIGN	Norm Christiansen
COVER ART	Heart figure/H.R. Muinos Chart figure/TechBooks
COPYEDITOR	Kathy Drasky
PROJECT MANAGER	Jane Shifflet/GTS Companies: PA/York Campus

This book was typeset in 10/12 Times Roman by GTS Companies and printed and bound by Von Hoffmann Press. The cover was printed by Von Hoffmann Press.

The paper in this book was manufactured by a mill whose forest management programs include sustained yield harvesting of its timberlands. Sustained yield harvesting principles ensure that the number of trees cut each year does not exceed the amount of new growth.

The procedures in this text are intended for use only by students with appropriate faculty supervision. In preparing the text, care has been taken to identify potentially hazardous steps and to insert safety precautions where appropriate. The authors and publisher believe the procedures to be useful tools if performed with the materials and equipment specified, in careful accordance with the instructions and methods in the text. However, these procedures must be conducted at one's own risk. The author and publisher do not warrant or guarantee the safety of individuals using these procedures and specifically disclaim any and all liability resulting directly or indirectly from the use or application of any information contained in this book.

This book is printed on acid free paper. ∞

To order books or for customer service please, call 1-800-CALL WILEY (225-5945).

ISBN 978-0-471-69122-4, 0-471-69122-4

Printed in the United States of America

10 9 8 7 6 5 4 3 06

I dedicate this book to our second grandchild, Jaxton James Allen.
—CONNIE ALLEN

I dedicate this book to my children, Scott and Kate, who have taught me so much, and to my husband Chuck who has supported me emotionally, made dinner, done laundry, and helped with the children.
—VALERIE HARPER

Preface

Anatomy and physiology is a challenging course, and this laboratory manual is written to help students meet that challenge. It is written for students interested in allied health fields, such as nursing; physical, respiratory, or occupational therapy; radiology; and dental hygiene. Although this manual incorporates many illustrations from *Principles in Anatomy and Physiology, 11th Edition,* by Gerard J. Tortora and Bryan Derrickson, it may be used with any two-semester anatomy and physiology textbook.

The design of this laboratory manual is based on the authors' experience as anatomy and physiology instructors and uses three learning styles: visual, auditory, and kinesthetic. When students label diagrams, they focus on the structure rather than just the dot at the end of a line. Writing out the structure's name and pronouncing it reinforces learning. Also, having students become subjects of laboratory exercises makes learning personal. Animal dissections give students an opportunity to physically manipulate structures, comparing location and texture, and to observe how structures are supported, protected, and attached by connective tissue.

Special features incorporated in this laboratory manual include:

- Just enough text is provided to introduce concepts in each section and to set up and support the laboratory section. The exercises are written so that students do not need their textbooks to complete the laboratory activities.
- New material is divided into small segments, starting with simple diagrams, illustrating the basic concepts and building up to more complex diagrams. Subsequent activities add to the students' knowledge in a stepwise fashion.
- Unlabeled four-color drawings, photographs, and photomicrographs are included for students to label either at home or in the laboratory. Students first write out the name of the structure to help memorize it. Then the completed diagrams will be used to identify structures on models.

- Word derivatives for bolded terms are given in the text.
- Phonetic pronunciation is included as new words are introduced. Students are directed to pronounce the name of a structure as they point to it on a model. Most students have difficulty pronouncing the anatomical terms, and need practice saying and hearing the words.
- Physiology experiments use students as subjects and can be completed with either simple, inexpensive equipment or more complex setups.
- Experimental report sections after physiology experiments where students are asked to make predictions, collect and analyze data, and write simple lab reports.
- Discussion Questions are within the activities to make the students think about the material presented.
- An Answer Key is provided at the end of the laboratory manual for the activities in each exercise. Students receive immediate feedback, and they are not dependent on the instructor for the correct answers.
- "Reviewing Your Knowledge" and "Using Your Knowledge" sections follow the activities at the end of each exercise. "Reviewing Your Knowledge" provides a thorough review of the material in the exercise, whereas "Using Your Knowledge" requires students to apply information learned. Either or both of these sections may be handed in to the instructors for a grade, because neither section has answers in the back of the laboratory manual. Answers to these two sections are in the online Instructor's Manual.

- Each exercise contains a list of objectives, materials needed for the exercise, and easily identifiable laboratory activity sections.

New to this Edition:

Helping Teachers Teach and Students Learn
www.wiley.com/college/allen

This title is available with eGrade Plus, a powerful online tool that provides instructors and students with an integrated suite of teaching and learning resources in one easy-to-use website. eGrade Plus is organized around the essential activities you and your students perform in class:

For Instructors

- **Prepare & Present:** Create class presentations using a wealth of Wiley-provided resources—such as an online version of the lab manual, images from the Anatomy & Physiology Visual Library, and interactive animations—making your preparation time more efficient. You may easily adapt, customize, and add to this content to meet the needs of your course.
- **Create Assignments:** Automate the assigning and grading of lab reports by using Wiley-provided question banks, or by writing your own. Student results will be automatically graded and recorded in your gradebook.
- **Track Student Progress:** Keep track of your students' progress via an instructor's gradebook, which allows you to analyze individual and overall class results to determine their progress and level of understanding.
- **Administer Your Course:** eGrade Plus can easily be integrated with another course management system, gradebook, or other resources you are using in your class, providing you with the flexibility to build your course, your way.

For Students

Wiley's eGrade Plus provides immediate feedback on student assignments and a wealth of support materials. This powerful study tool will help your students develop their conceptual understanding of the class material and increase their ability to answer questions.

- **A "Study and Practice"** area links directly to text content, allowing students to review the text while they study and answer questions. Additional resources can include interactive animations, lab reports and in-lab activities.
- **An "Assignment"** area keeps all the work you want your students to complete in one location, making it easy for them to stay "on task." Students will have access to a variety of interactive self-assessment tools, as well as other resources for building their confidence and understanding.
- **A Personal Gradebook** for each student will allow students to view their results from past assignments at any time.

Please view our online demo at www.wiley.com/college/egradeplus. Here you will find additional information about the features and benefits of eGrade Plus, how to request a "test drive" of eGrade Plus for this title, and how to adopt it for class use.

- New Exercises on Surface Anatomy (15), Human Development (40), and Heredity (41)
- Experimental Reports included in Activities devoted to physiological experiments
- PowerPhys CD-ROM
- Histology Atlas
- Word Roots Table
- Skeletal Muscle Origin, Insertion, and Action Table

Supplemental Materials:

- Instructor's Manual. Available on the book companion Website at www.wiley.com/college/allen. The Instructors' Companion Site contains Answer Keys for "Reviewing Your Knowledge" and "Using Your Knowledge," Interactions Review Sheets, and all Tables and Figures from the manual.
- Cat Dissection Manual, 2e. Contains dissection activities for use in the lab accompanied by full color photos and figures.
- Fetal Pig Dissection Manual, 2e. Contains dissection activities for use in the lab accompanied by full color photos and figures.
- Anatomy Drill and Practice exercises and other student exercises and resources are located at www.wiley.com/college/allen.

Acknowledgments

We deeply appreciate the support, instruction, and encouragement from the members of our editorial, production, and marketing team at Wiley: Bonnie Roesch, Mary

O'Sullivan, Hilary Newman, Anna Melhorn, Sarah Wolfman-Robichaud, Kevin Murphy, and Clay Stone. A special thanks goes to Mary O'Sullivan for all the encouragement, direction, and suggestions she gave us. Her steady, calm hand guided us through this whole project and we could not have done it without her. We also wish to thank Gerard Tortora and Bryan Derrickson for producing a wonderful textbook that provided many illustrations and ideas for our laboratory manual.

A special thank you to William Benyak for reviewing all the exercises, making suggestions, giving clinical guidance, and for photographing the fetal pig dissection. We also wish to thank Donalie Benyak for her technical assistance with the dissection photography. A special thanks to Charles Harper for answering many clinical questions.

Many thanks to Michael Ross and his assistant Todd Barnash from the University of Florida for answering our histology questions and providing photomicrographs for the laboratory manual. We also wish to thank Kierstan Hong at Imagineering Art for the artwork she provided for our laboratory manual.

Thank you to our colleagues at Edison College: Paul Monigan, Bob Clemence, Sue Pilbeam, Colleen Swanson, Jody Gootkin, Richard McCoy, Jeff Davis, Florence Levin, John Berte, Dick Felden, Lyman O'Neil, Kitty Gronlund, Ken Laser, Greg Stancel, and Tony Contino who encouraged us, answered our questions, and provided critiques of exercises. We also wish to thank Sheila Stancel and James Newton for their encouragement.

Thank you to Chaim Jay Margolin of Regional Radiology Associates and David Michie of Clinical Physiology Associates for providing images for this manual.

Reviewers

A special thank you to all the reviewers. Their suggestions were extremely helpful in developing this laboratory manual. Reviewers included the following:

Susan Baxley, *Troy State University, Montgomery*

Moges Bizuneh, *Ivy Tech State College, Indianpolis*

Juville Dario-Becker, *Central Virginia Community College*

Edward R. Fliss, *St. Louis Community College at Florissant Valley*

Molly Lee, *Harrisburg Community College*

Janet Lichti, *Ivy Tech Sstate College, Lafayette*

Karen K. McLellan, *Indiana University-Purdue University Fort Wayne*

Amy Ouchley, *University of Lousiana, Monroe*

John Pellegrini, *College of St.Catherine*

Beth Scaglione Sewell, *Valprariso University*

Mary Kay White, *Brevard College*

Vernon L. Wiersema, *Houston Community College*

Patricia Yokell, *New Hampshire Technical Institute*

Contents

Anatomical Language

Students often complain that they are not just learning anatomy but are learning a new language. This is quite true since most anatomical terms are derived from Latin and Greek word roots. In this exercise, you will learn anatomical terms for body structures that will be used throughout the course and will help you communicate effectively with other health care professionals.

A. Anatomical Position

The anatomical position is the reference position anatomists and people in medical fields use to describe the location of body parts or regions. In the anatomical position, the body is erect and facing forward; the arms are straight and at the sides of the body with the palms facing forward; the legs are straight with the feet facing forward and flat (Figure 1.1).

ACTIVITY 1 — Anatomical Position

1 Assume the anatomical position.
2 Have your laboratory partner determine if you are in the correct position.

B. Body Regions

The main body regions are described in Table 1.1. It is important that you learn the correct boundaries for each region. Two common misconceptions are that the arm is the area between the shoulder and wrist and that the leg includes the thigh. Actually, the arm is located between the shoulder and elbow, and the forearm is located from the elbow to wrist. The thigh is located between the groin and knee, and the leg is located between the knee and ankle.

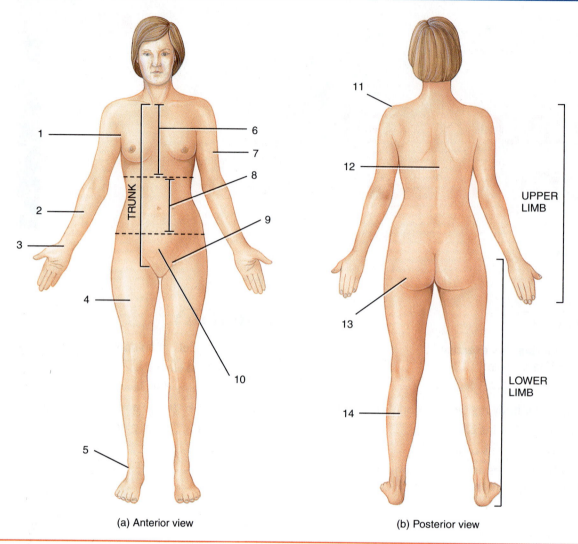

(a) Anterior view (b) Posterior view

FIGURE 1.1 Body regions.

(a) Anterior view

1 _Axillary (armpit)_

2 _ante brachial (forearm)_

3 _Carpal (wrist)_

4 _Femoral (thigh)_

5 _tarsal (ankle)_

6 _thorax (chest)_

7 _Brachial (arm)_

8 _abdomen_

9 _Inguinal (groin)_

10 _pubis (pubic)_

(b) Posterior view

11 _acromial (shoulder)_

12 _Spinal column (back)_

13 _Gluteal (buttock_

14 _sural (leg)_

TABLE 1.1 Body Regions

BODY REGION	DESCRIPTION
Head	
Skull	Bony portion of head
Face	Portion of skull not normally covered by scalp
Neck	Body area between head and trunk
Trunk	Central body area to which head and limbs are attached
Chest	Area of trunk between neck and abdomen; contains heart and lungs; diaphragm forms boundary between chest and abdomen
Abdomen	Area of trunk between chest and pelvis; contains digestive organs; hip bones form lower boundary of abdomen
Pelvis	Area of trunk below abdomen; contains internal reproductive organs and urinary bladder
Back	Posterior portion of trunk between neck and buttocks
Upper Limb	
Shoulder	Curved area where arm attaches to upper border of trunk
Armpit	Under-arm area where upper limb attaches to trunk
Arm	Area of upper limb between shoulder and elbow
Forearm	Area of upper limb between elbow and wrist
Wrist	Portion of hand that connects hand to forearm
Hand	Includes wrist and fingers
Lower Limb	
Buttocks	Rounded area on posterior surface where thigh attaches to trunk
Groin	Area where lower limb attaches to the pelvis
Thigh	Area of lower limb above the knee
Leg	Area of lower limb between knee and ankle
Ankle	Portion of foot that connects foot to leg
Foot	Includes ankle and toes

ACTIVITY 2 Body Regions

1 Label Figure 1.1 with the appropriate body regions. Use terms from Table 1.1.
2 Identify body regions on models or anatomical charts. Use terms from Table 1.1.

C. Anatomical Terms

Anatomical terms describe body regions, specific body areas, and landmarks. Most of these words are derived from Latin or Greek and are often part of the names of muscles, bones, nerves, and blood vessels. Learning these terms at this time will help you throughout the course.

Many anatomical terms have one or more word roots with a prefix and/or a suffix added. For example, in the word *antecubital, ante-* is a prefix meaning before or in front of; the word root *cubit-* means elbow; *-al* is a suffix meaning pertaining to. Table 1.2 contains anatomical terms with four different suffixes all of which mean pertaining to. These suffixes are: *-al, -ic, -ar,* and *-ary.* When suffixes like these are added to word roots, they form adjectives, whereas nouns have different endings, such as *-um, -us, -is,* and *-a.* For example, *stern-* is a word root meaning chest; *sternum* is the noun and *sternal* is the adjective. Anatomical terms and their definitions are found in Table 1.2. Word roots and their definitions are found in Appendix B, as well as nouns and adjectives formed from the word roots.

ACTIVITY 3 Anatomical Terms

1 Label Figure 1.2 with the appropriate anatomical terms for each body region or area. Refer to Table 1.2.
2 Use anatomical terms to identify the specific body regions or areas on models or anatomical charts.
3 Pronounce the anatomical terms as you point to them on models or anatomical charts.
4 Print out the glossary of prefixes, suffixes, and word roots from the book's companion web site (www. wiley.com/college/allen) for use as a reference.

TABLE 1.2 Anatomical Terms

TERM	DEFINITION	TERM	DEFINITION
AXIAL	Pertaining to the central part of the body, the head and trunk	**APPENDICULAR**	Pertaining to the extremities or limbs
Cephalic (se-FAL-ik)	Pertaining to the head	**Upper Limb (Appendage)**	
• Facial	Pertaining to the face	• Acromial (a-KROM-ee-al)	Pertaining to the highest point of the shoulder
• Frontal	Pertaining to the forehead		
• Orbital	Pertaining to the eye	• Axillary (AX-il-ary)	Pertaining to the armpit
• Otic (OH-tik)	Pertaining to the ear	• Brachial (BRAY-key-ul)	Pertaining to the arm
• Nasal	Pertaining to the nose	• Cubital	Pertaining to the elbow
• Buccal (BUCK-al)	Pertaining to the cheek	• Antecubital (an-teh-KYOO-bi-tul)	Pertaining to the anterior (front) surface of the elbow
• Oral	Pertaining to the mouth		
• Mental	Pertaining to the chin	• Olecranal (oh-LEK-ra-nul)	Pertaining to the posterior (back) surface of the elbow
• Cranial	Pertaining to the portion of the skull surrounding the brain	• Antebrachial	Pertaining to the forearm
		• Carpal	Pertaining to the wrist
• Occipital (ox-SIP-i-tal)	Pertaining to the back of head	• Manual	Pertaining to the hand
Cervical	Pertaining to the neck	• Palmar	Pertaining to the palm of the hand
Thoracic	Pertaining to the chest		
• Pectoral (PEK-tore-al)	Pertaining to the breast	• Digital	Pertaining to the digits (fingers)
• Sternal	Pertaining to the breast bone		
Abdominal	Pertaining to the abdomen	**Lower Limb (Appendage)**	
• Umbilical (um-BIL-ih-cal)	Pertaining to the navel	• Inguinal (ING-won-ul)	Pertaining to the groin where the thigh attaches to the pelvis
• Coxal (COX-al)	Pertaining to the hip		
Pelvic	Pertaining to the pelvis	• Gluteal (GLUE-tee-ul)	Pertaining to the buttocks
• Pubic (PYOO-bik)	Pertaining to the genital area	• Femoral (FEM-or-ul)	Pertaining to the thigh
Dorsal	Pertaining to the back	• Patellar (pa-TEL-ur)	Pertaining to the anterior (front) surface of the knee
• Scapular	Pertaining to the shoulder blade region	• Popliteal (pop-lih-TEE-ul)	Pertaining to the posterior (back) surface of the knee
• Vertebral (ver-TEE-brul)	Pertaining to the spinal column	• Crural (CROO-rul)	Pertaining to the anterior (front) surface of the leg
• Lumbar	Pertaining to the area of the back between the lowest rib and buttocks	• Sural (SIR-ul)	Pertaining to the posterior (back) surface of the leg
		• Fibular (FIB-you-lur) or peroneal (peh-RONE-ee-ul)	Pertaining to the lateral side of the leg
		• Tarsal (TAR-sul)	Pertaining to the ankle
		• Pedal	Pertaining to the foot
		• Calcaneal (kal-KANE-ee-ul)	Pertaining to the heel
		• Plantar	Pertaining to the sole of foot
		• Digital	Pertaining to the digits (toes)

(a) Anterior view

(b) Posterior view

FIGURE 1.2 Anatomical terms.

(a) Anterior view *(front)*

1. Cranial
2. facial
3. Cervical
4. acromial (shoulder)
5. axillary
6. brachial
7. antecubital (front of elbow)
8. antebrachial (forearm)
9. Carpal
10. Digital
11. palmar
12. femoral
13. patellar
14. Crural (leg)
15. tarsal
16. Digital
17. frontal (forehead)
18. Orbital (eye)
19. Otic (ear)
20. buccal (cheek)
21. Nasal (nose)
22. Oral (mouth)
23. Mental (chin)
24. sternal (breastbone)
25. pectoral (pertaining to the breast)
26. Umbilical (navel)
27. Coxal (hip)
28. Inguinal (groin)
29. Manual (hand)
30. pubic (pubis)
31. pedal (foot)

(b) Posterior view (back)

32. occipital (back of head)
33. scapular (shoulder blade)
34. vertebral (spinal column)
35. olecranal (back of elbow)
36. Gluteal (buttock)
37. popliteal (back of the knee)
38. sural (calf)
39. fibular (lateral side of the leg)
40. plantar (sole of foot)
41. Lumbar (loin)
42. Calcaneal (heel)

D. Directional Terms

Directional terms are used to describe the location of body structures relative to other structures. An example of a directional term is *inferior,* which means below. It would be correct to say that the neck is inferior to the head but incorrect to say that the neck is inferior. The directional terms are listed in Table 1.3, along with an example of how they are used. Note that opposite terms are paired.

The directional terms *proximal* and *distal* apply to the point of attachment of a limb to the torso or the point of origin of a structure such as a blood vessel or nerve. These terms refer to the location of structures relative to the point of attachment or point of origin, whether they are closer (proximal) or farther away (distal).

ACTIVITY 4 **Directional Terms**

1 Complete the sentences using the appropriate directional term from Table 1.3.

 a. The sternum is ___anterior___ to the vertebrae.

 b. The orbit is ___Superior___ to the oral cavity.

 c. The heart is ___Medial___ to the lungs.

 d. The carpus is ___distal___ to the brachium.

 e. The right lung and right kidney are ___ipSilateral___

 f. The skin is ___Superficial___ to the bones.

2 Label Figure 1.3 with directional terms.

FIGURE 1.3 Directional terms.

- anterior or ventral 1 ___Superior___
- distal 2 ___Posterior___
- inferior or caudal 3 ___anterior___
- posterior or dorsal 4 ___proximal___
- proximal 5 ___distal___
- superior or cephalic 6 ___inferior___

TABLE 1.3 **Directional Terms**

DIRECTIONAL TERM	DEFINITION	EXAMPLE OF USE
Superior (Cephalic)	Above	The head is superior to the neck.
Inferior (Caudal)	Below	The neck is inferior to the head.
Anterior (Ventral)	Closer to front of body	The lips are anterior to the teeth.
Posterior (Dorsal)	Closer to back of body	The teeth are posterior to the lips.
Medial	Closer to midline of body	The nose is medial to the eyes.
Lateral	Farther from midline of body	The eyes are lateral to the nose.
Intermediate	Between two structures	The elbow is intermediate between the shoulder and wrist.
Ipsilateral	On same side of body	The right arm and right leg are ipsilateral.
Contralateral	On opposite sides of body	The right arm and left arm are contralateral.
Proximal	Nearer to point of attachment of limb to trunk	The elbow is proximal to the wrist.
Distal	Farther from point of attachment of limb to trunk	The wrist is distal to the elbow.
Superficial	Closer to surface of body	The skin is superficial to the muscles.
Deep	Farther from surface of body	The muscles are deep to the skin.

E. Body Planes and Sections

Planes divide the body or organs in order to observe internal structures (Figure 1.4). The exposed surfaces produced by planes are called **sections**. Sagittal (*sagitta =* arrow) **planes** pass vertically through the body or organs and divide them into right and left sections (**sagittal sections**). If a plane passes vertically through the midline and divides the body into equal right and left halves, the plane is a **midsagittal plane,** but if a plane divides the body into unequal right and left portions it is a **parasagittal plane.** A **frontal** or **coronal plane** passes vertically through the body or organs and produces anterior and posterior sections (**frontal sections**). A **transverse plane** passes horizontally through the body and produces superior and inferior sections (**transverse sections** or **cross-sections**). **Oblique planes** pass through the body at an angle-forming oblique sections.

We often look at sections of individual organs, such as blood vessels, intestines, or long bones. Sections that are produced by a plane running along the long axis of a long narrow structure are called **longitudinal sections.** Sections that are produced by a plane running perpendicular to the long axis are called **cross-sections.** Because blood vessels and intestines twist and bend, one body plane may produce longitudinal sections, cross-sections, and oblique sections of these structures.

CLINICAL NOTE: *Transverse sections observed with computed tomography (CT) scans or magnetic resonance imaging (MRIs) are called* ***axial sections.***

ACTIVITY 5 Body Planes and Sections

1 Label the planes in Figures 1.4(a) and (b).
2 Observe sagittal, frontal, and transverse sections using an apple.
 • Working in a group, draw a face on the apple.
 • Cut sagittal, frontal, and transverse planes through the apple to make sagittal, frontal, and transverse sections.
 • Observe the appearance of the apple core in each section.
 • Keep sections together to form a whole apple to show to your instructor.

(a) Right anterolateral view

Longitudinal section Cross-section

(b) Longitudinal and cross-sections

FIGURE 1.4 Planes through the human body.
• frontal plane 1 _frontal plane_
• midsagittal plane 2 _Oblique plane_
• oblique plane 3 _transverse plane_
• parasagittal plane 4 _Parasagittal plane_
• transverse plane 5 _midsaggital plane_

3 Observe how one plane can produce both longitudinal sections and cross-sections through a tube.
- Observe a demonstration of a tube cut in a longitudinal section and in a cross-section.
- Twist a 1-foot section of plastic tubing so that one plane will divide one area of the tube into a longitudinal section and another area into a cross-section.
- Do not cut the tube unless instructed to do so.
- Show your instructor where a cut would produce both a longitudinal section and a cross-section.

4 Demonstrate sagittal, frontal, transverse, and oblique sections on sheep brains.
- Your instructor will display five sheep brains—one whole brain and four brains that have been cut into different sections.

- Determine the anterior, posterior, superior, and inferior surfaces of the brains.
- Decide which brain has been cut into sagittal, frontal, transverse, or oblique sections.
- Compare the appearance of the different sections.

Brain 1—Whole brain

Brain 2 _____ section

Brain 3 _____ section

Brain 4 _____ section

Brain 5 _____ section

Reviewing Your Knowledge

A. Anatomy Drill and Practice

Go to www.wiley.com/college/allen. Click on your lab manual, Student Companion Site, Anatomy Drill and Practice, and Exercise 1.

B. Anatomical Position

The person in Figure 1.5 is not in anatomical position. List four differences from anatomical position.

FIGURE 1.5 Person not in anatomical position.

1 _____ 3 _____

2 _____ 4 _____

C. Anatomical Terms for Body Structures

Write the anatomical terms that the phrase or word describes. Phrases or words referring to nouns are indicated. All other phrases refer to adjectives.

Umbilical **1.** Navel (noun).

thoracic **2.** Pertaining to the area between the neck and abdomen.

otic **3.** Pertaining to the ear.

palmar **4.** Pertaining to the palm of hand.

acromial **5.** Pertaining to the high point of the shoulder.

antecubital **6.** Pertaining to the anterior surface of the elbow region.

facial **7.** Pertaining to the face; anterior portion of the head.

nasal **8.** Pertaining to the nose.

cervical **9.** Pertaining to the neck.

popliteal **10.** Pertaining to the posterior surface of the knee.

Carpal **11.** Wrist (noun).

antebrachial **12.** Pertaining to the area between the elbow and wrist.

dorsal **13.** Back (noun).

Axillary **14.** Armpit area (noun).

Oral **15.** Pertaining to the mouth.

Patellar **16.** Pertaining to the anterior surface of the knee.

Sternal **17.** Breast bone (noun).

Coxal **18.** Pertaining to the hip.

fibular **19.** Pertaining to the side of the leg.

Sural **20.** Pertaining to the calf.

Axillary **21.** Pertaining to the area between the shoulder and elbow.

Digital **22.** Pertaining to the fingers or toes.

Manual **23.** Pertaining to the hand.

Pectoral **24.** Pertaining to the breast area.

buccal **25.** Pertaining to the cheek.

Calcaneal **26.** Pertaining to the heel.

plantar **27.** Pertaining to the sole of the foot.

Inguinal **28.** Pertaining to the groin where the thigh attaches to the pelvic region.

cephalic **29.** Pertaining to the head.

mental **30.** Pertaining to the chin.

pedal **31.** Pertaining to the foot.

Orbital **32.** Pertaining to the eye.

pubic **33.** Pertaining to the genital area.

buttock **34.** Pertaining to the area between the hip and knee.

Cranial **35.** Pertaining to the area that includes the bones enclosing the brain but excluding the facial bones.

frontal **36.** Pertaining to the forehead.

Vertebral **37.** Pertaining to the spinal column.

Occipital **38.** Pertaining to the inferior back of the head.

Crural **39.** Pertaining to the anterior surface of the leg.

Lumbar **40.** Pertaining to the area of the lower back or loin.

pelvic **41.** Pertaining to the anterior trunk below the abdomen.

scapular **42.** Pertaining to the area of the back that contains the shoulder blades.

Olecranal **43.** Pertaining to the posterior surface of the elbow region.

brachial **44.** Arm (noun).

D. Body Planes and Sections

Write the name of the plane that the phrase describes.

Parasaggital plane **1.** Divides body or organ into unequal right and left sections.

frontal plane **2.** Divides body or organ into anterior and posterior sections.

transverse plane **3.** Divides body or organ into superior and inferior sections.

Midsaggital Plane **4.** Divides body into right and left halves.

frontal Plane **5.**
midsaggital plane **6.** Which planes when passed through the body would result in two sections, with each section containing a piece of the heart and a piece of each lung?

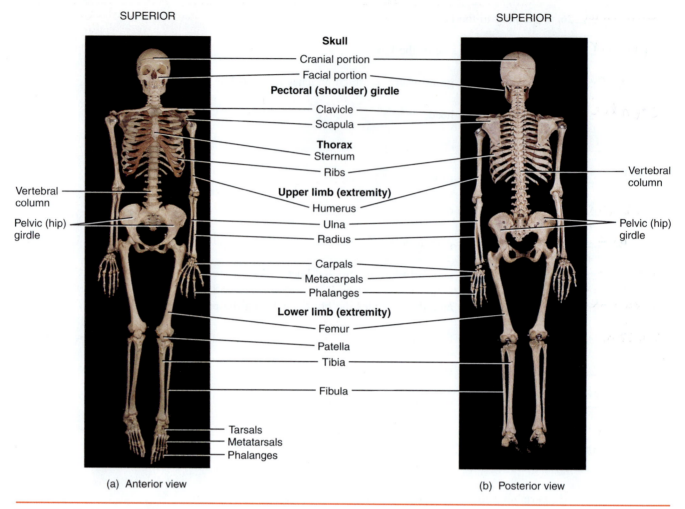

SUPERIOR

Skull
— Cranial portion
— Facial portion
Pectoral (shoulder) girdle
— Clavicle
— Scapula
Thorax
Sternum
— Ribs
Upper limb (extremity)
— Humerus
— Ulna
— Radius
— Carpals
Metacarpals
Phalanges
Lower limb (extremity)
— Femur
— Patella
— Tibia
— Fibula
Tarsals
Metatarsals
Phalanges

SUPERIOR

Vertebral column

Pelvic (hip) girdle

Vertebral column

Pelvic (hip) girdle

(a) Anterior view

(b) Posterior view

FIGURE 1.6 Human skeleton.

E. Directional Terms

Complete the sentences using directional terms. Use Figure 1.6 for reference.

1. The clavicle is _Superior_ to the ribs.

2. The ribs are _inferior_ to the sternum.

3. The humerus is _proximal_ to the radius.

4. The ulna is _proximal_ to the radius.

5. The tibia is _inferior_ to the femur.

6. The right humerus and the right radius are _ipsilateral_

7. The pelvic girdle is _distal_ to the ribs.

8. The sternum is _anterior_ to the vertebral column.

9. The scapula is _proximal_ to the clavicle.

10. The right fibula and left fibula are _ipsilateral_

Name _____ Date _____ Section _____

Using Your Knowledge

A. Body Regions, Anatomical Terminology, and Directional Terms

1. A 26-year-old female presented with an irregularly shaped and colored mole in the right lumbar region, just lateral to the vertebrae. Indicate on Figure 1.7 where this mole is likely to be found.

2. A 10-year-old male presented to the emergency room with a laceration on the left forearm just distal to the antecubital region. Indicate on Figure 1.7 where the laceration is likely to be found.

3. A 22-year-old female was identified by a tattoo on the fibular surface of the left leg just proximal to the tarsal region. Indicate on Figure 1.7 where the tattoo is likely to be found.

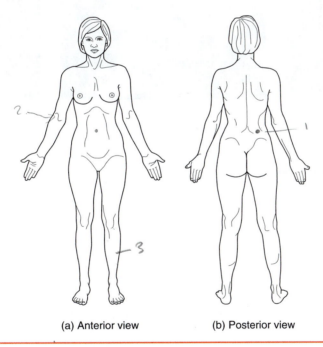

(a) Anterior view (b) Posterior view

FIGURE 1.7 Body regions, anatomical language, and directional terms.

Using the names of the structures in italics, answer questions 4–8.

4. Is the *pectoralis* major muscle anterior or posterior to the *subscapularis* muscle?

5. Is the *buccinator* muscle superior or inferior to the *frontalis* muscle?

6. Are the *thoracic* vertebrae medial or lateral to the *scapulae*?

7. Is the biceps *brachii* muscle proximal or distal to the extensor *digitorum* muscle?

8. The humerus bone is located in the arm. Is the *olecranon* process of the humerus on the distal or proximal end of the humerus?

B. Body Planes and Sections

Figure 1.8 contains three different sections through the thorax. Indicate which section (view a, b, or c) is a

9. Sagittal section ___a. or c___

10. Transverse (axial) section ___b.___

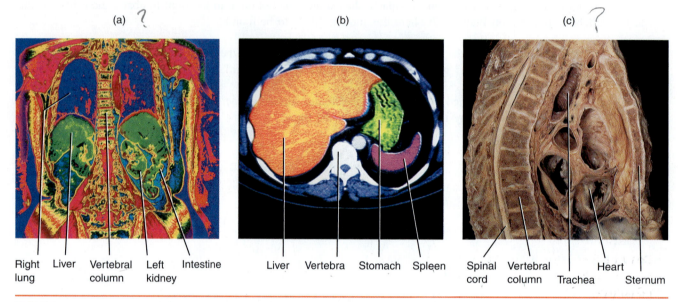

FIGURE 1.8 Sections through the thorax.

Organ Systems and Body Cavities

Objectives

After completing this exercise, you should be able to:

- Name the organ systems and describe the functions of each
- Name the major organs of each organ system and identify them on models or charts
- Describe the location of the body cavities and name the organs they contain
- Describe the structure and location of the serous membranes
- Identify the abdominopelvic quadrants and the organs found in each
- Identify the abdominopelvic regions and the major organs found in each

Materials

- human torso models or charts
- male and female human reproductive models or charts
- articulated skeleton
- one-gallon zippered plastic bags (2 per group)
- masking tape

A. Overview of Organ Systems

The body stays alive due to the interaction of different organ systems. An **organ system** is a group of organs performing a common function that helps provide the cells of the body with an optimal environment. All organ systems cooperate to maintain this optimal environment through a process called **homeostasis** (*homeo-* = same; *-stasis* = standing). Failure to maintain homeostasis results in disorders and disease. Organ system functions and major organs are summarized in Table 2.1.

ACTIVITY 1 Identifying Organ Systems

Identify the organ systems in Figure 2.1 using the organ system list provided. Refer to Table 2.1.

TABLE 2.1 Functions and Major Organs of the Organ Systems

ORGAN SYSTEM	FUNCTION
Cardiovascular	Transports nutrients, chemical messengers, gases, and wastes in blood. Major organs: heart and blood vessels
Respiratory	Adds oxygen to blood and removes carbon dioxide from blood; maintenance of carbon dioxide levels helps regulate pH. Major organs: nose, pharynx (throat), trachea, bronchi, lungs
Digestive	Breaks down food into units that can be absorbed into the body; eliminates wastes and nondigestible fiber in food. Major organs: mouth, pharynx, esophagus, stomach, intestines, pancreas, liver, gallbladder
Urinary	Removes nitrogenous wastes; maintains body fluid volume, pH, and electrolyte levels through urine production. Major organs: kidneys, ureters, urinary bladder, urethra
Integumentary	Provides a protective barrier for the body and aids in production of vitamin D; contains sensory receptors for pain, touch, and temperature; thermoregulation. Major organs: skin and skin structures (hair, nails, sweat glands, oil glands)
Lymphatic and Immune	Returns fluid to cardiovascular system; detects, filters, and eliminates disease-causing organisms, including cancer cells. Major organs: lymphatic vessels, lymph nodes, spleen, thymus, bone marrow
Skeletal	Protects major organs; provides levers and support for body movement. Major organs: bones
Muscular	Moves bones and maintains posture. Major organs: skeletal muscles
Nervous	Controls cell function with electrical signals; helps control body homeostasis. Major organs: brain, spinal cord, nerves
Endocrine	Controls cell function with hormones; helps control body homeostasis. Major organs: hypothalamus, pituitary, thyroid, pancreas, adrenal glands, ovaries, testes
Reproductive	Produces gametes; female uterus provides environment for development of fetus. Major organs in the male: testes, epididymis, ductus deferens, prostate, penis Major organs in the female: ovaries, uterine tubes, uterus, vagina

Hair

Skin and associated glands

Fingernails (and toenails)

1 _Integumentary_

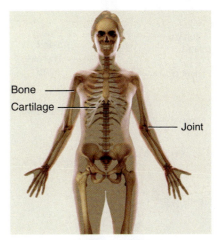

Bone

Cartilage

Joint

2 _Skeletal_

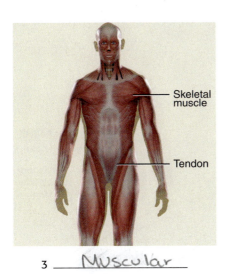

Skeletal muscle

Tendon

3 _Muscular_

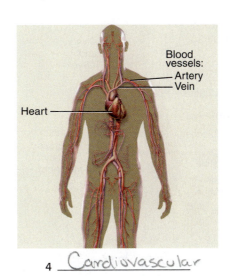

Blood vessels:
Artery
Vein

Heart

4 _Cardiovascular_

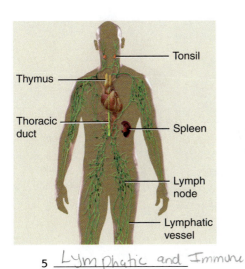

Tonsil

Thymus

Thoracic duct

Spleen

Lymph node

Lymphatic vessel

5 _Lymphatic and Immune_

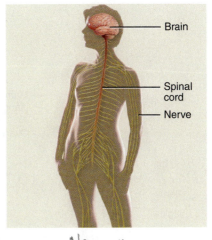

Brain

Spinal cord

Nerve

6 _Nervous_

FIGURE 2.1 Organ systems and selected organs.

7 _____Endocrine_____

8 _____Respiratory_____

9 _____Digestive_____

10 _____Urinary_____

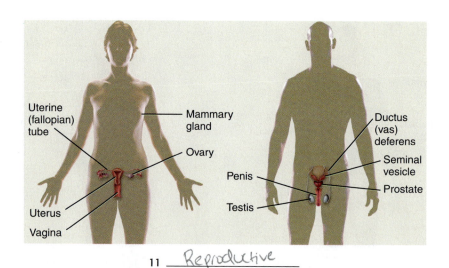

11 _____Reproductive_____

FIGURE 2.1 Organ systems and selected organs, *continued*.

B. Identification of Major Organs on a Torso Model

You will be identifying organs from anterior to posterior on a torso and answering questions concerning their position relative to the organs around them.

ACTIVITY 2 Identification of Organs

1 Identify the following organs on the anterior surface of a torso model. Identify all the organs without removing any organs from the model.
 • brain
 • trachea
 • heart
 • lungs
 • liver
 • stomach (left side of torso)
 • small intestine
 • large intestine (colon)

2 Remove the lungs, heart, liver, and stomach. Locate the gallbladder on the inferior surface of the liver.

3 Identify the following organs on the human torso model or chart.
 • esophagus
 • bronchi (right and left)
 • inferior vena cava
 • pancreas (posterior to stomach)
 • spleen

4 Remove the small intestine and large intestine. Locate the appendix at the inferior right end of the large intestine.

5 Identify the following organs on the human torso model.
 • abdominal aorta
 • adrenal glands (superior surface of kidneys)
 • kidneys
 • ureters
 • urinary bladder

6 Identify the female reproductive organs on a female reproductive model or chart. Observe the position of the urinary bladder relative to the uterus.
 • ovaries
 • uterus
 • urinary bladder

7 Identify the male reproductive organs on a male reproductive model or chart.
 • penis
 • scrotum (skin covering testes)
 • testes

8 Answer the questions about the position of each organ.

 1. The stomach is _____ to the small intestine.
 a. superior b. inferior c. medial d. lateral

 2. The liver is _____ to the lungs.
 a. superior b. inferior c. medial d. lateral

 3. The lungs are _____ to the heart.
 a. superior b. inferior c. medial d. lateral

 4. The trachea is _____ to the esophagus.
 a. medial b. inferior c. anterior d. posterior

 5. The pancreas is _____ to the stomach.
 a. superior and medial b. superior and anterior
 c. anterior and lateral d. posterior and inferior

 6. The gallbladder is on the _____ surface of the liver.
 a. superior b. inferior c. posterior d. lateral

 7. The stomach is _____ to the spleen.
 a. lateral b. medial c. superior d. inferior

 8. The abdominal aorta and inferior vena cava are _____ to the kidneys.
 a. medial b. lateral c. superior d. inferior

 9. The kidneys are _____ to the small intestine.
 a. anterior b. posterior c. superior d. inferior

 10. The urinary bladder is _____ to the uterus.
 a. posterior and superior b. anterior and inferior
 c. medial and superior d. lateral and posterior

C. Body Cavities

The two major cavities of the body are the dorsal and ventral body cavities. The **dorsal body cavity** is located near the posterior surface of the body and has two main subdivisions: the **cranial cavity** that contains the brain and the **vertebral** (*vertebra* = back) **canal** that contains the spinal cord. The cranial cavity and the vertebral canal are continuous.

The **ventral body cavity** is located near the anterior surface of the body and has two subdivisions, the thoracic cavity and the abdominopelvic cavity, which are separated by the **diaphragm.** The **thoracic cavity** is a space enclosed by the ribs, sternum, and vertebral column. This cavity contains three small cavities: the **pericardial cavity** (*peri-* = around; *-cardia* = heart) and two **pleural cavities** (*pleuro-* = side or rib). The pericardial cavity surrounds the heart, and each pleural cavity contains a lung. The **mediastinum** (*media-* = middle; *-stinum* = partition), a central area within the thoracic cavity, extends from the neck to the diaphragm and from the sternum to the vertebral column. The organs located in the mediastinum are the heart, thymus gland, esophagus, trachea and bronchi. The pleural cavities are located on either side of the mediastinum.

The **abdominopelvic cavity** consists of two continuous cavities: the abdominal cavity and the pelvic cavity. The **abdominal cavity** is the superior portion located between the diaphragm superiorly and the brim of the pelvis inferiorly. This cavity contains the stomach, liver, gallbladder, pancreas, spleen, small intestine, kidneys, appendix, and part of the large intestine. The **pelvic cavity,** located between the pelvic brim superiorly and the body wall inferiorly, is the inferior portion of the abdominopelvic cavity. The pelvic cavity contains part of the large intestine, rectum, urinary bladder, female reproductive organs (ovaries, uterine tubes, uterus, vagina), and male reproductive organs (prostate, part of ductus deferens). It is important to note that the testes and penis are not located in the ventral body cavity but are outside the body wall.

ACTIVITY 3 Body Cavities

1 Label the dorsal and ventral body cavities, their subdivisions, and the diaphragm on Figure 2.2(a) and (b).

2 Locate the dorsal body cavity and its subdivisions on a skeleton and torso model.

3 Locate the ventral body cavity, its subdivisions, and the diaphragm on a torso model.

4 Locate the mediastinum (meed-ee-uh-STINE-um) on a torso model.

5 Using a torso model, complete Table 2.2 by listing the major organ(s) found in each body cavity.

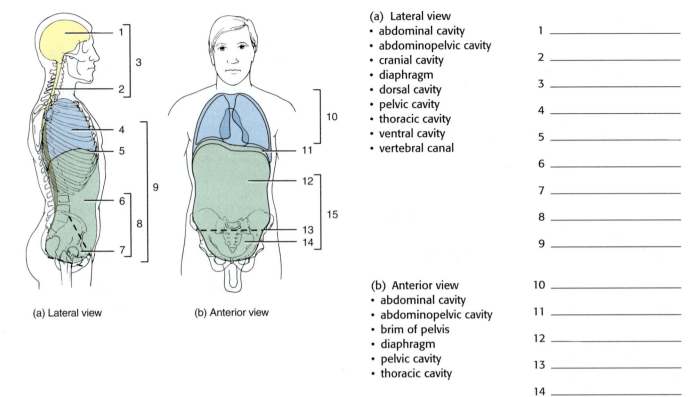

(a) Lateral view

(b) Anterior view

(a) Lateral view
- abdominal cavity
- abdominopelvic cavity
- cranial cavity
- diaphragm
- dorsal cavity
- pelvic cavity
- thoracic cavity
- ventral cavity
- vertebral canal

1 _____
2 _____
3 _____
4 _____
5 _____
6 _____
7 _____
8 _____
9 _____

(b) Anterior view
- abdominal cavity
- abdominopelvic cavity
- brim of pelvis
- diaphragm
- pelvic cavity
- thoracic cavity

10 _____
11 _____
12 _____
13 _____
14 _____
15 _____

FIGURE 2.2 Body cavities.

TABLE 2.2 Body Cavities

BODY CAVITY	ORGANS		
Dorsal body cavity			
Cranial cavity	1. _____		
Vertebral canal	2. _____		
Ventral body cavity			
Thoracic cavity			
Pleural cavities (2)	3. _____		
Mediastinum	4. (Within pericardial cavity) _____		5. _____
	6. _____	7. _____	8. _____
Abdominopelvic cavity			
Abdominal cavity	9. _____	10. _____	11. _____
	12. _____	13. _____	14. _____
	15. _____	16. _____	17. _____
Pelvic cavity	18. _____	19. _____	20. _____
	21. _____	22. _____	23. _____
	24. _____	25. _____	26. _____

D. Serous Membranes

Most of the organs in the ventral body cavity are covered with **serous** (*serum* = any clear, watery fluid) **membranes,** which are composed of two layers: a visceral layer and a parietal layer. The **visceral** (*viscera* = internal organs) **layer** covers the organ, whereas the **parietal** (*paries* = wall) **layer** attaches to and covers the ventral body wall. These two layers comprise one continuous sheet that folds to form a sac. Between the two layers is a small amount of serous fluid secreted by the membranes. The clear, watery **serous fluid** prevents friction as the organs move within the ventral body cavity. For example, the heart has movement within the thoracic cavity as it fills with and ejects blood.

Serous membranes are named for the cavities they surround. Thoracic serous membranes include the **pleura,** which covers the lungs, and the **pericardium,** which covers the heart. The serous membrane that covers the abdominal organs is the **peritoneum** (*peri-* = around; *teinein* = to stretch). Although most abdominal organs are positioned within the peritoneal cavity, a few organs are **retroperitoneal** (*retro-* = backward), or located posterior to the peritoneum. These organs are the pancreas, kidneys, adrenal glands, and portions of the large intestine, small intestine, aorta, and inferior vena cava. The peritoneum has three major double-layered folds called the mesentery, the greater omentum, and the lesser omentum. The **mesentery** encircles and holds the small intestine to the dorsal body wall. The **greater omentum** is suspended from the inferior portion of the stomach and covers the intestines in a similar manner to an apron. The **lesser omentum** suspends the superior portion of the stomach to the liver.

ACTIVITY 4 Serous Membranes

1 Label the serous membranes and cavities in the thoracic cavity in Figures 2.3(a) and (b).

2 Label the serous membranes and cavities in the abdominal cavity in Figures 2.4(a) and (b).

3 Make a replica or model of a serous membrane with your lab group. (This can be done as a demonstration.)
- Obtain 2 one-gallon zippered plastic bags.
- Make sure all the air is out of the first bag and then rezip the bag.
- In the second bag, add about 40 to 50 mL of water and push out the extra air before rezipping.
- Have a lab partner place a fist (simulating an organ) on the bottom edge of the first bag and push up into the bag.
- Now have the same lab partner place a fist on the bottom edge of the second bag and push up into the bag.

4 Visit www.wiley.com/college/allen to observe human serous membranes.

5 Answer the discussion questions with your lab partners.

? DISCUSSION QUESTIONS: SEROUS MEMBRANES

1 Was it easier to push a fist into the bag with no water or into the bag with water? Explain.

2 In the bag with water, what is the name of the simulated serous membrane layer that is touching the fist?

3 In the same bag, what is the name of the simulated outer serous membrane layer?

4 What does the water represent?

(a) Frontal section

(b) Transverse section

FIGURE 2.3 Serous membranes of the thoracic cavity.

(a) Frontal section
- parietal pericardium (pa-RYE-e-tul per-i-CAR-dee-um)
- parietal pleura (PLEUR-uh)
- pericardial cavity (peri-CAR-dee-ul)
- pleural cavity
- visceral pericardium (VIH-sir-ul)
- visceral pleura

1 _____

2 _____

3 _____

4 _____

5 _____

6 _____

(b) Transverse section
- parietal pericardium
- parietal pleura
- pericardial cavity
- pleural cavity
- visceral pericardium
- visceral pleura

7 _____

8 _____

9 _____

10 _____

11 _____

12 _____

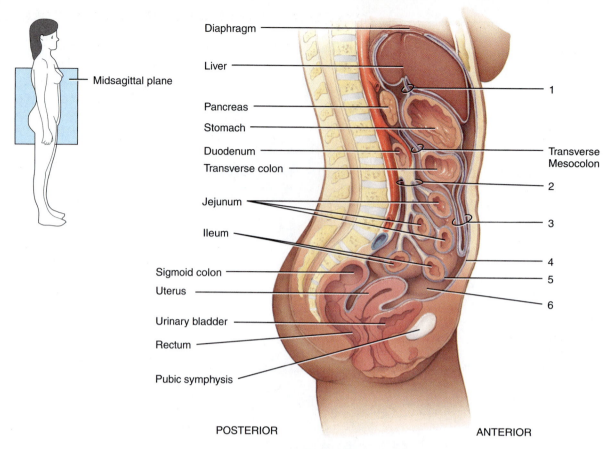

Midsagittal plane

Diaphragm

Liver

Pancreas

Stomach

Duodenum

Transverse colon

Jejunum

Ileum

Sigmoid colon

Uterus

Urinary bladder

Rectum

Pubic symphysis

1

Transverse
Mesocolon

2

3

4

5

6

POSTERIOR

ANTERIOR

(a) Sagittal section through the lower body

7

(cavity) 8

9

Small intestine

Vena cava

Liver

Kidney

Stomach

Pancreas

Aorta

Spleen

(b) Transverse section through abdomen

FIGURE 2.4 Serous membranes of the abdominal cavity.

(a) Sagittal section
• greater omentum
• lesser omentum
• mesentery
• parietal peritoneum
 (per-i-toe-NEE-um)
• peritoneal cavity
 (per-i-toe-NEE-ul)
• visceral peritoneum

1 _____

2 _____

3 _____

4 _____

5 _____

6 _____

(b) Transverse section
• parietal peritoneum
• peritoneal cavity
• visceral pertoneum

7 _____

8 _____

9 _____

E. Abdominopelvic Regions and Quadrants

Anatomists divide the abdominopelvic cavity into nine **regions** using two vertical and two horizontal lines in a tic-tac-toe grid so that the location of any organ is simple to describe. The two vertical lines are drawn midclavicular (mid-collar bone) and just medial to the nipples beginning at the diaphragm and extending inferiorly through the pelvic area. The upper horizontal line is drawn across the abdomen inferior to the ribs and across the inferior portions of the liver and stomach. The lower horizontal line is drawn a little inferior to the superior portion of the pelvic bones. These nine regions from the top right to the lower left are: **right hypochondriac** (*hypo-* = under; *chondro-* = cartilage), **epigastric** (*epi-* = upon; *gastro-* = stomach), **left hypochondriac, right lumbar** (*lumbar* = loin), **umbilical, left lumbar, right inguinal** or **iliac** (*inguinal* = groin), **hypogastric** or **pubic,** and **left inguinal** or **iliac.**

Clinicians are more apt to divide this cavity into four **quadrants** that are formed by transverse and sagittal planes running through the umbilicus (navel). These quadrants are useful clinically when one is trying to describe abnormalities or to determine which organ may be the cause of pain. The four quadrants are: **right upper quadrant (RUQ), left upper quadrant (LUQ), right lower quadrant (RLQ),** and **left lower quadrant (LLQ).**

ACTIVITY 5 Abdominopelvic Quadrants and Regions

Quadrants

1 Draw lines on Figure 2.5(a) separating the abdominopelvic cavity into quadrants and label the quadrants.

2 Using two pieces of masking tape, divide the abdominopelvic cavity of a human torso into quadrants.

3 Using the torso model (or chart), identify in which abdominopelvic quadrant(s) each organ is primarily located. Use the abbreviations RUQ, LUQ, RLQ, and LLQ.

a. appendix _____

b. large intestine or colon _____

c. liver _____

d. ovaries _____

e. pancreas _____

f. small intestine _____

g. spleen _____

h. stomach _____

Regions

1 Draw lines on Figure 2.5(b) separating the abdominopelvic cavity into regions and label the regions.

2 Using four pieces of masking tape, divide the abdominopelvic cavity of a human torso into regions.

3 Using the torso model (or chart), identify in which abdominopelvic region each organ is primarily located.

a. appendix _____

b. gallbladder _____

c. right ovary _____

d. bifurcation of the abdominal aorta _____

e. spleen _____

f. stomach (majority of) _____

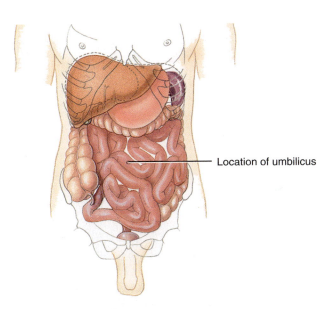

Location of umbilicus

(a) Quadrants

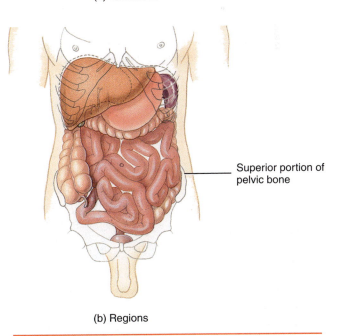

Superior portion of pelvic bone

(b) Regions

FIGURE 2.5 Abdominopelvic cavity.

Reviewing Your Knowledge

A. Functions of Organ Systems

Identify the organ system whose function is described below.

_____ **1.** Maintains blood oxygen and carbon dioxide levels.

_____ **2.** Controls muscles and glands by electrical impulses; helps control homeostasis.

_____ **3.** Causes movement of bones.

_____ **4.** Waterproof barrier that blocks the entrance of pathogens into the body and the loss of water from the body.

_____ **5.** Transports nutrients, oxygen, and carbon dioxide throughout the body.

_____ **6.** Changes food into absorbable nutrients; expels wastes.

_____ **7.** Regulates composition of blood by eliminating nitrogenous wastes, excess water, and minerals.

_____ **8.** Uses hormones to control cell function; helps control homeostasis.

_____ **9.** Provides framework for the body and protects body organs.

_____ **10.** Produces gametes (sperm and egg).

_____ **11.** Returns fluid to the bloodstream and provides protection against pathogens that have entered the body.

B. Organ Identification

Identify the correct organ system for the following organs.

Organ System **Organ**

_____ **1.** spleen

_____ **2.** liver

_____ **3.** trachea

_____ **4.** blood vessels

_____ **5.** hair

_____ **6.** kidney

_____ **7.** uterus

_____ **8.** pituitary gland

_____ **9.** spinal cord

_____ **10.** testes (2 systems)

_____ **11.** prostate gland

_____ **12.** large intestine

_____ **13.** pancreas (2 systems)

_____ **14.** adrenal gland

_____ **15.** thyroid

C. Body Cavities

Identify all the cavities for each organ as follows: dorsal (D), ventral (VN), cranial (C), vertebral (VR), thoracic (T), pleural (PL), pericardial (PC), peritoneal (PT), abdominal (A), or pelvic (P). All structures are present in more than one cavity, and some are in three cavities.

_____ **1.** brain

_____ **2.** small intestine

_____ **3.** heart

_____ **4.** lungs

_____ **5.** bronchi

_____ **6.** stomach

_____ **7.** spinal cord

_____ **8.** liver

_____ **9.** kidneys

_____ **10.** uterus

_____ **11.** urinary bladder

_____ **12.** ovaries

D. Serous Membranes

Write the term the phrase describes.

_____ **1.** Extends from the stomach and drapes over the intestines.

_____ **2.** Attaches the heart to the body cavity.

_____ **3.** Covers the surface of the lungs.

_____ **4.** Covers the surface of abdominal organs.

_____ **5.** The lubricating liquid in serous cavities.

_____ **6.** Extends between the superior part of the stomach and the liver.

_____ **7.** Attaches the small intestine to the posterior body wall.

_____ **8.** Circle the organs that are found within the peritoneal cavity: pancreas, liver, kidney, spleen, adrenal glands, abdominal aorta, inferior vena cava, stomach.

E. Abdominopelvic Quadrants and Regions

Quadrants

Name the quadrant(s) that the following organs predominantly occupy: RUQ, LUQ, RLQ, and LLQ.

_____ **1.** liver

_____ **2.** stomach

_____ **3.** spleen

_____ **4.** right adrenal gland

_____ **5.** gallbladder

_____ **6.** appendix

_____ **7.** left kidney

_____ **8.** right ovary

_____ **9.** uterus

_____ **10.** pancreas

Regions

Label the abdominopelvic regions in Figure 2.6.

1 _____

2 _____

3 _____

4 _____

5 _____

6 _____

7 _____

8 _____

9 _____

FIGURE 2.6 Abdominopelvic regions.

Using Your Knowledge

A. Homeostatic Imbalances of Organ Systems

Using your textbook, identify the organ system that is homeostatically imbalanced in the following diseases or disorders.

Organ System Disease

_____ **1.** muscular dystrophy

_____ **2.** multiple sclerosis

_____ **3.** myocardial ischemia

_____ **4.** infectious mononucleosis

B. Body Cavities

Identify *all* the cavities entered for each procedure, beginning with the largest cavity and ending with the most specific body cavity. Abdominal (A); cranial (C); dorsal (D); pelvic (P); pericardial (PC); pleural (PL); peritoneal (PT); thoracic (T); ventral (VN); and vertebral (VR).

_____ **5.** coronary bypass surgery

_____ **6.** cholecystectomy (gallbladder removal)

_____ **7.** spinal tap

C. Serous Membranes and Abdominopelvic Quadrants

8. A 44-year-old man went to the emergency room complaining of severe pain in his RLQ. The doctor palpated the area and determined that the pain was originating from an organ in that quadrant. Which organ might be involved?

(a) liver (b) appendix (c) gallbladder (d) spleen (e) stomach

9. A 23-year-old woman went to the doctor with the chief complaint of RLQ pain. Which organ is most likely the cause.

(a) adrenal gland (b) ovary (c) gallbladder (d) pancreas (e) kidney

D. Organ Identification

Identify the organs in the color enhanced medical images in Figure 2.7.

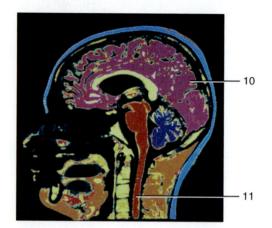

(a) MRI of head and neck.

(c) Radiograph of trunk.

(b) Radiograph of thorax.

(d) Radiograph of urinary tract organs.

FIGURE 2.7 Identification of organs on medical images.

10 _____ 14 _____ 18 _____

11 _____ 15 _____ 19 _____

12 _____ 16 _____ 20 _____

13 _____ 17 _____

Compound Light Microscope

Objectives

After completing this exercise, you should be able to:

- Describe and demonstrate how to carry, clean, use, and store a compound light microscope
- Identify the parts of a compound light microscope and describe their function
- Demonstrate how to view an object with the microscope using all magnifications
- Calculate total magnification
- Demonstrate how to measure the field of view

Materials

- compound light microscope, lens paper
- prepared slides of the letter "e"
- prepared slide of the trachea (or other organ)
- immersion oil
- thin, clear plastic ruler
- wet mount of cheek cells; clean microscope slides, coverslips, lens paper, flat toothpicks, and dropper bottle of dilute methylene blue, 0.9% saline solution, 10% bleach solution

A compound light microscope is used in the laboratory to observe small structures such as cells and tissues. The term *compound* refers to the set of lenses (one or two ocular lenses and one objective lens) used simultaneously to magnify the image. The term *light* refers to the necessity of using a light source to view the object. Most human cells must be magnified to be seen by the unaided human eye. The compound light microscope can magnify images up to approximately 1,000 times, depending on the magnifying power of the lenses.

Microscopic examination of cells and tissues allows students to observe how cell and tissue structure determines function. Changes in normal cell and tissue structure cause changes in organ function that lead to a disorder or disease. Tissue biopsies are performed to observe whether normal cellular structure has changed, which would indicate the absence or presence of a disorder or disease.

A. Transporting the Microscope

The compound light microscope is an expensive, precision instrument that must be handled appropriately. Demonstrate care in transporting, cleaning, using, and storing the microscope.

- Pick up the microscope with two hands, one holding the arm and the other supporting the base with the cord in a secure position.
- Carry the microscope upright so that a lens or eyepiece does not fall out, and carefully place the microscope on the lab table in front of you.

B. Parts of the Microscope

- **Base**—The widest bottom part that supports the microscope.
- **Arm**—The straight or curved vertical part that connects the base to the head.
- **Head** (or body tube)—The upper part of the microscope that extends from the arm and contains the ocular lens(es) and the rotating nosepiece with the objective lenses.

> **NOTE:** *All other microscope parts attach to the base, arm, and head—the three basic parts of the framework.*

- **Ocular lens(es)** (eyepieces)—One or two removable lenses that one uses to observe the microscope slide. Microscopes with one ocular lens are called **monocular** (*mono-* = one; *ocu-* = eye), and those with two ocular lenses are called **binocular** (*bi-* = two). Typically, these lenses magnify an object tenfold (10×). Look at an ocular lens and record the magnification power. _____ One of the ocular lenses may have a **pointer** used to identify a specific area on the slide. A **micrometer,** used to measure the field of view and object size, may also be present in one ocular lens.
- **Objective lenses**—A microscope will usually have three or four objective lenses mounted on a **revolving nosepiece.** Most microscopes have objective lenses that magnify an object 4× (scanning), 10× (low-power), 45× (high-dry), and 100× (oil immersion). List the magnification powers of the objective lenses on your microscope. _____ As the barrel of the objective lens increases in length, the magnifying power also increases.
- **Stage**—The flat platform located beneath the objective lenses on which the microscope slide is placed. The slide may be held onto the stage with either a mechanical stage clamp or two spring clips.
- **Mechanical stage**—Holds the slide securely in place with a spring clamp for viewing and can be moved with precision by using the **adjuster knobs.** One knob moves the slide side to side, and the other frontward and backward.
- **Coarse focus knobs**—On each side of the microscope toward the base is a large knob or dial that may or may not have a smaller knob in the middle. Only one of the large knobs needs to be used, depending on whether one is right- or left-handed. The large knob is used for coarse focusing and either knob moves the stage up and down quickly with the objective lens in place or moves the objective lenses up and down quickly with the stage in place. This knob is to be used with scanning or low-power lenses.
- **Fine focus knob**—The smaller knob on each side of the microscope that is used for precision focusing.
- **Condenser**—Located just below the stage is a lens that condenses light through the specimen on the slide above. If the condenser has an **adjustment knob** that raises and lowers the condenser, it usually needs to be in the highest position to focus the most light on the specimen.
- **Iris diaphragm**—Located beneath the condenser, the iris diaphragm works similarly to the iris of the eye. By adjusting its **lever,** the aperture changes diameter and regulates the amount of light that passes through the condenser. Decreasing the aperture size decreases the amount of light on the specimen and increases contrast. Some microscopes are equipped with a disk diaphragm instead.
- **Substage light**—The light source is usually built into the base of the microscope and typically has a dial or sliding bar on one side to control the light intensity.

C. Calculating Magnification

Total magnification is determined by multiplying the ocular lens power times the objective lens power. *Example:* Ocular lens power = 10×; Objective lens power = 4×; Total magnification = 40×.

> **ACTIVITY 1** **Parts of the Microscope and Calculating Magnification**

1. Identify the parts of your microscope as shown in Figure 3.1.
2. Compare your microscope with the one in Figure 3.1 and identify any differences with your lab partner.
3. Calculate the total magnification by multiplying the magnifying powers of your microscope lenses.

scanning lens _____ × ocular lens _____ = _____ total magnification

low-power lens _____ × ocular lens _____ = _____ total magnification

high-dry lens _____ × ocular lens _____ = _____ total magnification

oil immersion lens _____ × ocular lens _____ = _____ total magnification

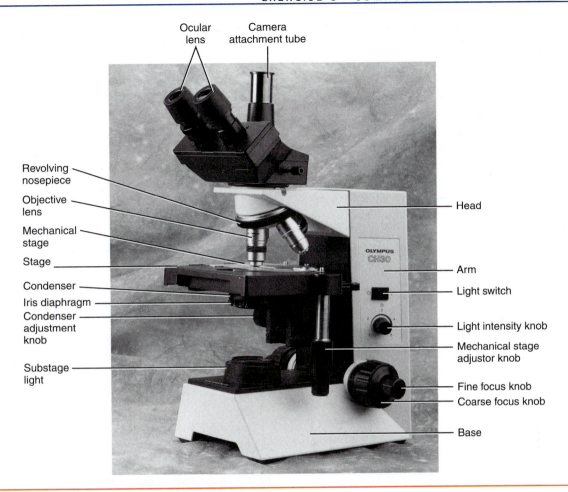

Ocular lens

Camera attachment tube

Revolving nosepiece

Objective lens

Mechanical stage

Stage

Condenser

Iris diaphragm

Condenser adjustment knob

Substage light

Head

Arm

Light switch

Light intensity knob

Mechanical stage adjustor knob

Fine focus knob

Coarse focus knob

Base

FIGURE 3.1 Parts of the microscope.

D. Using the Microscope

- Clean up your lab area and put nonessentials away so you will have plenty of room to use the microscope.
- Unwind the cord and plug it in.
- Clean the ocular, objective lenses, and condenser lenses only with the *special lens paper* (optical safe) provided by your instructor. Whenever the image on the slide cannot be focused clearly, it may be that the ocular, objective lens, or slide is dirty and needs additional cleaning. If all else has failed, consult your instructor.
- Turn on the light and adjust it to the lowest light setting feasible for good visibility and color to re-duce eye strain. The scanning (4×) and low-power (10×) lenses will not need as much light as the high-dry (40×) and oil immersion (100×) lenses.
- *Trouble-shooting:* If no light comes on initially, check two things before consulting your instructor. (a) Check the light dial to make sure it is not on

zero. (b) Check the safety switch on the electrical outlet by pushing in the reset button. If the light still does not work, plug your microscope into an outlet that you know works.

ACTIVITY 2 Using the Microscope

1 Move the **scanning objective lens** into place so it is over the light aperture on the stage. This objective lens has the shortest barrel. Make sure you feel the lens click into position, or your field of view will be black.

2 Obtain a prepared slide with the letter "e" from your instructor and place it on the stage, securing it with either the mechanical stage clamps or slide clips. Draw the letter "e" as it appears on the stage without looking in the ocular lens. _____

3 Without looking into the ocular lens, practice moving the slide from side to side in addition to backward and forward using the mechanical stage knobs (or your hands if your stage has slide clips).

4 Using the mechanical stage knobs, position the letter "e" over the light hole in the stage.

5 Check to see that the condenser lens is raised completely up to the stage.

6 If your *stage is moveable,* the coarse focus knob will move the stage. Raise the stage up as far as it will go. If your *objective is moveable,* use the coarse focus knob to lower the objective lens until it stops. The slide and the scanning objective lens will not actually touch.

7 If you have a binocular microscope, adjust the two ocular lenses as you would a pair of binoculars so that the two lenses are a comfortable distance apart for your eyes.

8 Looking through the ocular lens(es), adjust the light. Use the coarse focus knob to focus in the letter "e." Complete the focusing process by using the fine focus knob.

9 Using the mechanical stage knobs, bring the letter "e" directly into the center of the **field of view** (the lighted circular area you see as you look through the ocular lenses).

10 Draw the letter "e" as it appears through the microscope. Compare the appearance to your initial drawing. _____

11 While observing the letter "e" through the ocular lens(es), describe the movement that you observe as you move the slide:

- to the left _____

- to the right _____

- forward _____

- backward _____

12 Reposition the letter "e" directly in the middle of the field of view and switch to the **low-power objective lens.**

13 Most microscopes are **parfocal** so that when you move to a different magnification the specimen is almost, but not quite, in focus. You will need only the fine focus knob to focus the image. Center the specimen because the previously centered object is not in the center.

14 Use the iris diaphragm lever to adjust the amount of light and improve the contrast of your image.

> **CAUTION:** *Do not use the coarse focus knob with high-dry or oil immersion lenses.*

15 Repeat the above procedure with the **high-dry objective lens.** Be sure to focus *only* with the *fine focus knob.*

16 Describe the change in diameter of the field of view as one switches from the scanning lens to the low-power lens and then to the high-power lens.

> **NOTE:** Most slides used in anatomy and physiology do not need the magnification of the oil immersion lens. Your instructor will inform you if and when you will use this objective lens. *Only use oil immersion if instructed to do so.*

17 Center the area you want to view and then obtain a container of immersion oil made especially for the oil immersion lens.

18 Focus the slide with the high-power lens. Move the objective lens out of the way and apply a drop of oil directly on the part of the slide you wish to study.

19 Click the oil immersion lens into place, open the iris diaphragm as needed, adjust the light, and focus with only the *fine focus knob.*

20 Move the scanning power objective lens in place.

21 Remove the slide and clean the oil from the oil objective lens with lens paper. (Also clean the high-dry objective lens if you passed it through the oil.) Your instructor may ask you to use an additional cleaner.

22 Clean the slide with lens paper. If necessary, clean the stage as well.

E. Measuring the Field of View

The **field of view** is the area on the slide that is being observed and is inversely proportional to the magnification (the field of view decreases in size with increasing magnification). Once you know the diameter of the field of view in millimeters (mm) at various magnifications, you will be able to estimate the size of cells or other structures in the field of view. The object being viewed should be in the center of the field of view when you are switching to a higher objective lens (higher magnifying power).

Measuring the field of view at different magnifications demonstrates the advantage of scanning at a lower magnification to find a structure of interest before working up to a higher magnification.

ACTIVITY 3 **Measuring the Field of View**

1 Move the scanning objective lens in place.

2 Place a clear plastic ruler over the light opening in the stage, or use the micrometer in the ocular lens, or grid slide.

3 Look through the ocular lens and move the ruler or grid slide so that a line touches the left edge of the field and count the number of millimeter intervals that can be seen. Record. _____mm.

4 Switch to the low-power objective lens and repeat this procedure to count the number of millimeter intervals that can be seen. Record. _____mm.

5 Switch to the high-dry objective lens and repeat this procedure to count the number of millimeter intervals that can be seen. Record. _____mm. The closeness of this lens to the slide may not allow a ruler to be added.

6 Move the scanning power objective lens in place before removing the ruler.

F. Microscopic Structure of an Organ

An organ is comprised of a variety of cells and tissues. This activity starts with an observation of a section of whole organ with the scanning lens to get the "big picture," and then moves to higher magnifying powers to see additional detail.

ACTIVITY 4 **Observation of an Organ**

1 Obtain a prepared slide of the trachea or other organ supplied by your instructor.

2 Begin your observation using the scanning lens. Center and focus the organ and note that you can see several different tissues (stained different colors) present at this magnifying power.

3 Switch to the low-power lens. Center and focus and note the additional tissue detail that can be discerned at this magnifying power.

4 Switch to the high-power lens. Center and focus and note that you can now observe the cells that comprise the various tissues in this organ.

G. Wet Mount of Cheek Cells

Cells that line the interior of the mouth fit closely together like floor tiles and form a thick layer of thin cells that protect the underlying tissue from abrasion and microbes (bacteria and viruses). The superficial cells continually slough off and are replaced by underlying cells. Gently scraping the lining of the cheek removes the superficial cells that are about to slough off.

ACTIVITY 5 **Wet Mount of Cheek Cells**

1 Prepare a cheek smear slide and observe it under the compound microscope.
 * Obtain a toothpick, a clean microscope slide, and a coverslip.
 * Place a drop of 0.9% saline on the microscope slide.
 * Gently scrape the flat end of the toothpick (no blood, please!) on the inner lining of your cheek only one time. Do not scrape hard enough to hurt.

(a) Photomicrograph

(b) Student drawing

FIGURE 3.2 Cheek cells.

 * To apply the cells to the slide, rotate the toothpick between your thumb and forefinger to dislodge the cells into the saline.
 * Dispose of the toothpick as your instructor directs.
 * Add *one drop* of dilute methylene blue to the cells on your slide.
 * Cover the sample with a coverslip as directed by your instructor.
 * Using low-power, locate the blue-stained cells. Switch to high-power to observe more cellular detail.

2 Compare the cells on your slide with Figure 3.2(a). Draw a picture of your cells in Figure 3.2(b).

3 Move the scanning power objective lens in place and remove the slide.

4 Place microscope slides in a 10% bleach solution as your instructor directs. Clean your lab top with a 10% bleach solution.

5 Answer the discussion questions with your lab partners.

? DISCUSSION QUESTIONS: CHEEK SMEAR

1 Why was stain added to the cheek cells?

2 What cellular structures did you observe?

H. Storing the Microscope

It is important to put the microscope away properly.

 * Check that the scanning objective lens is in place and that the slide is removed from the stage.
 * Depending on the type of microscope, either lower the stage or raise the objective to put maximum distance between the objective and the stage.
 * Center the mechanical stage.
 * Turn off the substage light. The bulb life is extended if it cools before moving the microscope.
 * Clean the ocular and objective lenses with lens paper.
 * Coil the cord neatly according to your instructor's directions.
 * Place the dust cover over the microscope.
 * Using both hands and the proper carrying technique, return the microscope to the appropriate cabinet.

Reviewing Your Knowledge

A. Care and Use of the Microscope

Correct each statement by crossing out the incorrect word(s) and inserting the correct word(s).

1. Only one hand is needed to transport the microscope.

2. Tissue paper can be used to clean the microscope lenses and prepared slides.

3. The microscope should be put away with the high-dry lens in position.

4. The coarse focusing knob should be used when using the high-dry lens.

5. The iris diaphragm should be completely open to obtain maximum contrast.

6. The condenser should be in the lowest position (far from the stage) to focus the most light on the specimen.

B. Parts of the Microscope

Write the term that the phrase describes.

_____ **1.** Large knob that moves the stage or objective lens a great distance. Used with scanning or low-power objective lenses only.

_____ **2.** Flat platform beneath the objective lens on which the microscope slide is placed.

_____ **3.** Removable lenses that you look through to observe the microscope slide.

_____ **4.** Small knob that moves the stage or objective lens a very small distance and is used for precision focusing.

_____ **5.** Extends from the arm and contains the ocular lenses and rotating nosepiece.

_____ **6.** Lens that condenses light through the specimen and is located below the stage.

_____ **7.** Light from specimen passes through these lenses first. These lenses are located in the rotating nosepiece.

_____ **8.** Widest bottom part that supports the microscope.

_____ **9.** Regulates the amount of light passing through the condenser.

_____ **10.** Vertical portion that connects the base to the head.

C. Total Magnification and Field of View

1. Calculate the total magnification of an object viewed with a 10✕ ocular and a 60✕ objective lens.

2. Does the size of the field of view increase or decrease when going from a lower to higher power objective lens?

Cell Structure and Cell Cycle

Objectives

**After completing this exercise,
you should be able to:**

- Identify cellular components on a model or diagram
- Describe the function of the plasma membrane and cellular organelles
- Identify cells and observable cellular structures on prepared microscope slides
- Identify the stages of mitosis
- Describe the events of each stage of mitosis

Materials

- model or diagram of a cell
- compound microscopes and lens paper
- prepared slides of human skeletal muscle cells, pseudostratified ciliated columnar epithelium, motor neurons, sperm, and blood
- 3-dimensional models of mitosis
- whitefish blastula slides
- colored pencils

A. Cell Structure

Cells are the smallest structural and functional units of living organisms. They are enclosed by a plasma membrane which controls the movement of substances in to and out of the cell. The interior of the cell is filled with cytoplasm that contains cytosol (a viscous fluid) and organelles (little organs). Like an automobile, a cell has different parts or organelles that perform different functions. A "generalized" animal cell is shown in Figure 4.1, and functions of cellular organelles are described in Table 4.1.

ACTIVITY 1 Cell Structure

1 Using Table 4.1, label the cell structures in Figure 4.1.
2 Point to each cell structure shown in Figure 4.1 on a cell model or chart.

3 Pronounce the name of each organelle and describe its function as you point to it.
4 Go to www.wiley.com/college/allen and identify cellular organelles on electron micrographs of cells.
5 Answer the discussion question with your lab partners.

? DISCUSSION QUESTIONS: CELL ORGANELLES

1 Which cell structures from Table 4.1 are not found in most human cells?

FIGURE 4.1 Sectional drawing of a cell.

- centrioles
- chromatin
- cilia
- cytoplasm
- Golgi complex
- lysosome
- microfilament (cytoskeleton)

- microtubules (cytoskeleton)
- microvillus
- mitochondrion (singular)
- nuclear envelope
- nucleolus
- nucleus
- peroxisome

- plasma membrane
- ribosome
- rough endoplasmic reticulum
- secretory vesicle
- smooth endoplasmic reticulum

1 _____

2 _____

3 _____

4 _____

5 _____

6 _____

7 _____

8 _____

9 _____

10 _____

11 _____

12 _____

13 _____

14 _____

15 _____

16 _____

17 _____

18 _____

19 _____

TABLE 4.1 **Description and Function of Cell Structures**

STRUCTURE	DESCRIPTION	FUNCTION
PLASMA MEMBRANE	Phospholipid bilayer with proteins and carbohydrates (sugars) that forms the outer boundary of cell.	Protection; controls movement of substances in to and out of the cell; cell-to-cell communication.
Microvilli	Folds of the plasma membrane forming tiny extensions of cytoplasm.	Increases surface area of cell for increased absorption or secretion.
NUCLEUS	Large, spherical body that is composed of nucleoli and chromatin, enclosed by a nuclear envelope with pores.	Control center; DNA within the genes directs cellular activities and cellular structure.
Nucleolus	Small spherical body or bodies in the nucleus.	Synthesizes rRNA; assembles rRNA with protein to produce ribosomes.
Chromatin	Long, thin strands of DNA and proteins dispersed in nucleus; condenses to form chromosomes during cell division.	Carries hereditary information in genes and condenses to form chromosomes during cell division.
CYTOPLASM	Region of the cell that includes the cytosol and organelles.	Area where many components are involved in metabolism.
Cytosol	Fluid portion of cytoplasm that surrounds the organelles.	Allows diffusion of substances throughout the cell.
Organelles	Organized "little organs" within the cytoplasm.	Each organelle has specific cell functions.
• Mitochondria	Oval, double-membrane-bound organelle.	Site of aerobic cellular respiration for ATP production.
• Ribosomes	Tiny bodies composed of rRNA and protein; found in cytosol or attached to form RER.	Site of protein synthesis.
• Rough endoplasmic reticulum (RER)	Flattened membrane-bound channels with ribosomes attached, extending from nuclear envelope throughout cytoplasm.	Processes and transports proteins made at attached ribosomes; synthesizes phospholipids.
• Smooth endoplasmic reticulum (SER)	Extends from RER as a membrane-bound, tubular network without ribosomes.	Lipid and steroid synthesis; detoxifies toxic substances (drugs).
• Golgi complex	Curved, flattened membranous sacs with bulging edges and small vesicles nearby.	Receives and modifies proteins from RER; sorts and transports them.
• Secretory vesicles	Small membrane-bound vesicles.	Secretes substances outside the cell by exocytosis.
• Lysosomes	Membrane-bound vesicles formed from Golgi complex; contain digestive and hydrolytic enzymes.	Enzymes digest and recycle worn out organelles and substances entering the cell; can digest the cell.
• Peroxisomes	Small, membrane-bound vesicles containing enzymes.	Produces hydrogen peroxide, detoxifies harmful substances.
• Cytoskeleton Microfilaments Microtubules Intermediate filaments	Protein filaments throughout the cell. Thinnest filaments made of actin. Long, hollow tubes of tubulin. Intermediate diameter filaments.	Structural framework for the cell. Support and generate cellular movement. Cell shape; movement of organelles. Hold organelles in position; support.
• Centrosome (Centrioles)	Two perpendicular, cylindrical centrioles and material around the centrioles.	Forms mitotic spindle fibers and microtubules for cilia and flagella.
• Cilia	Short, abundant, hair-like cell projections formed of microtubules.	Move fluids and particles along cell surface by beating together.
• Flagella	Long cell projections with microtubules.	Whip-like motion moves sperm.

B. Cell Specialization

The human body contains many different types of cells with different functions. These differences in function are reflected in cell structure. Cells of the human body differ from the generalized animal cell in shape, size, or number and type of organelles present. In this activity you will observe cells of skeletal muscle, pseudostratified ciliated columnar epithelium, motor neurons, sperm, and blood. **Skeletal muscle** cells are long, cylindrical cells which contract (shorten in length) to move bones. **Pseudostratified ciliated columnar epithelial cells** have cilia which move substances like mucus along the surface of the cells. **Motor neurons** are nervous tissue cells with many processes (cell extensions) that receive information from other neurons and send electrical signals to muscle cells causing them to contract. **Sperm cells** (sperm) are small, oval cells with a flagellum which propels them through the female reproductive tract. **Red blood cells** do not have a nucleus (anucleate) but contain large amounts of hemoglobin, a red pigment that binds oxygen. **White blood cells** have nuclei with different shapes and defend the body from pathogens and cancerous cells.

ACTIVITY 2 Cell Specialization

1 Using a compound microscope, observe each prepared slide (skeletal muscle, pseudostratified ciliated columnar epithelium, motor neuron, sperm, and blood smear) and identify the cells and cell components shown in Figure 4.2.

2 Describe each cell's shape and list the cell structures that can be seen with the light microscope in each cell type.

a. skeletal muscle cell _____

b. pseudostratified ciliated columnar epithelial cell

c. motor neuron _____

d. sperm cell _____

e. red blood cell _____

f. white blood cell _____

3 Answer the discussion questions with your lab partners.

? DISCUSSION QUESTIONS: CELL SPECIALIZATION

1 What are the main differences you observed in the cells?

2 Speculate on the reasons for the anatomical differences you observed.

Width of skeletal
muscle cell Nucleus

(a) Skeletal muscle cells 400×

Goblet cell Nucleus Cilia

(b) Ciliated cells (Pseudostratified 400×
 ciliated columnar epithelium)

Cell body Nucleus Processes

(c) Motor neuron 350×

Nucleus Flagellum Red blood cells Nucleus of white
 blood cell

(d) Sperm cells 400× (e) Blood cells 400×

FIGURE 4.2 Cell specialization.

C. Somatic Cell Division: Mitosis and Cytokinesis

Somatic (*soma-* = body) **cell division** occurs when one cell divides to produce two genetically identical cells. Cell division is needed for growth of the individual and cell replacement.

The **cell cycle,** a period during which a cell grows and divides into two genetically identical cells (daughter cells), begins when a cell is produced by cell division and ends when the cell divides. The length of the cell cycle differs according to the type of cell, with some cells dividing more frequently than others. The cell cycle can be divided into two basic periods: **interphase,** a long period during which the cell conducts its normal activity, grows, and prepares for cell division; and the **mitotic phase,** when the cell is dividing. The mitotic phase consists of **mitosis,** or nuclear division, and **cytokinesis,** or cytoplasmic division. The four stages of mitosis are: prophase, metaphase, anaphase, and telophase (Table 4.2).

To observe interphase and the stages of mitosis, you will examine a prepared microscope slide containing several sections of a whitefish blastula. The **blastula** is an early embryonic stage in which cells are dividing rapidly, providing many cells in different stages of mitosis.

ACTIVITY 3 Mitotic Phases

1 Observe the 3-dimensional models of the mitotic phases, noting the changes in each phase.

2 Using Table 4.2, identify interphase, each phase of mitosis, and cytokinesis in Figure 4.3(a–e).

3 Obtain a prepared whitefish blastula slide and hold it up to the light. Notice that there are many blastula sections on each slide. It will be necessary to view several of these sections to find all of the phases.

4 Using a compound microscope, begin looking at your slide with the low-power objective lens. Use the high-power objective lens to identify interphase, the four stages of mitosis, and cytokinesis.

TABLE 4.2 Phases of Somatic Cell Cycle

PHASE	DESCRIPTION OF EVENTS
Interphase (*inter-* = between)	Normal cell work; cell metabolically active and growing, DNA replicates; chromosomes not visible.
Mitotic Phase	Cell division.
Mitosis (*mitos-* = thread)	Nucleus divides to form two genetically identical nuclei.
• Prophase (*pro-* = first)	Centrioles move to opposite poles; nucleolus and nuclear membrane disappear; chromatin condenses into chromosomes, each comprised of two identical chromatids; spindle fibers appear.
• Metaphase (*meta-* = next)	Chromosomes composed of identical chromatids line up at metaphasal plate; spindle fibers attach to centromeres of chromatids.
• Anaphase (*ana-* = apart)	Identical chromatids separate and move to opposite poles.
• Telophase (*telo-* = end)	Nucleolus and nuclear membrane reappear; chromosomes unravel into chromatin; spindle fibers disappear; centrioles move together.
Cytokinesis (*cyto-* = cell; *kinesio-* = movement)	Cytoplasmic division into two genetically identical daughter cells that usually occurs during telophase.

Chromatin in nucleus Nuclear membrane

(a)

Chromosomes forming Plasma membrane

(b)

Spindle fiber Chromosomes Centriole

(c)

Centriole Chromosomes Spindle fiber

(d)

Chromosomes Spindle fibers Cleavage furrow

(e)

FIGURE 4.3 Mitotic phases.

- anaphase (AN-a-faze)
- interphase (IN-ter-faze)
- metaphase (MEH-ta-faze)
- prophase (PRO-faze)
- telephase and cytokinesis (TELL-o-faze and cyto-kih-NEE-sis)

a _____

b _____

c _____

d _____

e _____

Reviewing Your Knowledge

A. Generalized Cell Anatomy with Organelles

Go to www.wiley.com/college/allen and click on the picture of your lab manual. Then click on Student Companion Site, Anatomy Drill and Practice, and Exercise 4.

B. Cellular Structure

Fill in the blank with the name of the cell structure that fits the description.

_____ **1.** short, hair-like projections for movement of substances along cell surface

_____ **2.** intracellular fluid

_____ **3.** site of energy production by cellular respiration

_____ **4.** site of protein synthesis

_____ **5.** site of carbohydrate and lipid synthesis

_____ **6.** small vesicle with digestive enzymes

_____ **7.** organelles needed to form cilia and flagella

_____ **8.** thread-like strand of DNA with associated proteins

_____ **9.** site of secretory and membrane protein synthesis

_____ **10.** site where protein products are stored, packaged, and exported

_____ **11.** contains genes that control cellular activities

_____ **12.** site of ribosome synthesis

_____ **13.** gives the cell shape, support, movement, and holds organelles in position

_____ **14.** controls movement of substances in to or out of the cell

_____ **15.** folds of the plasma membrane that increase the cell's surface area

_____ **16.** detoxifies harmful substances, produces hydrogen peroxide, and oxidizes amino acids

_____ **17.** double membrane that separates the nucleus from the cytoplasm

_____ **18.** a small membranous sac which delivers proteins to the plasma membrane to exit the cell

C. Phases of the Cell Cycle

Write the phase of the cell cycle that the phrase describes.

_____ **1.** cytoplasmic division

_____ **2.** cell performing normal functions; longest phase

_____ **3.** nuclear division

_____ **4.** chromatid pairs line up at equatorial plate

_____ **5.** chromatin condenses into chromosomes

_____ **6.** spindle fibers break up; nucleus reappears

_____ **7.** centromeres divide; chromosomes move to opposite poles

_____ **8.** nuclear membrane disassembles and disappears

_____ **9.** chromosomes unravel to form chromatin

_____ **10.** mitotic spindle forms

_____ **11.** DNA replicates

Using Your Knowledge

A. Cellular Organelles and Their Function

Write the letter for the correct answer in the blank.

____ 1. A cell makes and secretes a protein-based hormone. This particular cell would have a great amount of Golgi complex and:

(a) SER (b) RER (c) mitochondria (d) lysosomes

____ 2. Cells that make steroids (lipids) would have a larger amount of:

(a) SER (b) RER (c) mitochondria (d) lysosomes

____ 3. Cells that need large amounts of ATP would have many:

(a) Golgi complexes (b) ribosomes (c) SER (d) mitochondria

____ 4. Amoebas use what organelle for amoeboid movement and phagocytosis?

(a) ribosome (b) cytoskeleton (c) lysosome (d) Golgi complex

____ 5. Immune cells that destroy bacteria with chemicals need an abundance of:

(a) SER (b) ribosomes (c) lysosomes (d) centrioles

B. The Cell Cycle and Mitosis

Answer each question with a short answer.

6. Explain the role of somatic cell division advance from infancy to adulthood.

7. Explain the role of cell division in wound healing.

8. List the cell structures involved in mitosis.

9. Can red blood cells undergo mitosis? Explain.

10. Sperm and eggs have one-half the number of chromosomes of the somatic cells that divided to form them. Are sperm and eggs formed by mitosis? Explain.

Transport Across the Plasma Membrane

Objectives

After completing this exercise, you should be able to:

- Describe diffusion, osmosis, filtration, and phagocytosis
- Distinguish between and define hypotonic, hypertonic, and isotonic solutions

Materials

- Diffusion: 2 agar Petri dishes per group, small millimeter rulers, forceps, methylene blue crystals, potassium permanganate crystals
- Osmosis in Living Red Blood Cells: animal blood (uncoagulated), disposable gloves, clean microscope slides and coverslips, 4 medicine droppers per group, compound microscope, filter paper
- Osmosis and Diffusion Across a Dialysis Membrane: dialysis tubing 12 cm long, 2 dialysis clips, scissors, Congo red or red food coloring, 40% sucrose solution, 500 mL beakers, distilled water, gram scale
- Filtration: filter paper, funnel, ring stand with ring, 250 mL beaker, distilled water, 5% powdered charcoal solution, 1% copper sulfate solution

The plasma membrane with its unique design is responsible for selectively allowing substances in and out of the cell (selectively permeable membrane). Table 5.1 describes the variety of methods the cell uses to move substances across the cell membrane, either actively or passively. **Active processes** require ATP (energy) to get the work done. **Passive processes** do not require the cell to expend energy because the kinetic energy of the particles causes them to move from an area of their higher concentration to an area of their lower concentration.

TABLE 5.1 Movement of Substances across Cell Membranes

TRANSPORT PROCESS	DESCRIPTION
Passive Processes	Do not use ATP
• Diffusion	Movement of molecules or ions by kinetic energy from their greater concentration to their lesser concentration (down a concentration gradient) until equilibrium is reached
• Osmosis	Movement of *water molecules* by diffusion across a selectively permeable membrane; movement is from a hypotonic solution into a hypertonic solution; water moves until equilibrium is reached
• Facilitated diffusion	Movement of a substance down its concentration gradient using a transmembrane protein for transport
• Filtration	Flow of liquid (and solutes smaller than the pores) through the pores of a filter or plasma membrane; movement due to hydrostatic pressure
Active Processes	Cells use ATP to move substances across the plasma membrane
• Active transport	Movement of a substance against its concentration gradient using a transmembrane protein and ATP for transport
• Vesicular transport	Movement of substances in to or out of a cell using vesicles that form at or fuse with the plasma membrane
Endocytosis	Movement of substances into a cell via vesicles forming at the plasma membrane
Phagocytosis	A type of endocytosis; specialized movement of solid particles into a cell by pseudopodia forming a vesicle around the particles; "cell eating"
Pinocytosis	A type of endocytosis; drops of fluid brought into a cell by an invagination of the plasma membrane forming a small vesicle; "cell drinking"
Exocytosis	Movement of substances out of a cell via fusion of secretory vesicles with the plasma membrane

A. Passive Transport—Diffusion

In **diffusion,** there is a net movement of substances from a region of their greater concentration to a region of their lesser concentration. Diffusion can occur in solids, liquids, or gas, and across the plasma membrane of cells. In the following activity, diffusion of two crystals through a solid (agar) will be studied. Methylene blue has a molecular weight of 320, and potassium permanganate has a molecular weight of 158. The two crystals move at different rates through the agar, which is made up mostly of water.

ACTIVITY 1 Experiment: Diffusion

1 **Prediction:** With your lab group, predict which crystal will move faster. (*Hint: Use the molecular weights.*) Circle your answer: *methylene blue* or *potassium permanganate.*

2 **Materials:** Obtain the materials for diffusion.

3 **Data Collection:** Measure the diffusion rates of methylene blue and potassium permanganate.
 • Decide who will set up the experiment, who will time, who will measure, and who will record.

• Carefully using forceps, place a large crystal of methylene blue on the surface of an agar Petri dish (Figure 5.1). Be careful not to drop any extra crystals on the agar surface.

• Using the same technique, place a similar sized crystal of potassium permanganate on the other side of the Petri dish.

Methylene blue dye crystal

Potassium permanganate dye crystal

FIGURE 5.1 Diffusion in agar plate setup.

TABLE 5.2 Diffusion Results

TIME (MIN)	DIFFUSION OF METHYLENE BLUE (DIFFUSION DIAMETER IN MM)	DIFFUSION OF POTASSIUM PERMANGANATE (DIFFUSION DIAMETER IN MM)	DIFFUSION RATE MM/MIN
15			
30			
45			
60			
75			

- Using a millimeter (mm) ruler, measure the crystals' diameter of diffusion at 15-minute intervals for at least 1 hour (or longer if desired).
- After each observation, record the diffusion diameter of each crystal in millimeters (mm) in Table 5.2.

4 Clean up as directed by your instructor.

5 Data Analysis: Calculate the diffusion rates for each time period for the two crystals using Table 5.2.

6 Complete the Experimental Report with your lab group.

EXPERIMENTAL REPORT: DIFFUSION

Results: Which substance moved faster?

Discussion: Discuss why the diffusion rates of the two substances differed.

Conclusion: State how molecular weight affects diffusion rate.

B. Passive Transport—Osmosis

A **solvent** is a dissolving medium, typically a liquid and always water in living organisms. **Solutes** are the substances dissolved in a solvent. When two different solute concentrations are on either side of a selectively permeable membrane, there is a **concentration gradient** of the solute.

Osmosis is the net movement of water molecules through a selectively permeable membrane from a solution with a lower solute concentration (**hypotonic**) into a solution with a higher solute concentration (**hypertonic**). The terms hypotonic (*hypo-* = deficient; *-tonos* = stretching) solutions or hypertonic (*hyper-* = excessive) solutions are used only when two solutions are compared. Water will move into the hypertonic solution until the solute concentrations of the two solutions equalize to become **isotonic** solutions (*iso-* = same) or if enough pressure is applied to stop the flow of water.

Red blood cells (RBCs) are good examples to use for this osmosis experiment because their shape changes dramatically when they are exposed to hypotonic or hypertonic solutions. Their normal shape is a round biconcave disk with a smooth plasma membrane. Under the microscope, they look two-toned, with the center being lighter because of the concavity and thinness of the center. If the cell loses most of its water by osmosis when put in a hypertonic solution, it becomes **crenated** or shriveled with spiked edges. If the cell gains a significant amount of water by being placed in a hypotonic solution, it swells and may eventually burst—a process called **hemolysis** (*hemo-* = blood; *-lysis* = break down). As a basis for comparing various solutions to blood, the salt content (NaCl) of blood is 0.9% and is the same salt content as a physiologic saline solution.

NOTE: *Remember . . . salt shrivels and water swells (a cell).*

Safety Precautions

- Clean laboratory tabletops before and after conducting your experiments.
- Wear safety glasses and disposable gloves when handling body fluids or animal blood.

ACTIVITY 2 Experiment: Osmosis in Living RBCs

1 With your lab group, discuss and identify which solutions for this experiment are hypotonic, hypertonic, or isotonic to the RBCs. The isotonic solution is also called physiologic saline.

- 0.9% saline solution _____

- 5% saline solution _____

- distilled water _____

2 In Figure 5.2, identify the shape of the RBC (normal, swollen, or crenated) and the type of solution the RBC is in (isotonic, hypertonic, or hypotonic).

3 **Prediction:** With your lab group, predict the net movement of water in all three types of solutions. Circle your answer for all three situations:

- In the isotonic solution, the RBCs will *swell, crenate,* or *not change shape.*
- In the hypotonic solution, the RBCs will *swell, crenate,* or *not change shape.*
- In the hypertonic solution, the RBCs will *swell, crenate,* or *not change shape.*

4 **Materials:** Obtain materials to observe osmosis in RBCs.

5 **Data Collection:**

- Decide who will set up and initially observe each slide. Every member of your group should observe each slide.

Slide #1: 0.9% saline solution

- Place a drop of animal blood on a slide with a medicine dropper and cover with a coverslip. With a different medicine dropper, add a drop of 0.9% saline solution to the blood. Tilt the slide to intermix the two solutions, and cover with a coverslip.
- Using high power, observe the slide for changes in cell shape. Blood cells are very tiny and difficult to see under low power.

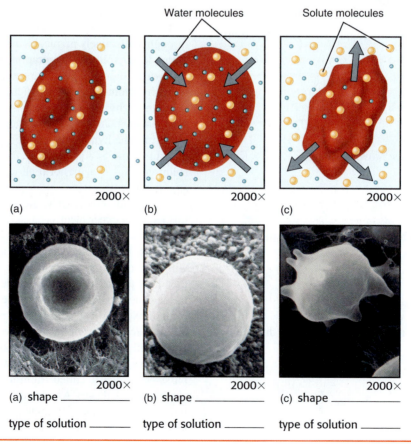

FIGURE 5.2 Osmosis in RBCs.

- Record shape of RBCs and type of solution in which they were placed.
 shape _____ type of solution _____

Slide #2: 5% saline solution
- Place a drop of animal blood on a second slide and cover with a coverslip. Observe the RBCs using high power. On one side of the coverslip, add one drop of 5% saline with a clean medicine dropper. Place filter paper or a small piece of paper towel on the opposite side of the coverslip to absorb the liquid and pull the 5% saline solution into the RBCs.
- Immediately observe the second slide under high power and watch for changes in cell shape. You may have to wait a few minutes for cell changes to occur.
- Record shape of RBCs and type of the solution in which they were placed.
 shape _____ type of solution _____

Slide #2 again: distilled water
- Using slide #2 again, add one drop of distilled water to the same edge of the coverslip that was used for the saline solution.
- Hold a piece of filter paper at the opposite edge of the coverslip, observe the paper absorb the saline solution, and watch as the distilled water is drawn under the coverslip and into the RBCs.
- Immediately observe the slide under high power and watch if the RBCs change shape.
- Record shape of RBCs and type of solution in which they were placed.
 shape _____ type of solution _____

6 Clean up your lab area.
- Place blood-stained items (slides, coverslips, and droppers) in a 10% bleach solution or as directed by your instructor.
- Place gloves in an autoclavable bag or location indicated by your instructor.
- Wash down the lab counters with 10% bleach solution in a squirt bottle and wipe with paper towels wet with 10% bleach solution. Allow to air dry.
- Wash your hands with soap and water before leaving the lab area.

7 Complete the Experimental Report with your lab group.

 ## EXPERIMENTAL REPORT: OSMOSIS IN LIVING RBCs

Results: Describe how the RBC shape changes when placed in each solution.

- 0.9% saline solution _____
- 5% saline solution _____
- distilled water _____

Discussion:

1. Describe how solute concentration affects osmosis.

2. Describe how water movement affects the cell shape.

Conclusion: State which solution(s) caused osmosis in RBCs. Compare what would happen if the RBCs were placed in tap water instead of distilled water.

C. Osmosis and Diffusion Across a Dialysis Membrane

In this activity, a sucrose solution is added to a dialysis membrane bag which is placed in distilled water. Dialysis membranes contain small pores which allow water and small solutes to cross. If water enters the bag, the weight of the bag will increase and if water leaves the bag, its weight decreases.

ACTIVITY 3 Experiment: Osmosis and Diffusion Across a Dialysis Membrane

NOTE: *This experiment should be set up at the beginning of class, so that changes can be observed and recorded throughout the lab time. The results become more dramatic the longer this experiment runs.*

1 Label Figure 5.3.

2 **Prediction:** With your lab group, predict the direction of water and sucrose movement in Figure 5.3 by circling the correct italicized choice.
- The net movement of water will be *into* or *out of* the dialysis bag (see Figure 5.3).
- The net movement of sucrose will be *into* or *out of* the dialysis bag (see Figure 5.3).

3 **Materials:** Obtain materials for osmosis and diffusion across a dialysis membrane.

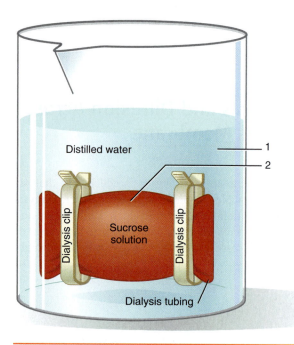

FIGURE 5.3 Dialysis bag setup.

- hypertonic solution 1 _____
 (hy-per-TOHN-ik)
- hypotonic solution 2 _____
 (hy-poh-TOHN-ik)

TABLE 5.3 Dialysis Bag Results

TIME (MIN)	COLOR OF BEAKER WATER	CHANGE IN WEIGHT OF SUCROSE SOLUTION (GRAMS)
0 min		Start weight:
15 min		
30 min		
45 min		
60 min		

4 **Data Collection:** Observe osmosis and diffusion through a dialysis membrane and record your results in Table 5.3.
- Decide who will set up the experiment, who will time, who will weigh and observe the color, and who will record.
- Cut a 12-cm piece of dialysis tubing and soak the bag in water for 3 minutes to make it easier to open.
- Fold over one end of the dialysis tubing and secure it with a dialysis clip.
- Rub the open end of the dialysis tubing between your thumb and finger to separate the sides, and fill the dialysis tubing more than three-quarters full with the red 40% sucrose solution.
- Pushing the air out of the bag, fold over the end of the bag and clip it with a dialysis clip. The dialysis bag now looks like a large "cell."
- Check to see that no liquid is leaking out of either end of the dialysis bag.
- Rinse the bag to remove excess sucrose solution, dry the bag, and weigh it. Record the beginning weight in Table 5.3.
- Submerge the dialysis bag in a beaker of distilled water.
- Record the time in Table 5.3.
- Dry and weigh the dialysis bag after 15-minute, 30-minute, and 45-minute intervals. If your lab lasts longer, you may also do a 60-minute measurement.
- Record each measurement in Table 5.3.

5 Clean up as directed by your instructor.

6 Complete the Experimental Report with your lab partners.

EXPERIMENTAL REPORT: DIALYSIS BAG

Results:

1. Describe how the weight of the bag changed over time.

2. Describe how the color of the beaker water changed over time.

Discussion:

1. Why was distilled water used instead of tap water?

2. Why was the red dye added to the sucrose solution?

3. Which direction did osmosis occur?

4. Which direction did diffusion of solutes occur?

5. Did net osmosis and diffusion of solutes stop? Did the two solutions become isotonic? Explain.

6. Explain why the bag gained weight since sucrose, Congo red, and water crossed the dialysis membrane. (*Hint: The molecular weight of sucrose is 342; Congo red, 697; and water, 18.*)

Conclusion: State what drives the osmosis and diffusion of solutes, and when does the net movement of both stop?

D. Passive Transport—Filtration

Filtration is the separation of substances passed through a filter or membrane from a region of higher pressure to a region of lower pressure. In the human body, filtration occurs in the kidneys and capillaries. Liquid and dissolved solutes flow through the pores of a filter or membrane, keeping solutes larger than the pores on the original side. In this activity, you will observe **selectivity** of the filtration membrane and a **change in filtration rate.** Keep in mind that the plasma membrane is part of a living cell and can have active processes going on simultaneously with passive processes, such as filtration.

ACTIVITY 4 Filtration

1 **Prediction:** With your lab partners, predict which substances will pass through the filter: water, copper sulfate (blue), or charcoal.

_____ will pass through the filter.

_____ will not pass through the filter.

2 **Materials:** Obtain materials for filtration.

3 **Data Collection:**
 • Decide who will set up the experiment, who will time, who will count, and who will record.
 • Fold the circular filter paper in half twice, and open it to form a cone (Figure 5.4). Place the cone in the funnel and place the funnel in a ring on the ring stand. Put the beaker under the funnel.

FIGURE 5.4 Filtration setup.

TABLE 5.4 Filtration Results

TIME (SEC)	#DROPS	FILTRATION RATE (DROPS/SEC)
15		
30		
45		
60		
90		

- Stir or mix the filtrate solution comprised of water, copper sulfate (blue), and powdered charcoal, and pour it into the funnel to just below the top of the filter paper. Watch so the edges of the filter paper do not capsize or that the solution doesn't go over the top.
- As soon as the liquid slows down enough to count individual drops, count number of drops in 15, 30, 45, 60, and 90 seconds.
- Record your information on Table 5.4.
- Observe the filtrate (liquid) in the beaker. If it is blue, the copper sulfate was filtered; if it is black, the charcoal was filtered.
- Compare your results with your predictions.

4 Clean up as directed by your instructor.

5 Complete the Experimental Report with your lab partners.

EXPERIMENTAL REPORT: FILTRATION

Results: Name the substance that passed through the filter paper.

Discussion:

1. What cell structure does the filter paper represent?

2. What determines which substances cross the filtration paper?

3. What is the force that drives the filtration?

4. Did you see a difference in the filtration rate during the activity? Explain.

Conclusion: State how cell membranes in kidneys control filtration.

E. Active Transport—Phagocytosis

Phagocytosis (*phage-* = to eat), or "cell eating," is an active process that involves bringing solid material into the cell via vesicles that form from cellular extensions of the plasma membrane called **pseudopodia** (*pseudo-* = false; *-podia* = feet). Although one-celled organisms like amoebas are involved in this activity, the human body also has cells that move in an amoeboid fashion and are involved in phagocytosis.

ACTIVITY 5 Phagocytosis in Amoebas

1 Observe phagocytosis in amoebas.
- Obtain a compound microscope, microscope slide, coverslip, amoeba culture, and amoeba food.
- Using a medicine dropper, place one drop of amoeba (starved) culture on a clean microscope slide.
- Place one drop of amoeba food in the drop of amoeba culture on the slide. Cover with a coverslip.
- Observe the amoebas first under low power and then under a higher power.
- Watch the amoebas change shape as they move. Observe them as they extend out their "false feet" to engulf the food.

2 Observe phagocytosis of amoebas at www.wiley.com/college/allen.

3 Answer the discussion question with your lab partners.

DISCUSSION QUESTION: PHAGOCYTOSIS

How would you describe the shape of an amoeba?

Reviewing Your Knowledge

A. Transport across the Plasma Membrane

Match the definition with the term.

 a. to shrink or shrivel
 b. water moving through selectively permeable membrane
 c. false feet of an amoeba or phagocytic white blood cell
 d. substance dissolved in a solution
 e. to burst a red blood cell
 f. difference between solute concentrations across a membrane
 g. same solute concentration on both sides of membrane
 h. cell eating of large solid particles
 i. lower concentration of solutes than in cytosol of cell
 j. a fluid that can contain dissolved substances
 k. higher concentration of solutes than in cytosol of cell
 l. random movement of particles from their greater concentration to their lesser concentration

_____ **1.** concentration gradient

_____ **2.** crenate

_____ **3.** diffusion

_____ **4.** hemolysis

_____ **5.** hypertonic solution

_____ **6.** hypotonic solution

_____ **7.** isotonic solution

_____ **8.** osmosis

_____ **9.** pseudopodia

_____ **10.** phagocytosis

_____ **11.** solute

_____ **12.** solvent

B. Diffusion

Select the correct lettered answer.

1. A larger molecular weight substance will diffuse _____ compared to a smaller molecular weight substance.

 (a) faster (b) slower (c) at the same rate

2. If there is no concentration gradient, a substance will not have net movement.

 (a) true (b) false

3. Passive processes use _____ to move substances across a cell membrane.

 (a) ATP (b) kinetic energy

4. Diffusion of a substance ceases when:

 (a) the cell runs out of ATP
 (b) too much pressure is applied to the cell
 (c) osmosis stops
 (d) equilibrium of the substance is reached

C. Osmosis

Select the correct lettered answer.

1. An isotonic solution will _____ an RBC.

 (a) crenate (b) hemolyze (c) cause no change to

2. A hypertonic solution will _____ an RBC.

 (a) crenate (b) hemolyze (c) cause no change to

3. A hypotonic solution will _____ an RBC.

 (a) crenate (b) hemolyze (c) cause no change to

Using Your Knowledge

Select the correct lettered answer in questions 1–8.

A. Diffusion

_____ **1.** White blood cells engulf bacteria. This is an example of diffusion.

 (a) true (b) false

_____ **2.** Food cooking on the stove in the kitchen can be smelled in the living room. This is an example of diffusion.

 (a) true (b) false

_____ **3.** Oxygen in the lungs moves into the bloodstream and carbon dioxide moves in the opposite direction. This is an example of diffusion.

 (a) true (b) false

B. Osmosis

_____ **4.** A test tube with blood in it has a particular solution added to it. After several minutes, the solution is not clear anymore, but becomes red. Which solution was added to the blood to obtain this result?

 (a) 0.9% saline (b) 5% saline (c) distilled water

_____ **5.** A 0.8% saline solution would be _____ to the cytosol of a cell.

 (a) hypotonic (b) hypertonic (c) isotonic

_____ **6.** If a 50% sugar solution had been used in the dialysis bag in Activity 3, there would be a faster rate of osmosis.

 (a) true (b) false

_____ **7.** If you placed a peeled apple or potato in a 5% salt solution, it would:

 (a) gain weight (b) lose weight (c) stay the same weight

_____ **8.** A person's hands become wrinkled after spending a long, relaxing time in the tub. Tub water does not have as many solutes in it compared with the human body. Explain why the hands look wrinkled.

(a) The tub water is *hypotonic* to body cells and water *enters* the cells.
(b) The tub water is *hypotonic* to body cells and water *leaves* the cells.
(c) The tub water is *hypertonic* to body cells and water *enters* the cells.
(d) The tub water is *hypertonic* to body cells and water *leaves* the cells.

C. Filtration

9. In the filtration experiment, hydrostatic pressure was the force driving filtration. What is the force driving filtration in the kidney?

10. What prevents blood cells from being filtered into the urine?

Tissues

6

There are four primary tissue types in the human body: epithelial, connective, muscle, and nervous. **Epithelial tissue** covers surfaces, lines cavities, and forms glands, and **connective tissue** supports and protects other tissues. **Muscle tissue** causes movement, and **nervous tissue** receives and generates nerve impulses. Organs are formed from two or more different tissues working together to perform a specific function.

A. Epithelial Tissue

Epithelial tissues or epithelia (epithelium, sing.) exhibit **cellularity;** that is, most of the tissue consists of cells packed together in an orderly fashion with little extracellular material (matrix) between cells. Epithelial tissue function is determined by the cell type and number of cell layers. Epithelia are supported by underlying connective tissue,

FIGURE 6.1 Epithelial tissue lining the mouth.

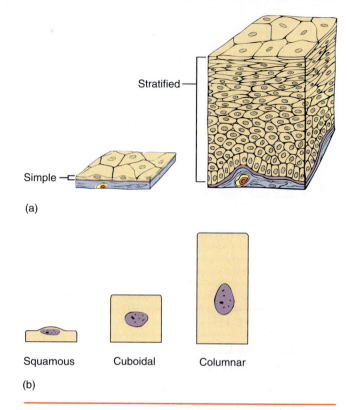

FIGURE 6.2 Epithelial tissue classification.

with the epithelial and connective tissues separated by a **basement membrane.**

There are two categories of epithelial tissues: covering and lining epithelia and glandular epithelia. **Covering and lining epithelia** cover the surface of the body and some organs, and line all hollow body structures. This type of epithelium forms barriers that protect and control what substances can cross into adjacent tissues. **Glandular epithelia** form glands that produce and secrete products needed by the body.

Covering and lining epithelia face exterior or interior spaces. The epithelial cell surface adjacent to the space is called the **apical** (*apex* = tip) **surface,** whereas the epithelial cell surface adjacent to the basement membrane is the **basal surface.** When viewed from the apical surface, one can observe how epithelial cells fit together like floor tiles to form a lining or barrier (Figure 6.1). When viewed in cross-section (side view of cells), one can observe differences in structure of the apical and basal surfaces. The differences in structure are associated with differences in function.

Epithelial tissue types are classified according to the shape of the cell in the apical cell layer and the number of epithelial cell layers.

Epithelial cells have four shapes: squamous, cuboidal, columnar, and transitional. Cell shapes are best seen in side views of the cells (Figure 6.2). **Squamous cells** are the thinnest cells, and they have a flattened nucleus. **Cuboidal cells** are cubelike with a round nucleus in the center of the cell. **Columnar cells** are tall with an oval nucleus close to

the base of the cell. **Transitional cells** change shape; the apical cells are cuboidal when the tissue is relaxed and squamous when the tissue is stretched.

Epithelial tissue with only one cell layer is **simple epithelium,** while epithelial tissue with two or more cell layers is **stratified epithelium** (Figure 6.2). **Simple epithelium** provides a selective barrier allowing diffusion, filtration, secretion, or absorption of selected substances. There are three types of simple epithelium: simple squamous, simple cuboidal, and simple columnar. **Stratified epithelium** is thicker, subject to wear and tear, and forms a protective barrier. Multiple cell layers make the tissue more resistant to damage, thereby preventing pathogens and foreign materials from crossing into underlying tissues. There are four types of stratified epithelium: stratified squamous, stratified cuboidal, stratified columnar, and transitional. **Pseudostratified** (*pseudo-* = false) **columnar epithelium** is only one cell layer thick because all the cells touch the basement membrane, although not all cells reach the apical surface. Therefore, there are cells of different shapes and heights, and their nuclei are at different levels. This gives the illusion of several different layers of cells. The tallest cells are narrow where they touch the basement membrane, but have a columnar shape toward the outer surface, while the shorter cells do not reach the apical surface.

NOTE: *Observe each tissue first with a scanning or low-power objective lens, focusing with the coarse focus knob. Before switching to a more powerful objective lens, move the slide so that the area you want to observe is in the center of the field of view. After changing to a more powerful lens, use the fine focus knob only to bring the tissue into focus. Use this procedure with all tissue slides.*

ACTIVITY 1 Microscopic Examination of Epithelia

1 Examine and color the stratified epithelium drawing in Figure 6.1. Color the apical and basal layers of the stratified squamous epithelium contrasting colors.

2 Examine the photomicrographs of the epithelial tissue types in Figures 6.3–6.8. Stratified cuboidal and stratified columnar are not included because they are less common.

Survey photomicrograph (40×) for each tissue

- Observe that the epithelial and connective tissue layers stain differently.
- Locate where the basement membrane forms the border between the epithelial and connective tissue layers.
- Note whether there is one epithelial cell layer (simple epithelium) or two or more epithelial cell layers (stratified epithelium).

Photomicrograph (400×) for each tissue

- Label the photomicrographs.
- Examine the shape of the epithelial cell at the apical surface. This determines the tissue type as squamous, cuboidal, or columnar epithelium.
- Note any structural differences between apical surface and basal surface of epithelial tissue.

3 Review the location and function for each epithelial tissue type in Tables 6.1–6.6.

4 Examine a prepared microscope slide of a whole mount of mesothelium (simple squamous epithelium).
- This slide shows cells viewed from the apical surface.
- Observe how the epithelial cells fit close together to form a good barrier.

5 Examine prepared microscope slides showing tissue cross-sections. Use the survey photomicrographs in Figures 6.3–6.8 to help you locate the epithelia at low-power. Use the photomicrograph at 400× to help you locate the major structures in each tissue.

6 Examine epithelial tissue photomicrographs in Histology Atlas Figures HA-1 to HA-8. Stratified cuboidal and stratified columnar epithelia are included.

Peritoneum

Intestinal wall

Peritoneum

40×
(a) Survey photomicrograph.

1 — 2

3
4

400×, H&E, human
(b) Simple squamous epithelium.

FIGURE 6.3 Sectional view of peritoneum.

- connective tissue
- nucleus of connective tissue cell
- nucleus of simple squamous epithelial cell
- simple squamous epithelium

1 _____

2 _____

3 _____

4 _____

TABLE 6.1 Location and Function of Selected Simple Squamous Epithelia

LOCATION	FUNCTION
Mesothelium (epithelial layer of peritoneum)	**Secretion** of serous fluid into peritoneal cavity.
Alveoli (air sacs of lungs)	Single layer of squamous cells creates a short distance for **diffusion** of oxygen and carbon dioxide.
Glomerular capsule (part of filtration membrane in kidney)	**Filtration** of blood to form urine filtrate (substance that is converted into urine).
Endothelium of capillaries	Single layer of squamous cells creates a short distance for **diffusion** of substances between blood and interstitial fluid (tissue fluid).

Kidney

Glomerulus

Lumen
of kidney
tubule

40×
(a) Survey photomicrograph.

1
2
3
4

Question: Explain why some tubule cross-sections are circular and others are oval.

400×, H&E, human
(b) Simple cuboidal epithelium.

FIGURE 6.4 Sectional view of kidney tubules.

- apical surface of simple cuboidal epithelial cell
- lumen of kidney tubule
- nucleus of simple cuboidal epithelial cell
- simple cuboidal epithelium

1 _____

2 _____

3 _____

4 _____

TABLE 6.2 Location and Function of Selected Simple Cuboidal Epithelia

LOCATION	FUNCTION
Walls of kidney tubules	Modify urine filtrate by **absorption** of substances from the filtrate and **secretion** of other substances into the filtrate.
Glands	**Secretion** of products made by the simple cuboidal epithelial cells.

Small intestine

Villus

Lumen

40×
(a) Survey photomicrograph.

Goblet cell

1

2

3

4

400×, H&E, human
(b) Simple columnar epithelium.

FIGURE 6.5 Sectional view of small intestine.

- connective tissue
- microvilli on apical surface of simple columnar epithelial cell
- nucleus of simple columnar epithelial cell
- simple columnar epithelium

1 _____

2 _____

3 _____

4 _____

TABLE 6.3 Location and Function of Selected Simple Columnar Epithelia

LOCATION	FUNCTION
Lining of stomach and intestines	**Secretion** of digestive juices by simple columnar cells and secretion of mucus by goblet cells. In the small intestine the simple columnar epithelial cells have **microvilli** (*micro-* = small; *villi* = shaggy hair) to increase surface area for **absorption.**
Uterine tubes (Fallopian tubes)	Simple columnar epithelial cells have **cilia** that help **move** the egg to the uterus.
Central canal of spinal cord	Ciliated cells **move** cerebrospinal fluid.

Mucous glands

Lumen

Epithelial tissue

Connective tissue

Smooth muscle

40×
(a) Survey photomicrograph.

200×, H&E, human
(b) Nonkeratinized stratified squamous epithelium.

FIGURE 6.6 Sectional view of esophagus.

- connective tissue
- nucleus of epithelial cell in basal layer of epithelium
- nucleus of squamous epithelial cell
- stratified squamous epithelium

1 _____

2 _____

3 _____

4 _____

TABLE 6.4 Location and Function of Selected Stratified Squamous Epithelia

LOCATION	FUNCTION
Surface of skin (keratinized stratified squamous epithelium)	Epithelial layer of skin is a tough, dry, waterproof outer surface that forms a **protective barrier.**
Lining of mouth, esophagus, anus, and vagina (nonkeratinized stratified squamous epithelium)	Moist epithelial layer that forms a **protective barrier** in areas subject to abrasion and friction.

Lumen

Connective tissue

Epithelial tissue

Smooth muscle

40×

(a) Survey photomicrograph.

Lumen

1

2

3

4

400×, H&E, human
(b) Transitional epithelium.

FIGURE 6.7 Sectional view of a relaxed urinary bladder.

- connective tissue
- nucleus of transitional epithelial cell in apical layer
- nucleus of transitional epithelial cell in basal layer
- transitional epithelium

1 _____

2 _____

3 _____

4 _____

TABLE 6.5 Location and Function of Transitional Epithelia

LOCATION	FUNCTION
Lining of urinary bladder and parts of the ureters and the urethra	Provides a **protective barrier** that permits **distension.**

Lumen

Epithelial
tissue

Connective
tissue

40× (a) Survey photomicrograph.

goblet cell 1

2

3

4

400×, H&E, human (b) Pseudostratified ciliated columnar epithelium.

FIGURE 6.8 Sectional view of the trachea.

- cilia on apical surface of columnar epithelial cell
- connective tissue
- nucleus of ciliated columnar epithelial cell
- pseudostratified ciliated columnar epithelium

1 _____

2 _____

3 _____

4 _____

TABLE 6.6 **Location and Function of Selected Pseudostratified Columnar Epithelia**

LOCATION	FUNCTION
Lining of nasal cavity, trachea, and bronchi	**Secretion** of mucus by goblet cells. The columnar cells have **cilia** which **move** mucus toward the pharynx.

B. Connective Tissue Overview

Connective tissue is the most abundant primary tissue in the body and has a variety of functions. It supports epithelial tissue and connects it to other tissues. Connective tissue forms tendons (connect muscle to bone) and ligaments (connect bone to bone). Connective tissue also provides coverings that support and protect muscle and nervous tissue. Bone, the hardest connective tissue, protects body organs and provides a framework for movement of muscles. Adipose tissue (fat) insulates body tissues and stores lipids, and blood provides a liquid medium for transportation of substances throughout the body.

Connective tissues are not very cellular, containing more extracellular matrix than cells (Figure 6.9). Connective tissue function is determined by the properties of the extracellular matrix components. **Extracellular matrix** consists of **fibers** and **ground substance** that are synthesized and secreted by connective tissue cells. The three major types of fibers are **collagen fibers** (which provide strength), **elastin fibers** (which provide strength and elasticity), and **reticular fibers** (fine, branching fibers which provide a netlike framework). The number and type of fibers present contribute to the strength, elasticity, and structure of the extracellular matrix. The **ground substance** may be fluid, semifluid, gelatinous, or hard. It secures connective tissue cells and fibers in the extracellular matrix and is the medium through which interstitial fluid diffuses between blood and cells and through which macrophages and white blood cells move.

Mature connective tissues include loose connective tissue, dense connective tissue, cartilage, bone, and blood. The connective tissue types have different connective tissue cells that form the extracellular matrix. The extracellular matrix of **loose and dense connective tissues** is formed by **fibroblasts** (-*blast* = early, developing stage); the extracellular matrix of **cartilage** is formed by **chondroblasts** (*chondro-* = cartilage) and maintained by **chondrocytes;** and the extracellular matrix of **bone** is formed by **osteoblasts** (*osteo-* = bone) and maintained by **osteocytes**. **Plasma** is the extracellular matrix in which blood cells and platelets are suspended; however, it is not made by these cells.

Loose Connective Tissue

Loose connective tissue includes areolar, reticular, and adipose tissues. In **loose connective tissue,** the fibers in the extracellular matrix are loosely arranged. Collagen, elastic, and reticular fibers in this tissue provide strength, elasticity, and support. Ground substance is semifluid (viscous), but interstitial fluid can easily diffuse through it. Fibroblasts and adipocytes (lipid-storing cells) permanently reside in the tissue. Other cells, such as macrophages, mast cells, and white blood cells, enter connective tissue from blood and are involved in body defenses.

Areolar connective tissue is the most abundant connective tissue. It contains all three fiber types, fibroblasts, a variety of cells involved in body defenses, and a viscous ground substance. **Reticular** (*reticulo-* = netlike) **fibers** are the dominant fiber type in **reticular connective tissue. Reticular cells** synthesize the reticular fibers and the ground substance. Reticular cells are actually fibroblasts, but they synthesize more reticular fibers than collagen fibers. **Adipose** (*adipo-* = pertaining to fat) **tissue,** like areolar connective tissue, contains fibroblasts, fibers, ground substance, and adipose cells. However, adipose tissue has a greater number of adipose cells or **adipocytes** that store lipid. The lipid within the cell occupies most of the cell volume and pushes cytoplasm and organelles to the periphery. Some slides do not have the lipid stained, or the lipid has been removed during tissue preparation. In those slides, the nucleus and cytoplasm are stained, and the space inside the cell normally occupied by lipid appears "empty."

Cell Fiber Blood vessel Cell

Fiber

Collagen fiber Cell Ground substance Cell

FIGURE 6.9 Connective tissue (areolar).

Dense Connective Tissue

The extracellular matrix of dense connective tissue is packed with fibers and contains very little ground substance and few fibroblasts. The extracellular matrix of **dense regular connective tissue** is packed with parallel bundles of collagen fibers that run in the direction of the pulling forces applied to the tissue. Fibroblasts are squeezed between the bundles, and the cytoplasm extends between and around collagen bundles. Fibroblasts secrete collagen fibers and ground substance. Collagen fibers give unstained tissue a silvery white appearance, which is why this tissue is often called *white fibrous tissue.*

Dense irregular connective tissue, like dense regular connective tissue, has little ground substance and few fibroblasts. The extracellular space is also packed with bundles of collagen fibers with fibroblasts squeezed between these bundles. In dense irregular connective tissue, however, the bundles of collagen fibers are irregularly arranged. This reflects the direction of the pulling forces to which this tissue is exposed, usually from many different directions. In **elastic connective tissue,** the extracellular matrix is packed with elastic fibers, and fibroblasts are found in the spaces between these fibers. Elastic fibers allow the tissue to be stretched and then regain its original size and shape (recoil).

Cartilage

Cartilage is a specialized form of connective tissue. Its extracellular matrix consists of collagen and elastic fibers embedded in a firm gelatinous ground substance. Collagen fibers give the tissue its strength, and the ground substance, which binds water to form a firm gel, gives cartilage its resiliency. **Chondroblasts** (*chondro-* = cartilage) secrete the fibers and ground substance, become isolated in spaces called **lacunae** (little lakes), and transform into **chondrocytes.** Since the tissue is avascular, the chondrocytes receive their nutrients through diffusion from adjacent vascular tissues.

There are three types of cartilage: hyaline (*hyalos* = glass), elastic, and fibrocartilage. These cartilage types differ in the amount and kinds of fibers and ground substance molecules present in their extracellular matrix. **Hyaline cartilage** is the most predominant cartilage in the body and contains fine collagen fibers in the extracellular matrix. This cartilage type appears glassy to the naked eye, and under the compound microscope its matrix looks smooth and homogeneous. It contains collagen fibers that are thin and not visible with a compound microscope. The matrix of hyaline cartilage is resilient and firm. **Elastic cartilage** is similar to hyaline cartilage, except the matrix is packed with elastic fibers. The many elastic fibers allow this cartilage to be much more flexible. **Fibrocartilage** has fewer lacunae and chondrocytes, and its extracellular matrix is packed with thick collagen fibers that give it tensile strength similar to dense connective tissue. Under the microscope, collagen fibers are the most prevalent structures seen.

Bone

Bone is the hardest of the connective tissues. It protects body organs and provides support for organs and skeletal muscles. The extracellular matrix is organized in layers called **lamellae** and consists of collagen fibers, ground substance, and inorganic salts. Inorganic salts, especially calcium salts, give bone its hardness; collagen fibers provide strength and flexibility. **Osteocytes** (mature bone cells) are trapped in spaces called **lacunae.** Small canals called **canaliculi** are two-way canals that connect lacunae to each other and to larger canals that contain blood vessels. Nutrients diffuse from a blood vessel through the canaliculi to the osteocytes, and waste materials diffuse back to the blood vessel for removal.

The two types of bone tissue are **compact bone** and **spongy bone.** These two types of osseous tissue differ in the amount and size of spaces present. The extracellular matrix of compact bone is organized into repeating structural units called **osteons.** In the center of each osteon is a large central canal. Spongy bone has large spaces compared to compact bone and does not have osteons. Instead, the extracellular matrix of spongy bone is arranged in **trabeculae** (little beams)—flat plates with a latticelike network of thin, bony columns.

Blood

Blood transports nutrients and chemical messengers to all body tissues and transports wastes from tissues to excretory organs. Blood is composed of red blood cells, white blood cells, platelets, and plasma. **Plasma** is the extracellular matrix, and its fibers are produced and observed only during blood clotting. Red blood cells obtain their red color from hemoglobin, and the mature cells do not contain a nucleus (anucleate). White blood cells, which appear white or clear on unstained slides, are nucleated. Platelets, anucleate fragments of cells that are much smaller than red blood cells, participate in blood clotting.

ACTIVITY 2 **Microscopic Examination of Connective Tissue Types**

1 Examine and color the areolar connective tissue drawing in Figure 6.9. Color the different fiber types with contrasting colors.

2 Label and examine the photomicrographs of the connective tissue types in Figures 6.10–6.21.
- Observe the extracellular matrix in each tissue.
- Compare the amount of fibers and ground substance.
- Identify the connective tissue cells.

3 Review the location and function for each connective tissue type in Tables 6.7–6.11.

4 Examine prepared microscope slides of the different connective tissue types. Use Figures 6.10–6.21 to help you locate the major structures in each tissue.

5 Examine connective tissue photomicrographs in Histology Atlas Figures HA-9 to HA-23.

400×, Verhoeff Orange Safrin, human

FIGURE 6.10 Areolar connective tissue spread of mesentery.

- collagen fiber
- connective tissue cells (fibroblasts or lymphocytes)
- elastic fiber

1 _____

2 _____

3 _____

Lymph node

400×, human

FIGURE 6.11 Sectional view of reticular tissue of the lymph node.

- reticulocyte
- reticular fiber

1 _____

2 _____

Heart

Fat

400×, H&E, human

FIGURE 6.12 Sectional view of adipose tissue.

- lipid storage area
- nucleus of adipocyte

1 _____

2 _____

TABLE 6.7 **Location and Function of Loose Connective Tissues**

LOOSE CONNECTIVE TISSUE TYPE	LOCATION	FUNCTION
Areolar	Beneath all epithelial tissues	Binds epithelium to underlying tissues and allows nutrients to diffuse to epithelial cells.
Reticular	Liver, spleen, lymph nodes	Forms support (framework) of these soft organs.
Adipose	Under skin and surrounding organs	Stores lipids for fuel and thermal insulation; cushioning organs.

400×, H&E, human

FIGURE 6.13 Sectional view of dense, regular connective tissue-forming tendons.

- collagen fiber bundle
- fibroblast

1 _____

2 _____

400×, H&E, human

FIGURE 6.14 Sectional view of dense, irregular connective tissue in skin.

- collagen fiber bundles running in different directions
- fibroblast
- parallel collagen fiber bundles

1 _____

2 _____

3 _____

Aorta

Heart

400×, human

FIGURE 6.15 Sectional view of elastic connective tissue in the wall of the aorta.

- bundle of elastic fibers
- fibroblast
- individual elastic fiber

1 _____

2 _____

3 _____

TABLE 6.8 **Location and Function of Dense Connective Tissues**

DENSE CONNECTIVE TISSUE TYPE	LOCATION	FUNCTION
Dense regular	Forms ligaments (connects bone to bone), tendons (connects muscle to bone), and aponeuroses (sheet-like tendons that connect muscle to muscle or muscle to bone)	Resists pulling forces at attachment points.
Dense irregular	Skin	Resists pulling forces from many different directions that would tear skin when skin is stretched.
Elastic	Lungs, trachea, and bronchi Aorta and other elastic arteries	Allows respiratory organs to recoil after inhalation. Recoil of elastic tissue helps push blood through cardiovascular system.

Trachea

400×, H&E, human

FIGURE 6.16 Sectional view of hyaline cartilage in the wall of the trachea.

- extracellular matrix
- lacuna (la-KOO-na)
- nucleus of chondrocyte

1 _____

2 _____

3 _____

400×

FIGURE 6.17 Sectional view of elastic cartilage of the ear.

- elastic fibers
- lacuna
- nucleus of chondrocyte

1 _____

2 _____

3 _____

Tendon of
quadriceps
femoris muscle

Patella
(knee cap)

Portion of
right lower limb

400×

FIGURE 6.18 Sectional view of fibrocartilage of tendon.

- chondrocytes
- collagen fibers
- lacuna

1 _____

2 _____

3 _____

TABLE 6.9 Location and Function of Cartilage

CARTILAGE TYPE	LOCATION	FUNCTION
Hyaline cartilage	Ends of long bones (articular cartilage)	Smooth surface that is resilient and reduces friction at joint.
	Trachea and bronchi	Provide support and flexibility to ensure an open airway.
	Anterior ends of ribs (costal cartilage)	Connect ribs to sternum with flexible joint (breastplate).
	Embryonic skeleton	Provides template for bone formation.
Fibrocartilage	Intervertebral discs	Provide strength to discs that form joints between vertebrae and act as shock absorbers.
	Cartilage pads in knee	Provide cushioning for bones forming knee joints.
	Pubic symphysis	Forms strong, flexible joint between hip bones.
Elastic cartilage	External ear	Provides support and maintains shape of external ear.
	Auditory tube	Provides support and elasticity to auditory tube as it changes diameter to equalize pressure in middle ear (ear pops).
	Epiglottis of larynx	Provides support and elasticity to epiglottis as it folds to block entrance to trachea while swallowing food and liquid.

400×, unstained, human

FIGURE 6.19 Sectional view of dried compact bone.

- canaliculus
- central canal
- lamella
- lacuna

1 _____

2 _____

3 _____

4 _____

End of long bone

Red marrow cell

400×

FIGURE 6.20 Sectional view of spongy bone.

- osteocyte
- trabecula

1 _____

2 _____

TABLE 6.10 Location and Function of Osseous Tissue

TYPE OF OSSEOUS TISSUE	LOCATION	FUNCTION
Compact	Exterior of bones	Support, protection, storage of minerals.
Spongy	Interior of bones	Support, protection, storage minerals.

Blood in blood vessels

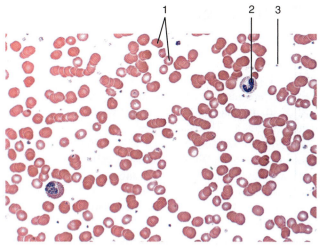

400×, H&E, human

FIGURE 6.21 Blood smear.

- red blood cells
- nucleus of white blood cell
- platelet

1 _____

2 _____

3 _____

TABLE 6.11 **Location and Function of Blood Components**

BLOOD COMPONENT	LOCATION	FUNCTION
Plasma	Liquid part of blood in arteries, capillaries, veins	Transport of nutrients, blood gases, wastes, chemical messengers, blood cells and platelets.
Red blood cells	Formed element of blood in arteries, capillaries, veins	Transport of blood gases.
White blood cells	Formed element of blood in arteries, capillaries, veins WBCs can also leave blood vessels and enter infected tissues.	Attack pathogens and other substances that invade the body.
Platelets	Formed element of blood in arteries, capillaries, veins	Participate in blood clotting.

C. Muscle Tissue Overview

Muscle tissue is very cellular, with most of the tissue consisting of muscle cells (Figure 6.22). All muscle tissues are highly vascularized and are innervated. Muscle cells are elongated cells called **fibers** that shorten (contract) when stimulated, causing movement. There are three types of muscle tissue—skeletal, cardiac, and smooth—with each type having a distinct appearance. **Skeletal muscle** cells are large, multinucleated, cylindrical cells. **Cardiac muscle** cells are smaller, branching cells with one or two centrally located nuclei per cell. **Intercalated discs** are dark bands where cardiac muscle cells connect end to end. Skeletal and cardiac muscle cells exhibit **striations** (light and dark bands). **Smooth muscle** cells are small, spindle-shaped (tapered at both ends) cells that are uninucleated and nonstriated.

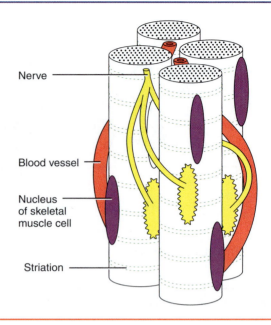

FIGURE 6.22 Skeletal muscle tissue.

ACTIVITY 3 Muscle Tissue

1 Examine and color Figure 6.22, which depicts skeletal muscle tissue. Color a muscle fiber and its striations using contrasting colors.

2 Label and examine the photomicrographs of muscle tissue in Figures 6.23–6.25.
 • Compare the size, shape, and number of nuclei of skeletal, cardiac, and smooth muscle fibers.
 • Look for striations in skeletal and cardiac muscle fibers.

3 Review the location and function for each muscle tissue type in Table 6.12.

4 Examine prepared microscope slides of the different muscle tissue types.

5 Examine muscle tissue photomicrographs in Histology Atlas Figures HA-24 to HA-26.

400×, H&E, human

FIGURE 6.23 Sectional view of skeletal muscle fibers.

• nucleus
• striation (stry-AY-tion)
• width of individual muscle fiber

1 _____

2 _____

3 _____

Heart

400×, iron H&E

FIGURE 6.24 Longitudinal view of cardiac muscle fibers.

- branches of cardiac muscle fiber
- intercalated discs
- nucleus
- width of cardiac muscle fiber

1 _____

2 _____

3 _____

4 _____

Smooth muscle layers
Connective tissue
Epithelial lining
Lumen

60×
(a) Survey photomicrograph

400×
(b) Smooth muscle fibers in cross-section
and longitudinal section

FIGURE 6.25 Sectional view of smooth muscle in wall of the ureter.

- nucleus of smooth muscle fiber in cross-section
- nucleus of smooth muscle fiber in longitudinal section

1 _____

2 _____

TABLE 6.12 Location and Function of Muscle Tissue Types

MUSCLE TISSUE TYPE	LOCATION	FUNCTION
Skeletal muscle	Attached to bones and skin	Movement of bones and skin. Contraction generates heat.
Cardiac muscle	Wall of heart	Movement of blood through cardiovascular system.
Smooth muscle	Walls of digestive tract organs	Movement of food through digestive tract.
	Walls of arteries and veins	Contraction and relaxation controls blood flow and blood pressure.
	Walls of ureters, urinary bladder, and urethra	Movement of urine through urinary tract.
	Intrinsic muscles of eye	Contraction and relaxation controls pupil size.

D. Nervous Tissue Overview

Nervous tissue, which forms the brain, spinal cord, and nerves, is very cellular (Figure 6.26). Two basic categories of nervous tissue cells are **neurons** and **neuroglia** (*neuro* = nerve; *glia* = glue). Neurons receive and send information, whereas neuroglia support the neurons and help them to function. Neurons have one or more processes (cellular extensions) that receive or send information as nerve impulses. **Dendrites** (*dendro* = tree) are processes that receive signals from sensory receptors or other neurons. An **axon** is a process that sends signals to other neurons, muscles, or glands. Neurons have one axon but may have many dendrites. The main part of the cell where the nucleus is located is called the **cell body.** In the following activity you will observe only **multipolar neurons,** which are neurons with many processes.

ACTIVITY 4 Nervous Tissue

1 Examine and color Figure 6.26, which depicts nervous tissue. Color the multipolar neuron cell bodies, nuclei, and processes using contrasting colors.

2 Label and examine the photomicrograph of nervous tissue in Figure 6.27.
 • Compare the sizes of the neuron and neuroglia.
 • Observe the cross-sections of processes surrounding the neuron and neuroglial cells.

3 Examine a prepared microscope slide of a multipolar neuron.

4 Examine nervous tissue photomicrographs in Histology Atlas Figures HA-27 to HA-29.

FIGURE 6.26 Nervous tissue.

Spinal cord

LM 260×

FIGURE 6.27 Photomicrograph of nervous tissue.

- cell body of multipolar neuron
- nucleus
- processes

1 _____

2 _____

3 _____

E. Body Membranes

The body contains synovial and epithelial membranes. These membranes cover or line body structures or cavities. **Synovial membranes** line synovial joint cavities and produce synovial fluid that lubricates the joints.

Epithelial membranes are layers of epithelial tissue that are supported by underlying connective tissue. The body contains three types of epithelial membranes: mucous, serous, and cutaneous membranes. **Mucous membranes** are wet or moist membranes that line body cavities opening to the exterior of the body, and are found lining part of the urinary tract and the digestive, respiratory, and reproductive tracts. Although mucous glands are found in some mucous membranes, not all mucous membranes are moistened by secretions of mucus (mucus, noun; mucous, adjective). The epithelial layer of the mucous membranes varies, but the underlying connective tissue layer, **lamina propria** (*propria* = one's own), is areolar connective tissue.

Serous membranes are moist membranes that line the ventral body cavities. They produce serous fluid and are composed of simple squamous epithelium (mesothelium) over areolar connective tissue.

The **cutaneous membrane** is the dry membrane that forms the skin and is composed of keratinized stratified squamous epithelium and a thick connective tissue layer.

ACTIVITY 5 **Body Membranes**

1 Examine the serous membrane in Figure 6.3.
2 Examine the mucous membranes in Figures 6.5–6.8.

Reviewing Your Knowledge

A. Primary Tissue Structure and Function

Name the primary tissue type (epithelial, connective, muscle, or nervous) that is described.

_____ **1.** Cells contain processes that receive and generate electrical signals to communicate with other cells.

_____ **2.** Tissue has elongated cells that shorten and cause movement.

_____ **3.** Tissue contains more extracellular matrix than cells.

_____ **4.** Cells either form a barrier that controls passage of molecules or form glands.

_____ **5.** ⎫

_____ **6.** ⎬ Primary tissue types that exhibit cellularity.

_____ **7.** ⎭

_____ **8.** ⎫

_____ **9.** ⎬ Properties of cells that determine function of tissue.

_____ **10.** ⎭

_____ **11.** Properties of extracellular matrix that determine function of tissue.

Identify the primary tissue types in Figure 6.28.

12 _____ 13 _____ 14 _____ 15 _____

FIGURE 6.28 Photomicrographs of primary tissues.

B. Connective Tissues

Write the name of the connective tissue type(s) that matches each description. A connective tissue can be used more than once.

_____ 1. Contains elastic fibers and is found in the lungs. This tissue allows the lungs to inflate during inhalation and return to their original shape after exhaling.

_____ 2. Packed with parallel bundles of collagen fibers and found in tendons. This tissue resists pulling forces applied by muscles.

_____ 3. Has a firm, gelatinous ground substance containing collagen fibers. This tissue is found in the tracheal wall to support and prevent the trachea from collapsing.

_____ 4. Found under covering and lining epithelium. Its extracellular matrix contains a loose arrangement of fibers, and its viscous ground substance facilitates the flow of interstitial fluid containing nutrients to epithelial tissues. It also cushions and supports epithelia.

_____ 5. Contains many elastic fibers in a firm gelatinous ground substance. Located in external ear, auditory tube, and epiglottis.

_____ 6. Hard extracellular matrix containing trabeculae.

_____ 7. Forms a framework in the spleen, bone marrow, and lymph nodes. It contains fine, branching fibers.

_____ 8. Is packed with bundles of collagen fibers running in different directions. It is found in skin and allows skin to resist pulling forces from many different directions.

_____ 9. Fluid extracellular matrix used to transport substances throughout the body.

_____ 10. Contains a large number of lipid-storing cells. It is found throughout the body, cushions and insulates organs, and stores lipids for future energy needs.

_____ 11. Hard extracellular matrix containing osteons; involved in protection and support.

_____ 12. Firm gelatinous ground substance packed with bundles of collagen fibers. This tissue is found in intervertebral discs, pubic symphysis, and knee meniscus.

_____ **13.**

_____ **14.**

_____ **15.**

_____ **16.** Fibroblasts produce extracellular matrix for these connective tissue types.

_____ **17.**

_____ **18.**

_____ **19.**

_____ **20.** Osteoblasts produce extracellular matrix.

_____ **21.**

_____ **22.** Chondroblasts produce extracellular matrix.

_____ **23.**

_____ **24.** Fibers not present unless injury occurs. Extracellular matrix not produced by cells present in this tissue.

C. Epithelial and Connective Tissue Identification

Identify the epithelial and connective tissue types in Figure 6.29.

1. _____

2. _____

3. _____

4. _____

5. _____

6. _____

7. _____

8. _____

9. _____

10. _____

11. _____

12. _____

13. _____

14. _____

15. _____

16. _____

17. _____

18. _____

1

2

3

4

5

6

FIGURE 6.29 Identification of epithelial and connective tissue types.

FIGURE 6.29 Identification of epithelial and connective tissue types, *continued*.

1 2 3

FIGURE 6.30 Muscle tissue identification.

D. Muscle Tissue Identification

Identify the muscle tissue types in Figure 6.30.

1. _____

2. _____

3. _____

E. Epithelial Membranes

Write the name of the epithelial membrane described by each statement.

_____ 1. Moist membranes that line body cavities that are open to the exterior.

_____ 2. Dry membrane.

_____ 3. Membrane composed of simple squamous epithelium over areolar connective tissue.

_____ 4. Membrane that lines the digestive tract.

_____ 5. Membrane covering the soles of the feet.

_____ 6. Membrane that lines the respiratory tract.

_____ 7. Membrane that lines the urinary and reproductive tracts.

_____ 8. Membrane forming pericardium and pleura.

Using Your Knowledge

A. Genetic Diseases and Tissue Function

Each of the following diseases is due to a gene mutation and the production of an abnormal protein. Look up the diseases in your textbook or other resource. For each disease (1) identify the abnormal protein and (2) state how the abnormal protein affects the function of the tissue.

Marfan's syndrome and connective tissue

1. Abnormal protein:

2. The effect of the abnormal protein on tissue function:

Hemophilia A and blood

3. Abnormal protein:

4. The effect of the abnormal protein on tissue function:

B. Nutritional Deficiencies and Tissue Function

Nutritional deficiencies interfere with cell metabolism and tissue structure. Look up rickets in your textbook or other resource and, (5) describe how rickets affects the extracellular matrix of bone and, (6) describe how this change in extra cellular matrix causes the symptoms of rickets.

Rickets and bone

5. Effect on extracellular matrix:

6. Extracellular matrix changes and symptoms:

C. Tissues Forming Organs

Name the tissue types that form the wall of each organ in questions 7–10. Use your textbook as a reference and list the tissues in order from lumen to outer surface.

7. esophagus _____

8. small intestine _____

9. trachea _____

10. ureter _____

The Integumentary System Structure and Function

Objectives

After completing this exercise, you should be able to:

- Describe the structure and function of the integumentary system
- Identify the layers of the epidermis and dermis, and locate the hypodermis on models or microscope slides
- Identify the accessory structures of the skin on models or microscope slides
- Describe the function of the epidermal accessory structures
- Compare eccrine sweat gland density on the forehead, forearm, palm, and anterior leg

Materials

- integumentary system model or chart
- compound microscopes and lens paper
- prepared microscope slides: thick skin, thin skin with hair
- ink pad for fingerprinting or demonstrating of fingerprints
- Eccrine Gland Density: 1 cm × 1 cm squares of bond paper (4 per student); Betadine-soaked cotton swabs; surgical tape

The **integumentary system** (*inte-* = whole; *-gument* = body covering) consists of the skin, hair, nails, sweat and sebaceous glands, and associated muscle and nervous tissue. This system provides a protective barrier for the body, contains sensory receptors, aids in the production of vitamin D, is important in regulating body temperature, and plays a minor role in excretion and absorption.

A. Major Divisions of the Skin

The skin has two major divisions: the superficial epidermis and the deep dermis. The **epidermis** (*epi-* = above), the outer layer of the skin, is composed of epithelial tissue. The **dermis** is the connective tissue layer that is firmly attached to the epidermis by a basement membrane, supports the epidermis, and connects it to the underlying hypodermis. Although it is not a part of the integumentary system, the **hypodermis** (*hypo-* = below) or **subcutaneous layer** (subQ) is located below the skin and is usually discussed with the skin. The hypodermis is a major storage site for adipose tissue.

Epidermis

The epidermis is **keratinized stratified squamous epithelium.** This is a very thick epithelial layer in comparison with other epithelial layers of the body. Keratinization is a process that results in tough, waterproof, dead cells in the superficial layer of the epithelium.

Four types of cells are found in the epidermis: keratinocytes, melanocytes, Langerhans cells, and Merkel cells. **Keratinocytes** (*keratino* = hornlike; *cytes* = cells) comprise 90% of the cells of the epidermis and produce keratin, a fibrous protein important for waterproofing the epidermis. **Melanocytes** (*melano* = black) comprise 8% of the epidermal cells and produce and secrete the pigment melanin. **Langerhans cells** are immune system cells that attack pathogens that enter the skin. **Merkel cells** are the least abundant cell type and are found only in the deepest layer of the epidermis. These cells function as touch receptors and are associated with sensory neurons.

The layers of the epidermis are in order from deepest to most superficial: stratum basale, stratum spinosum, stratum granulosum, stratum lucidum, and stratum corneum. The **stratum basale** (*strata* = layers; *basa-* = base) or stratum germinativum (*germ* = sprout), a single row of cells attached to the basement membrane, contains stem cells that divide to form new keratinocytes. As new cells are formed, the older cells are pushed toward the surface and undergo a process called keratinization, which produces tough, water-repellent cells in the superficial layer. Melanocytes, Langerhans cells, and Merkel cells are also found within the stratum basale.

The **stratum spinosum** (*spinos-* = thornlike) contains 8 to 10 rows of cells, mainly keratinocytes. In prepared slides, cells in this layer have thornlike projections caused by the tissue preparation process.

Granules are observed in the **stratum granulosum** (*granulos-* = little grains). This layer contains 3 to 5 rows of flattened keratinocytes that are beginning to die. No dividing cells are present in this layer or in more superficial layers.

The **stratum lucidum** (*lucid-* = clear) contains 3 to 5 rows of flat, dead keratinocytes. This layer is translucent in specimens of fresh skin, but in prepared slides it may be clear or stained.

The outermost layer, the **stratum corneum** (*corn-* = hard or hooflike) is a very thick layer containing 25 to 30 or more rows of dead, squamous-shaped keratinocytes. This layer is tough and water-repellent. These cells continually slough off and are replaced by cells in the adjacent layer.

The epidermis differs in thickness in thin and thick skin. **Thick skin** is found on the palms of the hands and the soles of the feet and has all five strata. **Thin skin,** which covers the rest of the body, does not have a visible stratum lucidum and has a thinner stratum corneum than thick skin.

Dermis

The dermis consists of two regions: the papillary (*papilla* = nipples) region and the reticular (*reticul-* = netlike) region. The **papillary region** is a thin layer of areolar connective tissue that is deep to the stratum basale of the epidermis and the basement membrane. **Dermal papillae** are fingerlike projections of the papillary region that extend into the epidermis. In the palms, fingers, soles, and toes, the dermal papillae cause genetically determined whorls in the epidermis called **epidermal ridges** that increase surface area, friction, and grip. Sweat glands deposit their secretions onto these ridges, resulting in fingerprints when these ridges touch surfaces. The **reticular layer** is the deeper and much thicker region of the dermis. It is composed mainly of dense, irregular connective tissue whose collagen fibers provide the skin with strength and whose elastic fibers provide elasticity. Some adipose tissue is also found in the reticular region.

The dermis is highly vascularized, allowing nutrients to diffuse from the dermis into the epidermis. Hair follicles, sweat glands, and sebaceous glands are all derived from epithelial tissue and extend into the dermis. Nerves and lymphatic vessels are also found in this layer.

ACTIVITY 1 Major Divisions of the Skin

1 Label the skin structures in Figure 7.1. Observe where the basement membrane separates the epidermis and dermis.

2 Label the layers of the epidermis in thick skin in Figure 7.2.

3 Identify the skin structures in Figures 7.1 and 7.2 on a model or chart.

4 Pronounce the terms as you point to them.

5 Examine a microscope slide of thick skin.
- Using the scanning or low-power objective, identify the structures in Figure 7.1.
- Using the high-power objective, identify the layers of the epidermis in Figure 7.2. Observe the different cell shapes in the various layers of the epidermis.
- Observe the areolar connective tissue in the papillary layer of the dermis, the dense irregular connective tissue in the reticular layer of dermis, and the adipose tissue in the hypodermis.

30×

FIGURE 7.1 Photomicrograph of the skin.

- epidermis (EPI-derm-is)
- dermal papillae (puh-PILL-ee)
- hypodermis (HY-poh-der-mis)
- papillary layer (PAP-il-lary) of dermis
- reticular layer of dermis

1 _____
2 _____
3 _____
4 _____
5 _____

110×

FIGURE 7.2 Photomicrograph of the epidermal layers in thick skin.

- stratum basale (bay-SAL)
- stratum corneum (kor-NEE-um)
- stratum granulosum (gran-you-LOW-sum)
- stratum lucidum (LOU-sih-dum)
- stratum spinosum (spy-NO-sum)

1 _____
2 _____
3 _____
4 _____
5 _____

B. Accessory Structures of the Skin

The accessory structures of the skin include hairs, hair follicles, nails, sweat glands, sebaceous glands, ceruminous glands, and mammary glands. All of these are derived from epidermal tissue and extend into the dermis.

Sudoriferous and Sebaceous Glands

Sudoriferous glands (*sudori* = sweat; *-ferous* = bearing) or sweat glands secrete a watery substance that is important in excretion and body temperature regulation. There are two types of sudoriferous glands: eccrine glands and apocrine glands. **Eccrine glands** (*eccrine* = sweating outwardly) are the most common type of sudoriferous gland and are found on most areas of the body. Ducts from the eccrine glands deposit their secretions, called sweat, on the epithelial surface. **Apocrine glands** are found only in the axilla, genital area, and pigmented area around the nipples (areolae). Apocrine glands produce a secretion similar in composition to sweat but more viscous. This secretion, which is deposited on the distal end of the hair root, is odorless until broken down by bacteria. Ceruminous glands and mammary glands are modified sudoriferous glands. **Ceruminous glands** (*ceri-* = wax) are found in the ear canal and secrete a waxy substance called cerumen that prevents foreign substances (including insects) from entering the auditory canal. **Mammary glands** are found in the breasts and synthesize and secrete milk after appropriate hormonal stimulation.

Sebaceous glands (*sebace-* = greasy) or oil glands are found surrounding hair follicles and deposit sebum, an oily substance that lubricates the skin and hair, onto the hair shaft.

Hair and Hair Follicles

Hairs are found all over the body with the exception of the palms, soles, lips, and parts of the external genitalia. Hair consists of dead, keratinized epithelial cells and has two main sections: the **shaft,** which projects from the skin surface, and the **root,** which extends into the dermis of the skin. The **hair follicle,** which surrounds the hair root, is formed from epidermal layers that project into the dermis. The expanded base of the hair follicle, the **hair bulb,** contains the papilla of the hair and the matrix. The **papilla** of the hair is a projection of connective tissue into the hair follicle and contains blood vessels that provide nutrients to the dividing cells of the matrix. The **matrix,** which is derived from the stratum basale of the epidermis, forms new hair cells that are added to the base of the hair root. Surrounding the hair follicle is a **connective tissue sheath** comprised of dermal tissue.

The **arrector pili** (*arrect* = to raise; *pili* = hair) is a bundle of smooth muscle cells attached to the connective tissue sheath around the hair follicle. Contraction of the arrector pili muscle moves the hair from its normal angle to a 90° angle (perpendicular) with the skin surface, elevating the skin surrounding the hair shaft and causing goosebumps. Arrector pili muscles contract in response to stress (including cold temperature).

Examination of the hair in cross-section shows three layers: the cuticle (outer layer), the cortex (middle layer), and the medulla (inner layer). The **cuticle,** the layer of hair that we see, is a thin layer of dead, flattened, keratinized cells that overlap each other like shingles on a roof. Split ends occur when the free ends of these cells are pulled away from each other. Cells in the **cortex** and **medulla** contain pigment granules that give hair its color. The shape of hair in cross-section indicates whether the hair is curly, wavy, or straight. Curly hair is flat in cross-section, wavy hair is oval, and straight hair is round.

ACTIVITY 2 Hair, Hair Follicles, Sebaceous Glands, and Sudoriferous Glands

1 Label the diagram of the skin and accessory structures in Figure 7.3.

2 Point to the structures on a model or chart of the integumentary system.

3 Pronounce the names of the structures as you point to them.

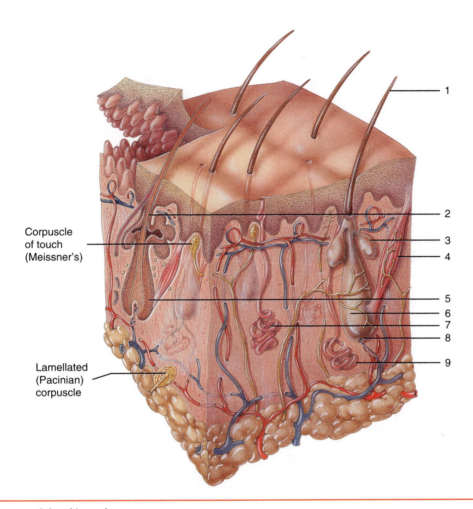

Corpuscle of touch (Meissner's)

Lamellated (Pacinian) corpuscle

FIGURE 7.3 Diagram of the skin and accessory structures.

- apocrine sweat gland (AP-oh-krin)
- arrector pili muscle (PIE-lee)
- eccrine sweat gland (EK-rin)
- hair bulb
- hair follicle
- hair root
- hair shaft
- papilla of hair (puh-PILL-uh)
- sebaceous gland (se-BAY-shus)

1 _____

2 _____

3 _____

4 _____

5 _____

6 _____

7 _____

8 _____

9 _____

4 Label the photomicrograph in Figure 7.4.

5 Examine a slide of hairy skin and identify the structures in Figure 7.4.

10×

FIGURE 7.4 Photomicrograph of the skin and accessory structures.

- hair bulbs
- hair follicle
- hair root
- papilla of hair
- sebaceous gland

1 _____

2 _____

3 _____

4 _____

5 _____

6 Observe the fingerprints your instructor has as a demonstration or use an ink pad to make your own fingerprints. Examine the fingerprints of several students to observe the variety of whorl patterns.

7 Examine a strand of hair with the compound microscope.
- Pull out one strand of scalp hair and lay the hair horizontally on the stage over the light source.
- Observe the dark cuticle and lighter medulla of the hair shaft with the 4× objective lens. Switch to the low- and high-power lenses and look for the cells of the cuticle overlapping each other. Observe the root end of the hair. What differences do you see between the root end and the shaft area?
- If someone is willing to donate hair with "split ends," compare the cuticle of a hair with split ends with undamaged hair.

ACTIVITY 3 Experiment: Eccrine Gland Density

1 Prediction: List the eccrine gland density of the following body areas from highest to lowest: forehead, anterior forearm, palm, anterior leg.

- _____ (highest)

- _____

- _____

- _____ (lowest)

2 Materials: Obtain materials for the eccrine gland density experiment (see Materials list).

3 Data Collection: Count eccrine gland density in different body areas and record your results in Table 7.1.
- Decide who will be the subject and who will do each of the following tasks: apply Betadine and tape, time experiment, remove tape, count blue-black dots, and record data.
- Find areas on your forehead, right anterior forearm, right palm, and right anterior leg that do not have large crease lines.
- For each location, apply Betadine to an area slightly larger than the paper square.
- Securely tape a paper square over the Betadine-covered area at each location.
- After 20 minutes, remove the tape and paper squares, and count the blue-black dots on each square. Each blue-black dot is an active sweat gland. Remove Betadine from skin by washing the area with soap and water.
- Record the number of dots per cm^2 in Table 7.1.

4 Clean up as directed by your instructor.

5 Data Analysis:
- Collect the individual eccrine gland density value for each body area from the subject for each group.
- Calculate the average eccrine gland density for each body area by adding the number of glands/cm^2 for each subject dividing by the number of subjects.
- Record the average value in Table 7.1.

6 Complete the Experimental Report with your lab group.

EXPERIMENTAL REPORT: ECCRINE GLAND DENSITY

Results:

1 State which area has the highest average eccrine gland density and which has the lowest.

2 State whether variation in eccrine gland density per body area was observed among subjects.

Discussion:

1 Discuss the variation in sweat gland density for a given area among individuals.

2 Discuss how eccrine gland density correlates with the amount of sweat that appears on these body areas during exercise or nervousness. Base this on your own observations and experiences.

Conclusion:

1 Write a sentence that postulates a correlation between average eccrine gland density and the amount of sweat produced by an area.

2 Write a sentence that postulates why eccrine gland density per area varied or did not vary among individuals.

TABLE 7.1 Eccrine Sweat Gland Density

BODY AREA	ECCRINE SWEAT GLAND DENSITY (NO. OF GLANDS/CM²)	
	Individual Value	Class Average
Forehead		
Right anterior forearm		
Right palm		
Right anterior leg		

Nails

Nails, which are found on the distal ends of the digits, assist in grabbing and manipulating objects, and protect the digits. The nail consists of a nail body, a free edge, and a root. The **nail body** is the part of the nail that is visible, the **free edge** extends beyond the digit, and the **root** is underneath the skinfold at the proximal end of the nail body. The **lunula** (*lunula* = little moon) is the crescent-shaped area of the nail body distal to the nail root. The **cuticle** or **eponychium** (*epi-* = above; *-onych* = nail) is the thickened epithelial tissue along the proximal border of the nail body. The **hyponychium** (*hypo-* = below) is deep to the free edge and forms the nail bed. The **nail matrix** is the epithelial tissue within the nail root that divides to produce new cells that are added to the nail body as it grows.

ACTIVITY 4 Nail Structure

1 Identify the nail structures on the diagram in Figure 7.5.
2 Identify nail structures on your own nails.

(a) Dorsal view (b) Sagittal section through finger

FIGURE 7.5 Nail structures.

(a) Dorsal view
• eponychium (eh-poh-NICK-ee-um)
• free edge
• lunula
• nail body

(b) Sagittal section through finger
• eponychium
• free edge
• hyponychium (hypo-NICK-ee-um)
• lunula (Loon-you-luh)
• nail root
• nail body
• nail matrix

1 _____

2 _____

3 _____

4 _____

5 _____

6 _____

7 _____

8 _____

9 _____

10 _____

11 _____

Reviewing Your Knowledge

A. Identification of Skin Layers and Accessory Structures

Go to www.wiley.com/college/allen and click on the picture of your lab manual. Then click on Student Companion Site, Anatomy Drill and Practice, and Exercise 7.

B. Skin Layers and Structures

Write the name of the skin layer or structure that fits the description.

_____ **1.** Layer of epidermis where there is the most rapid cell division.

_____ **2.** Tough, water-repellent epidermal layer; contains dead squamous-shaped cells.

_____ **3.** Areolar connective tissue layer beneath basement membrane.

_____ **4.** Projections of dermis that cause epidermal ridges.

_____ **5.** Translucent layer found in thick skin, absent in thin skin.

_____ **6.** Appears to have thornlike projections in prepared slides.

_____ **7.** Thick dermal layer containing dense, irregular connective tissue.

_____ **8.** Epidermal layer containing visible granules.

C. Cells of the Epidermis

Write the name of the epidermal cell type that fits the description.

_____ **1.** Defends skin against microbes.

_____ **2.** Produces keratin.

_____ **3.** Produces the pigment melanin which shields cell nuclei from UV radiation.

_____ **4.** Associated with sensory neurons and functions in sensation of touch.

D. Accessory Structures of the Skin

Write the name of the accessory structure that best fits the description.

_____ **1.** Sudoriferous glands located in axillary and genital areas; become active after puberty.

_____ **2.** Connective tissue projection; provides blood supply for hair matrix.

_____ **3.** Secures nail to digit.

_____ **4.** Epithelial layer that surrounds hair root.

_____ **5.** Secretes sebum onto hair and skin.

_____ **6.** Part of nail that overlies nail bed.

_____ **7.** Part of nail that extends beyond digit.

_____ **8.** Part of hair that contains the matrix.

_____ **9.** Sudoriferous glands that deposit sweat onto epidermal ridges causing fingerprints.

_____ **10.** Part of hair in epidermis and extending beyond skin surface.

_____ **11.** Moves hair shaft perpendicular to skin; causes goosebumps.

_____ **12.** Thickened epithelial tissue at proximal end of nail body.

_____ **13.** Crescent-shaped area of nail body near cuticle.

_____ **14.** Part of hair within dermis.

_____ **15.** Part of nail within skin.

Using Your Knowledge

Identification of Skin from Different Body Locations

Observe diagrams of skin from different body locations in Figure 7.6. Based on number of epithelial layers and skin accessory structures, determine if skin is from the **axillary area, forearm,** or **sole of foot.**

1. _____

2. _____

3. _____

1 2 3

FIGURE 7.6 Identification of skin from different body locations.

4. Cortisone is a steroid that is applied to the skin to reduce inflammation. Cortisone acts on cells within the dermis and can travel through unbroken epidermis to reach cells in the dermis. If the epidermis is such a good barrier, how can cortisone easily travel through it?

5. Explain how dandruff is formed.

6. Joey has a splinter in his finger. His mother pulled the skin away from the splinter with a sterile needle and removed the splinter. Joey did not feel pain and did not bleed. Explain.

7. Identify the epidermal layer(s) in which the following cancers arise: (a) basal cell carcinoma; (b) malignant melanoma; (c) squamous cell carcinoma.

8. Name the part(s) of the skin that "peel(s)" off after a minor sunburn.

9. As we age, our skin wrinkles due to changes in collagen and elastic fibers in the dermis and a decrease in their production. Explain why topical application of collagen and elastic fibers would not eliminate wrinkles.

10. Permanent tattoos are made by injecting pigment into the skin. Into which part of the skin is the pigment injected, the epidermis or dermis? (You can answer this without research, just use your knowledge.)

Bone Structure and Function

Objectives

After completing this exercise, you should be able to:

- Identify bones as either long, short, flat, irregular, or sesamoid
- Describe the gross structure of a long bone and identify the parts on a real bone or chart
- Describe the difference between compact and spongy bone
- Identify microscopic structures of compact and spongy bone on drawings and slides
- Describe the effect of collagen and mineral salts on bone hardness and flexibility

Materials

- disarticulated bones—for bone classification
- human long bone cut longitudinally and transversely
- dissecting microscope
- fresh animal bones sectioned longitudinally
- 1-inch cross-section cut from fresh long bone
- dissecting tray and instruments (blunt probe and forceps); disposable gloves
- compound microscope and lens paper
- prepared slides of cross-section of ground compact bone and cancellous (spongy) bone
- Collagen and Mineral Salts: 3 chicken thigh or leg bones: 1 raw, 1 baked for 2 hours or more at 250° F, and 1 soaked in an acidic solution (vinegar or nitric acid) for 5–7 days

Bones are organs composed of a complex arrangement of several tissues. A typical bone has compact and spongy osseous tissues, connective tissues, cartilage, and adipose tissue. In addition, bones contain blood vessels and nerves.

A. Classification of Bones

Human bones have different shapes and distinct gross anatomical features. Bones are placed in five classifications according to their shapes: long, short, flat, irregular, and sesamoid. **Long bones** are longer than they are wide and contain a shaft and contain spongy bone with a thick compact bone exterior. Distribution of spongy and compact bone in long bones is covered in detail later in this exercise. **Short bones** are almost equal in length and width and contain a thick interior of spongy bone covered by a thin veneer of compact bone. **Flat bones** are relatively flat, but may be curved, and contain a thin, spongy bone interior covered by a thin veneer of compact bone. **Irregular bones** are self-explanatory and do not easily fit into any of these categories, and **sesamoid bones** (*sesame* = seed) are small bones that develop in tendons (e.g., patella) for protection against wear and tear.

ACTIVITY 1 Classification of Bones According to Shape

1 Identify the types of bones in Figure 8.1 according to shape.

2 Classify each of the bones your instructor has displayed according to shape.

3 Discuss with the rest of the class which bones are difficult to classify and why.

B. Gross Features of Long Bones

The enlarged proximal and distal ends of long bones are called **epiphyses** (*epi-* = above, over; *physis* = growing), and the middle shaft area is called the **diaphysis** (*dia-* = through). The **metaphyses** are the areas in an adult bone where the epiphyses and diaphysis join. In a growing bone, the metaphyses contain a layer of hyaline cartilage called the **epiphyseal plate.** Division of cartilage allows the bone to grow in length. Bone growth stops when the epiphyseal

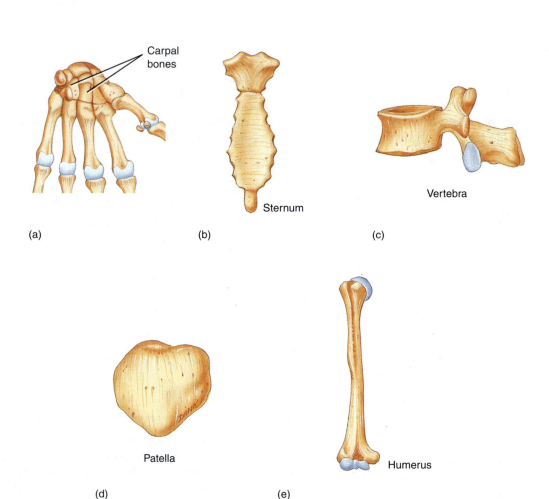

FIGURE 8.1 Classification of bones.

- flat
- irregular
- long
- sesamoid (SES-a-moid)
- short

(a) ___irregular___

(b) ___flat___

(c) ___short___

(d) ___sesamoid___

(e) ___long___

plate cartilage becomes ossified and forms a bony structure called the **epiphyseal line.** The cavities within the spongy bone of the epiphyses contain **red marrow,** a tissue that produces blood cells. **Articular cartilage** (*articul-* = joint), composed of hyaline cartilage, covers both epiphyses, and the rest of the bone exterior is covered with a tough, connec-tive tissue membrane, the **periosteum** (*peri-* = around; *osteo* = bone). The diaphysis has an exterior of thick compact bone over a thin layer of spongy bone. The hollow center of the bony diaphysis is called the **medullary cavity** (*medulla* = marrow, pith), and a small amount of spongy bone is found in this cavity. The medullary cavity is lined with a connective tissue membrane called the **endosteum** (*endo-* = within) and is filled with a fatty substance called **yellow marrow.** The endosteum also lines the cavities within the spongy bone of the epiphyses. Both the periosteum and the endosteum contain **osteoblasts** (*-blast* = builder) and **osteoclasts** (*klasis* = breaking) for bone formation, bone tissue repair, and bone remodeling.

ACTIVITY 2 Gross Features of a Long Bone

1 Label Figure 8.2.

2 Identify the items from Figure 8.2 on a long bone that is partially sectioned longitudinally and transversely.

3 Use a dissecting microscope to observe the medullary cavity with spongy bone at a higher magnification.

4 Pronounce the terms as you point to them on the bone.

Longitudinal section

FIGURE 8.2 Features of an adult long bone.

- articular cartilage
- compact bone
- diaphysis (die-AF-ih-sis)
- distal epiphysis (e-PIF-ih-sis)
- endosteum (en-DOS-tee-um)
- epiphyseal line (ep-i-PHY-zee-al or ee-PIF-ih-seal)
- medullary cavity (MED-yoo-lar-y)
- nutrient artery
- periosteum (peri-OS-tee-um)
- proximal epiphysis
- spongy bone
- yellow marrow

1 Proximal epiphysis
2 diaphysis
3 distal epiphysis
4 articular cartilage.
5 epiphyseal line
6 compact bone
7 medullary cavity
8 yellow marrow
9 periosteum
10 nutrient artery
11 endosteum
12 spongy bone.

ACTIVITY 3 Freshly Cut Section of Long Bone

1 Using gloves, examine a freshly cut longitudinal section of beef or other bone.

2 Identify as many features as you can from Activity 2.

3 Note the shiny appearance of the articular cartilage.

4 Use a blunt probe and forceps to loosen the periosteum, ligaments, or tendons.

5 Find the epiphyseal line.

6 Note the consistency of the yellow and red bone marrow.

7 Observe a 1-inch cross-section from fresh long bone.

8 Remove the yellow bone marrow from this section to observe the spongy bone with a dissecting microscope.

9 In this cross-section, identify compact bone, spongy bone, and the periosteum.

C. Microscopic Structure of Compact and Spongy Bone Tissue

Compact bone is composed of repeating units of **osteons,** with each unit having a **central** (Haversian) **canal** running longitudinally. The central canal contains blood vessels, lymphatic vessels, and nerves that serve compact bone tissue. The blood vessels, lymphatic vessels, and nerves travel from the **periosteum,** dense regular connective tissue covering the bone surface, to the central canal through **perforating (Volkmann) canals.** These canals run horizontally in compact bone and connect with the central canal. The main feature of each osteon is the concentric rings, or **concentric lamellae** (*lamella* = small plate or ring), which look similar to the rings of a tree trunk cut in cross-section. When viewed on a stained slide of compact bone, there are dark areas with thin lines extending between the lamellae. The dark areas are **lacunae** (*lacuna* = little lake) that are found between concentric lamellae, and the thin lines are

canaliculi (small channels) that connect the lacunae. **Osteocytes** are mature bone cells that reside in the lacunae, and osteocyte processes extend through the canaliculi. Canaliculi allow nutrients from the blood vessels in the central canal to diffuse to the osteocytes embedded in the solid bone material. The canaliculi are also the route by which waste materials are removed from these cells. **Interstitial lamellae** (*inter-* = between; *-stitial* = to stand) fill in the spaces between the osteons.

Spongy bone does not contain osteons but instead has **trabeculae** (little beams)—flat plates with a latticelike network of thin, bony columns. The trabeculae have **lamellae, lacunae, osteocytes,** and **canaliculi.** Spongy bone has many spaces filled with red marrow. Blood vessels within the red marrow provide the osteocytes with nutrients. This fragile spongy bone needs the protection of an outer layer of compact bone. Spongy bone is found in the epiphyses of long bones and in the interior of short, flat, and irregular bones.

ACTIVITY 4 Microscopic Structure of Ground Compact Bone and Spongy Bone

1 Label the structures in Figures 8.3(a) and (b), and Figure 8.4.

2 Examine a prepared slide of a cross-section of ground compact bone. Blood vessels, osteocytes, and periosteum cannot be observed on ground bone slides.
 - Using the low-power objective lens, identify an osteon, a central canal, concentric lamellae, and interstitial lamellae. Count the number of osteons in one field of view.
 - Using the high-power objective lens, identify canaliculi and lacunae.

3 Examine a prepared microscope slide of spongy bone. Note that on slides of spongy bone, the preparations are generally demineralized (mineral salts removed) and do not show the lamellae and canaliculi as depicted in the drawing in Figure 8.4(b).
 - Using the low-power objective lens, identify the trabeculae.
 - Using the high-power objective lens, identify the lacunae, which appear as darker areas in the trabeculae.

4 Go to www.wiley.com/college/allen to observe osteoblasts, osteoclasts, and bone formation.

(a) Osteons (Haversian systems) in compact bone and trabeculae in spongy bone

(b) Enlarged aspect of spongy bone trabeculae

FIGURE 8.3 Microscopic features of bone.

(a)
- blood vessels
- canaliculus
 (can-a-LIK-yoo-lus)
- central canal
- compact bone
- concentric lamellae
 (la-MEL-lee)
- lacuna (la-COO-na)
- osteocyte (OS-tee-o-site)
- perforating canal
- periosteum
 (per-ee-OS-tee-um)
- spongy bone
- trabeculae (trah-BEK-
 yoo-lee) of spongy bone

1 _Spongy bone_
2 _trabeculae of spongy bone_
3 _Concentric lamellae_
4 _blood vessels_
5 _Canaliculus_
6 _Lacuna_
7 _Osteocyte_
8 _Periosteum_
9 _Central Canal_
10 _Perforating Canal_
11 _Compact bone_

(b)
- interstitial lamellae
- osteocyte in lacuna
- trabeculae

12 _trabeculae_
13 _Osteocyte in lacuna_
14 _interstitial lamellae_

LM 285×

(a) Cross-section of ground compact bone

Red marrow cell

(b) Photomicrograph of spongy bone

FIGURE 8.4 Microscopic features of compact and spongy bone.

(a)
• canaliculi
• central canal
• concentric lamella
• lacuna

(b)
• osteocyte
• trabeculae

1 _____

2 _____

3 _____

4 _____

5 _____

6 _____

D. Role of Collagen and Mineral Salts in Osseous Tissue

The properties of osseous tissue are determined by its extracellular matrix, which contains approximately 25% water, 25% protein fibers, and 50% mineral salts. Collagen fibers provide tensile strength and flexibility so that bone does not break with normal stress. The mineral salts consist mainly of calcium phosphate and calcium carbonate salts, giving the "backbone" or hardness to bone. As we age, the collagen content of osseous tissue decreases, causing bones to become brittle and break more easily. Decreased mineral content of bone, as occurs with rickets, causes bones to be soft and to bend due to body weight.

ACTIVITY 5 **Role of Collagen and Mineral Salts in Osseous Tissue**

1 With your group, examine a chicken thigh or leg bone baked at about 250° F for a minimum of 2 hours or until brittle. Compare what happens when you try to bend a treated and untreated bone.

2 Examine a chicken thigh or leg bone that has been soaked in an acidic solution (vinegar or nitric acid) for 5–7 days. Try bending the bone and compare this with an untreated bone and the baked bone.

3 Answer the Discussion Questions with your lab group.

? DISCUSSION QUESTIONS: COLLAGEN AND MINERAL SALTS

1 Compare the flexibility of the untreated, baked, and acid-treated bones.

2 Which bone is brittle?

3 What substance has been damaged in the brittle bone? _____ Why would this increase bone breakage?

4 Which bone is the softest?

5 What substance has been leached from the soft bone? _____ What clinical disorder is this bone simulating?

Reviewing Your Knowledge

A. Identification of Gross and Microscopic Bone Structures

Go to www.wiley.com/college/allen and click on the picture of your lab manual. Then click on Student Companion Site, Anatomy Drill and Practice, and Exercise 8.

B. Gross Features of Long Bones

Complete the sentences with the correct word or words.

1. What area of the long bone is covered with cartilage? _____

2. What type of cartilage is articular cartilage? _____

3. What area (epiphysis or diaphysis) is made up of a *thin* layer of compact bone and a *thick* spongy bone?

4. What area (epiphysis or diaphysis) is made up of a *thick* layer of compact bone and a *very thin* layer of spongy bone?

5. The long bone in Figure 8.2 is from an adult. If this figure were from a child, what structure would be present

instead of the epiphyseal line? _____

C. Microscopic Features of Long Bones

Identify the term that describes the phrase about long bones.

_____ **1.** bone cell that is found in lacunae

_____ **2.** vertical canal in an osteon

_____ **3.** cavity that contains yellow marrow in adults

_____ **4.** bone shaft

_____ **5.** spaces where the osteocytes are located

_____ **6.** horizontal canal in an osteon

_____ **7.** small canal connecting lacunae

_____ **8.** membrane lining medullary cavity

_____ **9.** membrane covering surface of bone

_____ **10.** thin bony columns in spongy bone

D. Comparison of Compact and Spongy Bone

Identify whether the statements below describe compact bone, spongy bone, or both.

_____ **1.** composed of osteons

_____ **2.** contains osteocytes and lacunae

_____ **3.** has lamellae

_____ **4.** has trabeculae

_____ **5.** has red marrow

_____ **6.** located in the epiphyses

_____ **7.** located in the diaphysis

_____ **8.** has a central canal

_____ **9.** spaces filled with red marrow

_____ **10.** has canaliculi

E. Chemical Composition of Bone

_____ **1.** The hardness of bone is due to _____.

_____ **2.** The flexibility and tensile strength of bone are due to _____.

_____ **3.** What type of a macromolecule (carbohydrate, lipid, protein) is collagen?

_____ **4.** A bone that has the collagen removed is flexible or inflexible?

_____ **5.** A bone that has calcium removed is flexible or inflexible?

Using Your Knowledge

A. Bone Tissue

Fill in the blank with the correct term.

1. Two bone cells located in the periosteum and endosteum are _____ and _____.

2. Which type of bone tissue, compact bone or spongy bone, significantly degenerates first in osteoporosis?

3. As we age, the amount of collagen in the extracellur matrix of bone decreases and bones become more brittle.

Identify the osseous tissue cell that secretes collagen. _____

4. Explain the importance of the integumentary system to bone formation.

B. Bone Images

Identify the structures indicated on Figure 8.5 and Figure 8.6.

FIGURE 8.5 Longitudinal section of a long bone.

FIGURE 8.6 X-ray of a child's knee joint.

5 _____

6 _____

7 _____

8 _____

9. Observe the bones in Figure 8.7(a) and (b). Identify the child's hand and the adult's hand.

(a) _____ (b) _____

(a) (b)

FIGURE 8.7 X-rays of the child and adult hand.

10. Observe Figure 8.8(a) and (b). Identify the normal bone and osteoporotic bone.

(a) _____ (b) _____

(a) (b)

FIGURE 8.8 Normal and osteoporotic bone tissue.

Axial Skeleton

Objectives

After completing this exercise, you should be able to:

- Identify the three main parts of the axial skeleton
- Identify the major bones of the axial skeleton
- Identify selected bone surface markings and sutures on the axial skeleton
- Identify paranasal sinuses and bones of the nasal septum, hard palate, and orbit of eye
- Identify the fontanels in the fetal skull
- Identify the parts of a typical vertebra
- Compare the three individual types of vertebrae and the identifying features of each
- Compare and contrast the normal and abnormal curvatures of the vertebral column
- Identify the bones of the thoracic cage and selected surface markings

Materials

- articulated skeleton
- disarticulated skeleton
- articulated skulls, Beauchene (disarticulated) skull, and fetal skull
- articulated vertebral column with intervertebral discs
- small, plastic straws (coffee stirrers) or pipe cleaners for pointers

There are 206 named bones in the adult skeleton, which can be separated into the **axial** and the **appendicular** divisions. The **axial skeleton** is composed of 80 bones located along a vertical line, the longitudinal axis of the body. Its bones support and protect the organs of the head, neck, and torso. The **appendicular** skeleton (*appendere* = to hang upon) permits a greater variety of movements than the axial skeleton. This division of the skeleton has four main parts and is composed of 126 bones that make up the upper limbs or extremities, lower limbs, and the bones of the girdles that attach the limbs to the axial skeleton. The appendicular skeleton will be studied in the next exercise.

A. Parts of the Axial Skeleton

The axial skeleton includes the **skull, hyoid bone, vertebral column,** and **thoracic cage** (rib cage). Smaller bones included in this division are the **ear ossicles** (bones).

Skull

- cranial portion
- facial portion
- ear ossicles

Hyoid Bone

Vertebral Column

- vertebrae
- intervertebral discs (*inter-* = between)
- sacrum
- coccyx

Thoracic Cage

- sternum
- ribs
- costal cartilages
- thoracic vertebrae

ACTIVITY 1 The Axial Skeleton

1 Identify the major components of the axial skeleton and the bones from Figure 9.1 on an articulated skeleton.

2 Pronounce the terms as you point to them on the skeleton.

3 Palpate the following bones on your body: cranial and facial portion of the skull, sternum, costal cartilages, ribs, and vertebrae.

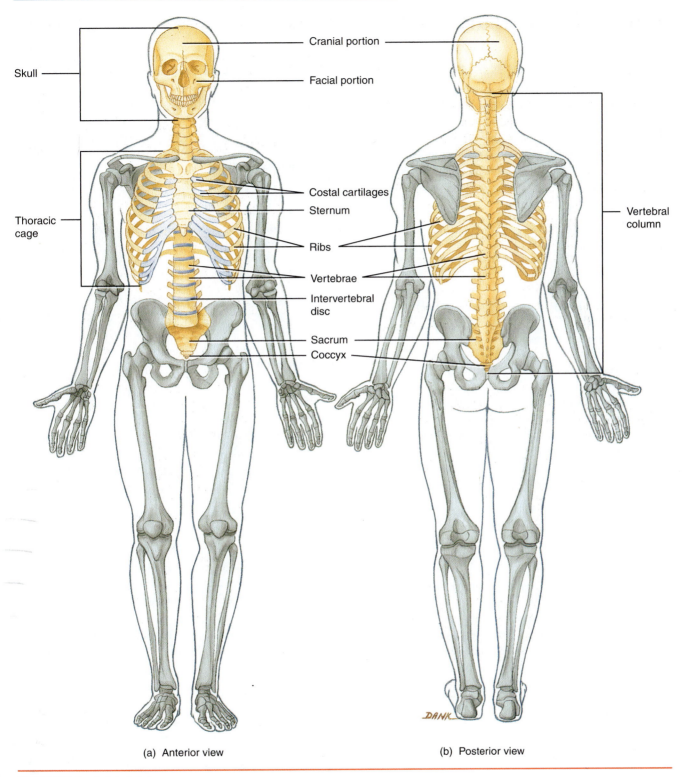

(a) Anterior view

(b) Posterior view

FIGURE 9.1 Axial skeleton.

B. The Skull and Sutures

The **cranium** (cranial bones) harbors and protects the brain and houses organs of hearing and equilibrium. **Facial bones** provide the shape of the face, house the teeth, and provide attachments for all the muscles of facial expression. Other major features of the skull include **sutures, bone markings, orbit of eye, paranasal sinuses, nasal septum, hard palate,** and **fontanels** in the fetal skull.

Cranial Bones

There are a total of eight bones in the cranium, two paired and four single bones.

- **(2) parietal bones** (*paries* = wall)—superior lateral wall
- **(2) temporal bones**—inferior lateral wall; house organs of inner ear
- **(1) frontal bone**—anterior portion
- **(1) occipital bone** (*occipit-* = atlas)—posterior wall
- **(1) sphenoid bone** (*sphen-* = wedge; *-eidos* = form)—floor posterior to ethmoid
- **(1) ethmoid bone** (*ethmos* = sieve)—floor and anterior wall

Sutures

Sutures are immovable joints between cranial and facial bones. The four main sutures in the skull are the:

- **coronal** (*corona* = crown)—joins frontal and parietal bones
- **sagittal** (*sagitta-* = arrow)—joins parietal bones
- **lambdoid** (shape of the Greek letter *lambda*)—joins both parietal bones with occipital bone
- **squamous**—joins temporal and parietal bones

Facial Bones

There are a total of 14 facial bones, six paired and two single bones.

- **(2) maxillae** (*mala* = jaw)—upper jaw and hard palate
- **(2) zygomatic bones** (*zygoma* = yoke or bar)—cheek bone

- **(2) lacrimal bones** (*lacri-* = tear)—orbit of eye near nasal bone
- **(2) nasal bones**—bridge of nose
- **(2) inferior nasal conchae** (*concha* = shell or scroll-shaped) or turbinate—lateral wall of nasal cavity
- **(2) palatine bones** (*palatum* = palate)—posterior part of hard palate
- **(1) mandible** (*mandere* = to chew)—lower jaw
- **(1) vomer** (*vomer* = plowshare)—inferior portion of nasal septum

NOTE: *BE ACCOUNTABLE FOR THE CARE OF SKULLS AND BONES. Please be careful not to use pencils, pens, or markers as pointers while you are studying the skull and other bones. Thin bones can be broken, and other bones can be permanently marked. Your instructor will provide a small plastic straw, pipe cleaner, or other suitable pointer.*

ACTIVITY 2 **Cranial Bones, Facial Bones, and Sutures**

1 Label the cranial bones, facial bones, and sutures on the photographs in Figures 9.2–9.6.
2 Identify these bones and sutures on an articulated skull and a Beauchene (disarticulated) skull.
3 Pronounce the terms as you point to them on a skull.
4 Palpate the cranial and facial bones on your own skull and say their names.
5 Locate the six bones that comprise the orbit of the eye.
6 Palpate your temporomandibular joint (TMJ) by putting your fingers on either side of your jawbone just anterior to the ears. Open and close your mouth to feel this joint.
7 Go to www.wiley.com/college/allen to observe individual cranial and skull bones.

CLINICAL NOTE: *A dislocated or defective TMJ has a history of causing clinical problems characterized by pain, tenderness, dysfunction, and clicking noises.*

- coronal suture (cor-RONE-al)
- ethmoid bone
- frontal bone (FRON-tal)
- lacrimal bone (LAK-rih-mal)
- lambdoid suture (LAMB-doid)
- mandible
- maxilla (ma-XIL-la)
- nasal bone
- occipital bone (oc-CI-pi-tal)
- parietal bone (pa-RYE-e-tal)
- sphenoid bone (SFEE-noid)
- squamous suture (SQUAY-mus)
- temporal bone (TEM-por-ul)
- zygomatic bone (zy-go-MA-tic)

FIGURE 9.2 Lateral view of skull.

1 _____ 6 _____ 11 _____

2 _____ 7 _____ 12 _____

3 _____ 8 _____ 13 _____

4 _____ 9 _____ 14 _____

5 _____ 10 _____

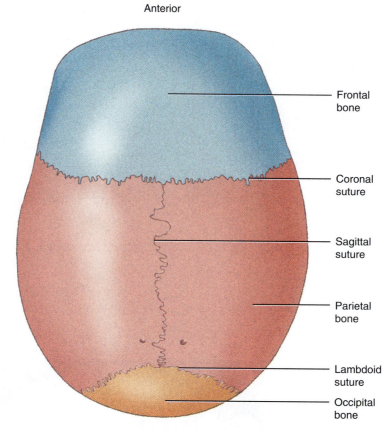

Anterior

Frontal
bone

Coronal
suture

Sagittal
suture

Parietal
bone

Lambdoid
suture

Occipital
bone

Posterior

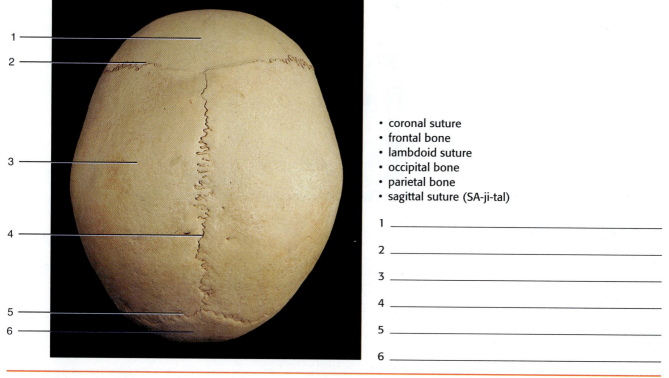

- coronal suture
- frontal bone
- lambdoid suture
- occipital bone
- parietal bone
- sagittal suture (SA-ji-tal)

1 _____

2 _____

3 _____

4 _____

5 _____

6 _____

FIGURE 9.3 Superior view of skull.

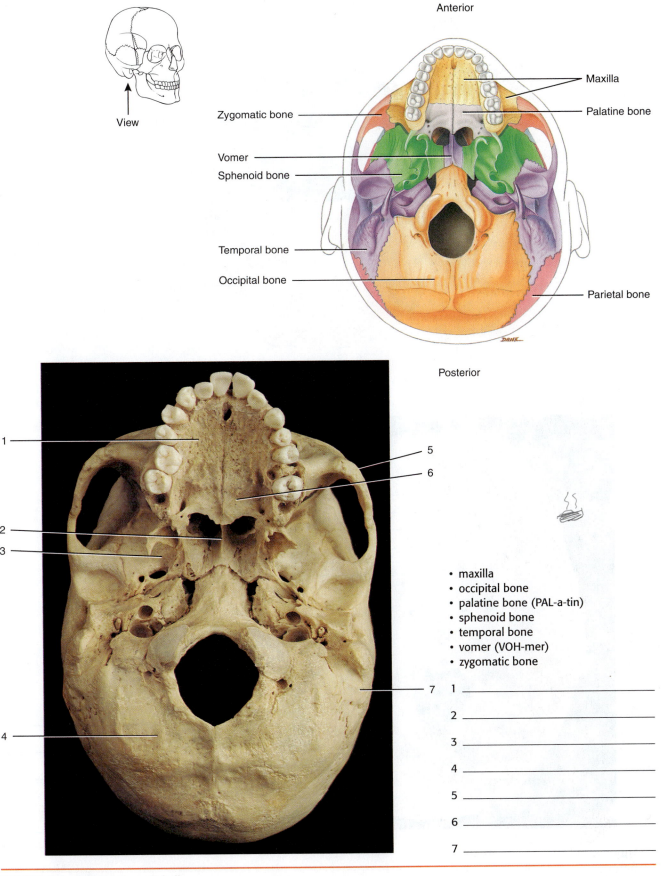

Anterior

Maxilla

Zygomatic bone

Palatine bone

Vomer

Sphenoid bone

Temporal bone

Occipital bone

Parietal bone

Posterior

View

1

2

3

4

5

6

7

• maxilla
• occipital bone
• palatine bone (PAL-a-tin)
• sphenoid bone
• temporal bone
• vomer (VOH-mer)
• zygomatic bone

1 _____

2 _____

3 _____

4 _____

5 _____

6 _____

7 _____

FIGURE 9.4 Inferior view of skull.

View

Transverse plane

Anterior

Frontal bone

Ethmoid bone

Sphenoid bone

Temporal bone

Parietal bone

Lambdoid suture

Occipital bone

Posterior

- ethmoid bone
- frontal bone
- lambdoid suture
- occipital bone
- parietal bone
- sphenoid bone
- temporal bone

1 _____

2 _____

3 _____

4 _____

5 _____

6 _____

7 _____

FIGURE 9.5 Superior view of floor of cranium.

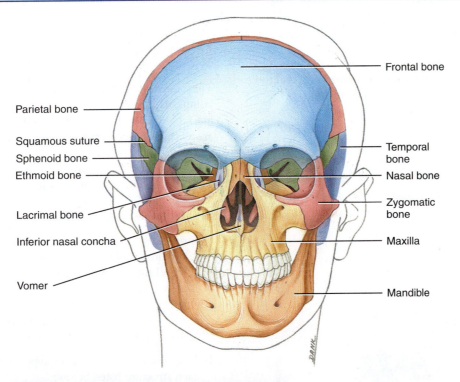

- Frontal bone
- Parietal bone
- Squamous suture
- Sphenoid bone
- Ethmoid bone
- Temporal bone
- Nasal bone
- Lacrimal bone
- Zygomatic bone
- Inferior nasal concha
- Maxilla
- Vomer
- Mandible

- ethmoid bone
- frontal bone
- inferior nasal concha or turbinate (CON-cha)
- lacrimal bone (LAC-ri-mal)
- mandible (MAN-di-ble)
- maxilla
- nasal bone
- parietal bone
- sphenoid bone
- squamous suture
- temporal bone
- vomer
- zygomatic bone

1 _____
2 _____
3 _____
4 _____
5 _____
6 _____
7 _____
8 _____
9 _____
10 _____
11 _____
12 _____
13 _____

FIGURE 9.6 Anterior view of skull.

Orbit of the Eye

The orbit of the eye is a mosaic of six different bones, three cranial bones (**frontal, sphenoid,** and **ethmoid**), and three facial bones (**maxilla, zygomatic,** and **lacrimal**).

ACTIVITY 3 Orbit of the Eye

1 Label the bones forming the orbit of the eye in Figure 9.7.

2 Pronounce each term as you point to it on a skull.

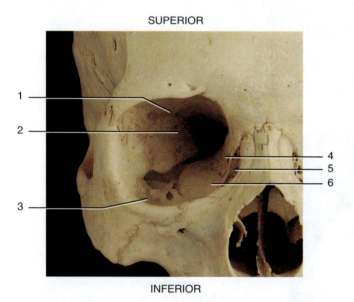

SUPERIOR

INFERIOR

FIGURE 9.7 Orbit of the eye.

- ethmoid bone
- frontal bone
- lacrimal bone
- maxilla
- sphenoid bone
- zygomatic bone

1 _____

2 _____

3 _____

4 _____

5 _____

6 _____

Selected Bone Surface Markings of the Skull

The bones of the skull are not totally smooth but have bumps, depressions, grooves, holes, and projections. Bone markings have specific names and distinct purposes and are classified according to their function. Several bone markings found on the axial skeleton are listed in Table 9.1 along with their descriptions and functions. Selected bones and their markings are listed below.

Cranial Bones: Selected Bone Surface Markings

1. *Frontal bone* (1)
 - **supraorbital foramina** (*supra-* = above; *orbit* = wheel rut; *foram-* = opening)—openings located above the orbit of the eye for supraorbital nerve and artery
 - **supraorbital ridges or margins**—thickening of frontal bone superior to orbit of the eye

2. *Temporal bone* (2)—Each temporal bone has one of each structure listed below:
 - **external auditory meatus** (*meatus* = passageway)—tubelike openings for the ear canal
 - **mastoid process** (*mastoid* = breastlike)—rounded projection posterior to external auditory meatus; attachment for muscles
 - **styloid process** (*stylo-* = point)—long, thin projection on inferior skull surface; attachment for muscles and ligaments of tongue and neck
 - **zygomatic process**—projection that articulates with the zygomatic bone
 - **mandibular fossa**—depression for articulation with condylar process (mandibular condyle)
 - **foramen lacerum** (*lacerare* = torn or lacerated)—jagged opening filled with cartilage in a living person
 - **carotid foramen (canal)**—foramen for internal carotid artery
 - **jugular foramen** (*jugular* = throat)—foramen for jugular vein and cranial nerves IX, X, and XI
 - **stylomastoid foramen**—opening for an artery and cranial nerve VII
 - **internal auditory meatus**—opening for cranial nerve VIII

TABLE 9.1 Bone Markings

MARKING	DESCRIPTION	PURPOSE
Depressions or Openings		
1. Fissure (FISH-er)	Narrow slit or cleft in a bone	Opening for blood vessels and nerves
2. Foramen (for-AY-men)	Opening or hole (foramina, pl.)	Opening for blood vessels and nerves
3. Fossa (FOS-sa)	Shallow depression (fossae, pl.)	Muscle attachment or articulation (joint)
4. Meatus (me-AY-tus)	Tubelike passageway or opening (meati, pl.)	Passageway or canal for blood vessels and nerves
Processes		
5. Condyle (CON-dile)	Smooth, rounded articular process	Articulation
6. Ramus (RAY-mus)	A small branch (rami, pl.)	Articulation
7. Spine	Pointed process	Articulation

3. *Occipital bone* (1)
 - **foramen magnum** (*magnum* = large)—opening for lower brain to spinal cord connection
 - **hypoglossal foramina**—openings for cranial nerve XII
 - **occipital condyles**—rounded processes that articulate with the atlas (C1)

4. *Ethmoid bone* (1)
 - **cribriform plates** (*cribr-* = sieve)—one on either side of crista galli; form roof of nasal cavity
 - **crista galli** (*crist-* = crest; *galli* = rooster)— projection for attachment of membranes covering brain
 - **olfactory foramina** (*olfact-* = smell)—tiny holes in cribriform plates for cranial nerve I
 - **perpendicular plate**—forms superior part of nasal septum
 - **middle nasal conchae**—scroll-like projections on each lateral wall of nasal cavity
 - **superior nasal conchae**—scroll-like projections on each lateral wall of nasal cavity

5. *Sphenoid bone* (1)
 - **foramina ovale** (*ovale* = oval)—openings for mandibular branch of cranial nerve V
 - **foramina rotundum**—openings for maxillary branch of cranial nerve V
 - **sella turcica** (*sella* = saddle; *turcica* = Turkish)— bony projection that surrounds pituitary gland
 - **greater and lesser wings**—form anterior and lateral floor of cranium
 - **optic foramina**—openings for cranial nerve II
 - **inferior orbital fissures**—openings for blood vessels and nerves

 - **superior orbital fissures** (*fissure* = cleft)— openings for blood vessels and cranial nerves III, IV, V, and VI (ophthalmic branch)
 - **pterygoid processes (medial and lateral)** (*pterygoid* = wing-shaped)—winglike projections on the base of the skull in the middle section of the sphenoid bone

Facial Bones: Selected Bone Surface Markings

1. *Maxillae* (2)
 - **alveoli** (*alve-* = socket)—tooth sockets (alveolus, sing.)
 - **palatine process**—fused processes that form the anterior part of hard palate

2. *Mandible* (1)
 - **alveoli**—tooth sockets
 - **body**—curved, anterior portion of mandible
 - **mental foramina** (*menta* = chin)—openings in chin for nerves and blood vessels
 - **rami**—posterior branches, one on either side of the body of mandible
 - **condylar processes (mandibular condyles)**— rounded processes on rami that articulate with temporal bone at the mandibular fossa
 - **coronoid processes**—triangular projections of rami anterior to the condylar processes

3. *Lacrimal bone* (2)
 - **lacrimal fossa**—canal that houses lacrimal sac; formed from the maxilla and lacrimal bone

4. *Zygomatic bone* (2)
 - **temporal process**—projects posteriorly; temporal process of zygomatic bone and zygomatic process of temporal bone form the **zygomatic arch**

ACTIVITY 4 Bone Surface Markings

1 Label the bone markings listed in Figures 9.8–9.11.

2 Identify these structures on an articulated skull and an unarticulated (Beauchene) skull, if available.

3 Pronounce the terms as you point to them on a skull.

4 Using two small, plastic straws, poke each straw into an orbit of the eye and through an optic foramen. Observe the straws in the superior view of the cranial floor and notice that the straws cross above the sella turcica. This simulates the crossing of the optic nerves called the optic chiasma.

- inferior orbital fissure
- mental foramen
- middle nasal concha
- orbit of eye
- perpendicular plate of ethmoid
- superior orbital fissure
- supraorbital foramen (supra-OR-bi-tal)
- supraorbital margin

1 _____
2 _____
3 _____
4 _____
5 _____
6 _____
7 _____
8 _____

FIGURE 9.8 Anterior view of skull.

Zygomatic process
of temporal bone

External auditory
meatus

Mastoid process

Styloid process

Lacrimal fossa

Coronoid process

Condylar process
(Mandibular condyle)

Body of mandible

Ramus of mandible

Hyoid bone

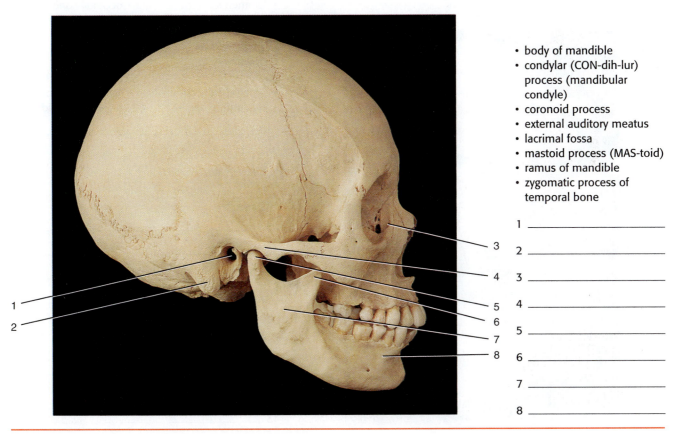

- body of mandible
- condylar (CON-dih-lur)
 process (mandibular
 condyle)
- coronoid process
- external auditory meatus
- lacrimal fossa
- mastoid process (MAS-toid)
- ramus of mandible
- zygomatic process of
 temporal bone

1 _____

2 _____

3 _____

4 _____

5 _____

6 _____

7 _____

8 _____

FIGURE 9.9 Lateral view of skull.

Anterior

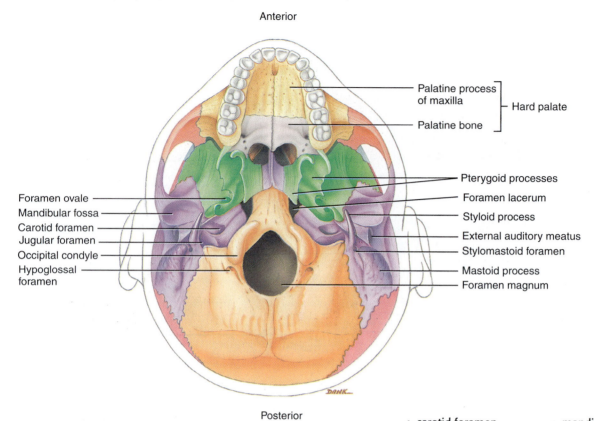

Palatine process
of maxilla
Hard palate
Palatine bone

Pterygoid processes
Foramen lacerum
Styloid process
External auditory meatus
Stylomastoid foramen
Mastoid process
Foramen magnum

Foramen ovale
Mandibular fossa
Carotid foramen
Jugular foramen
Occipital condyle
Hypoglossal
foramen

DANK

Posterior

ANTERIOR

POSTERIOR

• carotid foramen
 (canal)
• foramen lacerum
 (LA-sir-um)
• foramen magnum
• foramen ovale
 (o-VAL-ee)
• hard palate
• hypoglossal foramen
• jugular foramen

• mandibular fossa
• mastoid process
• occipital condyle
 (ox-CI-pi-tal)
• palatine bone
• palatine process of
 maxilla
• pterygoid processes
• stylomastoid foramen
 (sty-lo-MAS-toid)

1 _____

2 _____

3 _____

4 _____

5 _____

6 _____

7 _____

8 _____

9 _____

10 _____

11 _____

12 _____

13 _____

14 _____

FIGURE 9.10 Inferior view of skull.

View

Transverse plane

Anterior

Crista galli
Olfactory foramina
Cribriform plate
Lesser wing of sphenoid

Foramen rotundum

Foramen ovale

Internal auditory meatus

Hypoglossal foramen

Foramen magnum

Optic foramen
Sella turcica
Greater wing of sphenoid

Foramen lacerum

Jugular foramen

DANK

Posterior

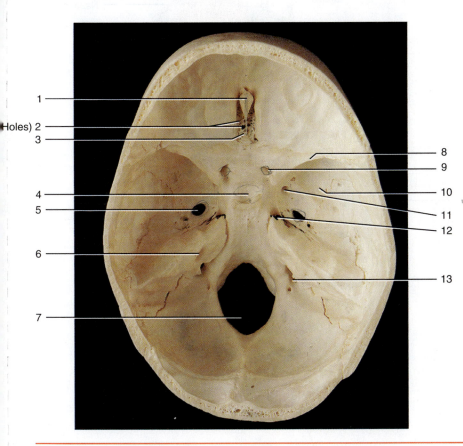

Holes) 2

1
2
3

4
5

6

7

8
9

10

11
12

13

- cribriform plate (CRIB-ri-form)
- crista galli (CRIS-ta GAL-li)
- foramen lacerum
- foramen magnum
- foramen ovale
- foramen rotundum
- greater wing of sphenoid
- internal auditory meatus
- jugular foramen
- lesser wing of sphenoid
- olfactory foramina
- optic foramen
- sella turcica (SEL-la TUR-si-ca)

1 _____
2 _____
3 _____
4 _____
5 _____
6 _____
7 _____
8 _____
9 _____
10 _____
11 _____
12 _____
13 _____

FIGURE 9.11 Superior view of floor of cranium.

Paranasal Sinuses

Paranasal (*para-* = next to; *nasal* = nose) **sinuses** (*sinu-* = hollow) are cavities lined with mucous membranes that are located near and have openings into the nasal cavities. The four paranasal sinuses, the **ethmoidal, frontal, maxillary,** and **sphenoidal,** are named for their locations in bones of the same name.

ACTIVITY 5 Paranasal Sinuses

Label the four paranasal sinuses in Figure 9.12.

(a) (b)

FIGURE 9.12 Paranasal sinuses.

(a) Anterior frontal view (b) Sagittal view

- ethmoidal sinus
- frontal sinus
- maxillary sinus
- sphenoidal sinus

1 _____

2 _____

3 _____

4 _____

Nasal Septum

The **nasal septum** (*saeptum* = wall or partition) is comprised of bone and cartilage. The inferior bone is the **vomer,** and the **perpendicular plate** of the **ethmoid** bone is superior. The **septal cartilage** is anterior to the two bones.

ACTIVITY 6 Nasal Septum

1 Label the bones and cartilage of the nasal septum in Figure 9.13.

2 Identify these structures on an articulated skull (and a Beauchene skull, if available).

3 Palpate your nasal septum to find the junction of the nasal bones and nasal cartilage.

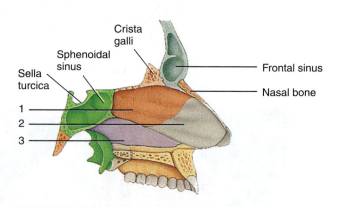

FIGURE 9.13 Nasal septum.

- perpendicular plate of ethmoid
- septal cartilage
- vomer

1 _____

2 _____

3 _____

Hard Palate

The **hard palate,** or roof of the mouth, is formed by the fusion of four bones: two **palatine processes of the maxillary bones** and two **palatine bones.** Approximately three-fourths of the hard palate is composed of the palatine processes of the maxillary bones, and one-fourth is the palatine bones.

> **CLINICAL NOTE:** *A **cleft palate** is a common congenital malformation of the hard palate. This defect is the result of bones of the palate failing to fuse along the midline during embryonic development. The cleft can be partial or complete.*

ACTIVITY 7 Hard Palate

1 Identify the bones of the hard palate in Figure 9.10.
2 Locate these bones on a skull.
3 Palpate your own palate with your tongue or finger, noting the junction of the hard and soft palate.

Fetal and Newborn Skull

The newborn skull is not completely ossified but has membranous sections composed of fibrous connective tissue called **fontanels** (*fontaine* = fountain) or "soft spots." Fontanels allow the cranial bones to compress during the journey through the birth canal, and they also permit rapid growth of the brain and skull during the early years. At birth, the physician or midwife can tell how the baby is presenting itself in the birth canal by palpating the fontanel. If the fontanel is a diamond shape, the baby is presenting the frontal bone first and is face down. If the fontanel is triangular in shape, the baby is presenting the occipital bone and is face up. These membrane templates that are present at birth eventually ossify.

ACTIVITY 8 Fetal Skull and Major Fontanels

1 Label the fontanels in Figures 9.14(a) and (b).
2 Identify the fontanels on a fetal skull (if available).
3 Compare the location of the fontanels and sutures in the fetal skull with adult sutures.

(a) Superior view

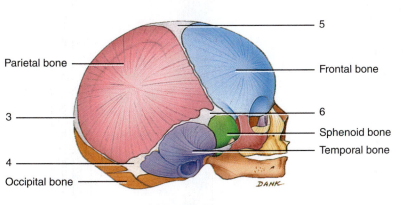

Parietal bone

Frontal bone

Sphenoid bone
Temporal bone

Occipital bone

DANK

(b) Lateral view

(a) Superior View
• anterior (frontal) fontanel
• posterior (occipital) fontanel

1 _____

2 _____

(b) Lateral View
• anterior (frontal) fontanel (fon-ta-NEL)
• anterolateral (sphenoidal) fontanel
• posterior (occipital) fontanel
• posterolateral (mastoid) fontanel

3 _____

4 _____

5 _____

6 _____

FIGURE 9.14 Major fontanels of the fetal skull.

C. Hyoid Bone

The hyoid bone is not attached to the axial skeleton but is included with the axial skeleton because of its midline location and proximity to the mandible and vertebral column. This bone is U-shaped and has the distinction of not articulating with any other bones. It is secured in place by ligaments and muscles, including many muscles of the tongue and neck. It is located in the anterior neck region between the mandible and larynx. (This is the bone that is crushed during strangulation.)

ACTIVITY 9 Hyoid Bone

1 Locate the hyoid bone in Figure 9.9.
2 Identify the hyoid bone on an articulated skeleton and on a disarticulated skeleton.
3 Palpate your hyoid bone by placing the thumb and middle finger of one hand on either side of the neck about 1 inch inferior to the mandible. This bone can be moved laterally from side to side.

D. Vertebral Column

The vertebral column protects the spinal cord and provides attachment points for back and abdominal muscles. The curved **vertebral column** (backbone) is a flexible structure that can be bent, twisted, and rotated, especially in the cervical region. The vertebral column consists of **cervical** (*cervic-* = neck), **thoracic** (*thorac-* = chest), and **lumbar** (*lumb-* = loin) **vertebrae,** the **sacrum** (*sacer* = sacred), and **coccyx** (*coccyx* = cuckoo's beak). The sacrum consists of five sacral vertebrae that are fused in the adult. The coccyx (tailbone) is usually composed of four small, fused coccygeal vertebrae. An infant has 33 individual vertebrae, and the adult has 26 due to fusion of the sacral and coccygeal vertebrae. The vertebral column articulates with the skull, the ribs, and the pelvis.

Regions and Normal Curvatures of the Vertebral Column

The five regions of the vertebral column and the number of vertebrae in the adult, from superior to inferior, are: **cervical** (7), **thoracic** (12), **lumbar** (5), **sacral** (1), and **coccygeal** (1). To remember the number of vertebrae in the first three groups, students say they eat "breakfast at 7, lunch at 12, and dinner at 5." There are four normal curvatures that correspond with the regions of the vertebral column: **cervical, thoracic, lumbar,** and **sacral** (pelvic) **curvatures.** A newborn has a single anteriorly concave primary curve that will become the thoracic curve and sacral curve. Two anteriorly convex secondary curves, the cervical and lumbar, develop several months later. The cervical curve develops when the baby can hold its head erect, while the lumbar curve develops when the baby can stand.

ACTIVITY 10 The Vertebral Column

1 Label the five regions of the vertebral column in Figure 9.15.
2 Label the four normal curvatures in Figure 9.15.
3 Identify the five regions on an articulated skeleton or vertebral column model.
4 Identify the four curvatures on an articulated skeleton or vertebral column.
5 Pronounce the terms as you point to them on a skeleton or vertebral column.

FIGURE 9.15 Normal spinal curvatures of the vertebral column.

- cervical curve
- cervical vertebrae (VER-te-bray)
- coccyx (COCK-six)
- lumbar curve
- lumbar vertebrae
- sacral curve
- sacrum (SAY-crum)
- thoracic curve
- thoracic vertebrae

1 _____

2 _____

3 _____

4 _____

5 _____

6 _____

7 _____

8 _____

9 _____

Parts of a Typical Vertebra

The cervical, thoracic, and lumbar vertebrae have similar as well as different structures. The similarities are studied in this activity, and the differences will be studied in the next activity.

- **body**—located anteriorly; is the largest part of the vertebra
- **pedicle** (*pediculus* = little foot)—attached to and extends posteriorly on either side of the body
- **transverse process**—extends laterally from each pedicle
- **lamina** (*lamin-* = plate)—connects transverse processes to the spinous process
- **spinous process**—projects posteriorly from fused lamina
- **vertebral arch**—formed by the fusion of pedicles and laminae
- **vertebral foramen**—large opening formed by the vertebral arch that protects the spinal cord
- **superior** and **inferior articular processes** (with **facets**)—extend from the vertebra at the junction of the pedicle and lamina to articulate with a superior and inferior vertebra, respectively

ACTIVITY 11 **Parts of a Typical Vertebra**

1 Locate the parts of a vertebra shown in Figure 9.16.
2 Identify the parts on a disarticulated thoracic vertebra.
3 Pronounce the terms as you point to them on the vertebra.
4 Palpate the spinous processes of your vertebral column.

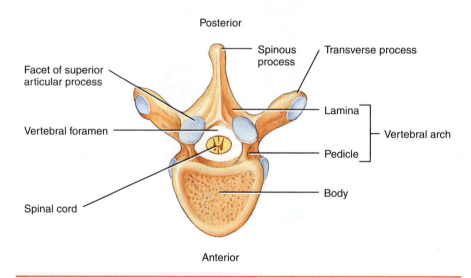

FIGURE 9.16 Parts of a typical vertebra.

Comparison of Cervical, Thoracic, and Lumbar Vertebrae

The first two of the seven cervical vertebrae look different from the others. The **atlas** is the first cervical vertebra (C_1) and is named for the mythical Greek god who held up the world on his shoulders. The large **superior articular facets** of the atlas articulate with the **occipital condyles** of the skull and allow the head to tilt up and down in a nodding motion. The **axis,** C_2, has a superior toothlike protuberance called the **dens** (*dens* = tooth in Latin) or **odontoid process** (*odous* = tooth in Greek). The dens extends superiorly into the vertebral foramen of the atlas and allows the atlas to pivot laterally in a "no" type motion. Some head traumas push the head onto the vertebral column, and the dens is shoved into the medulla oblongata (brain stem). This is the cause of death in such head injuries. The easiest way to distinguish any **cervical vertebra** is by its three foramina (pl.; foramen, sing.)—a **vertebral foramen** plus two **transverse foramina.** The transverse foramina, found only in cervical vertebrae, allow passage of the vertebral arteries, veins, and sympathetic nerves to and from the brain. Since the cervical vertebrae do not carry much weight, they are lightweight and have small bodies. C_3 through C_6 have bifurcated (forked) spinous processes. The **vertebra prominens** (C_7) has a prominent single spinous process that protrudes at the base of the neck and can be seen and palpated. A **whiplash,** which occurs in some rear-ended auto accidents, causes partial or complete dislocation of the cervical vertebrae.

A **thoracic vertebra** has a medium-sized body and a long, narrow spinous process that commonly slants inferiorly at a sharp angle. Thoracic vertebrae also have facets (*facette* = little face) on the transverse processes and demifacets (*demi-* = half) on the bodies, both of which articulate with ribs.

Lumbar vertebrae have the largest bodies to support more weight and thick, hatchet-shaped spinous processes that extend horizontally.

Intervertebral discs can be found between the bodies of vertebrae from C_2 to the sacrum. As you recall, they are made of fibrocartilage to give strength, permit movement, and absorb shock. The outer portion of the disc is called the **annulus fibrosus** (*annulus* = ringlike) and consists of tough fibrocartilage. The inner part, the **nucleus pulposus** (*pulposus* = pulplike), is soft and pulpy. A *herniated* or *slipped disc* occurs when the fibrocartilage of the annulus is stressed and torn or cracked. In this situation, the inner pulplike center protrudes outward and causes pressure on the spinal cord or a spinal nerve. A *laminectomy* (removal of the lamina and sometimes spinous process of the vertebra) relieves the pressure.

Intervertebral foramina are formed between adjacent vertebrae when vertebrae are stacked on one another. Spinal nerves exit the vertebral column through these foramina.

ACTIVITY 12 **Comparison of Cervical, Thoracic, and Lumbar Vertebrae**

1 Identify the specific parts of an atlas, axis, and cervical vertebra in Figure 9.17(a), (b), and (c).

2 Identify the specific parts of a thoracic vertebra in Figure 9.18(a) and (b) and a lumbar vertebra in Figure 9.19(a) and (b).

3 Using disarticulated vertebrae, place all three types side by side and note their differences. Mix them up and identify each type of vertebra with a partner.

4 Identify the parts of articulated vertebrae in Figure 9.20(a) and a herniated disc in 9.20(b).

5 Pronounce each term as you point to it.

6 Palpate your own spinous process of the vertebra prominens (C_7).

Transverse foramen

(a) Superior view of the atlas (C₁)

Bifurcated spinous process

Transverse process

Dens

(b) Superior view of the axis (C₂)

Bifurcated spinous process

Lamina

Superior articular facet

Pedicle

(c) Superior view of a typical cervical vertebra (C₃)

FIGURE 9.17 Cervical vertebrae.

(a) Atlas
• superior articular facet
• transverse foramen
• transverse process

(b) Axis (*Note:* The transverse foramen cannot be seen in this photo.)
• dens or odontoid process
• lamina
• spinous process

(c) Typical cervical vertebra
• bifurcated spinous process
• body
• pedicle
• transverse process

1 _____

2 _____

3 _____

4 _____

5 _____

6 _____

7 _____

8 _____

9 _____

10 _____

POSTERIOR

ANTERIOR

(a) Superior view

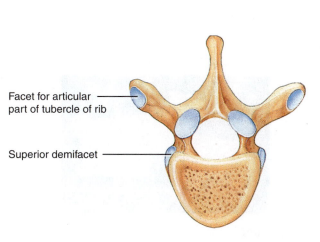

Facet for articular part of tubercle of rib

Superior demifacet

(a) Superior view

SUPERIOR

POSTERIOR ANTERIOR

(b) Right lateral view

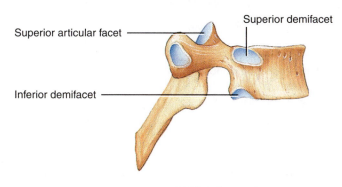

Superior articular facet

Superior demifacet

Inferior demifacet

(b) Right lateral view

FIGURE 9.18 Thoracic vertebrae.

(a) Superior view
• facet for articular part of tubercle of rib
• superior articular facet
• superior demifacet
• transverse process

(b) Right lateral view
• facet for articular part of tubercle of rib
• inferior demifacet
• slanted spinous process
• superior demifacet

1 _____

2 _____

3 _____

4 _____

5 _____

6 _____

7 _____

8 _____

POSTERIOR

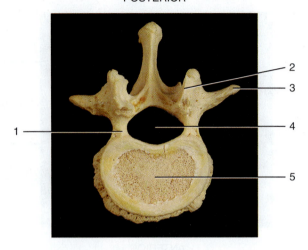

ANTERIOR

(a) Superior view

SUPERIOR

POSTERIOR ANTERIOR

(b) Right lateral view

FIGURE 9.19 Lumbar vertebrae.

(a) Superior view
- body
- pedicle
- superior articular process
- transverse process
- vertebral foramen

1 _____

2 _____

3 _____

4 _____

5 _____

(b) Right lateral view
- hatchet-shaped spinous process
- inferior articular facet

6 _____

7 _____

(a) Right lateral view of articulated vertebrae with intervertebral disc

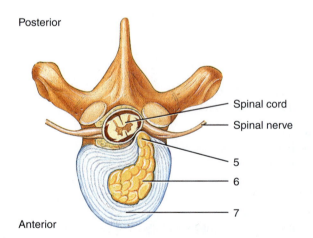

Posterior

Spinal cord

Spinal nerve

5

6

7

Anterior

(b) Superior view of herniated disc

FIGURE 9.20 Intervertebral discs.

(a) Articulated vertebrae
- annulus fibrosus
- intervertebral disc
- intervertebral foramen
- nucleus pulposus

1 _____

2 _____

3 _____

4 _____

(b) Herniated disc
- annulus fibrosus
- herniation
- nucleus pulposus

5 _____

6 _____

7 _____

Sacrum and Coccyx

The **sacrum** has a slightly curved, triangular shape and has two lateral winglike projections called **alae** (ala, sing.). The **sacral promontory** (*promontory* = a projecting part) which protrudes anteriorly is an important landmark in females during labor and delivery. The sacrum has **sacral foramina** that provide exits for spinal nerves.

From the posterior view, the opening of the **sacral canal** is located just posterior to the body of the sacrum and is open all the way through the sacrum. (*Note:* On many plastic models, the sacral canal is closed.) Two **superior articular processes** with facets that articulate with the fifth lumbar vertebra are located on either side of the opening to the sacral canal. The lateral surfaces of the sacrum, the

auricular surfaces, are roughened to articulate with the iliac portion of the os coxa on each side, forming **sacroiliac joints.** The vertebral column ends with a small **coccyx,** or tailbone. The tiny, fused vertebrae of the coccyx are attached to the sacrum with ligaments.

ACTIVITY 13 **Sacrum and Coccyx**

1 Locate the parts of the sacrum and coccyx in Figure 9.21.

2 Identify these structures on a model of the sacrum and coccyx.

3 Pronounce the terms as you point to them on the model.

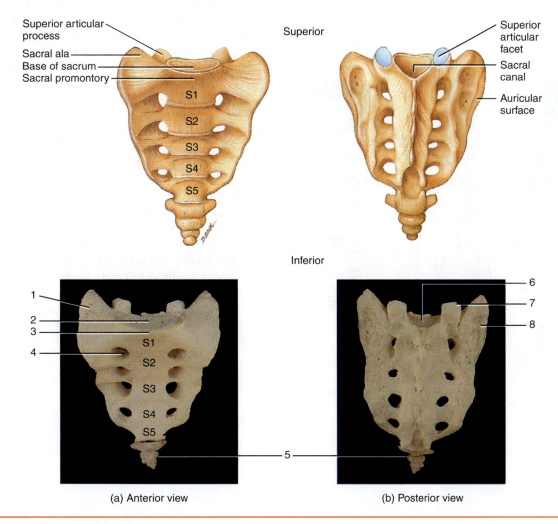

(a) Anterior view (b) Posterior view

FIGURE 9.21 Sacrum and coccyx.

(a)
- base
- coccyx
- sacral ala
- sacral foramen
- sacral promontory

(b)
- auricular surface (for sacroiliac joint)
- sacral canal
- superior articular facet

1 _____ 6 _____

2 _____ 7 _____

3 _____ 8 _____

4 _____

5 _____

Abnormal Vertebral Curvatures

Three abnormal curves of the vertebral column are scoliosis, kyphosis, and lordosis. The vertebral column bends laterally in **scoliosis** (*scolio-* = crooked). **Kyphosis** is an exaggerated thoracic curve that results in a hunched back with rounded shoulders. **Lordosis** is an exaggerated lumbar curve that appears as a swayback with the abdomen protruding anteriorly.

ACTIVITY 14 **Abnormal Curves of the Vertebral Column**

Identify the three different types of abnormal curves in Figure 9.22(a), (b), and (c).

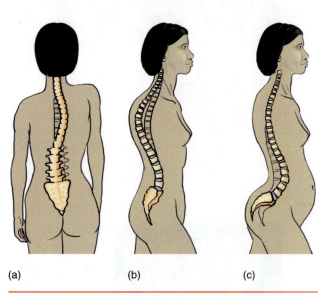

(a) (b) (c)

FIGURE 9.22 Abnormal curves of the vertebral column.

- kyphosis
- lordosis
- scoliosis

a _____

b _____

c _____

E. Thoracic Cage (Rib Cage)

The bony cage that encircles the chest is called the **thoracic cage,** or rib cage, and is composed of the sternum, ribs, costal cartilages, and thoracic vertebrae.

The **sternum** is a narrow flat bone that is composed of three fused bones: the manubrium, the body of the sternum, and the xiphoid process. The **manubrium** (handle), the superior portion of the sternum, has a concave superior surface called the **suprasternal notch** or **jugular notch.** Between the **body** of the sternum and the manubrium is the **sternal angle,** an important clinical landmark indicating the attachment of the second rib, below which is the second intercostal space. The manubrium and body articulate (*articul-* = pertaining to a joint) with the **costal cartilages** of the **ribs.** The **xiphoid process** (swordlike) is the inferior portion of the sternum that is shaped like a small sword.

There are 12 pairs of **ribs** in both males and females. The first 7 pairs are called the **true ribs** or **vertebrosternal ribs** because their **costal (rib) cartilages** have a direct attachment to the sternum. The last 5 rib pairs (8–12) are called **false ribs.** Rib pairs 8 through 10 are **vertebrochondral ribs** (*vertebro* = ribs; *chondral* = cartilage) because their costal cartilages do not have a direct attachment to the sternum, but attach to the costal cartilage of the seventh rib instead. Rib pairs 11 and 12 do not have any attachment to the sternum or cartilage and are called **floating ribs** or **vertebral ribs.** The space between the ribs is called the **intercostal space.**

The main parts of a rib are the head, neck, tubercle, and body. The **head** projects from the posterior part of the rib and articulates with **demifacets** of the bodies of thoracic vertebrae. The **neck** is the constricted part lateral to the head. The **tubercle** is a small, knoblike projection close to the neck that articulates with the facet of a transverse process. The **body** is the main part of the rib.

ACTIVITY 15 **Thoracic Cage and Rib Articulations**

1 Locate the parts of the thorax in Figure 9.23.

2 Identify these parts on an articulated thorax or skeleton.

3 Locate the parts of a rib in Figure 9.24(a) and its articulation with vertebrae in 9.24(b) and (c).

4 Identify the four parts of a rib and its articulations on a skeleton.

5 Pronounce the name of each part as you point to it.

6 Palpate as many of these parts as you can on your body.

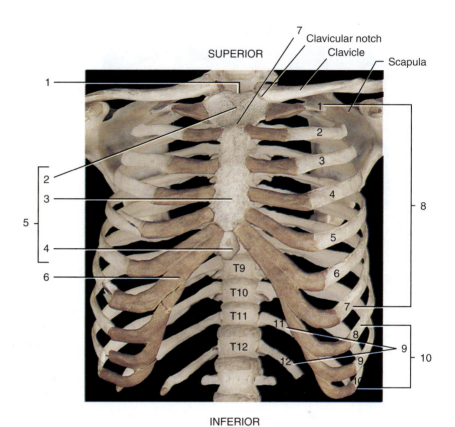

FIGURE 9.23 Thoracic cage.

- body of sternum
- costal cartilage
- false ribs
- floating ribs
- manubrium
- sternal angle
- sternum
- suprasternal notch (jugular notch)
- true ribs
- xiphoid process

1 _____ 6 _____

2 _____ 7 _____

3 _____ 8 _____

4 _____ 9 _____

5 _____ 10 _____

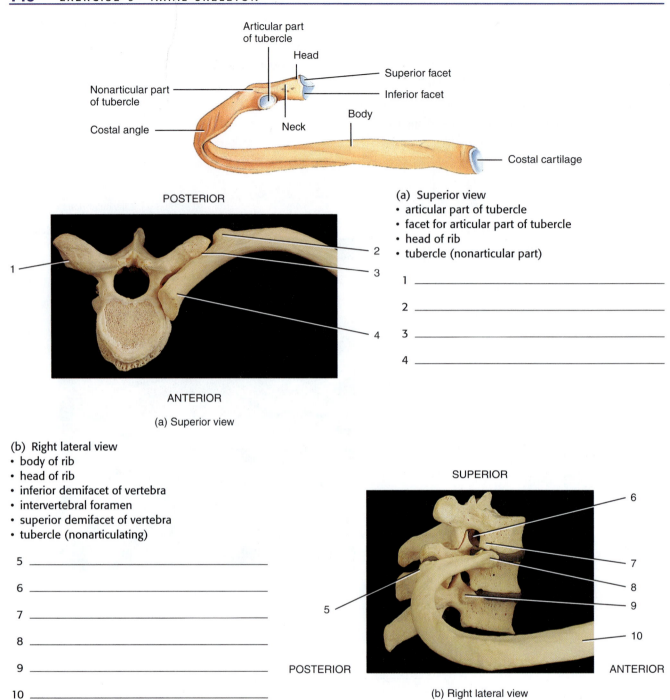

Articular part
of tubercle

Head

Superior facet

Nonarticular part
of tubercle

Inferior facet

Body

Costal angle

Neck

Costal cartilage

POSTERIOR

1

2

3

4

ANTERIOR

(a) Superior view

(a) Superior view
• articular part of tubercle
• facet for articular part of tubercle
• head of rib
• tubercle (nonarticular part)

1 _____

2 _____

3 _____

4 _____

(b) Right lateral view
• body of rib
• head of rib
• inferior demifacet of vertebra
• intervertebral foramen
• superior demifacet of vertebra
• tubercle (nonarticulating)

5 _____

6 _____

7 _____

8 _____

9 _____

10 _____

SUPERIOR

6

7

8

9

10

5

POSTERIOR

ANTERIOR

(b) Right lateral view

FIGURE 9.24 Rib and its articulation with thoracic vertebrae.

Reviewing Your Knowledge

A. Bone Surface Markings

Name the depression, opening, process, or projection that describes the definition.

_____ **1.** Tubelike passageway

_____ **2.** Rounded articular process

_____ **3.** Opening or hole through a bone (oval or round)

_____ **4.** Shallow depression

_____ **5.** Branchlike process

_____ **6.** Narrow slit or cleft in bone

_____ **7.** Pointed projection

B. Cranial Bones and Bone Surface Markings

Identify the cranial bones that have the following bone markings:

_____ **1.** Middle nasal concha

_____ **2.** Foramen magnum

_____ **3.** Alveoli (2 bones)

_____ **4.** Sella turcica

_____ **5.** External auditory meatus

_____ **6.** Crista galli

_____ **7.** Greater wing

_____ **8.** Mastoid process

_____ **9.** Occipital condyle

_____ **10.** Optic foramen

C. Skull Bones and Bone Markings

Identify the bones, bone markings, and sutures in Figure 9.25, anterior view of skull.

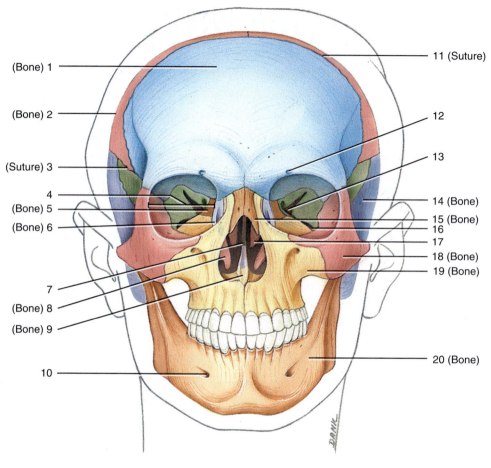

Anterior view

FIGURE 9.25 Anterior view of skull.

1 _____ 8 _____ 15 _____

2 _____ 9 _____ 16 _____

3 _____ 10 _____ 17 _____

4 _____ 11 _____ 18 _____

5 _____ 12 _____ 19 _____

6 _____ 13 _____ 20 _____

7 _____ 14 _____

D. Skull Bones and Bone Markings, Lateral View

Identify the bones, bone markings, and sutures in Figure 9.26, lateral view of skull.

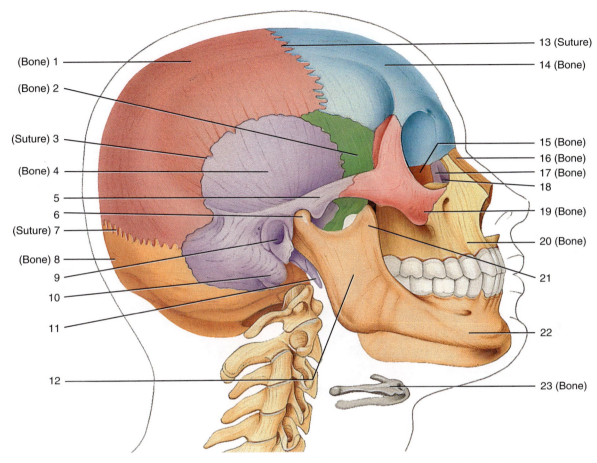

FIGURE 9.26 Right lateral view of skull.

1 _____	9 _____	17 _____
2 _____	10 _____	18 _____
3 _____	11 _____	19 _____
4 _____	12 _____	20 _____
5 _____	13 _____	21 _____
6 _____	14 _____	22 _____
7 _____	15 _____	23 _____
8 _____	16 _____	

E. Skull Bones and Bone Markings, Inferior View of Skull

Identify the bones, bone markings, and sutures in Figure 9.27, inferior view of skull.

Anterior

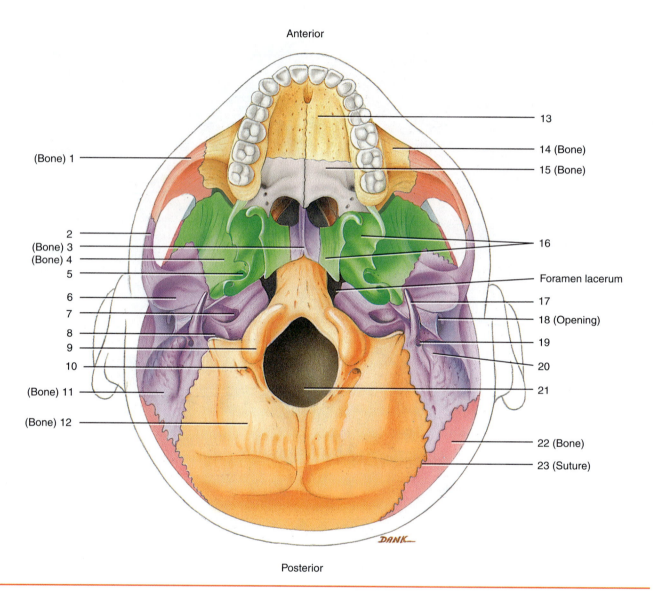

(Bone) 1

2
(Bone) 3
(Bone) 4
5
6
7
8
9
10
(Bone) 11
(Bone) 12

13
14 (Bone)
15 (Bone)
16
Foramen lacerum
17
18 (Opening)
19
20
21
22 (Bone)
23 (Suture)

DANK

Posterior

FIGURE 9.27 Inferior view of skull.

1 _____	9 _____	17 _____
2 _____	10 _____	18 _____
3 _____	11 _____	19 _____
4 _____	12 _____	20 _____
5 _____	13 _____	21 _____
6 _____	14 _____	22 _____
7 _____	15 _____	23 _____
8 _____	16 _____	

F. Skull Bones and Bone Markings, Superior View of Floor of Cranium

Identify the bones, bone markings, and sutures in Figure 9.28, superior view of floor of cranium.

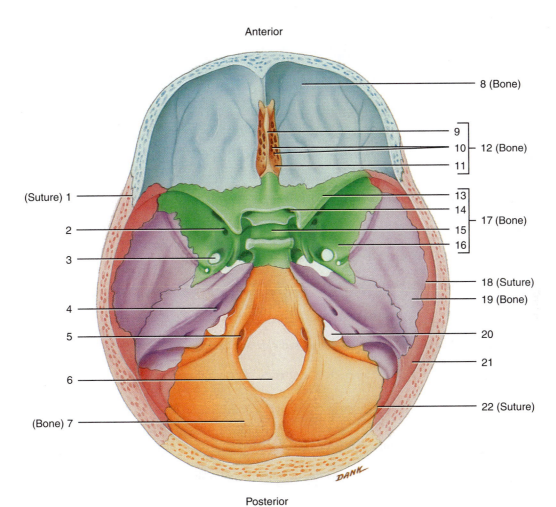

Anterior

8 (Bone)

9
10 ⊦ 12 (Bone)
11

(Suture) 1

13
14
15 ⊦ 17 (Bone)
16

2

3

18 (Suture)
19 (Bone)

4

5

20

21

6

22 (Suture)

(Bone) 7

DANK

Posterior

Superior view of floor of cranium

FIGURE 9.28 Superior view of floor of cranium.

1 _____	9 _____	17 _____
2 _____	10 _____	18 _____
3 _____	11 _____	19 _____
4 _____	12 _____	20 _____
5 _____	13 _____	21 _____
6 _____	14 _____	22 _____
7 _____	15 _____	
8 _____	16 _____	

G. Anatomy Drill and Practice

Go to www.wiley.com/college/allen and click on the picture of your lab manual. Then click on Student Companion Site, Anatomy Drill and Practice, and Exercise 9.

H. Vertebral Column and Thorax

Write the name of the structure that is described.

_____ **1.** Softens jolts to the vertebral column

_____ **2.** Opening for spinal nerve exit

_____ **3.** Bony protection for thoracic organs

_____ **4.** First cervical vertebra

_____ **5.** Second cervical vertebra

_____ **6.** Articulates with rib posteriorly

_____ **7.** Tailbone

_____ **8.** Encloses and protects the spinal cord

_____ **9.** Primary curvatures

_____ **10.** Secondary curvatures

_____ **11.** Formed by the fusion of pedicles and lamina

_____ **12.** Passageway for vertebral arteries in cervical vertebrae

I. Other Major Features of the Axial Skeleton

Identify the major skull feature or structure that is described.

_____ **1.** The suture that joins the occipital bone to the parietal bones.

_____ **2.** The facial bone that contains a paranasal sinus.

_____ **3.** The articulation between the mandible and skull.

_____ **4.** Neck and tongue muscles are attached to this bone.

_____ **5.** This structure allows the baby's cranium to compress during childbirth.

_____ **6.** A special feature present on the transverse process of cervical vertebrae.

_____ **7.** Special features present on the thoracic vertebrae for rib articulations.

_____ **8.** A part of the rib that articulates with the body of the thoracic vertebrae.

_____ **9.** A part of the rib that articulates with the transverse process of the thoracic vertebrae.

_____ **10.** The odontoid process is a part of which bone?

_____ **11.** Name the superior bone of the nasal septum.

_____ **12.** Name the inferior bone of the nasal septum.

Using Your Knowledge

A. Identifying Bones and Bone Markings on Radiographs

Identify the structures in Figure 9.29(a).

1 _____ (bone marking)

2 _____ (bone)

3 _____ (bone)

4 _____ (bone)

5 _____ (bone)

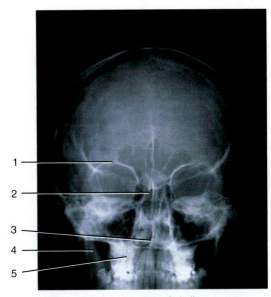

(a) Anterior view of skull.

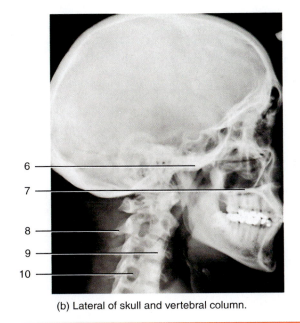

(b) Lateral of skull and vertebral column.

Identify the structures in Figure 9.29(b).

6 _____ (bone marking)

7 _____ (bone marking)

8 _____ (bone marking)

9 _____ (bone marking)

10 _____ (space between bones)

FIGURE 9.29 Radiographs.

B. Identifying Bones and Bone Markings on CT Scans

Identify the structures in Figure 9.30(a), CT of lateral view of skull.

11 _____ (bone)

12 _____ (bone)

13 _____ (sinus)

14 _____ (bone marking in sphenoid)

15 _____ (opening)

(a) Lateral view of skull.

ANTERIOR

Occipital condyles Dens

POSTERIOR

(b) Transverse section through skull at the level of the atlas.

Identify the structures in Figure 9.30(b).

16 _____ (bone marking)

17 _____ (bone marking)

18 _____ (bone sinus)

19 _____ (bone marking)

20 _____ (bone)

FIGURE 9.30 CT scans.

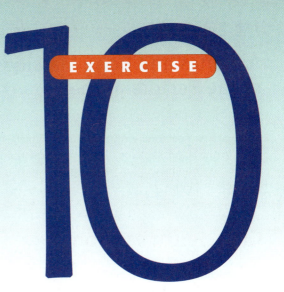

Appendicular Skeleton

Objectives

**After completing this exercise,
you should be able to:**

- Identify the bones of the appendicular skeleton and be able to identify selected bones as right or left
- Identify the principal bone markings of the pectoral (shoulder) girdle and the upper limb
- Describe the articulation of the pectoral girdle with the humerus
- Describe the articulation of the humerus with the forearm bones
- Identify the bone markings of the pelvic girdle and the lower limbs
- Describe the articulation of the bones of the pelvic girdle with the femur
- Describe the articulation of the femur with the leg bones

Materials

- articulated skeleton
- disarticulated skeleton

The appendicular skeleton (*appendere* = to hang upon) has larger bones than the axial skeleton and bears more weight. The bones of this division are separated into four main areas: the **pectoral girdles** (*pectus* = breast; *girdle* = encircles), the **upper limbs** or extremities, the **pelvic girdle** (*pelvis* = basin), and the **lower limbs.** The bone markings include sites of muscle attachment and articulations with other bones to form a joint. Bone surface markings that pertain to the appendicular skeleton are given in Table 10.1.

A. Bones and Bone Markings of the Pectoral Girdle

There are two pectoral girdles, and each attaches an upper limb to the axial skeleton. Each **pectoral** (or shoulder) **girdle** is composed of a **scapula** and a **clavicle** (*clavicle* = key).

Clavicle (collar bone)

- **sternal end**—blunt, medial end
- **acromial end** (*acrom-* = topmost)—broader, flat, roughened, lateral end

Scapula (shoulder blade)

- **spine**—sharp ridge located on posterior side
- **acromion**—flattened process at lateral end of spine
- **glenoid cavity** (*glene* = joint socket) or **fossa**—depression inferior to acromion
- **coracoid process** (*coracoid* = crow's beak)—superior and medial to glenoid cavity; projects anteriorly
- **supraspinous fossa** (*supra-* = above; *spinous* = spine)—depression superior to spine

- **infraspinous fossa** (*infra-* = below)—depression inferior to spine
- **subscapular fossa** (*sub-* = under)—depression on anterior surface of scapula
- **lateral** or **axillary border**—margin near axilla
- **medial** or **vertebral border**—margin near vertebral column

The acromial end of the clavicle articulates with the acromion (acromial process) of the scapula laterally, forming the pectoral girdle. The sternal end of the clavicle articulates with the manubrium of the sternum medially, attaching the pectoral girdle to the axial skeleton. The scapula does not articulate directly with the axial skeleton but is attached to it with muscles.

ACTIVITY 1 Bones of the Pectoral Girdle

1 Identify the parts of the clavicle and scapula in Figure 10.1(a), (b), and (c).

2 Locate these structures on disarticulated bones.

3 Pronounce the terms as you point to them.

4 Palpate these structures on your own body: clavicle, acromion (process), spine of scapula, and muscles located in the supraspinous and infraspinous fossae.

TABLE 10.1 Selected Bone Surface Markings of the Appendicular Skeleton

MARKING	DESCRIPTION	PURPOSE
Depressions or Openings		
1. Foramen	Opening or hole	Opening for blood vessels and nerves
2. Fossa	Shallow depression	Muscle attachment or articulation
Processes		
3. Crest	Prominent ridge	Muscle attachment
4. Condyle	Smooth, rounded articular process	Articulation
5. Epicondyle	Projection above a condyle	Muscle attachment
6. Head	Rounded articular projection supported on the neck of a bone	Articulation
7. Line	Long, narrow ridge (less prominent than a crest)	Muscle attachment
8. Trochanter	Very large projection	Muscle attachment
9. Tubercle	Small, rounded projection	Muscle attachment
10. Tuberosity	Large, roughened projection	Muscle attachment

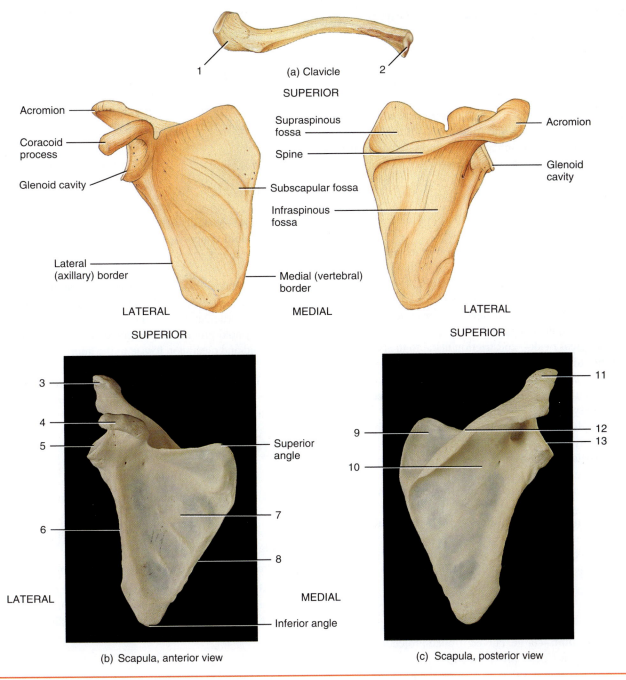

1 _____ (a) Clavicle 2 _____

SUPERIOR

Acromion

Supraspinous fossa

Acromion

Coracoid process

Spine

Glenoid cavity

Glenoid cavity

Subscapular fossa

Infraspinous fossa

Lateral (axillary) border

Medial (vertebral) border

LATERAL MEDIAL LATERAL

SUPERIOR SUPERIOR

3

4

5

6

Superior angle

7

8

9

10

11

12
13

LATERAL MEDIAL

Inferior angle

(b) Scapula, anterior view (c) Scapula, posterior view

FIGURE 10.1 The right pectoral girdle.

(a) Clavicle
• acromial end (a-CROHM-ee-al)
• sternal end

(b) Scapula, anterior view
• acromion or acromial process
 (a-CROW-mee-yon)

• lateral (axillary) border
• coracoid process (COR-a-coid)
• glenoid cavity or fossa (GLEN-oid)
• subscapular fossa (sub-SCAP-u-lar)
• medial (vertebral) border
 (ver-TEE-bral)

(c) Scapula, posterior view
• acromion or acromial process
• glenoid cavity
• infraspinous fossa
 (in-fra-SPINE-us FOS-sa)
• spine of scapula
• supraspinous fossa (su-pra-SPINE-ous)

1 _____

2 _____

3 _____

4 _____

5 _____

6 _____

7 _____

8 _____

9 _____

10 _____

11 _____

12 _____

13 _____

B. Bones and Selected Bone Markings of the Arm, Forearm, and Hand

The upper limb consists of the **humerus, ulna, radius, carpals, metacarpals,** and **phalanges.** Of the 30 bones in each upper limb, 1 is in the arm, 2 in the forearm, and the other 27 are in the hand (includes wrist).

Humerus (arm bone)

- **head**—rounded, proximal end
- **anatomical neck**—constriction immediately distal to head
- **greater tubercle** (*tuber* = swelling)—lateral projection distal to anatomical neck
- **lesser tubercle**—smaller, anterior projection distal to anatomical neck
- **surgical neck**—constriction distal to the tubercles
- **deltoid tuberosity**—raised area on lateral side between the proximal and distal ends of humerus
- **trochlea** (*trochlea* = pulley)—spool-shaped medial condyle on distal end
- **capitulum** (*caput* = head)—rounded, knoblike condyle lateral to trochlea
- **medial epicondyle** (*epi-* = upon)—rough projection above trochlea
- **lateral epicondyle**—rough projection above capitulum; smaller than medial epicondyle
- **coronoid fossa** (*corona* = crown)—shallow anterior depression on distal end
- **olecranon fossa** (*olekranon* = tip of the elbow)—largest depression on posterior, distal end

Ulna (medial bone of forearm)

- **olecranon**—large, curved, liplike projection on posterior side of proximal end
- **coronoid process**—smaller, curved, liplike projection on anterior side of proximal end; distal to olecranon
- **trochlear notch**—deep, curved area between olecranon and coronoid process
- **styloid process** (*stylos* = pole; *-oid* = like)—slender, pointed projection on distal end
- **radial notch**—depression on proximal end where head of radius articulates with ulna

Radius (lateral bone of forearm)

- **head**—flat, disc-shaped proximal end
- **styloid process**—slender, pointed projection; distal end

The shoulder joint that connects the upper limb to the pectoral girdle is formed by the head of the humerus (*humeri* = shoulder) articulating with the glenoid cavity of the scapula. The coronoid process and olecranon of the ulna fit into the coronoid and olecranon fossae of the humerus. The trochlea of the humerus articulates with the trochlear notch of the ulna. The head of the radius articulates with the capitulum of the humerus at the elbow joint and with the radial notch of the ulna at the proximal radioulnar joint.

ACTIVITY 2 **The Upper Limb Bones and Bone Markings**

1 Identify the parts of the humerus in Figure 10.2(a) and (b) and the ulna and radius in Figure 10.3(a) and (b).

2 Locate these bones and their bone markings on disarticulated bones.

3 Pronounce the terms as you point to them.

4 Palpate the following bone parts on your own body: medial and lateral epicondyle of the humerus, olecranon of the ulna, and styloid processes of the radius and ulna.

FIGURE 10.2 Right humerus.

(a) Anterior view
- anatomical neck
- capitulum (ca-PIT-u-lum)
- coronoid fossa (COR-a-noid)
- deltoid tuberosity (DEL-toid tu-ber-OS-ity)
- greater tubercle (TU-ber-cul)
- head
- intertubercular groove (in-ter-tu-BER-cue-lar)
- lateral epicondyle (epi-CON-dile)
- lesser tubercle
- medial epicondyle
- trochlea (TRO-clee-a)

1 _____
2 _____
3 _____
4 _____
5 _____
6 _____
7 _____
8 _____

9 _____
10 _____
11 _____

(b) Posterior view
- lateral epicondyle
- medial epicondyle
- olecranon fossa (o-LEH-cra-non)

12 _____
13 _____
14 _____

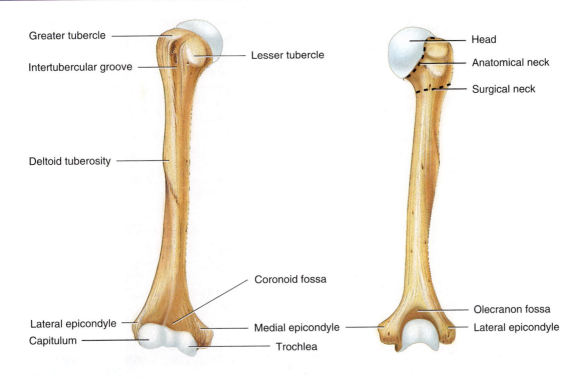

Greater tubercle

Lesser tubercle

Intertubercular groove

Head

Anatomical neck

Surgical neck

Deltoid tuberosity

Coronoid fossa

Olecranon fossa

Lateral epicondyle

Medial epicondyle

Lateral epicondyle

Capitulum

Trochlea

SUPERIOR

SUPERIOR

1

2

3

6

7

Surgical neck

8

Radial fossa

4

5

9

10

11

LATERAL

MEDIAL

12

13

14

LATERAL

(a) Anterior view

(b) Posterior view

FIGURE 10.2 Right humerus, *continued.*

Olecranon (process)

Trochlear notch

Head of radius

Coronoid process

Radial notch

Radial tuberosity

Olecranon (process)

Head of radius

Interosseous membrane

Styloid process of radius

Styloid process of ulna

Styloid process of radius

(a) Anterior view
• coronoid process
• head of radius
• olecranon (process)
• radial notch
• styloid process of radius (STY-loid)
• styloid process of ulna
• trochlear notch (semilunar)

1 _____

2 _____

3 _____

4 _____

5 _____

6 _____

7 _____

(b) Posterior view
• radius
• ulna

8 _____

9 _____

SUPERIOR

SUPERIOR

Radial tuberosity

LATERAL

MEDIAL

LATERAL

(a) Anterior view

(b) Posterior view

FIGURE 10.3 Right ulna and radius.

Carpus (Wrist)

The carpus is composed of eight short bones of the wrist, the **carpal** bones, which are lined up to form a proximal and a distal row of bones. Two of the carpal bones articulate with the radius, but there is no articulation of the carpal bones with the ulna.

Metacarpus (Palm of Hand)

The metacarpus is composed of five **metacarpal** bones that make up the palm of the hand. They are numbered as Roman numerals I–V from the metacarpal of the thumb (lateral side) to the little finger side. The metacarpals articulate with the carpals proximally and with the phalanges distally.

Phalanges (Fingers)

The phalanges (phalanx, sing.) make up the fingers or digits. The fingers are also numbered I–V from the thumb (*pollex*) to the little finger. The thumb has two phalanges, proximal and distal. Digits II through V each have **proximal, middle,** and **distal phalanges** (*phalanx* = closely knit row).

ACTIVITY 3 **Bones of the Hand**

1 Identify the parts of the hand in Figure 10.4. For each phalanx, include the Roman numeral.

2 Locate the bones of the hand on an articulated hand or skeleton.

3 Pronounce the terms as you point to them.

4 Locate and palpate these bones on yourself.

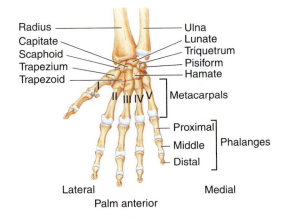

Radius — Ulna
Capitate — Lunate
Scaphoid — Triquetrum
Trapezium — Pisiform
Trapezoid — Hamate
II III IV V — Metacarpals
Proximal
Middle] Phalanges
Distal
Lateral Medial
Palm anterior

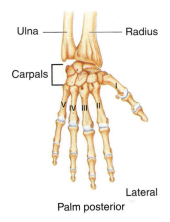

Ulna — Radius
Carpals [
V IV III II
Lateral
Palm posterior

I II III IV V
1
2
3
4
5
LATERAL MEDIAL
Anterior view

- carpals (CAR-puls)
- distal phalanx V (FAY-lanx)
- metacarpals (meta-CAR-pul)
- middle phalanx V
- proximal phalanx V

1 _____

2 _____

3 _____

4 _____

5 _____

FIGURE 10.4 Right hand and wrist.

C. Bones and Selected Bone Markings of the Pelvic Girdle

The **pelvic girdle** (*pelvis* = basin) is composed of two hip (coxal) bones called the **os coxae** (*os-* = bone; *cox-* = hip) that attach the lower limb to the axial skeleton. Each **os coxa** is formed by the fusion of three separate bones: the **ilium** (*ilia* = flank), **ischium** (*ischion* = hip joint), and **pubis** (*pub-* = grown up or adult) bones. These three bones are identifiable as separate bones in children.

Os Coxa

ilium—largest and most superior of the three components of os coxa

- **iliac crest**—superior border of ilium
- **anterior superior iliac spine**—anterior end of iliac crest
- **anterior inferior iliac spine**—below the anterior superior iliac spine
- **posterior superior iliac spine**—posterior end of iliac crest
- **posterior inferior iliac spine**—below the posterior superior iliac spine
- **greater sciatic notch**—large notch on posterior side
- **iliac fossa**—depression on anterior surface

ischium—inferior, posterior portion of os coxa

- **ischial tuberosity**—large, roughened projection on posterior and inferior edge
- **ischial spine**—posterior projection between greater and lesser sciatic notches
- **lesser sciatic notch**—smaller indentation between ischial spine and ischial tuberosity

pubis—anterior, inferior portion of os coxa

- **pubic symphysis** (*symphysis* = growing together)—joint where the two pubic bones join anteriorly

bone markings formed by ilium, ischium, and pubis

- **acetabulum** (*acetabulum* = little saucer)—deep indentation, or cup, for head of the femur
- **obturator foramen**—largest foramen in the skeleton

Pelvis

- **pelvic brim**—divides the false pelvis from the true pelvis; begins at the sacral promontory and extends laterally and inferiorly to end at the pubic symphysis
- **false pelvis**—portion of pelvis superior to pelvic brim; wide area extending to top of iliac crest
- **true pelvis**—portion of pelvis inferior to pelvic brim; surrounds the pelvic cavity
- **pelvic inlet**—superior opening of true pelvis; bordered by pelvic brim
- **pelvic outlet**—inferior opening of true pelvis; bordered by the coccyx, ischial spine, and ischial tuberosities

The **hip** or **coxal joint** is formed by the acetabulum articulating with the head of the femur to form a ball-and-socket joint. There is a strong ligament that connects these two structures deep inside the joint itself. When articulated os coxae are united posteriorly with the sacrum at the **sacroiliac joints** (*sacro-* = sacrum; *iliac* = ilium), they form the bony **pelvis** and the connection of the lower extremity with the axial skeleton.

There are definite anatomical differences between the male and female pelves (pl.). The bones of the male are typically heavier and rougher, with larger bone markings than the female. The **male pelvis** is more vertical and narrower, has a pelvic inlet that is heart-shaped, and has a 90° or less pubic arch angle. The **female pelvis** generally has more space in the true pelvis for childbirth, and is tilted backward and flared. The pelvic inlet is round or oval, and the angle of the pubic arch is generally greater than 90°.

ACTIVITY 4 Os Coxa and Pelvis

1 Identify the bone markings of an os coxa in Figure 10.5.
2 Locate the markings on a disarticulated os coxa and on an articulated pelvis.
3 Pronounce the terms as you point to them.
4 Palpate the iliac crest, the anterior superior iliac spine, and the pubic symphysis on your own body.
5 Identify the bone markings of the pelvis in Figure 10.6.
6 Locate these markings on an articulated pelvis or skeleton.
7 Identify the differences between male and female pelves on models or articulated skeletons.

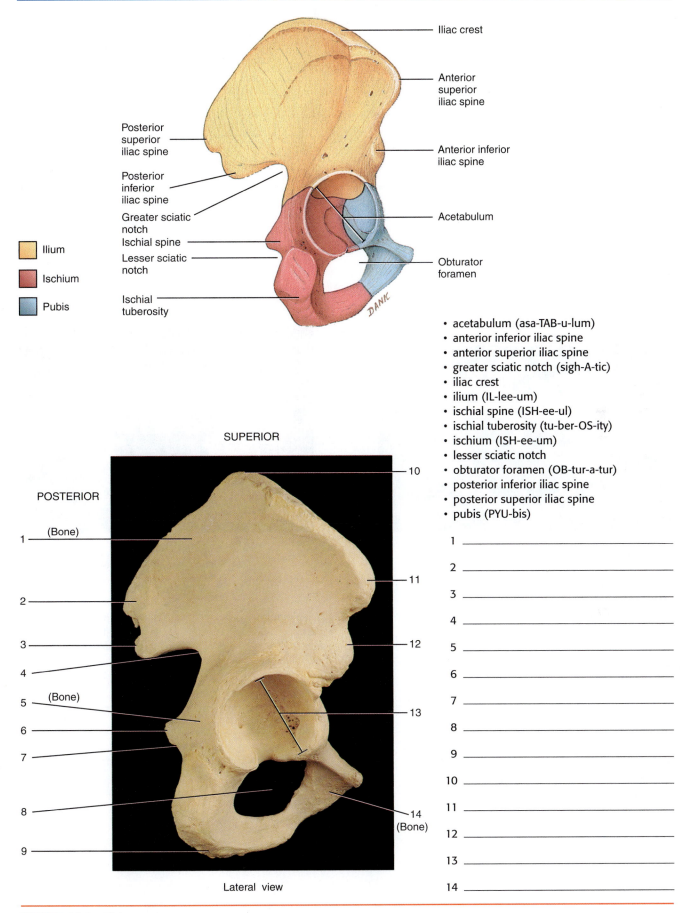

Ilium

Ischium

Pubis

Iliac crest

Anterior superior iliac spine

Posterior superior iliac spine

Anterior inferior iliac spine

Posterior inferior iliac spine

Greater sciatic notch

Acetabulum

Ischial spine

Lesser sciatic notch

Obturator foramen

Ischial tuberosity

SUPERIOR

POSTERIOR

1 (Bone)

2

3

4

5 (Bone)

6

7

8

9

10

11

12

13

14 (Bone)

Lateral view

- acetabulum (asa-TAB-u-lum)
- anterior inferior iliac spine
- anterior superior iliac spine
- greater sciatic notch (sigh-A-tic)
- iliac crest
- ilium (IL-lee-um)
- ischial spine (ISH-ee-ul)
- ischial tuberosity (tu-ber-OS-ity)
- ischium (ISH-ee-um)
- lesser sciatic notch
- obturator foramen (OB-tur-a-tur)
- posterior inferior iliac spine
- posterior superior iliac spine
- pubis (PYU-bis)

1 _____

2 _____

3 _____

4 _____

5 _____

6 _____

7 _____

8 _____

9 _____

10 _____

11 _____

12 _____

13 _____

14 _____

FIGURE 10.5 Right os coxa.

POSTERIOR

(a) Superior view of female pelvis

(b) Superior view of male pelvis

FIGURE 10.6 Pelvis.

(a) Female pelvis
• false pelvis
• iliac crest
• ilium
• ischial spine

• pelvic brim
• pubic symphysis
 (PYU-bic SYM-fah-sis)
• pubis (PYU-bis)
• true pelvis

(b) Male pelvis
• coccyx (COCK-six)
• false pelvis
• ischial spine
• pubis

• sacroiliac joint
 (say-crow-ILL-ee-ac)
• sacrum (SAY-crum)
• true pelvis

1 _____

2 _____

3 _____

4 _____

5 _____

6 _____

7 _____

8 _____

9 _____

10 _____

11 _____

12 _____

13 _____

14 _____

15 _____

D. Bones and Selected Bone Markings of the Thigh, Leg, and Foot

The lower limb consists of bones of the **femur, patella, tibia, fibula, tarsals, metatarsals,** and the **phalanges.** Of the 30 bones in each lower limb, 4 are in the thigh and leg, and the other 26 are in the foot (including ankle).

Femur (Thigh Bone)

The femur is the largest and strongest bone in the human skeleton. This bone is bowed anteriorly in a slight curve. Its head articulates with the acetabulum, and the distal end articulates with the tibia and patella.

- **head**—large, rounded, knoblike proximal end
- **neck**—narrower, constriction distal to head
- **greater trochanter** (*trochanter* = runner)—large and roughened superior projection; lateral to neck
- **lesser trochanter**—smaller, posterior-medial prominence distal to greater trochanter
- **medial condyle**—rounded, medial process on posterior side of distal end
- **lateral condyle**—similar to medial condyle on lateral side
- **intercondylar fossa**—deep fossa between medial and lateral condyles
- **medial epicondyle**—bumplike projection superior to medial condyle
- **lateral epicondyle**—bumplike projection superior to lateral condyle; a little smaller
- **gluteal tuberosity**—posterior surface of body of femur; roughened projection inferior to lesser trochanter
- **linea aspera**—vertical ridge on posterior surface

Patella (Kneecap)

This small, triangular bone has an anterior surface that is smoother than the posterior surface. Shallow, irregular-shaped articular facets are on the posterior surface that articulate with the condyles of the femur.

Tibia (Leg Bone)

The tibia is the weight-bearing bone of the two leg bones and is medially located. Laterally, the tibia forms a joint with the fibula.

- **medial condyle**—flattened, expanded medial projection on proximal end
- **lateral condyle**—similar to medial condyle on lateral side
- **tibial tuberosity**—large, roughened projection on anterior surface, inferior to condyles
- **anterior border (crest)**—slender ridge on anterior surface
- **medial malleolus**—medial process on distal end, forms medial bump of ankle

Fibula (Leg Bone)

The slender fibula is the lateral leg bone that is important for muscle attachment but not for bearing weight. The **head** of the fibula is proximal and articulates with the tibia but not the femur. The **lateral malleolus** is distal and articulates with the talus laterally.

- **head**—proximal end
- **lateral malleolus**—distal end, forms lateral bump of ankle

ACTIVITY 5 **Bones and Bone Markings of the Thigh and Leg**

1 Identify the bones and bone markings of the femur in Figure 10.7(a) and (b), and the patella, tibia, and fibula on Figure 10.8(a) and (b).

2 Locate the bones and bone markings on disarticulated bones as well as on an articulated skeleton.

3 Pronounce the terms as you point to them.

4 Palpate these bone markings on your own body: greater trochanter, medial and lateral epicondyles, patella, head of the fibula, tibial tuberosity, anterior crest (shin) of the tibia, medial malleolus of the tibia, and lateral malleolus of the fibula.

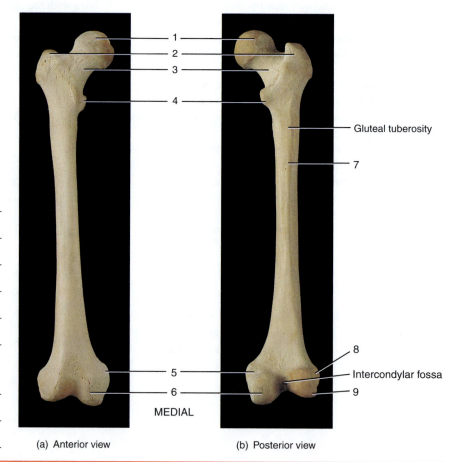

- greater trochanter (tro-CAN-ter)
- head of femur
- lateral condyle (CON-dile)
- lateral epicondyle (epi-CON-dile)
- lesser trochanter
- linea aspera (LIN-ee-uh ASP-er-uh)
- medial condyle
- medial epicondyle
- neck

(a) Anterior view

1 _____

2 _____

3 _____

4 _____

5 _____

6 _____

(b) Posterior view

7 _____

8 _____

9 _____

FIGURE 10.7 Right femur.

FIGURE 10.8 Right tibia, fibula, and patella.

- fibula (FIB-u-la)
- head of fibula
- lateral condyle
- lateral malleolus (mal-LAY-e-lus)
- medial condyle
- medial malleolus
- tibia
- tibial tuberosity

1 _____ 5 _____

2 _____ 6 _____

3 _____ 7 _____

4 _____ 8 _____

Tarsus (Ankle)

The tarsus is composed of seven tarsal bones of the foot, with two of them being larger than the rest. The largest tarsal bone is the **calcaneus** (*calcaneum* = heel), also known as the heel bone. The other large tarsal bone is the **talus** (*talus* = ankle), which articulates with the tibia and fibula.

Metatarsus

The metatarsus is composed of five metatarsal bones (*meta-* = after or next) that are analogous to the metacarpals in the hand. They are numbered the same way, I to V, from the great toe to the little toe.

Phalanges (Toes)

The phalanges (toes or digits) are similar to the phalanges in the hand. The toes are numbered I to V from the great toe

(*hallux*) to the little toe. The great toe is made of two phalanges (**proximal** and **distal**), and digits II to V have three bones each—**proximal, middle,** and **distal phalanges.**

ACTIVITY 6 **Bones of the Foot: Tarsals, Metatarsals, and Phalanges**

1 Label Figure 10.9(a) and (b). For each phalanx, include the Roman numeral.

2 Locate the bones of the foot on an articulated foot.

3 Pronounce the names as you point to them.

4 Palpate these parts on your own body: lateral malleolus, calcaneus, and talus.

(a) Superior view
- calcaneus (cal-CANE-ee-us)
- distal phalanx II (FAY-lanx)
- middle phalanx II
- metatarsals (meta-TAR-suls)
- phalanges
- proximal phalanx II
- talus (TA-lus)
- tarsals (TAR-suls)

1 _____ 5 _____

2 _____ 6 _____

3 _____ 7 _____

4 _____ 8 _____

(b) Lateral view
- calcaneus
- lateral malleolus of fibula
- tibia
- metatarsals
- phalanges
- talus
- tarsals

9 _____ 13 _____

10 _____ 14 _____

11 _____ 15 _____

12 _____

FIGURE 10.9 Bones of the right foot.

ACTIVITY 7 **Assembly of a Complete Disarticulated Skeleton**

1 Obtain a disarticulated skeleton from your instructor.

2 With your lab partners, assemble the skeleton.

3 Have your instructor check it to see that you have assembled it properly.

Reviewing Your Knowledge

A. Identification of Bones and Bone Markings of the Appendicular Skeleton

Go to www.wiley.com/college/allen and click on the picture of your lab manual. Then click on Student Companion Site, Anatomy Drill and Practice, and Exercise 10.

B. Bone Surface Markings

Name the process that is described.

_____ **1.** Projection above a condyle

_____ **2.** *Very* large projection

_____ **3.** Large, roughened projection

_____ **4.** Rounded articular projection supported on a bone neck

_____ **5.** Long, narrow ridge (less prominent than a crest)

_____ **6.** Small, rounded projection

_____ **7.** Prominent, thicker ridge

_____ **8.** Opening or hole

_____ **9.** Smooth, rounded articular process

_____ **10.** Shallow depression

C. Pectoral Girdle (Shoulder)

Fill in the blank with the correct term.

_____ **1.** The acromion (process) articulates with what bone?

_____ **2.** Is the clavicle anterior or posterior compared to the scapula?

_____ **3.** The clavicle articulates medially with which bone?

_____ **4.** The humerus articulates with what bone marking of the scapula?

_____ **5.** Is the subscapular fossa located anterior or posterior to the supraspinatus and infraspinatus fossae?

_____ **6.** Name the two bones that make up the pectoral girdle.

D. Upper Limb

Fill in the blank with the correct term.

_____ **1.** What part of the radius articulates with the humerus?

_____ **2.** What part of the ulna fits into the olecranon fossa of the humerus?

_____ **3.** The coronoid process articulates with what depression on the distal end of the humerus?

_____ **4.** Is the ulna medial or lateral compared with the radius?

_____ **5.** What are the bones called that make up the fingers?

_____ **6.** What are the bones called that make up the palm of the hand?

E. Pelvic Girdle and Pelvis

Fill in the blank with the correct term.

_____ **1.** When you put your hands on your hips, which bone marking of each os coxa are you touching?

_____ **2.** With your hands on your hips, you can feel a point of the pelvis protruding out anteriorly just above your thigh. Name this bone marking.

_____ **3.** Name the two bones that form the pelvic girdle.

_____ **4.** What prominent bone marking on each os coxa do you sit on?

_____ **5.** Name the bones of the os coxae that articulate anteriorly.

_____ **6.** The female pelvis has smoother bone markings than the male. True or False?

F. Lower Limb

Fill in the blank with the correct term.

_____ **1.** Is the fibula medial or lateral to the tibia?

_____ **2.** What is the correct term for the medial process at the distal end of the tibia that forms the anklebone?

_____ **3.** What is the heaviest and strongest bone of the leg (not thigh)?

_____ **4.** Name the thigh bone.

_____ **5.** The fibula is a weight-bearing bone. True or False?

_____ **6.** Name the tarsal bone that articulates with the tibia.

_____ **7.** Name the heelbone.

_____ **8.** Name the bones of the distal part of the instep.

Using Your Knowledge

A. Upper Limb

Identify the bones and bone markings of the upper limb as shown in Figure 10.10(a), (b), and (c).

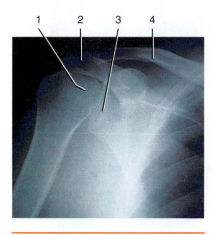

FIGURE 10.10a Radiograph of the shoulder joint, posterior view.

1 _____ (bone marking)

2 _____ (bone marking)

3 _____ (bone marking)

4 _____ (bone)

5 6 7 8 9 Epiphyseal plate

FIGURE 10.10b Radiograph of the elbow joint.

5 _____ (bone)

6 _____ (bone)

7 _____ (bone)

8 _____ (bone marking)

9 _____ (bone marking)

10 11

FIGURE 10.10c Radiograph of the left hand.

10 _____ (bone marking)

11 _____ (bone)

12 _____
Draw a band on the specific bone where a wedding ring is worn.

13 _____
Name the bone the ring is on.

B. Surface Anatomy

Identify the bone and bone markings of the surface anatomy of the upper limb (Figure 10.11a).

Identify the bone and bone markings of the surface anatomy of the hip and lower limb (Figure 10.11b).

FIGURE 10.11a Surface anatomy of the upper limb and pectoral girdle.

14 _____ (bone marking)

15 _____ (bone marking)

16 _____ (bone marking)

FIGURE 10.11b Surface anatomy of the hip and lower limb.

17 _____ (bone marking)

18 _____ (bone)

19 _____ (bone marking)

20 _____ (bone marking)

Joints and Synovial Joint Movements

Objectives

After completing this exercise, you should be able to:

- Name the three major structural categories of joints and give an example of each
- Describe the cartilage type and location of joint cartilages
- Distinguish between the three major functional categories of joints and give an example of each
- Describe the basic structure of a typical synovial joint
- Describe the structure of the knee joint
- List six different types of synovial joints and give an example of each
- Describe the types of movements of synovial joints and demonstrate them

Materials

- articulated skeleton
- whole and longitudinally cut fresh, frozen, or glycerinated mammalian synovial joint
- dissection tray, instruments, disposable gloves
- model or chart of a knee joint
- compound microscope, lens paper, prepared slides of hyaline cartilage and fibrocartilage

A **joint or articulation** (*articulare* = to divide into joints) connects a bone with another bone, cartilage, or tooth. Joints are commonly classified according to their structure and function.

A. Structural Classification of Joints

The structural classification of joints depends on the type of connective tissue forming the joint and on whether or not there is a space or synovial cavity between the bones.

Fibrous joints have dense fibrous connective tissue with strong collagen fibers that hold the joints firmly together with no synovial cavity. This type of joint permits little to no movement. Examples are the skull joints, teeth in sockets, and the distal joint between the tibia and fibula. **Cartilaginous joints** have either hyaline cartilage or fibrocartilage connecting the bones with no synovial cavity. Usually, there is a small degree of movement with this type of joint. Examples are the intervertebral joints, the pubic symphysis, and the joint between the manubrium and body of the sternum. **Synovial joints** (*syn-* = together; *ovum* = egg) have a small synovial cavity (space) between the two bones that permits a greater amount of movement than fibrous or cartilaginous joints. The term *synovial* comes from the synovial fluid present in the synovial cavity that resembles the albumin of an uncooked egg, only more

viscous. Dense fibrous connective tissue on the exterior of the joint holds the bones together. The majority of the joints in the human body are synovial joints—for example, the shoulder, elbow, hip, and knee joints.

ACTIVITY 1 **Structural Classification of Joints**

1 Label the structural category of each joint in Figure 11.1.

2 Identify these joints on an articulated skeleton.

3 Pronounce the terms as you point to them on the articulated skeleton.

4 Point out these joints and palpate the ones you can on your own body.

B. Histology and Location of Joint Cartilages

Cartilage is distinctly different from bone tissue in that cartilage has no blood vessels or nerves and therefore takes longer to heal than bone. Similar to bone, cartilage has oval-shaped **lacunae** with cartilage cells or **chondrocytes** (*chondro-* = cartilage) inside. The **perichondrium,** a connective tissue covering, surrounds cartilage similar to the periosteum of bone.

Cartilage is composed of both living cells and nonliving **matrix.** The matrix is made up of a gel-like ground substance, fibers, and water secreted by chondrocytes. Two fiber types, **collagen** and **elastin,** give the three types of cartilage their different microscopic appearances. **Hyaline cartilage** is the predominant type of cartilage in the body and is found at the end of bones **(articular cartilage),** attaching ribs to the sternum **(costal cartilage),** in the nose,

(a) Anterior view (b) Tooth in alveolus

• cartilaginous joint (car-tih-LA-jih-nous)
• fibrous joint (FY-brous)
• synovial joint (sih-NO-vi-ul)

1 _____

2 _____

3 _____

4 _____

5 _____

6 _____

7 _____

8 _____

9 _____

10 _____

11 _____

FIGURE 11.1 Structural classification of joints.

larynx, trachea, and bronchi. This cartilage type appears glassy to the naked eye, and under the compound microscope its matrix looks smooth and homogeneous. It contains collagen fibers that are thin and not visible with a compound microscope. The matrix of hyaline cartilage is resilient and firm. **Fibrocartilage** has fewer lacunae and chondrocytes, and its extracellular matrix is packed with thick collagen fibers that give it tensile strength similar to dense connective tissue. Under the microscope, collagen fibers are the most prevalent structures seen. This cartilage type is found in intervertebral discs (pads between the vertebrae), the knee menisci, and the pubic symphysis.

(a) Hyaline cartilage 512×

(b) Fibrocartilage 742×

FIGURE 11.2 Microscopic structure of joint cartilages.

(a) Hyaline cartilage
• chondrocyte
 (CHON-dro-cyte)
• extracellular matrix
• lacuna

(b) Fibrocartilage
• chondrocyte
• collagen fiber
• lacuna

1 _____

2 _____

3 _____

4 _____

5 _____

6 _____

ACTIVITY 2 The Histology and Location of Joint Cartilages

1 Label the two types of cartilage in Figure 11.2.

2 Pronounce each term as you write it in the blank.

3 Examine prepared microscope slides or photomicrographs of hyaline cartilage, elastic cartilage, and fibrocartilage, and identify the structures that were labeled in the photomicrographs.

4 Identify the cartilage type for each location in Figure 11.3 as fibrocartilage or hyaline cartilage.

5 Identify the location of each cartilage from Figure 11.3 on the articulated skeleton.

FIGURE 11.3 Location of joint cartilages in the skeleton.

1. invertebral disc _____

2. meniscus of knee _____

3. nasal cartilage _____

4. articular cartilage _____

5. coastal cartilage _____

6. pubic symphysis _____

C. Functional Classification of Joints

The functional classification of joints is made on the basis of the amount of movement the joint allows. Immovable joints or **synarthroses** (*syn-* = union; *arthro-* = joint) include the sutures between the skull bones and the teeth sockets. Intervertebral joints, the tibiofibular joint, and the pubic symphysis are examples of slightly movable joints or **amphiarthroses** (*amphi-* = on both sides). Most of the joints in the body, about 90%, are freely movable joints, or **diarthroses** (*di-* = apart; away from).

As you can see from the description of structural and functional joints, there are similarities in the amount of movement and certain types of structural joints. Most of the fibrous joints such as sutures and teeth sockets are nonmovable joints. However, the fibrous tibiofibular joint is a slightly movable joint. Most of the cartilaginous joints are slightly movable joints, such as the intervertebral discs and the pubic symphysis. The cartilaginous epiphyseal plates of long bones, however, are nonmovable joints.

ACTIVITY 3 Functional Classification of Joints

1 Identify the functional category of each joint in Figure 11.4.

2 Identify these joints on an articulated skeleton.

3 Pronounce the terms as you point to them on the articulated skeleton.

D. Basic Structure of Synovial Joints

Although synovial joints vary in structure, they have several common features as follows:

- **synovial cavity**—small space between the two articulating bones
- **articular cartilage**—hyaline cartilage covering the ends of the bones in the synovial cavity
- **articular capsule**—structure that encloses the synovial joint and synovial cavity; has two layers—the fibrous capsule and synovial membrane
- **fibrous capsule**—outer dense fibrous connective tissue layer of the articular capsule that is continuous with the periosteum of the bone; also forms ligaments when fibrous bundles are parallel
- **synovial membrane**—inner layer of the articular capsule; composed of areolar connective tissue containing elastic fibers and adipocytes
- **synovial fluid**—secreted by the synovial membrane; lubricates the articular cartilages to reduce friction

 - amphiarthrosis (amphi-ar-THROW-sis)
 - diarthrosis (die-ar-THROW-sis)
 - synarthrosis (syn-ar-THROW-sis)

1 _____

2 _____

3 _____

4 _____

5 _____

6 _____

7 _____

8 _____

9 _____

10 _____

(a) Anterior view (b) Tooth in alveolus

FIGURE 11.4 Functional classification of joints.

ACTIVITY 4 **Typical Synovial Joint**

1 Label the simple synovial joint in Figure 11.5.

2 Pronounce each term as you complete the labeling exercise.

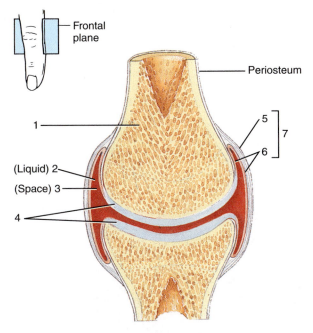

Frontal plane

Periosteum

5
7
6

(Liquid) 2
(Space) 3

1

4

Frontal section

FIGURE 11.5 Typical synovial joint.

- articular bone
 (ar-TIH-ku-lar)
- articular capsule
- articular cartilage
- fibrous capsule
- synovial cavity
- synovial fluid
- synovial membrane

1 _____

2 _____

3 _____

4 _____

5 _____

6 _____

7 _____

E. Accessory Structures of Major Synovial Joints

In addition to the common joint features, many synovial joints also have one or more of the following accessory structures.

- **ligaments** (*liga-* = bound or tied)—tough, dense tissue that holds bone to bone; **extracapsular ligaments**—external to articular capsule; **intracapsular ligaments**—internal to articular capsule
- **bursa** (*byrsa* = wineskin or purse)—saclike structure lined with a synovial membrane and containing synovial fluid; located in friction areas to cushion a joint
- **tendon sheath**—tubelike bursa that wraps around tendons; forms a cushion that reduces friction
- **meniscus** (*meniskos* = crescent)—fibrocartilage cushioning pad between articulating surfaces of bones; found only in the knee

The knee joint is specifically studied here because it has the distinction of being the most complex and highly stressed joint, as well as being the location of many joint injuries. The knee joint is classified as a hinge joint, but when flexed it also demonstrates gliding and rotation movements.

- **medial meniscus**—inside of joint cavity; cushions knee joint
- **lateral meniscus**—inside of joint cavity; cushions knee joint
- **medial** (tibial) **collateral ligament**—extracapsular ligament; adds strength to joint medially
- **lateral** (fibular) **collateral ligament**—extracapsular ligament; adds strength to joint laterally
- **anterior cruciate ligament** (*cruci-* = cross) or **ACL**—intracapsular ligament; attaches the femur and tibia anteriorly
- **posterior cruciate ligament** or **PCL**—intracapsular ligament; stabilizes joint posteriorly
- **patellar ligament**—extension of tendon from quadriceps muscle; connects patella to tibial tuberosity and stabilizes the joint anteriorly
- **infrapatellar fat pad** (*infra-* = beneath)—cushion between patellar ligament and synovial capsule
- **bursa**—reduces friction; 13 bursae in knee

ACTIVITY 5 Major Synovial Joints

1 Label the parts of the knee joint in Figure 11.6.

2 Identify the parts on a model or chart of the knee.

3 Pronounce the terms as you point to them.

4 Use your knowledge of bones and synovial joint structures to identify the parts of the shoulder, elbow, hip, and knee joints on the Wiley Website. Go to www.wiley.com/college/allen. Click on the student site, choose Anatomy Drill and Practice, and then choose the exercise number.

(a) Anterior deep view

(b) Sagittal section

FIGURE 11.6 Knee joint.

(a) Anterior view
- articular cartilage of femur
- anterior cruciate ligament (KRU-she-ate)
- lateral (fibular) collateral ligament
- lateral meniscus (meh-NIS-cus)
- medial (tibial) collateral ligament
- medial meniscus
- patellar ligament
- posterior cruciate ligament

(b) Sagittal section
- articular capsule
- articular cartilage
- bursae (BUR-see)
- infrapatellar fat pad
- patellar ligament
- synovial cavity with synovial fluid

1 _____	5 _____	9 _____	13 _____
2 _____	6 _____	10 _____	14 _____
3 _____	7 _____	11 _____	
4 _____	8 _____	12 _____	

F. Types of Movement at Synovial Joints

Skeletal muscle contraction causes bone movement at synovial joints. Professionals in kinesiology and physical therapy use particular terms to describe the movements of these joints. Many movements are grouped to demonstrate opposite movements. The four main categories of synovial joint movements are gliding, angular, rotation, and special movements (Table 11.1).

ACTIVITY 6 **Types of Movements at Synovial Joints**

1 Label the movements illustrated in Figures 11.7–11.11.
2 Demonstrate each movement with a partner(s).
3 Use an articulated skeleton to demonstrate each movement. During pronation, observe the radius crossing over the ulna.

TABLE 11.1 **Types of Movement at Synovial Joints**

MOVEMENT	DESCRIPTION
A. GLIDING	Nearly flat bone surfaces slide or glide over each other.
B. ANGULAR	
Flexion (*flex-* = to bend)	Decrease in the angle of a joint.
Extension (*exten-* = to stretch out)	Increase in the angle of a joint.
Hyperextension (*hyper-* = excessive)	Excessive extension movement beyond normal anatomical position.
Abduction (*ab-* = away; *-duct* = to lead)	Move appendage away from the midline.
Adduction (*ad-* = toward)	Move appendage toward midline.
Circumduction (*circ-* = circle)	Move a distal part of an appendage in a circular motion.
C. ROTATION (*rota-* = revolve)	Turn on a pivot with a circular motion.
D. SPECIAL JOINT MOVEMENTS	
Elevation	Upward movement raising body part vertically.
Depression	Downward movement lowering body part vertically.
Protraction (*pro-* = in front of; *trahere* = to draw)	Move a body part forward or anterior on a horizontal plane.
Retraction (*retractare* = to draw back)	Move a body part backward or posterior.
Supination (*supine* = lying on the back)	Turn palm of the hand to face forward, or, if arm is outstretched, turn palm upward.
Pronation (*pronate* = lying face downward)	Turn palm of the hand to face backward, or, if arm is outstretched, turn palm downward.
Inversion	Turn the sole of foot inward.
Eversion	Turn the sole of foot outward.
Dorsiflexion	Point your toes upward; stand on your heels.
Plantar flexion	Point your toes downward and raise your heel.

(a) Atlanto-occipital and cervical intervertebral joints

(b) Shoulder joint

(c) Elbow joint

(d) Wrist joint

(e) Hip joint

(f) Knee joint

FIGURE 11.7 Angular joint movements of flexion, extension, and hyperextension.

- extension
- flexion
- hyperextension

(a)

1 _____

2 _____

3 _____

(b)

4 _____

5 _____

6 _____

(c)

7 _____

8 _____

(d)

9 _____

10 _____

11 _____

(e)

12 _____

13 _____

14 _____

(f)

15 _____

16 _____

(a) Shoulder joint

(b) Hip joint

Lateral Medial

(c) Right wrist joint

(d) Shoulder joint

(e) Hip joint

(f) Metacarpophalangeal joints of the fingers (not the thumb)

FIGURE 11.8 Angular joint movements of abduction, adduction, and circumduction.

- abduction (ab-DUK-shun)
- adduction (ad-DUK-shun)
- circumduction (sir-cum-DUC-shun)

(a)

1 _____

2 _____

(b)

3 _____

(c)

4 _____

5 _____

(d)

6 _____

(e)

7 _____

8 _____

(f)

9 _____

10 _____

(a) Atlanto-axial joint (b) Shoulder joint

FIGURE 11.9 Joint movements of rotation.

- lateral rotation
- medial rotation
- rotation

1 _____

2 _____

3 _____

(a) Temporomandibular joint (b) (c) Temporomandibular joint (d)

FIGURE 11.10 Special joint movements.

Palm posterior Palm anterior

5 6

(e) Radioulnar joint

- depression
- elevation
- pronation (pro-NAY-shun)
- protraction (pro-TRAC-shun)
- retraction (re-TRAC-shun)
- supination (soup-in-NAY-shun)

1 _____

2 _____

3 _____

4 _____

5 _____

6 _____

(a) Intertarsal joint (b) (c) Ankle joint

FIGURE 11.11 Special joint movements of the foot.

- dorsiflexion
- eversion
- inversion
- plantar flexion

1 _____ 3 _____

2 _____ 4 _____

G. Six Types of Synovial Joints

There are six types of synovial joints based on the structure of the articulating bone surfaces at the joints (Table 11.2). The joint structure determines the movement of the joint.

ACTIVITY 7 Palpation of Synovial Joint Movements

1 Palpate the following synovial joints on your body and use an articulated skeleton to observe the structure of the joint.
- **Planar joint (gliding)**—flex your arm up and feel the gliding movement at the acromioclavicular joint.
- **Hinge joint**—flex and extend your elbow and your knee as you feel this joint type. Note how this joint type only moves by flexion and extension.

- **Pivot joint**—feel the proximal part of your forearm until you locate the head of the radius. Rotate the radioulnar joint as you palpate this rotation movement.
- **Condyloid joint**—feel the joint between the 2nd metacarpal and the 2nd proximal phalanyx. Extend and flex, abduct and adduct, and circumduct this joint.
- **Saddle joint**—the thumb joint is the only saddle joint in the body. Feel the movement of trapezius bone (carpal bone) with the 1st metacarpal as you move this joint on two different axes.
- **Ball-and-socket joint**—feel your shoulder joint as you move your arm in different motions. Make a full circle with your shoulder joint.

2 Identify the joints listed above on an articulated skeleton and note the shape of the articulating surfaces. Refer to Table 11.2.

3 Identify the bones involved in these joints with your lab group.

TABLE 11.2 Types of Synovial Joints

NAME OF JOINT	ARTICULAR SURFACE DESCRIPTION	MOVEMENT
Planar (gliding)	Flat or slightly curved plane	Gliding motion back and forth and/or side to side
Hinge	Convex bone surface articulates with a concave bone surface	Flexion and extension
Pivot	Rounded or pointed projection articulates with ring formed by bone and ligament	Rotation
Condyloid	Oval convex projection articulates with oval concave depression	Flexion and extension, abduction and adduction, circumduction
Saddle	Saddle-shaped depression articulates with projection that fits into the saddle	Same as condyloid joint, except more exaggerated
Ball-and-socket	Ball-shaped head articulates with cup-shaped socket	Freely movable joint; flexion and extension; abduction and adduction; circumduction; rotation

H. Dissection of a Synovial Joint

NOTE: *If you are using fresh or frozen joints, wear protective gloves and wash your hands thoroughly with soap and water after the dissection.*

ACTIVITY 8 **Synovial Joint Dissection**

1 Dissect a whole synovial joint following the dissection instructions.
 - After placing your **synovial joint** in a dissecting tray, observe the external white **articular capsule** that holds the bones together. Note that the articular capsule is continuous with the periosteum of the bone.
 - Using a scalpel, cut open the joint capsule. Feel the slippery **synovial fluid** between your fingers.
 - Try to detect the difference between the outer **fibrous capsule** layer and the inner **synovial membrane** layer of the articular capsule.
 - Note the **synovial cavity** or space between the **articular bones.**
 - Observe the ends of the joint bones for the **articular cartilage.** Cut out a small piece of hyaline cartilage to observe its thickness.

2 Pronounce the terms as you point to them on the joint.

3 Observe a longitudinally cut synovial joint. Repeat the dissection steps above, but do not cut the joint with a scalpel.

CLEAN UP: *Thoroughly wash your dissection tray, instruments, and work surface with soap and water. Be careful not to wash bits of tissue down the drain. Dispose of the joint and cut tissue as directed by your instructor.*

Reviewing Your Knowledge

A. Synovial Joint Review

Go to www.wiley.com/college/allen and click on the picture of your lab manual. Then click on Student Companion Site, Anatomy Drill and Practice, and Exercise 11.

B. Structural and Functional Classification of Joints

Fill in the blank with the most appropriate term: **cartilaginous, fibrous, synovial, amphiarthrosis, diarthrosis,** or **synarthrosis.**

_____ **1.** Functional category that has the greatest amount of movement.

_____ **2.** Functional category that has the least amount of movement.

_____ **3.** Functional category that is slightly movable.

_____ **4.** Structural category that has an articular (joint) capsule.

_____ **5.** Structural category that has cartilage joining the ends of the articulating bones.

_____ **6.** Structural category with a joint cavity.

_____ **7.** Structural category that is tightly held together by fibrous connective tissue.

C. Basic Structure of Synovial Joints

Fill in the blank with the correct answer.

_____ **1.** Name the type of cartilage that covers the articular ends of bones.

_____ **2.** Name the fluid that lubricates, reduces friction, and gives nutrition to a joint.

_____ **3.** Name the tissue at the end of a bone that reduces friction in a joint.

_____ **4.** Identify the saclike structure in a synovial joint that is sometimes present to reduce friction.

_____ **5.** Identify the structure that secretes synovial fluid.

D. Types of Synovial Joints

There are six different types of synovial joints. Fill in the blank with the correct type.

_____ **1.** The type of synovial joint between the atlas and axis.

_____ **2.** The type of synovial joint at the shoulder.

_____ **3.** The type of synovial joint at the knee.

_____ **4.** The type of synovial joint between the scapula and clavicle.

_____ **5.** The type of synovial joint between the thumb and the 1st metacarpal.

_____ **6.** The type of synovial joint between the metacarpal and proximal phalanx.

E. Movement at Synovial Joints

Fill in the blank with the correct term for the type of movement of the synovial joint.

_____ **1.** Moves appendage toward the midline.

_____ **2.** Increases the angle of a joint.

_____ **3.** Downward movement, lowering the body part vertically.

_____ **4.** Palm of hand faces forward or upward.

_____ **5.** Pointing toes downward; raising the heel on your tiptoes.

_____ **6.** Decreases the angle of a joint.

_____ **7.** Moves an appendage away from the midline.

_____ **8.** Turns on a pivot with a circular motion.

_____ **9.** Movement that raises mandible.

_____ **10.** Move body part forward along horizontal plane.

_____ **11.** Turn sole of foot inward.

Using Your Knowledge

A. Synovial Joints

Identify the synovial joint structures and the bone markings of the synovial joint in Figure 11.12(a) and (b).

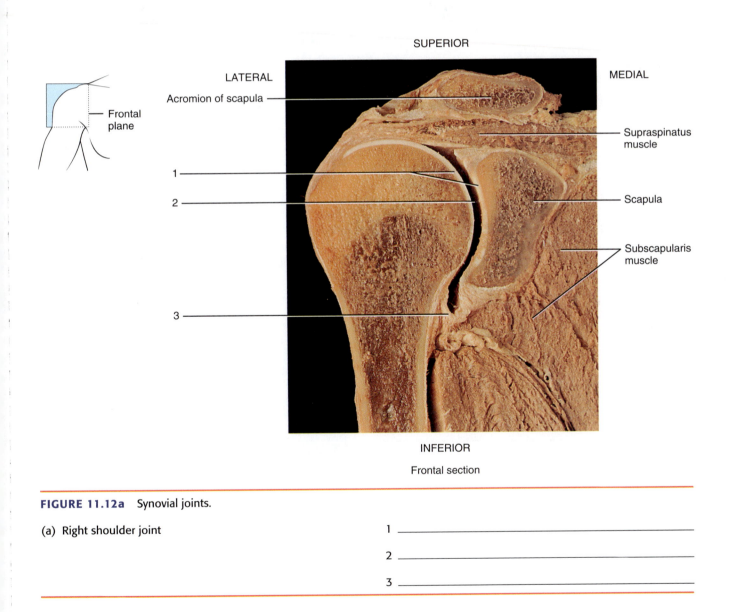

SUPERIOR

LATERAL

MEDIAL

Acromion of scapula

Supraspinatus muscle

1

2

Scapula

Subscapularis muscle

3

INFERIOR

Frontal section

Frontal plane

FIGURE 11.12a Synovial joints.

(a) Right shoulder joint

1 _____

2 _____

3 _____

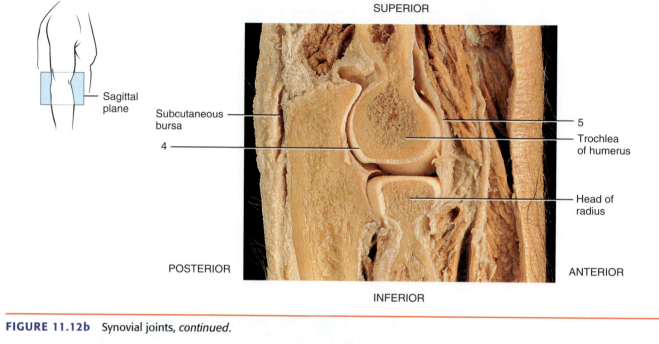

FIGURE 11.12b Synovial joints, *continued.*

(b) Right elbow joint

4 _____

5 _____

B. Synovial Joint Movements

Identify the joint movements involved in walking that are described in the sentences below. Use the following terms **dorsiflex, plantar flex, flex,** and **extend** to describe the movement of right leg.

First you flex the thigh, (6) the leg at the knee, and (7) the foot, then you (8) the leg at the knee, (9) the thigh, and (10) the foot.

6. _____

7. _____

8. _____

9. _____

10. _____

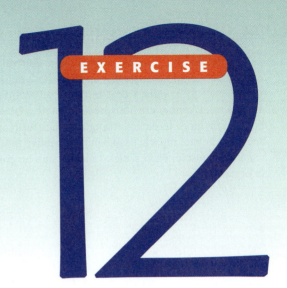

Skeletal Muscle Structure

**After completing this exercise,
you should be able to:**

- Describe the structure of skeletal tissue and skeletal muscle fibers
- Identify the connective tissue structures in skeletal muscle on a microscope slide or photomicrograph
- Describe the structure of the sarcomere
- Describe the structure of the neuromuscular junction

- prepared microscope slides of skeletal muscle tissue and neuromuscular junction
- compound microscopes and lens paper

Skeletal muscles are organs comprised of skeletal muscle tissue and connective tissue. These organs also contain nerves and blood vessels. The skeletal muscle fibers within skeletal muscles contract (shorten) and cause movement of our skeleton or skin. The signal for contraction is carried by neurons that innervate each skeletal muscle fiber. We consciously control contraction of skeletal muscles, so the contraction is called voluntary.

A. Skeletal Muscle Tissue and Connective Tissue Coverings

Skeletal muscle fibers (cells) are striated and multinucleated. The striations are light and dark stripes along the muscle cell. A skeletal muscle fiber is actually many embryonic cells that have fused together to form one large cell with multiple nuclei.

Individual skeletal muscle fibers are surrounded by a layer of areolar connective tissue called **endomysium** (*endo* = within; *mys* = muscle). Skeletal muscle fibers are grouped into bundles called **fascicles** that are surrounded by a layer of dense regular connective tissue called **perimysium** (*peri* = around). A muscle is formed from a number of fascicles that are surrounded by a dense regular connective tissue layer called **epimysium** (*epi* = on; upon). **Tendons,** connective tissues that attach the muscle to bone, are formed from endomysium, perimysium, and epimysium that extend beyond each skeletal muscle fiber.

The connective tissues surrounding skeletal muscle fibers separate and electrically insulate these cells. Therefore, electrical impulses that initiate contraction in one skeletal muscle fiber are not spread to another fiber. Recall that two or more tissues working together form an organ. Because skeletal muscle tissue and connective tissues form a skeletal muscle, each skeletal muscle can be considered an organ.

ACTIVITY 1 **Skeletal Muscle Tissue and Connective Tissue Coverings**

1 Label Figures 12.1 and 12.2.

2 Examine a prepared slide of skeletal muscle.
- Using the low-power objective, locate a cross-section through the muscle showing a **fascicle.** Identify the **perimysium** surrounding the fascicle and an **endomysium** surrounding a **skeletal muscle fiber.**
- Locate a section of skeletal muscle fiber in longitudinal section.
- Using the high-power objective, identify **nuclei** and **striations.**

Transverse plane

Periosteum covering the bone

Tendon

Deep fascia

Skeletal muscle

1

2

3

4

5

FIGURE 12.1 Skeletal muscle.

- endomysium (endo-MY-zee-um)
- epimysium (epi-MY-zee-um)
- fascicle (FAS-i-kul)
- muscle fiber
- perimysium (peri-MY-zee-um)

1 _____ 4 _____

2 _____ 5 _____

3 _____

(a) Cross-section 400×, H&E, human

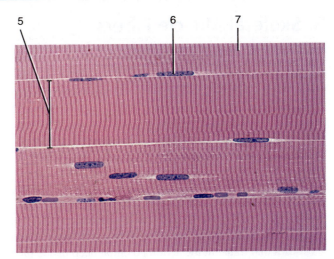

(b) Longitudinal section 400×, H&E, human

FIGURE 12.2 Sectional views of skeletal muscle fibers.

(a) Cross-section
• endomysium
• fascicle
• perimysium
• skeletal muscle fiber in cross-section

(b) Longitudinal section
• nucleus
• striation (stry-AY-tion)
• width of skeletal muscle fiber

1 _____

2 _____

3 _____

4 _____

5 _____

6 _____

7 _____

B. Skeletal Muscle Fibers

Skeletal muscle tissue contains elongated cells called **muscle fibers.** Muscle fibers have specialized features important to their function. The plasma membrane, called the **sarcolemma** (*sarco* = flesh; *lemma* = sheath) in muscle fibers, can conduct an electrical signal called an action potential. The **sarcoplasmic reticulum** is endoplasmic reticulum that is specialized to store calcium ions. Muscle cells contain **thick** and **thin filaments** or **myofilaments** (*myo* = muscle) that contain different contractile proteins. Movement of the myofilaments causes muscle shortening or contraction.

The sarcolemma has tubelike invaginations called **T tubules** that transmit the action potential deep inside the fiber to widened areas of the sarcoplasmic reticulum called **terminal cisternae.** The arrangement of two terminal cisternae with a T tubule between them is called a **triad.** An action potential traveling down the T tubules causes calcium to be released from the terminal cisternae into the cytoplasm of the muscle fiber.

Thick and **thin filaments,** which are bundled into **myofibrils,** occupy 80% of the volume of the fiber. Each myofibril is a chain of **sarcomeres,** the smallest structural and contractile unit of a muscle myofibril. Sarcomeres are composed of an orderly arrangement of thin and thick filaments and shorten when the muscle fiber is stimulated to contract.

Thin filaments contain actin, tropomyosin, and troponin molecules. **Actin** molecules comprise the majority of the thin filament and contain a **binding site** for the myosin molecules of the thick filament. A strand of **tropomyosin** molecules is nestled between the actin molecules. Several troponin molecules bind at precise intervals along the tropomyosin strand.

Thick filaments are composed of **myosin** molecules. A myosin molecule resembles two golf clubs twisted together. The tails of many myosin molecules form the thick filament, whereas the **heads** or **crossbridges** project toward the thin filaments. When the myosin-binding sites on the actin molecules are uncovered, myosin crossbridges attach and pull the thin filaments past the thick filaments.

ACTIVITY 2 **Skeletal Muscle Fibers**

1 Label the skeletal muscle fiber in Figure 12.3.

FIGURE 12.3 Skeletal muscle fiber.

(a) Muscle fiber
- myofibril (myo-FY-bril)
- sarcolemma (sar-co-LEM-ma)
- sarcoplasmic reticulum (sar-co-PLAZ-mic re-TIC-u-lum)
- terminal cisternae of sarcoplasmic reticulum (cis-TER-nee)
- triad
- T tubule

(b) Thick and thin filaments
- sarcomere
- thick filament
- thin filament

(c) Contractile proteins
- actin
- myosin crossbridges (MY-oh-sin)
- myosin tails
- tropomyosin (tro-poh-MY-o-sin)
- troponin

1 _____

2 _____

3 _____

4 _____

5 _____

6 _____

7 _____

8 _____

9 _____

10 _____

11 _____

12 _____

13 _____

14 _____

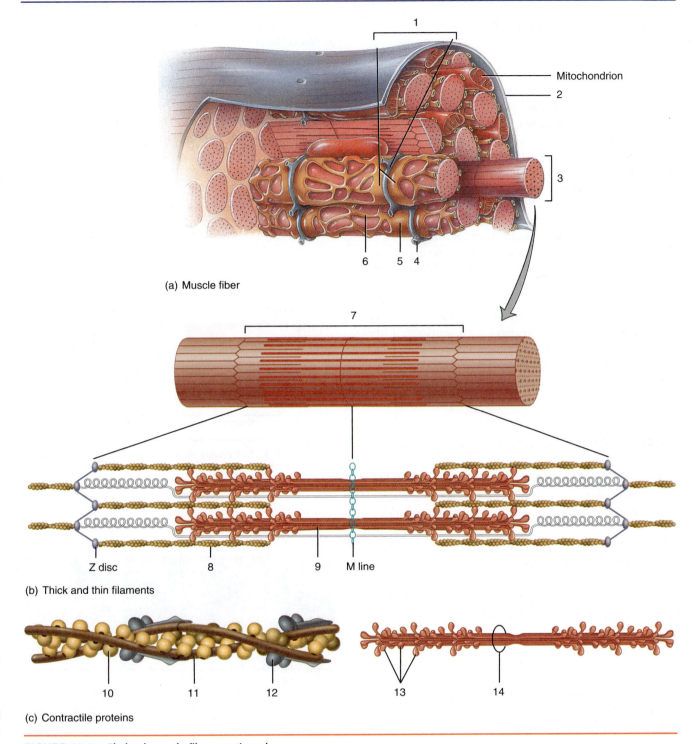

(a) Muscle fiber

(b) Thick and thin filaments

(c) Contractile proteins

FIGURE 12.3 Skeletal muscle fiber, *continued*.

C. Sarcomere Structure

The regular arrangement of thin and thick filaments in the sarcomeres gives the myofibrils and the skeletal muscle fibers light and dark bands (Figure 12.4). Lighter color bands with a dark stripe down the middle are the **I bands.** The I bands contain only thin filaments, and the stripes are **Z discs** which secure the thin filaments. **A bands** are dark-colored bands that extend the length of the thick filaments. In the A band of relaxed muscle, thin filaments overlap the thick filaments except in the middle of the A band, the **H zone,** which appears lighter. The **M line** is found in the middle of the H zone and secures the thick filaments.

Muscle fibers shorten because myosin heads (cross-bridges) on the thick filaments attach to the thin filaments and pull the thin filaments past the thick filaments toward the M line. This action increases the overlap of thin and thick filaments (Figure 12.4). As these filaments slide past each other in each sarcomere within each myofibril, the ends of the muscle fiber are brought closer together, the I bands decrease in length, and the H zone (lighter area) disappears. The increase in overlap of thin and thick filaments can be readily observed in sarcomeres of skeletal muscle.

ACTIVITY 3 Sarcomere Structure

1 Label the electron micrograph of the sarcomere in Figure 12.5.

2 Answer discussion questions with your lab partners.

? DISCUSSION QUESTIONS: SARCOMERE STRUCTURE

1 Do the lengths of the thin or thick filaments change when a muscle contracts? Explain.

2 Does the A band length change when a muscle contracts? Explain.

3 Does the I band length change when a muscle contracts? Explain.

4 Does the H zone length change when a muscle contracts? Explain.

5 Is the sarcomere in Figure 12.5 most likely from a fully contracted or fully relaxed skeletal muscle cell? Explain.

FIGURE 12.4 Sarcomere structure.

TEM 21,600×

FIGURE 12.5 Transmission electron micrograph illustrating sarcomere structure.

- A band
- H zone
- I band
- M line
- Z disc

1 _____

2 _____

3 _____

4 _____

5 _____

D. The Neuromuscular Junction

Each skeletal muscle fiber is stimulated to contract by a **motor neuron.** A motor neuron and all the skeletal muscle fibers it innervates are a **motor unit.** Within the muscle, the axon of a motor neuron divides into many branches or **axon terminals,** each of which forms a **neuromuscular junction** with a skeletal muscle fiber. At the neuromuscular junction, each axon terminal divides into **synaptic end bulbs.** Within the synaptic end bulbs are **synaptic vesicles** filled with acetylcholine (neurotransmitter molecules). When a nerve impulse reaches the synaptic end bulb, **acetylcholine** is released and diffuses across the **synaptic cleft** (the space between the synaptic end bulb and the sarcolemma). **Acetylcholine** molecules bind to receptors in the **motor end plate,** the region of the sarcolemma directly across from the synaptic end bulb. If enough acetylcholine binds, an action potential is generated, stimulating the skeletal muscle fiber to contract.

ACTIVITY 4 Neuromuscular Junction

1 Label the neuromuscular junction in Figure 12.6.
2 Label the photomicrograph in Figure 12.7.
3 Examine a microscope slide of the neuromuscular junction and identify the structures listed in Figure 12.7.

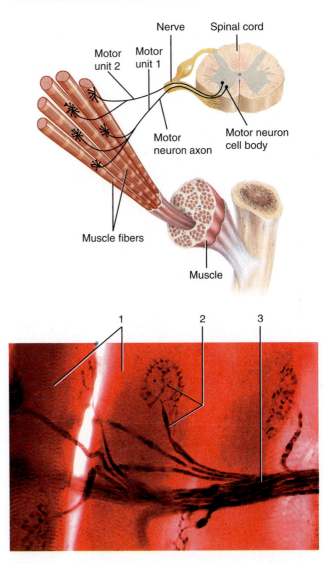

FIGURE 12.6 Structure of the neuromuscular junction.

- axon terminal
- motor end plate
- receptors
- synaptic cleft
- synaptic end bulb
- synaptic vesicle with acetylcholine

1 _____
2 _____
3 _____
4 _____
5 _____
6 _____

365×

FIGURE 12.7 Section of skeletal muscle showing axon terminals and synaptic bulbs.

- skeletal muscle fibers
- axon terminal with synaptic end bulbs
- motor neuron axon

1 _____
2 _____
3 _____

Reviewing Your Knowledge

A. Identification of Skeletal Muscle Structures

Go to www.wiley.com/college/allen and click on the picture of your lab manual. Then click on Student Companion Site, Anatomy Drill and Practice, and Exercise 12.

B. Skeletal Muscle Structure

Number the following nestled, cylindrical structures in order from largest (1) to smallest (5).

____ myofibril ____ muscle fiber

____ fascicle ____ muscle

____ filaments

C. Skeletal Muscle Structural Terms

Identify the structure that matches the description.

_____ **1.** Connective tissue covering surrounding a fascicle.

_____ **2.** Fingerlike invaginations of plasma membrane; extend into interior of fiber and surround myofibrils.

_____ **3.** Plasma membrane of skeletal muscle fiber.

_____ **4.** Connective tissue covering surrounding the muscle.

_____ **5.** Smallest contractile unit within individual muscle fibers.

_____ **6.** Stores calcium within muscle fiber.

_____ **7.** Connective tissue covering surrounding individual muscle fibers.

_____ **8.** Terminal cisternae, T tubule, terminal cisternae.

_____ **9.** Rodlike structures within skeletal muscle fiber that contain thin and thick filaments organized into sarcomeres.

D. Sarcomere Structure and the Sliding Filament Mechanism of Contraction

Write the name of the sarcomere structure defined. A structure may be used more than once, and a definition may apply to more than one structure.

_____ **1.** Length does not change when sarcomere shortens.

_____ **2.** This area is the length of thick filaments.

_____ **3.** Center point of attachment for thick filaments.

_____ **4.** Length decreases when sarcomere shortens.

_____ **5.** This area only contains thin filaments.

_____ **6.** Point of attachment for thin filaments.

_____ **7.** This area only contains thick filaments.

_____ **8.** This area contains overlapping thin and thick filaments.

_____ **9.** The area from Z disc to Z disc.

_____ **10.** This area disappears in a fully contracted muscle.

E. The Neuromuscular Junction

Write the name of the structure(s) described.

_____ **1.** Found in synaptic end bulbs of axon terminal; contains neurotransmitter molecules.

_____ **2.** Area of sarcolemma across from synaptic end bulbs of axon terminal; contains neurotransmitter receptors.

_____ **3.** Space between synaptic end bulbs of axon terminal and sarcolemma.

_____ **4.** Divides into synaptic end bulbs at neuromuscular junction.

_____ **5.** Parts of axon terminal that form neuromuscular junction.

Using Your Knowledge

1. With age, the collagen-containing connective tissue coverings of skeletal muscles increase, and the number of muscle fibers decreases. Is meat from an older or a younger animal more tender? Explain.

2. Weight training increases muscle fiber size by increasing the number of myofibrils. Explain.

3. The diaphragm is a muscle that controls inspiration. Is control of the diaphragm voluntary, involuntary, or both? Explain.

Using your textbook or another reference, for each condition below, indicate which part of the neuromuscular junction is affected: the motor end plate or the axon terminal.

4. Myasthenia gravis

5. Curare poisoning

6. Botulinum toxin poisoning

Put a check mark next to the muscles that are skeletal muscles. **Hint:** Consider location and whether muscle contraction is voluntary or involuntary.

_____ **7.** arector pili

_____ **8.** tongue

_____ **9.** muscle in gallbladder

_____ **10.** muscles that control movement of the eyeball

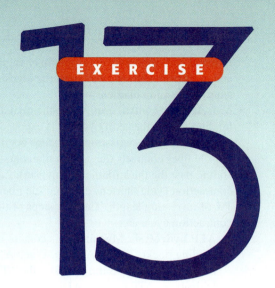

Contraction of Skeletal Muscle

Objectives

After completing this exercise, you should be able to:

- Describe the role of ATP in skeletal muscle contraction
- Describe the three muscle fiber types and their influence on contraction
- Describe how skeletal muscles achieve a smooth, sustained contraction
- Describe how skeletal muscles vary the force of contraction
- Define threshold of contraction, maximal stimulus, recruitment, and fatigue, and explain how to observe them

Materials

- Role of ATP in Skeletal Muscle Contraction: ATP-glycerinated muscle kit from biological supply house and the following items per group: 1 Petri dish, 5 test tubes, marker, 2 teasing needles or straight pins, watchmaker forceps, sharp scissors, 5 microscope slides, 1 millimeter ruler, and 5 plastic transfer pipettes. Dissecting microscope, compound microscope and coverslips are optional.
- Rulers and ankle weights (students may bring them from home)
- *Optional*—Recording Force Generated by Frog Skeletal Muscle Contraction: frogs, needle probes, dissecting equipment, frog Ringer's solution in squeeze bottle, femur clamp, recorder, stimulator, force transducer.
- *Optional*—Recording Human Skeletal Muscle EMGs: Alcohol wipes or cotton balls and 70% alcohol, electrodes and electrode paste, equipment for recording EMGs.

A. Role of ATP in Skeletal Muscle Contraction

There are three main events that occur in skeletal muscle contraction—electrical excitation of a muscle fiber, excitation-contraction coupling, and muscle fiber contraction due to the sliding filament mechanism.

- **Electrical excitation of a muscle fiber.** Skeletal muscle fibers (cells) can be stimulated either by a motor neuron in the body or by a voltage stimulator in the lab. Stimulation given by either method results in a depolarization of the sarcolemma. If the depolarization reaches threshold, an action potential (electrical signal) is initiated.

- **Excitation-contraction coupling.** The action potential is transmitted along the sarcolemma and down the T-tubules. This action causes calcium ions to be released from the terminal cisternae of the sarcoplasmic reticulum. Calcium ions couple electrical excitation to muscle fiber contraction by binding to troponin, which is attached to the actin

filament and tropomyosin. Troponin changes shape and pulls tropomyosin away from the myosin-binding sites on the actin filament.

- **Muscle fiber contraction.** When the myosin-binding site on the actin filament is exposed, a contraction cycle consisting of 4 steps begins. A muscle fiber contracts due to thin filaments (actin) sliding past the thick filaments (myosin). Each contraction cycle shortens each muscle fiber about 1% of its resting length. The four steps of the contraction cycle are:

 1. **ATP hydrolysis.** Myosin heads contain an ATP-binding pocket and an ATPase. When ATP binds to the ATP-binding pocket, the ATPase hydrolyzes ATP forming ADP and inorganic phosphate. Hydrolysis of ATP energizes the myosin head.

 2. **Attachment of myosin to actin to form crossbridges.** Energized myosin heads bind to the unblocked myosin-binding site on actin and the inorganic phosphate is released from the myosin head.

 3. **Power stroke.** The release of the inorganic phosphate starts the power stroke, which is the rotation of the myosin head that pulls the thin filament toward the center of the sarcomere. During the power stroke, ADP is released from the mysosin head, but the myosin head remains attached.

 4. **Detachment of myosin from actin.** Another ATP molecule binds to the myosin ATP-binding pocket, releasing the myosin head from actin, and the contraction cycle begins again.

The contraction cycle continues as long as intracellular calcium levels remain high. As the intracellular calcium levels drop, tropomyosin blocking of the myosin-binding sites on actin return, energized myosin is prevented from binding, and the muscle fiber relaxes.

The amount of shortening that occurs when a muscle is stimulated to contract depends on how long the contraction cycle continues. With each contraction cycle, sarcomeres shorten a little more until maximal contraction of a sarcomere is reached. Muscle fibers can shorten up to 40% of their resting length.

Exposure of skeletal muscle fibers to glycerin creates holes in the sarcolemma allowing ions and ATP to diffuse into and out of the muscle fiber. **Glycerination** also disrupts the troponin-tropomyosin complex so that *calcium is not needed* to bind to troponin and pull tropomyosin to unblock the myosin-binding sites on the actin molecules.

In this activity, you will be observing the effect of varying concentrations of ATP on contraction in glycerinated skeletal muscle fibers.

ACTIVITY 1 Experiment: Role of ATP in Skeletal Muscle Contraction

1 **Prediction:** With your lab group, predict the % muscle fiber shortening each solution will cause by circling a choice in italics.

0.25% ATP in distilled water—*50%, 40%, 30%, 20%, 10%, 0%*

0.25% ATP in salt solution—*50%, 40%, 30%, 20%, 10%, 0%*

0.10% ATP in salt solution—*50%, 40%, 30%, 20%, 10%, 0%*

0.05% ATP in salt solution—*50%, 40%, 30%, 20%, 10%, 0%*

salt solution, no ATP—*50%, 40%, 30%, 20%, 10%, 0%*

2 **Materials:** Obtain materials for Role of ATP in Skeletal Muscle Contraction.

3 **Data Collection:** Measure muscle length of glycerinated skeletal muscle fibers before and after exposure to four different ATP solutions and a solution that does not contain ATP. Decide who will: mix the solutions, prepare the muscle strands, apply the solutions to the muscle strands, time, measure, and record.

- Label test tubes and microscope slides 1, 2, 3, 4, and 5,

- Place the following in the appropriate test tube.

 Test tube 1 (0.25% ATP in distilled water)—5 drops from dropper bottle labeled 0.25% ATP in distilled water

 Test tube 2 (0.25% ATP in salt solution)—5 drops from dropper bottle labeled 0.25% ATP, 0.05M KCl and 0.001M $MgCl_2$ in distilled water

 Test tube 3 (0.10% ATP in salt solution)—2 drops from dropper bottle labeled 0.25% ATP, 0.05M KCl and 0.001M $MgCl_2$ in distilled water and 3 drops from dropper bottle labeled 0.05M KCl and 0.001M $MgCl_2$ in distilled water

 Test tube 4 (0.05% ATP in salt solution)—1 drop from dropper bottle labeled 0.25% ATP, 0.05M KCl and 0.001M $MgCl_2$ in distilled water and 4 drops from dropper bottle labeled 0.05M KCl and 0.001M $MgCl_2$ in distilled water

 Test tube 5 (salt solution, no ATP)—5 drops from dropper bottle labeled 0.05M KCl and 0.001M $MgCl_2$ in distilled water

- From your instructor, obtain a Petri dish containing a 2 cm length piece of glycerinated skeletal muscle in a small amount of glycerin.
- Using teasing needles or straight pins, separate the skeletal muscle (in a Petri dish) into at least 15 strands not more than 0.2 mm in diameter (2-4 muscle fibers per strand).

Effect of Varying Concentrations of ATP Solutions on Muscle Fiber Contraction

- Using forceps, place 3 (of the 15) thin strands of muscle fibers on a microscope slide. Arrange the strands so they are straight and parallel. Do not cover the strands with a cover slip. *Note:* The amount of glycerol that is transferred with the strands should be enough to keep them moist. Only add a small drop of glycerol if the strands are exposed to heat from the microscope lamp for a sustained period.
- Measure the length of each muscle strand with a millimeter ruler and record the value in Table 13.1. Measurements can be made using a dissecting microscope, if desired.
- Using a clean transfer pipette, transfer all the solution from test tube 1 to the microscope slide and measure the length of the strands after 40 seconds. Record the results in Table 13.1.

- Repeat the above steps for the solutions in Test tubes 2-5 using 3 new muscle strands, transfer pipettes, and microscope slides for each solution. Record the results in Table 13.1.
- *Optional: Save all 5 slides to observe striations. See instructions below.*

Observation of Striations in Resting and Contracted Muscle Fibers (optional)

- Place a cover slip over each microscope slide.
- Using a compound microscope, observe one microscopic slide with the low-power objective lens and note the striations on each strand.
- Switch to the high-dry objective lens. Identify the light-colored I bands and the dark-colored A bands.
- Compare the distance between the bands.
- Repeat for each slide.

4 Clean up as directed by your instructor

5 **Data Analysis**
- Calculate the percent of contraction by dividing the length after exposure to ATP solution by the resting length.
- Average the values for each solution and record in Table 13.1.

6 Complete the Experimental Report with your lab group

TABLE 13.1 Role of ATP in Muscle Contraction

	TEST TUBE 1 0.25% ATP, DISTILLED WATER	TEST TUBE 2 0.25% ATP, SALT SOLUTION	TEST TUBE 3 0.10% ATP, SALT SOLUTION	TEST TUBE 4 0.05% ATP, SALT SOLUTION	TEST TUBE 5 0 ATP, SALT SOLUTION
Muscle Strand 1					
Length before solution					
Length after solution					
% Contraction					
Muscle Strand 2					
Length before solution					
Length after solution					
% Contraction					
Muscle Strand 3					
Length before solution					
Length after solution					
% Contraction					
Average % Contraction					

EXPERIMENTAL REPORT: ROLE OF ATP IN MUSCLE CONTRACTION

Results:

- List the ATP solutions that exhibited muscle strand contractions from the greatest to the least contraction.

- State whether there were variations in contractions of muscle strands from the same ATP solutions.

- State the average percent contraction for each solution.

Discussion:

- Discuss the reasons why there might be variations in contraction of muscle strands from the same ATP solutions.

- Discuss whether the sarcomere length increased, decreased, or stayed the same in muscle fibers from the four ATP solutions.

- Discuss how ATP concentration affects the amount of shortening that occurs in a muscle fiber. (*Hint:* Think number of contraction cycles.)

Conclusion:

- State how ATP concentration affects the amount of muscle fiber shortening.

B. Influence of Muscle Fiber Type on Muscle Contraction

There are 3 types of muscle fibers: **slow oxidative, fast oxidative-glycolytic,** and **fast glycolytic.** These muscle fiber types differ in diameter, maximum contractile force generated, myosin ATPase, speed of contraction, methods of ATP production, size of glycogen stores, blood supply, and myoglobin content. Myoglobin content also determines the color of the fiber.

- **Diameter and Force.** Fast glycolytic fibers have the largest diameter of the muscle fiber types, and slow-oxidative fibers have the smallest diameter. Muscle fibers with a larger diameter develop more force. The large diameter fibers have more myofibrils and therefore more myosin heads that can attach to actin. The amount of force a muscle fiber develops is dependent on the number of myosin heads attached to actin.

- **Myosin ATPase and Speed of Contraction.** Fast glycolytic fibers and fast oxidative-glycolytic fibers have a myosin ATPase that breaks down ATP faster than the myosin ATPase in slow oxidative fibers. The speed at which the contraction cycle can occur is determined by how fast myosin ATPase can break down ATP. The faster the myosin ATPase, the greater the speed at which the contraction cycle can occur, and the faster the muscle can contract.

- **Methods of ATP Generation, Glycogen Stores, Blood Supply, and Myoglobin Content.** Muscle fibers with fast myosin ATPase need ATP to be produced quickly to achieve a fast contraction cycle. Fast glycolytic fibers produce the majority of ATP by glycolysis, while slow oxidative fibers use aerobic cellular respiration to produce ATP. Fast oxidative-glycolytic fibers use both ATP production methods.

 Glycolysis produces ATP quickly, but cannot produce large amounts of ATP over time. Therefore, muscle fibers that primarily use glycolysis fatigue quickly. Glycolysis requires large glycogen (glucose source) stores, and muscle fibers that primarily use glycolysis to make ATP are paler due to lower myoglobin content and less blood supply.

Aerobic cellular respiration produces ATP slowly, but can produce large amounts of ATP over time if there is a sufficient blood supply and myoglobin (red pigment that stores oxygen) stores. Therefore, muscle fibers that use aerobic respiration take longer to fatigue. These muscle fibers have lower glycogen stores, are darker due to higher myoglobin content, and have a greater blood supply. These fibers also have greater numbers of mitochondria because the enzymes for aerobic cellular respiration are located in the mitochondria.

ACTIVITY 2 **Influence of Muscle Fiber Type on Muscle Contraction**

1 Label the muscle fiber types in Figure 13.1.
2 Complete Table 13.2 by choosing the correct choice in parentheses.
3 Answer the discussion questions with your lab partners.

LM 440×

FIGURE 13.1 Cross-section of skeletal muscle showing all three fiber types.

- fast glycolytic fiber
- fast oxidative-glycolytic fiber
- slow oxidative fiber

1 _____
2 _____
3 _____

? **DISCUSSION QUESTIONS: INFLUENCE OF MUSCLE FIBER TYPE ON MUSCLE CONTRACTION**

1 Discuss why slow oxidative fibers are more fatigue-resistant than fast glycolytic fibers.

2 Discuss why slow oxidative fibers have more capillaries and mitochondria than fast glycolytic fibers.

3 Discuss why fast glycolytic fibers have more glycogen stores than slow oxidative fibers.

4 Weight training causes fast glycolytic fibers to hypertrophy (increase in size). Explain why this increases strength.

5 Identify which muscle fiber type would be primarily involved in each of the following functions.
 a. walking
 b. standing
 c. lifting a heavy object and immediately putting it down.

TABLE 13.2 **Characteristics of Skeletal Muscle Fiber Types**

	SLOW OXIDATIVE	FAST OXIDATIVE-GLYCOLYTIC	FAST GLYCOLYTIC
Fiber diameter (smallest, intermediate, largest)			
Force (lowest, intermediate, greatest)			
Myosin ATPase (slowest, fastest)			
Contraction velocity (slow, fast)			
Methods of ATP generation (glycolysis, aerobic respiration)			
Glycogen stores (low, intermediate, high)			
Capillaries (many, few)			
Myoglobin content (low, high)			
Color (white, red-pink, red-brown)			

C. Contraction of Skeletal Muscle Fibers

A motor neuron stimulates all the skeletal muscle fibers in a motor unit simultaneously. The amount of force generated by a motor unit is determined by the number of muscle fibers within the motor unit, the size of the individual muscle cells, and the number of crossbridges attached in the muscle fibers.

A **twitch contraction** is the quick shortening observed in a skeletal muscle when a single action potential traveling down a motor neuron stimulates the skeletal muscle fibers of the motor unit to contract. The minimal stimulus that results in a muscle twitch is called the **threshold stimulus.** The three phases of a twitch contraction are the latent period, the contraction period, and the relaxation period. The **latent period** lasts about 2 msec (milliseconds) and is the time between stimulation of muscle cells and force generation. The **contraction period** lasts about 10–100 msec and is the period during which force (measured in grams) is increasing, whereas the **relaxation period,** which lasts 10–100 msec, is the period when force is decreasing. Normal muscle contractions are not twitch contractions, but are sustained contractions of varying force.

If the muscle fibers of a motor unit are stimulated before the relaxation phase of a muscle twitch is complete, then the next contraction will produce a greater force. This is called **wave summation.** Increasing the frequency of muscle stimulation produces sustained force generation. **Unfused tetanus** (*tetan* = rigid) occurs when there is a partial relaxation between muscle twitches. **Fused tetanus** is a sustained contraction with no relaxation observed between twitches. Increasing the number of motor units contracting at the same time, **motor unit recruitment,** also increases force generated. Lifting a feather requires fewer motor units than lifting your anatomy and physiology textbook. **Maximal force** development occurs when all motor units of a muscle are stimulated and all muscle fibers are contracting. In the lab, the stimulus that produces maximal force is called the **maximal stimulus.** Therefore, a stimulus to the muscle greater than maximal does not produce a greater force.

Most sustained voluntary skeletal muscle contractions are unfused tetanic contractions with different motor units stimulated at different times (**asynchronous contractions**). The asynchronous contractions delay **muscle fatigue,** which is an inability to contract caused by long periods of muscle contraction.

Muscle contraction results in development of tension or force usually measured in grams. Muscles will shorten if they develop more force than the force that is opposing them. For example, contracting muscles in our arm will shorten and allow us to lift a book if the force developed by the muscles is greater than the weight (force) of the book. The weight of the book is also called load because it is a force that the muscle is contracting against.

If a muscle is generating a constant force to move a load, that contraction is called an **isotonic contraction.** There are two types of isotonic contraction, concentric and eccentric isotonic contractions. **In concentric isotonic contractions,** the muscle is shortening while it is contracting. An example of this is using the biceps brachii (large anterior muscle of arm) to lift a book off the table. In **eccentric isotonic contractions,** the muscle is lengthening while it is contracting. This occurs when you slowly lower your arm to return the book to the table. The biceps brachii is still contracting, but it is lengthening while it is contracting. This enables you to lower the book in a controlled manner.

Isometric contractions are contractions in which the muscle is developing force but not shortening and no visible movement is seen. In this case, the force developed by the muscle equals the force (load) it is contracting against. An example of this would be holding a book in the same position. Isometric muscle contractions are maintaining the position against the weight (load) of the book, but the biceps brachii muscle is not moving the book because it is not generating a force greater than the weight (force) of the book.

ACTIVITY 3 Contraction of Skeletal Muscle Fibers

1 Draw a twitch contraction that has a latent period of 2 msec, a contractile period of 10 msec with a maximum force of 2 g, and a relaxation period of 10 msec. Use the graph in Figure 13.2.

2 Observe unfused tetanus, motor unit recruitment, fatigue, isotonic, and isometric contractions in the muscles that cause bending of the knee.
- In your lab group, decide who will be the subject, observer, and recorder.
- Have the subject stand while holding onto the lab bench for support. Have subject bend the left knee

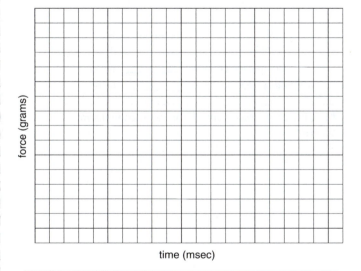

force (grams)

time (msec)

FIGURE 13.2 Student drawing of twitch contraction.

to a 90-degree angle and hold this position. The hamstring muscles are used to bend the knee.
- Observe the unfused tetanic contraction. Start timing.
- Time how long it takes for the muscle to fatigue, which is demonstrated by any vertical movement in the leg.
- Time until fatigue: _____ min.
- Place a 5-lb ankle weight on the subject's right ankle and repeat the steps.
- Time until fatigue: _____ min.

3 Answer discussion questions with your lab partners.

? DISCUSSION QUESTIONS: OBSERVATION OF UNFUSED TETANUS, RECRUITMENT, AND FATIGUE IN HAMSTRING MUSCLES

1 Explain why you can maintain contraction of the hamstring muscles over time.

2 Explain why you can sustain the same contraction with a 5-lb weight attached to the ankle.

3 Explain why the leg with the 5-lb ankle weight fatigued faster.

4 As the subject bends the knee and holds it at 90 degrees, state which muscle movement causes isometric contractions and which causes isotonic contractions.

5 Is the isotonic contraction observed when the hamstrings contract a concentric isotonic contraction or an eccentric isotonic contraction? What would you ask the subject to do to observe the other type of isotonic contraction?

D. Recording Force Generated by Muscle Contractions and Measuring Threshold and Maximal Stimulus

Polygraphs amplify and record physiological changes that are detected by various sensors. Examples of **sensors** include those that detect changes in pH, force, temperature, and pressure. A **myograph** is a type of polygraph used to record the force generated by muscle contractions along with the following equipment:

1. A **stimulator** stimulates skeletal muscle fibers to contract, and the stimulus strength is measured in volts. The **frequency of stimulation** is the number of stimuli per second. Increasing the frequency of stimulation increases the frequency of contractions.

2. A **transducer,** which is a sensor that converts the force generated by a contracting muscle into an electric signal (current), is attached to the muscle.

3. An **amplifier** increases the amplitude to the transducer signal and transmits it to a recorder channel.

4. The **recorder** uses the electric signal from the transducer to move a pen across a sheet of paper that is moving at preset speed. This allows students to observe if the force changes over time. Computers can also act as recorders and display the pen deflections as lines moving across the computer screen. The visual displays can be saved and printed.

To calculate the force recorded from a contracting muscle, the transducer must be calibrated before attaching it to the muscle. Different gram weights are attached to the transducer to observe the electric signal or pen deflection for each weight (or force). Pen deflection is adjusted so that each gram of force will result in a pen deflection of a known length.

Figure 13.3 illustrates the use of equipment to record a **myogram** (recording of force generated by a frog gastrocnemius muscle). A myogram is illustrated in Figure 13.4.

Recorders can have more than one **channel,** allowing several different measurements to be recorded at one time. For example, an electrode can be attached to the muscle and channel 1 of a recorder to record the electric signal or action potential of a stimulated muscle, while a force transducer connected to channel 2 records the force of contraction (myogram). Time and event markers of recorders are pens that mark the time when events occur, such as stimulation of a muscle.

FIGURE 13.3 Apparatus used to measure force generated in frog gastrocnemius muscle.

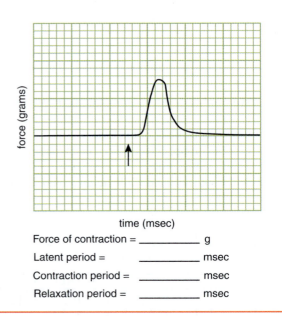

Force of contraction = _____ g

Latent period = _____ msec

Contraction period = _____ msec

Relaxation period = _____ msec

FIGURE 13.4 Myogram of frog gastrocnemius muscle contraction.

ACTIVITY 4 Recording Muscle Contractions

1 Measure the force generated in grams and the length of the latent, contraction, and relaxation periods in milliseconds on the myogram in Figure 13.4 and record your results on the figure.
 - The arrow indicates when a stimulus was applied to the muscle.
 - Each small square is 0.2 cm^2 and each large square is 1 cm^2.
 - An upward deflection of 1 cm represents generation of 1 g of force.
 - Paper speed was 10 cm/sec. Therefore, a horizontal distance of 10 cm represents 1 sec and a horizontal distance of 1 cm = 0.1 sec (100 msec).

2 With your group, discuss how you would measure threshold stimulus, maximal stimulus, and maximal force in the frog gastrocnemius preparation.

3 Optional: Myograms of frog gastrocnemius muscle. Follow the instructions supplied by your instructor to prepare and conduct the experiment.

4 Optional: Myograms of human skeletal muscle contractions. Follow the instructions supplied by your instructor to prepare and conduct the experiment.

Reviewing Your Knowledge

A. Role of ATP in Muscle Contraction

Write T for true or F for false for the following statements.

_____ **1.** There is only one type of myosin ATPase involved in muscle contraction.

_____ **2.** ATP causes the detachment of myosin from actin.

_____ **3.** Glycerinated muscle fibers need calcium to couple the electrical excitation with the contraction cycle.

_____ **4.** Skeletal muscle fibers need ATP in order to contract.

_____ **5.** A solution of 0.05% ATP causes more skeletal muscle contraction than a solution of 0.25% ATP.

B. Influence of Muscle Fiber Type on Muscle Contraction

Match the correct fiber type to the fiber characteristic. Questions may have more than one answer.

 a. fast glycolytic
 b. fast oxidative-glycolytic
 c. slow oxidative

_____ **1.** largest diameter

_____ **2.** is pink in color

_____ **3.** myoglobin content is low

_____ **4.** has the fastest contraction velocity

_____ **5.** has many capillaries

_____ **6.** is red in color

_____ **7.** generates ATP by aerobic respiration

_____ **8.** has the slowest myosin ATPase

_____ **9.** has glycogen stores

_____ **10.** lowest force

C. Contraction of Skeletal Muscle Fibers

Match structure(s) to definition.

a. contractile period g. muscle fatigue
b. fused tetanus h. relaxation period
c. latent period i. threshold stimulus
d. maximal stimulus j. twitch
e. motor unit k. unfused tetanus
f. motor unit recruitment

_____ **1.** Inability of muscle fibers to contract after a long period of contraction.

_____ **2.** Phase of contraction; time between muscle fiber stimulation and measurement of force generation.

_____ **3.** Motor neuron and all the muscle fibers it innervates.

_____ **4.** Phase of contraction; period of force generation.

_____ **5.** Type of wave summation with partial relaxation observed between twitches.

_____ **6.** Increasing the number of motor units that are stimulated to contract.

_____ **7.** Stimulus that results in maximal force generation.

_____ **8.** Lowest stimulus that results in force generation.

_____ **9.** Single contractile event in response to single action potential.

_____ **10.** Phase of contraction; period during which more crossbridges detach than reattach to thin filaments.

_____ **11.** Type of wave summation with no observable relaxation between twitches.

D. Recording Force Generated by Muscle Contractions

Match the term to the definition.

a. amplifier e. sensors
b. myograph f. stimulator
c. polygraph g. transducer
d. recorder

_____ **1.** Examples are: pH, pressure, temperature.

_____ **2.** Increases the amplitude to the transducer signal and transmits it to a recorder channel.

_____ **3.** Makes skeletal muscle fibers contract.

_____ **4.** A machine that amplifies and records physiological changes detected by a variety of sensors.

_____ **5.** Uses the electric signal from transducer to make a tracing on moving paper.

_____ **6.** A machine that records force generated by muscle contractions.

_____ **7.** Sensor that converts force generated by a contracting muscle into an electric signal.

EXERCISE

13

Using Your Knowledge

Answer the following questions with a short answer.

1. Explain why muscles are stiff (contracted) when rigor mortis occurs.

2. Large muscles, such as the muscles of the leg, have more muscle fibers than small muscles, such as the muscles of the finger. Explain why the muscles of the finger cannot develop as much force as the muscles of the leg.

3. Which type of muscle would fatigue faster, one that has many blood vessels, or one that has fewer blood vessels? Explain.

Using your textbook or another reference, for each contraction below, indicate whether it is an example of a muscle twitch (mt), unfused tetanus (ut), or fused tetanus (ft).

_____ **4.** muscle spasm

_____ **5.** facial muscle tic

_____ **6.** cardiac fibrillation

_____ **7.** smiling

The percentage of each muscle fiber type in any given muscle of your body is determined genetically. Also, physical activity can cause slight changes in muscle fiber type. Discuss the relative amount of each muscle type in the thigh muscles of Olympic athletes participating in the following events:

8. marathon

9. weight lifting

10. 100 meter dash

Skeletal Muscles and Their Actions

A. Criteria for Naming Skeletal Muscle

Learning the criteria used to name skeletal muscles will assist you in identifying and learning the function of these muscles. Skeletal muscles are named for their orientation relative to the midline of the body, size, shape, action, number of origins, location, or origin and insertion. The muscle attachment points, the origin and insertion, are usually skin or bone. The **origin** is the nonmoving point of attachment when a muscle contracts, and the **insertion** point moves toward the origin. For characteristics used to name skeletal muscles, refer to your textbook or the table located at www.wiley.com/college/allen. Click on the picture of your lab manual and then click on Student Companion Site, Anatomy Drill and Practice, and Exercise 14.

B. Skeletal Muscle Identification and Action

When skeletal muscles contract, they pull their point of insertion toward their point of origin. For the muscles of facial expression, this action moves the skin on the face and allows us to express emotion. For most other skeletal muscles, contraction results in the movement of a bone at a joint. In the appendages, muscles that perform similar functions are surrounded by fascia, forming **compartments** that are served by the same major blood vessels and nerves. Also, within a location some muscles are grouped according to whether they are **superficial** or **deep.**

Muscles of the Head and Neck

Muscles of Facial Expression and Mastication

Muscles located in the head include those involved in facial expression, mastication, blinking, and eye movements. Their origins are found in the superficial fascia beneath the skin and attached to skull bones, while their insertions are in the skin of the face. Muscles that insert on the exterior of the eyeball and move the eye will be studied in the exercise on Special Senses—The Eye.

Muscles That Move the Head and Neck

Muscles that flex the head and neck are found in pairs on the anterior and lateral surface of the neck, while muscles that extend the head and neck are found on the posterior surface of the neck. However, the levator scapulae, a posteriolateral neck muscle, elevates the scapula but does not move the neck.

The **sternocleidomastoid muscle,** a muscle that runs diagonally from the sternum and the clavicle to the mastoid process of the temporal bone, divides each side of the neck into an anterior and posterior triangle. The **anterior triangle** includes the carotid artery and the internal jugular vein, while the **posterior triangle** includes the external jugular vein and part of the subclavian artery.

Muscles That Move the Hyoid Bone and Larynx

Muscles that move the hyoid bone and larynx include the suprahyoid muscles and the infrahyoid muscles. The **suprahyoid muscles** are located superior to the hyoid bone and elevate the hyoid bone, floor of the oral cavity, and tongue while swallowing. The **infrahyoid muscles** are located inferior to the hyoid bone and depress the hyoid bone and larynx while swallowing and talking.

ACTIVITY 1 Muscles of the Head and Neck

1 Review muscle location and action in Table 14.1.
2 Label Figure 14.1(a), (b), (c), and (d) and Figure 14.2(a), (b), and (c).
3 Use Figures 14.1 and 14.2, and Table 14.2 to point to the muscles on models, diagrams, and/or cadaver photographs. Pronounce the names as you point to them.
4 Locate the superficial muscles on yourself and perform the action indicated in Table 14.1. Palpate the muscle while performing the action.
5 Optional: Color black and white muscles in contrasting colors. Color in the same direction as the muscle fibers.
6 Optional: Use Appendix C to locate the origin and insertion on an articulated skeleton. Pull the insertion toward the origin to simulate muscle action.

TABLE 14.1 Muscles of the Head and Neck

MUSCLE	LOCATION AND *FUNCTION*
MUSCLES OF FACIAL EXPRESSION	**(FUNCTION IN ITALICS)**
Occipitofrontalis muscle	Combination of frontalis and occipitalis muscle connected by an aponeurosis (a flat, broad tendon).
Frontalis (*frontalis* = in front)	Lies over frontal bone. *Raises eyebrows and wrinkles forehead.*
Occipitalis (*occipit-* = base of skull)	Lies over occipital bone. *Pulls scalp posteriorly.*
Orbicularis oculi (*orbicularis* = little circle; *oculi* = eye)	Circular muscle that encircles the eye. *Closes eye.*
Zygomaticus minor	Between zygomatic bone and corner of mouth. *Raises upper lips, exposing upper teeth.*
Zygomaticus major (*Zygoma* = bar)	Between zygomatic bone and corner of mouth. *Raises corners of mouth (smiling).*
Orbicularis oris (*oris* = mouth)	Circular muscle that encircles the mouth. *Closes and purses lips.*
Platysma	Wide, flat muscle that covers lower mandible and entire anterior neck, and ends on chest. *Tenses neck skin; depresses mandible (pouting).*

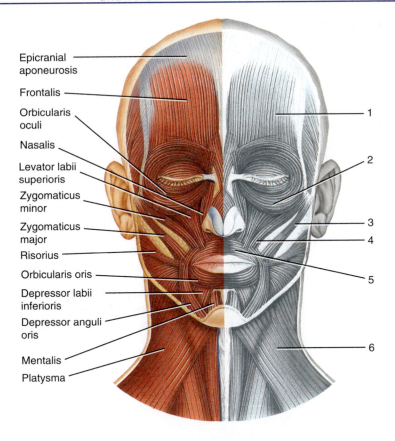

Epicranial
aponeurosis

Frontalis

Orbicularis
oculi

Nasalis

Levator labii
superioris

Zygomaticus
minor

Zygomaticus
major

Risorius

Orbicularis oris

Depressor labii
inferioris

Depressor anguli
oris

Mentalis

Platysma

1

2

3

4

5

6

(a) Anterior view

FIGURE 14.1 Selected muscles of the head.

(a)
- frontalis (fron-TA-lis)
- orbicularis oculi (or-bi-kyoo-
 LAR-is OK-yoo-lie)
- orbicularis oris (OR-is)
- platysma (pla-TIZ-ma)
- zygomaticus major (zy-go-MAH-ti-kus)
- zygomaticus minor

1 _____

2 _____

3 _____

4 _____

5 _____

6 _____

TABLE 14.1 *(Continued)*

MUSCLE	LOCATION AND *FUNCTION*
MUSCLES OF MASTICATION	
Superficial Muscles	
Temporalis (*tempora* = temples)	Lies over temporal bone. *Elevates and retracts mandible.*
Masseter (*maseter* = chewer)	Anterior to ear between zygomatic arch and posterior portion of mandible. *Elevates and retracts mandible.*
Deep Muscles	
Buccinator (*bucca* = cheek)	Deep to masseter. Fibers run transversely and form fleshy part of cheek. *Presses cheeks inward to whistle, blow, and suck. Helps hold food between teeth while chewing.*
Lateral pterygoid (*pterygoid* = like a wing)	Deep to masseter and superior to medial pterygoid. Transverse muscle fibers between pterygoid process of sphenoid and mandible. *Protracts mandible, depresses mandible (opening mouth), and moves it sideways.*
Medial pterygoid	Deep to masseter and inferior to lateral pterygoid. Verticle muscle fibers between pterygoid process of sphenoid and mandible. *Elevates and protrudes mandible and moves it sideways.*

Frontalis
Orbicularis oculi
Occipitalis
Zygomaticus minor
Zygomaticus major
Masseter
Buccinator
Orbicularis oris
Platysma

7
8
9
10
11
12
13

(b) Lateral superficial view

Temporalis
Lateral pterygoid
Medial pterygoid
Buccinator
Orbicularis oris

14
15
16
17
18

(c) Lateral deep view

FIGURE 14.1 Selected muscles of the head, *continued.*

(b)
- buccinator
- frontalis
- masseter
- occipitalis (ok-si-pi-TA-lis)
- orbicularis oculi
- zygomaticus major
- zygomaticus minor

7 _____

8 _____

9 _____

10 _____

11 _____

12 _____

13 _____

(c)
- buccinator
- lateral pterygoid (TER-ih-goid)
- medial pterygoid
- orbicularis oris
- temporalis (tem-por-A-lis)

14 _____

15 _____

16 _____

17 _____

18 _____

TABLE 14.1 (*Continued*)

MUSCLE	LOCATION AND *FUNCTION*
MUSCLES THAT MOVE THE HEAD AND NECK	
Anterior and Lateral Muscles	
Sternocleidomastoid (*sterno-* = breastbone; *cleido* = clavicle; *mastoid* = mastoid process)	Straplike muscle on anterior and lateral neck. Fibers run diagonally across the neck from the sternum and clavicle to the mastoid process. *Both muscles contracting flex head (prayer muscle). One muscle contracting laterally flexes head and rotates head to opposite side (as in saying no).*
Scalenes (*scalenos* = uneven)	Three muscles—anterior, middle, and posterior scalenes—located in lateral neck deep to sternocleidomastoid muscle. *Flex head and elevate 1st and 2nd ribs when muscles on both sides contract during deep inspiration. When muscles on one side contract they laterally flex head and rotate head to opposite side (as in saying no).*
Posterior Muscles	
Splenius capitis	Posterior neck deep to trapezius. *Extends head when both muscles contract. Laterally flexes and rotates head to same side as contracting muscle when only one muscle contracts.*
Trapezius (superior portion)	Superficial muscle of the upper back and posterior neck. *Extends head and elevates scapula.*
MUSCLE THAT ONLY MOVES THE SCAPULA	
Levator scapulae (*levare* = to raise)	Posterior to sternocleidomastoid; runs diagonally from posterior head to scapula. *Elevates scapula and rotates it downward; stabilizes scapula.*
MUSCLES THAT MOVE HYOID BONE	
Suprahyoid Muscles	
Digastric (*di* = two; *gastric* = belly)	
Anterior belly	Straplike muscle under chin that is parallel to midline of chin and extends from posterior belly. *Elevates hyoid bone and depresses mandible (as in opening mouth).*
Posterior belly	Runs along posterior border of chin. *Elevates hyoid bone and depresses mandible (as in opening mouth).*
Stylohyoid	Chin muscle, medial to posterior belly of digastric. *Elevates hyoid bone and moves it posteriorly.*
Mylohyoid	Deep muscle of chin extending from right side of mandible to left side. *Elevates hyoid bone and floor of oral cavity and depresses mandible.*
Infrahyoid Muscles	
Omohyoid	Lateral to sternohyoid. *Depresses hyoid bone.*
Sternohyoid (*sterno* = sternum)	Runs along midline of anterior neck. *Depresses hyoid bone.*

Sternocleidomastoid

Splenius capitis

Trapezius

Levator scapulae

Platysma

Middle scalene

(a) Lateral superficial view

FIGURE 14.2 Selected muscles of the chin and neck.

(a)
- levator scapulae (le-VAY-tor SKA-pyoo-lee)
- middle scalene
- platysma (pla-TIZ-ma)
- splenius capitis
- sternocleidomastoid (ster-no-kli-do-MAS-toid)
- trapezius (tra-PEE-zee-us)

1 _____

2 _____

3 _____

4 _____

5 _____

6 _____

(b) Anterior view with platysma removed

(c) Anterior deep view

FIGURE 14.2 Selected muscles of the chin and neck, *continued.*

(b)
- digastric, anterior belly
- digastric, posterior belly
- mylohyoid
- omohyoid
- sternocleidomastoid
- sternohyoid
- stylohyoid

7 _____

8 _____

9 _____

10 _____

11 _____

12 _____

13 _____

(c)
- levator scapulae
- mylohyoid
- scalenes (SKAY-leens)

14 _____

15 _____

16 _____

Muscles of the Trunk

Muscles That Move the Arm

These muscles originate on the anterior or posterior trunk and insert on the humerus. Muscle contraction pulls the humerus toward the insertion, resulting in movement of the humerus at the shoulder joint. Because of the structure of the shoulder joint and the way these muscles surround the humerus, contraction can cause a full range of motion at a synovial joint—flexion, extension, abduction, adduction, circumduction, and rotation.

The rotator cuff muscles are deep muscles that move the arm and surround the shoulder joint like a sleeve cuff and help strengthen and stabilize the shoulder joint. They include the **subscapularis, supraspinatus, infraspinatus,** and **teres minor.**

Muscles That Move the Scapula

Movements of the scapula include elevation, depression, abduction (protraction), adduction (retraction), and rotation. Shrugging the shoulders or lifting an object over the head elevates the scapula. Depression of the scapula occurs when doing a pullup. Performing a "pushup" abducts the scapula, and standing at attention adducts the scapula.

Muscles That Move the Abdominal Wall and Vertebral Column

Muscles on the anterior surface of the abdomen (abdominal muscles) flex the vertebral column, while the muscles of the back extend it. However, the transversus abdominis muscle cannot flex the vertebral column because it runs horizontally.

Muscles Used in Breathing

Muscles of inspiration increase the diameter or length of the thoracic cavity when they contract, while muscles of expiration decrease the diameter or length of the thoracic cavity. In this activity we will be examining the diaphragm, external intercostals, and internal intercostals. All muscles involved in breathing will be examined together when studying the respiratory system.

ACTIVITY 2 Muscles of the Trunk

1 Review muscle location and action in Tables 14.2 and 14.3.

2 Label Figure 14.3(a) and (b) and Figures 14.4(a), (b), and (c).

3 Use Figures 14.3 and 14.4 and Tables 14.2 and 14.3 to point to the muscles on models, charts, and/or cadaver photographs. Pronounce the names as you point to them.

4 Locate the muscle on yourself and perform the action indicated in Tables 14.2 and 14.3. Palpate the muscle while performing the action.

5 Optional: Color black and white muscles in contrasting colors. Color in the same direction as the muscle fibers.

6 Identify the rotator cuff muscles on Figures 14.3 and 14.4.

7 Optional: Use Appendix C to locate the origin and insertion on an articulated skeleton. Pull the insertion toward the origin to stimulate muscle action.

TABLE 14.2 Muscles of the Anterior Trunk

MUSCLE	LOCATION AND *FUNCTION*
MUSCLES THAT MOVE THE ARM AT THE SHOULDER JOINT	
Superficial Muscles	
Deltoid (*delta* = triangle)	Large, rounded, triangular shoulder muscle. **Anterior fibers:** _flex and medially rotate arm._ **Lateral fibers:** _abduct arm._ **Posterior fibers:** _extend and laterally rotate arm._
Pectoralis major (*pectus* = chest)	Large muscles on superior, anterior chest between sternum and arm. _Whole muscle adducts and medially rotates arm, clavicular head only flexes arm, and sternocostal head only extends arm._
Deep Muscle	
Subscapularis	On ventral side of the scapula in the subscapular fossa. _Medially rotates arm._
MUSCLES THAT MOVE THE SCAPULA	
Serratus anterior (*serratus* = saw-toothed)	Serrated-looking muscles on the lateral trunk inferior to arms and rib. _Abducts scapula and rotates it upward. This occurs when throwing a punch and is often called the boxer's muscle._
Pectoralis minor	Deep to pectoralis major attaching to third through fifth ribs. _Abducts scapula and rotates it downward; elevates third through fifth ribs during forced inspiration._
MUSCLES THAT MOVE THE ABDOMINAL WALL	
Superficial Muscles	
Rectus abdominis (*rectus* = parallel fibers)	Midline abdominal muscles located between sternum and inguinal ligament. _Flexes vertebral column and compresses abdomen. Does not laterally flex vertebral column._
External oblique	Lateral and anterior sheetlike abdominal muscles whose fibers run obliquely toward the midline (linea alba). _Flexes vertebral column and compresses abdomen when both muscles contract and laterally flexes and rotates vertebral column when only one muscle contracts._
Deep Muscles	
Internal oblique	Abdominal muscles deep to external oblique whose fibers run obliquely toward linea alba. _Flexes vertebral column and compresses abdomen when both muscles contract and laterally flexes and rotates vertebral column when only one muscle contracts._
Transversus abdominis	Abdominal muscles deep to internal oblique whose fibers run transversely. _Compresses abdomen only._
MUSCLES USED IN BREATHING	
Deep Muscles	
External intercostals (*inter-* = between; *costal* = ribs)	Oblique muscles located between ribs superficial and lateral to internal intercostals. _Increase thoracic diameter by elevating ribs during inspiration._
Internal intercostals	Oblique muscles located between ribs near sternum. Deep and medial to external obliques. _Decrease thoracic diameter by depressing ribs during forced expiration._
Diaphragm	Dome-shaped muscle inside the ventral body cavity separating it into the thoracic and abdominal cavities. Contains openings for the esophagus, aorta, and vena cava. _Flattens, increasing length of thoracic cage, during normal inspiration._

Platysma

Clavicle

Pectoralis major

Latissimus dorsi

Serratus anterior

Linea alba

External oblique

Rectus sheath

Inguinal ligament

Inguinal ring (superficial)

- external oblique
- pectoralis (pek-tor-A-lis) major
- serratus (ser-RAY-tus) anterior

1 _____

2 _____

3 _____

1

2

3

(a) Anterior superficial view

Sternocleidomastoid

Trapezius

Deltoid (cut)

Pectoralis minor

Subscapularis

External intercostal

Internal intercostal

Rectus abdominis

Internal oblique (cut)

External oblique (cut)

Transverse abdominis

- external intercostal
- internal intercostal
- internal oblique
- pectoralis minor
- rectus abdominis (REK-tus ab-DOM-in-is)
- transverse abdominis

4 _____

5 _____

6 _____

7 _____

8 _____

9 _____

4

5

6

7

8

9

(b) Anterior deep view

FIGURE 14.3 Muscles of the anterior trunk.

TABLE 14.3 Muscles of the Posterior Trunk

MUSCLE	LOCATION AND *FUNCTION*
MUSCLES THAT MOVE THE ARM AT THE SHOULDER JOINT	
Superficial muscles	
Deltoid (*delta* = triangle)	Large, rounded, triangular shoulder muscle. *Posterior portion of deltoid extends and laterally rotates arm.*
Infraspinatus (*infra-* = below; *spinatus* = spine of scapula	Located in infraspinous fossa (inferior to spine of scapula) and partially covered by deltoid. *Laterally rotates and adducts arm.*
Teres minor (*teres* = long and round)	Inferior to infraspinatus and partially covered by deltoid. *Laterally rotates, extends, and adducts arm.*
Teres major	Inferior to teres minor and partially covered by deltoid. *Extends, adducts, and medially rotates arm.*
Latissimus dorsi (*latissimus* = widest; *dorsum* = back)	Main large, flat muscle of middle and lower back that runs laterally and attaches to superior portion of humerus. *Extends, adducts, and medially rotates arm (if arm is elevated over head, it brings it down).*
Deep muscles	
Supraspinatus (*supra* = above)	Located in supraspinous fossa (superior to spine of scapula) and totally covered by deltoid. *Assists deltoid muscle in abducting arm.*
MUSCLES THAT MOVE THE SCAPULA	
Superficial muscle	
Trapezius (*trapezoides* = trapezoid shape)	Diamond-shaped muscle of posterior neck and upper back that extends from the skull and vertebral column to spine of scapula and lateral clavicle. *Superior portion elevates scapula and extends head, middle portion adducts scapula, inferior portion depresses scapula.*
Deep muscles	
Levator scapulae (*levare* = to raise)	Posterior to sternocleidomastoid and runs diagonally from posterior head to scapula. *Elevates scapula and rotates it downward.*
Rhomboideus minor	Deep to trapezius in upper back located between the vertebrae and scapula, and superior to rhomboideus major. *Elevates and adducts scapula and rotates it downward, stabilizes scapula.*
Rhomboideus major	Deep to trapezius in upper back located between the vertebrae and scapula, and inferior to rhomboideus minor. *Elevates and adducts scapula and rotates it downward, stabilizes scapula.*

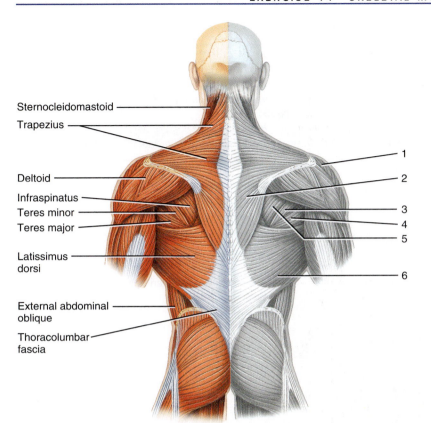

Sternocleidomastoid

Trapezius

Deltoid

Infraspinatus

Teres minor

Teres major

Latissimus dorsi

External abdominal oblique

Thoracolumbar fascia

(a) Posterior superficial view

- deltoid
- infraspinatus (in-fra-spi-NAY-tus)
- latissimus dorsi (la-TIS-i-mus DOR-sigh)
- teres (TEH-rez) major
- teres minor
- trapezius (tra-PEE-zee-us)

1 _____

2 _____

3 _____

4 _____

5 _____

6 _____

- infraspinatus
- rhomboid major (ROM-boid)
- rhomboid minor
- supraspinatus (soo-pra-spi-NAY-tus)
- teres major
- teres minor

7 _____

8 _____

9 _____

10 _____

11 _____

12 _____

Splenius capitis

Levator scapulae

Rhomboid minor

Rhomboid major

Supraspinatus

Infraspinatus

Teres minor

Teres major

Serratus posterior inferior

(b) Posterior deep view

FIGURE 14.4 Muscles of the posterior trunk.

TABLE 14.3 (*Continued*)

MUSCLE	LOCATION AND *FUNCTION*
MUSCLES THAT MOVE THE VERTEBRAL COLUMN	
Erector spinae (*erector* = erect posture; *spinae* = spine) consists of iliocostalis, longissiumus, and spinalis groups	Group of muscles next to the vertebral column deep to the trapezius, latissimus dorsi, and scapula. *Extend vertebral column and maintain erect posture when both muscles contract. Laterally flexes vertebral column when only one muscle.*

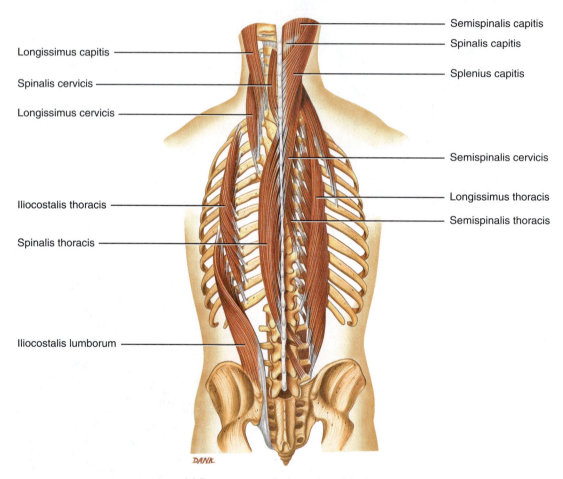

Longissimus capitis

Spinalis cervicis

Longissimus cervicis

Iliocostalis thoracis

Spinalis thoracis

Iliocostalis lumborum

Semispinalis capitis

Spinalis capitis

Splenius capitis

Semispinalis cervicis

Longissimus thoracis

Semispinalis thoracis

DANK

(c) Deep erector spinal muscles of the back

FIGURE 14.4 Muscles of the posterior trunk, *continued.*

Muscles of the Arm

These muscles have their origin on the humerus or scapula, cross the elbow joint, and insert on the bones of the forearm. When these muscles contract, they flex or extend the forearm at the elbow joint or cause movement at the radioulnar joints. The muscles that have origins on the scapula can also move the arm at the shoulder joint.

Muscles of the limbs that perform similar functions often are surrounded by fascia-forming compartments. Muscles in the **anterior compartment** of the arm, such as the biceps brachii and brachialis muscles, flex the forearm, whereas those in the **posterior compartment,** the triceps brachii, extend the forearm.

Muscles of the Forearm

Muscles located in the forearm flex or extend the hand or cause movement at the radioulnar joint. The muscles that move the hand cross the wrist and insert on the carpals, metacarpals, or phalanges. These muscles are found in the anterior (flexor) compartment and posterior (extensor) compartment of the forearm. The muscles of the superficial anterior compartment of the forearm from lateral to medial are: flexor carpi radialis, palmaris longus, flexor carpi ulnaris, and flexor digitorum superficialis. The flexor digitorum superficialis is actually deep to the other three muscles. These muscles flex the hand or digits and originate on the medial epicondyle of the humerus. Muscles of the superficial posterior compartment of the forearm from medial to lateral are: extensor carpi ulnaris, extensor digitorum

minimi, extensor digitorum, extensor carpi radialis brevis, extensor carpi radialis longus. These muscles extend the hand and digits and originate on the lateral epicondyle of the humerus. Flexor and extensor muscles near the lateral surface of the forearm may also abduct the wrist, whereas those near the medial surface may adduct it. Other muscles located in the forearm include the brachioradialis, pronator teres, and supinator, which move the forearm at the radioulnar joint.

ACTIVITY 3 — Muscles of the Arm and Forearm

1 Review muscle location and action in Tables 14.4 and 14.5.

2 Label Figure 14.5(a) and (b) and Figures 14.6(a), (b), (c), and (d).

3 Use Figures 14.5 and 14.6 and Tables 14.4 and 14.5 to point to the muscles on models, charts, and/or cadaver photographs. Pronounce the names as you point to them.

4 Locate the muscle on yourself and perform the action indicated in Tables 14.4 and 14.5. Palpate the muscle while performing the action.

5 Optional: Color black and white muscles in contrasting colors. Color in the same direction as the muscle fibers.

6 Optional: Use Appendix C to locate the origin and insertion on an articulated skeleton. Pull the insertion toward the origin to stimulate muscle action.

TABLE 14.4 Muscles of the Arm

MUSCLE	LOCATION AND *FUNCTION*
ANTERIOR SURFACE	
Biceps brachii (*biceps* = 2 heads; *brachii* = arm)	Large muscle with 2 heads. *Flexes and supinates forearm, and flexes arm.*
Brachialis	Deep to biceps brachii and anterior to humerus. *Flexes forearm.*
POSTERIOR SURFACE	
Triceps brachii	Large muscle with three heads. **Long head:** Superficial muscle on medial side. **Medial head:** Deep muscle on medial side. **Lateral head:** Lateral side. *All three heads extend forearm and help extend arm.*

Biceps brachii

Brachialis

1

2

DANK DANK

(a) Anterior view

Triceps brachii
long head

Triceps brachii
lateral head

Triceps brachii
medial head

3

4

5

(b) Posterior view

FIGURE 14.5 Muscles of the arm.

(a)
• biceps brachii (BI-ceps
 BRAY-key-eye)
• brachialis (bray-key-AL-is)

1 _____

2 _____

(b)
• triceps brachii, lateral head
• triceps brachii, long head
• triceps brachii, medial
 head

3 _____

4 _____

5 _____

TABLE 14.5A Muscles of the Anterior Forearm

MUSCLE	LOCATION AND *FUNCTION*
MUSCLES THAT MOVE THE RADIUS AND ULNA	
Superficial muscles	
Brachioradialis (*brachi-* = arm; *radialis* = radius)	Large, superficial lateral muscle. *Flexes forearm and pronates and supinates forearm to neutral position.*
Pronator teres (*pronare* = to bend forward; *teres* = long, cylindrical)	Lateral to brachioradialis with oblique muscle fibers that cross antecubital fossa. *Pronates and weakly flexes forearm.*
Deep muscle	
Supinator (*supinate* = supine; palm upward)	Anterior portion is deep to brachioradialis. *Supinates forearm.*
MUSCLES THAT MOVE THE HAND	
Superficial muscles of anterior (flexor) compartment *(Lateral to medial)*	
Flexor carpi radialis (*carpi-* = wrist)	Medial and inferior to pronator teres and medial to brachioradialis in forearm. *Flexes and abducts hand.*
Palmaris longus (*palma* = palm)	Medial to flexor carpi radialis. *Weakly flexes wrist.*
Flexor carpi ulnaris	Small part of muscle is seen anteriorly; most medial muscle of posterior forearm. *Flexes and adducts hand.*
Flexor digitorum superficialis	Deep to tendons of flexor carpi radialis, palmaris longus, and flexor carpi ulnaris. *Flexes hand and proximal and middle phalanx of each finger.*
Deep muscle of anterior (flexor) compartment	
Flexor pollicis longus	Inferior and lateral to flexor digitorum superficialis. *Flexes distal phalanx of thumb.*

Biceps brachii
Brachialis
Medial epicondyle of humerus
Pronator teres
Brachioradialis
Flexor carpi radialis
Palmaris longus
Flexor carpi ulnaris
Flexor digitorum superficialis
Flexor pollicis longus
Pisiform bone
Abductor digiti minimi
Flexor digiti minimi brevis

Abductor
pollicis brevis
Flexor
pollicis brevis
Adductor
pollicis

1
2
3
4
5

(a) Anterior superficial view

Supinator
Flexor
digitorum
profundus
Flexor
pollicis
longus
Carpal tunnel
Opponens
digiti minimi
Dorsal interossei

6
7
8

(b) Anterior deep view

FIGURE 14.6 Muscles of the forearm, wrist, and hands.

(a)
• brachioradialis
• flexor carpi (KAR-pee) radialis
• flexor carpi ulnaris
• palmaris (pall-MARE-is) longis
• pronator (PRO-nay-tor) teres

1 _____
2 _____
3 _____
4 _____
5 _____

(b)
• flexor digitorum profundus
• flexor pollicis (POL-ih-kis) longus
• supinator

6 _____
7 _____
8 _____

TABLE 14.5B Muscles of the Posterior Forearm

MUSCLE	LOCATION AND *FUNCTION*
MUSCLE THAT MOVES THE RADIUS AND ULNA	
Supinator	Posterior portion is deep to lateral muscles of posterior forearm. *Supinates forearm.*
MUSCLES THAT MOVE THE HAND	
Superficial muscles of posterior (extensor) compartment (Medial to lateral)	
Extensor carpi ulnaris	Lateral to flexor carpi ulnaris. *Extends and adducts hand.*
Extensor digitorum minimi	Lateral to extensor carpi ulnaris. *Extends hand and proximal phalanx of fifth digit.*
Extensor digitorum	Lateral to extensor digitorum minimi. *Extends hand and proximal, middle, and distal phalanges.*
Extensor carpi radialis brevis	Lateral to extensor digitorum. *Extends and abducts hand.*
Extensor carpi radialis longus	Lateral to extensor radialis brevis. *Extends and abducts hand.*
Deep muscle of posterior (extensor) compartment	
Extensor pollicis longus	Deep to lateral extensor muscles on posterior forearm. *Extends distal phalanx of thumb and first metacarpal of thumb and abducts hand.*

Medial epicondyle

Flexor carpi ulnaris

Extensor retinaculum

Extensor carpi radialis longus

Extensor carpi ulnaris

Extensor carpi radialis brevis

Extensor digitorum

Extensor digiti minimi

Abductor pollicis longus

Extensor pollicis brevis

9

10

11

12

13

14

(c) Posterior superficial view

- extensor carpi radialis brevis
- extensor carpi radialis longus
- extensor carpi ulnaris
- extensor digitorum
- extensor digiti minimi
- flexor carpi ulnaris

9 _____

10 _____

11 _____

12 _____

13 _____

14 _____

Supinator

Abductor pollicis longus

Extensor pollicis longus

Extensor pollicis brevis

(d) Posterior deep view

15

16

- extensor pollicis longus
- supinator (SOUP-ih-nay-tor)

15 _____

16 _____

FIGURE 14.6 Muscles of the forearm, wrist, and hands, *continued.*

Muscles of the Thigh

Muscles of the thigh flex, extend, abduct, adduct, circumduct, and rotate the thigh. Muscles that move the thigh have origins on the pelvis or vertebrae and insertions on the femur or tibia. Those with origins on the pelvis move the thigh at the hip joint and the leg at the knee joint, whereas those with origins on the femur only move the leg at the knee joint. Muscles of the **anterior compartment** of the thigh, the quadriceps femoris group, extend the leg. The rectus femoris muscle of this group also flexes the thigh. The sartorius which runs diagonally from the hip to the medial tibia marks the boundary between the anterior and medial compartments. Muscles of the **medial compartment** are: pectineus, adductor magnus, adductor longus, adductor brevis, and gracilis. These muscles adduct and flex the thigh. Muscles of the **posterior compartment** are the hamstring muscle group, which flex the leg and extend the thigh.

ACTIVITY 4 **Muscles of the Thigh**

1 Review muscle location and action in Table 14.6.

2 Label Figure 14.7(a), (b), (c), and (d).

3 Use Figure 14.7 and Table 14.6 to point to the muscles on models, diagrams, and/or cadaver photographs. Pronounce the names as you point to them.

4 Locate the muscles on yourself and perform the actions indicated in Table 14.6. Palpate the muscles while performing the action.

5 Optional: Color black and white muscles in contrasting colors. Color in the same direction as the muscle fibers.

6 Optional: Use Appendix C to locate the origin and insertion on an articulated skeleton. Pull the insertion toward the origin to simulate muscle action.

TABLE 14.6 **Muscles of the Thigh**

MUSCLE	LOCATION AND *FUNCTION*
ANTERIOR SURFACE (*Lateral to medial*)	
Tensor fasciae latae (*tensor* = to stretch; *fascia* = band; *latus* = wide)	Small lateral hip muscle with fascia lata (deep fascia) that extends down thigh. *Flexes and abducts thigh.*
Quadriceps femoris (*quad-* = four; *cep-* = head or origins)	Four-part muscle on anterior thigh.
Rectus femoris	Located along midline of thigh. *Extends leg and flexes thigh.*
Vastus lateralis	Lateral to rectus femoris. *Extends leg.*
Vastus medialis	Medial to rectus femoris. *Extends leg.*
Vastus intermedius	Deep to rectus femoris and intermediate to vastus laterali and vastus medialis. *Extends leg.*
Sartorius (*sartor* = tailor)	Diagonal muscle. Extends from anterior superior iliac spine to medial surface of tibia. *Flexes leg and flexes, abducts, and laterally rotates thigh (allows us to flex and cross our legs).*
Iliopsoas	Combination of psoas major and iliacus.
Psoas major (*psoas* = loin) **Iliacus** (*iliacus* = ilium)	A small portion of these muscles are observed in anterior/medial compartment of thigh inferior to inguinal ligament. *Together flex thigh, rotate thigh laterally, and flex trunk.*
Pectineus (*pecten* = comb)	Medial to iliopsoas. *Adducts and flexes thigh.*
Adductor longus	Medial to pectineus. *Adducts and flexes thigh and medially rotates thigh.*
Adductor magnus	Observed medial and inferior to adductor longus. A larger portion of this muscle is deep to the adductor longus. *Adducts and medially rotates thigh. Anterior portion flexes thigh and posterior portion extends thigh.*
Adductor brevis	Deep to adductor longus, it is totally covered by superficial muscles in figure. *Adducts and flexes thigh and medially rotates thigh.*
Gracilis	Medial to adductor magnus; straight muscle on inside of thigh. *Adducts thigh and medially rotates thigh. Flexes leg.*

Iliopsoas
- Psoas major
- Iliacus

Tensor fasciae latae
Sartorius
Rectus femoris (cut)
Iliotibial tract
Vastus lateralis
Vastus intermedius
Vastus medialis

Pectineus
Adductor longus
Gracilis
Adductor magnus

(a) Anterior superficial view

- adductor longus
- adductor magnus
- gracilis (grah-SIL-us)
- iliacus (il-ee-AK-us)
- pectineus (pek-TIN-ee-us)
- psoas major (SO-us)
- rectus femoris (FEM-or-is)
- sartorius (sar-TOR-ee-us)
- tensor fasciae latae (FA-schee LA-tee)
- vastus intermedius
- vastus lateralis
- vastus medialis

1 _____ 7 _____

2 _____ 8 _____

3 _____ 9 _____

4 _____ 10 _____

5 _____ 11 _____

6 _____ 12 _____

Inguinal ligament
Obturator externus
Gracilis
Adductor brevis
Adductor magnus
Adductor longus (cut)

- adductor brevis
- adductor magnus
- gracilis

13 _____

14 _____

15 _____

(b) Anterior deep view

FIGURE 14.7 Muscles of the thigh.

TABLE 14.6 (*Continued*)

MUSCLE	LOCATION AND *FUNCTION*
POSTERIOR SURFACE	
Gluteus maximus	Largest buttocks muscle. *Extends and laterally rotates thigh.*
Gluteus medius	Observed superior to gluteus maximus but is partially covered by it. *Abducts and medially rotates thigh.*
Gluteus minimus	Deep muscle covered by gluteus medius. *Abducts and medially rotates thigh.*
Piriformis (*pirum* = pear, *-forma* = shape)	Deep muscle inferior to gluteus minimus; important landmark for sciatic nerve which passes deep to piriformis. *Laterally rotates and abducts thigh.*
Hamstrings	Three muscles on posterior thigh.
Biceps femoris	Most lateral hamstring. *Flexes leg and extends thigh.*
Semitendinosus (*semi-* = half or part; *tendinosus* = long tendon)	Medial to biceps femoris and superficial to semimembranosus. *Flexes leg and extends thigh.*
Semimembranosus (*membranosus* = partly membrane)	Most medial of the hamstrings; (remember 'm' for medial). *Flexes leg and extends thigh.*

- biceps femoris
- gluteus maximus
- gluteus medius
- semimembranosus
- semitendinosus

16 _____

17 _____

18 _____

19 _____

20 _____

(c) Posterior superficial view

Coccyx

- gluteus minimus
- piriformis

21 _____

22 _____

(d) Posterior deep view

FIGURE 14.7 Muscles of the thigh, *continued.*

Muscles of the Leg and Foot

Palpate the tibia of your leg to find the anterior crest and the lateral and medial borders of the tibia. Observe that the muscles on the anterior surface of the tibia are located lateral to the anterior crest of the tibia. The anterior and lateral superficial muscles are in order starting at the anterior crest just below the knee and moving laterally: tibialis anterior, extensor digitorum longus, and the fibularis (peroneus) longus. The tibialis anterior and extensor digitorum are muscles of the **anterior compartment** of the leg. These muscles dorsiflex the foot and also extend the toes if they insert on the phalanges.

Muscles of the **posterior compartment** of the leg include the gastrocnemius, soleus, and flexor digitorum longus. These muscles plantar flex the foot and also flex the toes if they insert on the phalanges. Eversion of the foot is due to contraction of muscles near the lateral surface of the leg, whereas inversion is due to contraction of muscles near the medial surface of the leg.

ACTIVITY 5 Muscles of the Leg and Foot

1 Label Figure 14.8(a), (b), (c), and (d).

2 Use Figure 14.8 and Table 14.7 to point to the muscles on models, diagrams, and/or cadaver photographs. Pronounce the names as you point to them.

3 Locate the muscle on yourself and perform the action indicated in Table 14.7. Palpate the muscle while performing the action.

4 Optional: Color black and white muscles in contrasting colors. Color in the same direction as the muscle fibers.

5 Optional: Use Appendix C to locate the origin and insertion on an articulated skeleton. Pull the insertion toward the origin to simulate muscle action.

TABLE 14.7 Muscles of the Leg and Foot

MUSCLE	LOCATION AND *FUNCTION*
LATERAL AND ANTERIOR SURFACES (*Superficial muscles*)	
Tibialis anterior	Lateral to the anterior crest of the tibia. *Dorsiflexes and inverts foot.*
Extensor digitorum longus	Lateral to tibialis anterior. *Dorsiflexes foot and extends toes.*
Fibularis (peroneus) longus	Lateral to the extensor digitorum longus. *Plantar flexes and everts foot.*
POSTERIOR SURFACE (*Superficial muscles*)	
Gastrocnemius	Large superficial muscle on posterior leg. *Plantar flexes foot and flexes leg.*
Soleus	Deep to gastrocnemius and posterior to fibularis (peroneus) longus. *Plantar flexes foot only.*
(*Deep muscles*)	
Flexor digitorum longus	Deep to soleus, medial muscle (not shown on figure). *Plantar flexes foot and flexes toes.*
Flexor hallucis longus	Deep to soleus, lateral muscle. *Plantar flexes foot and flexes great toe.*

- extensor digitorum longus
- flexor digitorum longus
- gastrocnemius (gas-trok-
 NEE-mee-us)
- fibularis (peroneus) longus
- soleus (SOLE-ee-us)
- tibialis anterior

1 _____

2 _____

3 _____

4 _____

5 _____

6 _____

Tibialis anterior

Extensor digitorum longus

Fibularis (peroneus) longus

Fibularis (peroneus) brevis

Gastrocnemius

Soleus

Flexor digitorum longus

Tendon of extensor hallucis longus

1
2
3

4
5

6

(a) Anterior superficial view

- extensor digitorum longus
- gastrocnemius
- fibularis (peroneus) longus
- soleus
- tibialis anterior

7 _____

8 _____

9 _____

10 _____

11 _____

Plantaris

Gastrocnemius

Soleus
fibularis (peroneus) longus

Extensor digitorum longus

Tibialis anterior

7
8
9
10
11

(b) Lateral superficial view

FIGURE 14.8 Muscles of the leg, ankle, and foot.

Plantaris

Gastrocnemius

Soleus

Fibularis (peroneus) longus

Flexor hallucis longus

Calcaneal tendon

(c) Posterior superficial view

- gastrocnemius
- soleus

12 _____

13 _____

12

13

Tibialis posterior

Flexor digitorum longus

Flexor hallucis longus

(d) Posterior deep view

- flexor digitorum longus
- flexor hallucis longus (HAL-a-kis)

14 _____

15 _____

14

15

FIGURE 14.8 Muscles of the leg, ankle, and foot, *continued*.

Reviewing Your Knowledge

A. Skeletal Muscle Identification

Go to www.wiley.com/college/allen and click on the picture of your lab manual. Then click on Student Companion Site, Anatomy Drill and Practice, and Exercise number 14.

B. Muscle Function

Identify the muscles that perform each function.

Muscles of the Head and Neck

a. masseter
b. orbicularis oculi
c. orbicularis oris
d. sternocleidomastoid
e. temporalis
f. trapezius
g. zygomaticus major

____ **1.** Smiling muscle.

____ **2.** Kissing muscle.

____ **3.** Closes eyelid.

____ **4.**
⎫
⎬ Two muscles that close mouth.
____ **5.** ⎭

____ **6.** Paired muscle that flexes head and rotates head to side.

____ **7.** Extends head.

Muscles of the Anterior Trunk

a. deltoid
b. external oblique
c. internal oblique
d. pectoralis major
e. rectus abdominis
f. serratus anterior
g. transversus abdominis

_____ **8.** Adducts and flexes arm.

_____ **9.** Anterior portion flexes arm; lateral portion abducts arm.

_____ **10.** Abducts scapula and rotates it upward (boxer's muscle).

_____ **11.** Flexes vertebral column and compresses abdomen.

_____ **12.**
_____ **13.** } Two muscle pairs that flex vertebral column, compress abdomen, and laterally flex vertebral column.

_____ **14.** Compresses abdomen only.

Muscles of the Posterior Trunk

a. deltoid
b. erector spinae
c. latissimus dorsi
d. teres major
e. trapezius

_____ **15.**
_____ **16.** } Extends, adducts, and medially rotates arm.

_____ **17.** Posterior portion extends arm; lateral portion abducts arm.

_____ **18.** Superior portion elevates scapula, middle portion adducts scapula, inferior portion depresses scapula.

_____ **19.** Paired muscle that extends vertebral column, maintains erect posture, and laterally flexes vertebral column.

Muscles of the Arm and Forearm

a. biceps brachii
b. brachialis
c. brachioradialis
d. extensor carpi radialis
e. extensor carpi ulnaris
f. extensor digitorum
g. flexor carpi radialis
h. flexor carpi ulnaris
i. palmaris longus
j. triceps brachii

_____ **20.** Extends forearm at elbow and extends arm.

_____ **21.** Flexes forearm at elbow and flexes arm.

_____ **22.** Flexes forearm.

_____ **23.** Flexes forearm and pronates and supinates forearm.

_____ **24.** Flexes and abducts hand.

_____ **25.** Flexes and adducts hand.

_____ **26.** Weakly flexes hand.

_____ **27.** Extends hand and extends phalanges.

_____ **28.** Extends and adducts hand.

_____ **29.** Extends and abducts hand.

Muscles of the Thigh

a. adductor magnus, longus, brevis
b. biceps femoris
c. gluteus maximus
d. gluteus medius
e. gracilis
f. rectus femoris
g. sartorius
h. semimembranosus
i. semitendinosus
j. tensor fasciae latae
k. vastus intermedius
l. vastus lateralis
m. vastus medialis

_____ **30.** Extends leg at knee and flexes thigh at hip.

_____ **31.**
_____ **32.** } Three muscles that extend leg only.
_____ **33.**

_____ **34.** Flexes leg and flexes, abducts, and laterally rotates thigh (allows us to flex and cross our legs).

_____ **35.** Adducts thigh and flexes leg.

_____ **36.** Group of muscles that adducts and flexes thigh.

_____ **37.**
} Abduct thigh.
_____ **38.**

_____ **39.** Extends thigh.

_____ **40.**
_____ **41.** } Three muscles that flex leg.
_____ **42.**

Muscles of the Leg

a. extensor digitorum longus
b. fibularis (peroneus) longus
c. flexor digitorum longus
d. gastrocnemius
e. soleus
f. tibialis anterior

_____ **43.** Plantar flexes and everts foot.

_____ **44.** Plantar flexes foot and flexes toes.

_____ **45.** Plantar flexes foot only.

_____ **46.** Plantar flexes foot and flexes leg.

_____ **47.** Dorsiflexes foot and extends toes.

_____ **48.** Dorsiflexes and inverts foot.

Using Your Knowledge

Identify the muscles and/or actions in the following questions.

1. Identify the *muscles of the head* that you use when you *whistle.*

2. Identify the *muscles of the head* used to *blink.*

3. Identify the major *thigh* and *leg* muscles you use when you *kick* a ball.

4. Give the action of each muscle listed in question 3.

5. Identify the *forearm* muscles you use when you *turn* a key in a lock.

6. Give the action of each muscle listed in question 5.

7. Identify the major *arm* muscles you use when you *lift* a glass to drink from it.

8. Give the action of each muscle listed in question 7.

9. Identify the *neck* muscles you use when you *look both ways* to cross a street.

10. Give the action of each muscle listed in question 9.

11. Identify the *abdominal* muscles you use when you do the dance called the *"twist".*

12. Give the action of each muscle listed in question 11.

13. Identify the muscles used to move the *upper limb* when you do *jumping jacks.*

14. Give the action of each muscle listed in question 13.

15. Identify the muscles that move the *lower limb* when you do *jumping jacks.*

16. Give the action of each muscle listed in question 15.

17. Identify the muscles used to bend the *trunk* when you *touch your toes.*

18. Give the action of each muscle listed in question 17.

19. Identify the *upper limb* muscles used to *play* the cymbals in a band.

20. Give the action of each muscle listed in question 19.

Surface Anatomy

Objectives

After completing this exercise, you should be able to:

- Define and describe the importance of surface anatomy
- Locate and identify important anatomical landmarks on the external surface of the body
- Identify the borders of the anterior and posterior triangles of the neck and the femoral triangle

Materials

- hand mirror
- articulated skeleton
- skeletal muscle models or charts

Surface anatomy is the study of anatomical landmarks observed on the external surface of the body. These landmarks can be used to locate underlying structures by palpation. **Palpation** (*palpare* = to touch gently) is a technique that uses the hands or fingers to locate internal body structures and to determine the size and texture of the structures.

ACTIVITY 1 Surface Anatomy of the Head and Neck

1 Read the location and description of each structure.
2 Locate and label each structure in Figure 15.1(a), (b), (c), and (d).
3 Palpate the designated structures on your body.

A. Anterior Surface of the Head

Structures located in Figure 15.1(a).

- **frontalis muscle**—Lies over the forehead. Palpate the frontalis muscle with fingers while raising eyebrows.
- **supraorbital margins**—Superior borders of the frontal bone that border the eye orbits. Palpate the supraorbital margins.
- **nasal bone**—Forms the bridge of the nose. Place fingers along the bridge of the nose to feel nasal bones. Find the anterior border of the nasal bones and palpate the **nasal cartilage** inferior to nasal bones.
- **zygomaticus major muscle**—Originates on the zygomatic bone and inserts on the corner of the mouth. Palpate this muscle near the zygomatic bone while smiling.
- **body of mandible**—Horizontal portion of the lower jawbone. Palpate this main part of the mandible that includes the chin.
- **mental protuberance**—Palpate this anterior tip of the chin.

B. Lateral and Posterior Surfaces of the Head

Structures located in Figure 15.1(b).

- **temporomandibular joint** (TMJ)—Located anterior to the external auditory meatus of the ear. Palpate the joint while opening and closing the mouth.
- **ramus of mandible**—Vertical process of mandible. Palpate ramus inferior to the TMJ. The rounded **condylar process** (mandibular condyle) at the posterior portion of the ramus articulates with the mandibular fossa of the temporal bone to form the TMJ.
- **masseter muscle**—Located anterior to ramus of the mandible. Palpate the masseter muscle while closing the mouth and clenching your teeth.
- **temporalis muscle**—Located superior to the ear. Palpate the temporalis muscle while closing the mouth and clenching your teeth.
- **mastoid process**—Rounded projection on the inferior portion of the temporal bone posterior to the ear. Palpate area of the skull inferior and posterior to external auditory meatus of ear.
- **external occipital protuberance**—Rounded projection superior to foramen magnum of the occipital bone. Palpate the base of the skull near the midline. Lateral to the external occipital protuberance are two curved ridges called the **superior nuchal lines** which mark the boundary between the head and neck.
- **occipitalis muscle**—Lies over the inferior portion of the occipital bone. Firmly palpate the posterior surface of the skull immediately above the neck as you raise and lower eyebrows.

Orbicularis oculi muscle

Depressor labii inferioris muscle

Orbicularis oris muscle

(a) Anterior view of the head

(b) Right lateral view of the head

FIGURE 15.1 Surface anatomy of the head and neck.

(a) Anterior view of head
• body of mandible
• frontalis muscle
• mental protuberance
• nasal bone
• supraorbital margin
• zygomaticus major muscle

1 _____

2 _____

3 _____

4 _____

5 _____

6 _____

(b) Right lateral view of head
• external occipital protuberance
• masseter muscle
• mastoid process
• occipitalis muscle
• ramus of mandible
• temporalis muscle
• temporomandibular joint

7 _____

8 _____

9 _____

10 _____

11 _____

12 _____

13 _____

C. Anterior Surface of the Neck

Structures located in Figure 15.1(c).

- **hyoid bone**—Located in the anterior neck between the mandible and larynx. With the head in anatomical position, palpate the hyoid bone by placing the thumb and middle finger of one hand on either side of the neck about 1 inch inferior to the mandible. This bone can be moved laterally from side to side.
- **thyroid cartilage**—Largest cartilage of the larynx that has a prominence called the Adam's apple. Located in anterior neck inferior to hyoid bone. Move fingers 1 inch inferiorly from the hyoid bone until a firm structure is reached. Swallow to feel this cartilage move superiorly.
- **cricoid cartilage**—Located inferior to the thyroid cartilage and used as a landmark when locating the trachea during a tracheostomy. Move your fingers down the thyroid cartilage. There is a depression between the thyroid cartilage and the cricoid cartilage. Palpate the cricoid cartilage inferior to this depression.
- **thyroid gland**—Located inferior to the larynx on either side of trachea. Palpate by placing fingers on neck inferior and lateral to the thyroid cartilage and feeling for a soft mass.
- **sternocleidomastoid muscle**—Originates on the sternum and clavicle and inserts on the mastoid process of the temporal bone. Look into a mirror while turning your head to the side. Observe the sternal and clavicular heads of the sternocleidomastoid muscle and palpate the muscle.
- **trapezius muscle**—Located in the posterior and lateral neck. Inflammation of trapezius muscle may result in "stiff neck." Place fingers on the posterior portion of the lateral neck and palpate while flexing and extending the neck.

- **common carotid arteries**—Located in the lateral neck between the trachea and the sternocleidomastoid muscle. Branch into the external and internal carotid arteries at the superior border of the larynx. Palpate the right common carotid artery by placing your fingers on the right side of the neck just lateral to the trachea.
- **external jugular vein**—Located lateral to the sternocleidomastoid muscle. To observe this vein, look in the mirror while clenching your teeth as in anger or placing fingers on the skin above the clavicle and pressing firmly to prevent blood from draining the external jugular vein.
- **suprasternal (jugular) notch**—Located at the base of the neck between the sternal heads of the sternocleidomastoid muscles and just superior to the sternum. Palpate the sternum and move the fingers superiorly to feel suprasternal notch.

D. Anterior and Posterior Triangles of the Neck

Structures located in Figure 15.1(d).

- **anterior triangle**—Superiorly bounded by the inferior margin of the mandible, anteriorly by the midline of the neck, and posteriorly by the sternocleidomastoid. The **carotid artery** and **internal jugular vein** are located in this triangle. Palpate to identify the triangle's borders and palpate your carotid pulse in the anterior triangle.
- **posterior triangle**—Anteriorly bordered by the sternocleidomastoid, posteriorly by the trapezius, and inferiorly by the clavicle. Palpate to identify the triangle's borders. The **brachial plexus** and **external jugular vein** are located in this triangle.

(c) Anterior view of the neck

1 ——————————

Thyroid gland ——————
2 ——————
Subclavian artery ——————
Clavicle ——————

3 ——————
External jugular vein
4 ——————
5 ——————
6 ——————
7 ——————

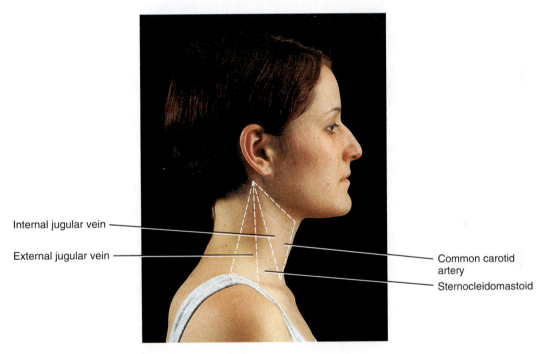

Internal jugular vein ——————

External jugular vein ——————

Common carotid artery
Sternocleidomastoid

(d) Anterior and posterior triangles of the neck

FIGURE 15.1 Surface anatomy of the head and neck, *continued.*

- common carotid artery
- cricoid cartilage
- hyoid bone
- sternocleidomastoid muscle
- suprasternal (jugular) notch
- thyroid cartilage
- trapezius muscle

1 _____

2 _____

3 _____

4 _____

5 _____

6 _____

7 _____

1 Read the location and description of each structure.

2 Locate and label each structure in Figure 15.2(a), (b), and (c).

3 Palpate the designated structures on your body.

E. Surface Anatomy of the Chest

Structures located in Figure 15.2(a).

- **manubrium of the sternum**—Superior part of the sternum between the suprasternal notch and body of the sternum. Palpate by placing fingers at suprasternal notch and moving them inferiorly until a ridge (sternal angle) is reached.
- **sternal angle**—Raised area that can be felt at border of manubrium and sternum. The trachea divides into the **right** and **left bronchi** at the level of the sternal angle. The sternal angle is located at the level of the second rib. Palpate the sternal angle and move fingers laterally until the **second rib** can be felt. Move fingers inferiorly counting each rib and each intercostal space between the ribs.
- **ribs**—The ribs can be palpated lateral to the sternum. The location of the second, third, fifth, and sixth ribs is used to identify the location of the **heart.**
- **costal margin**—Anterior edge of costal cartilage of ribs 7–10. The **liver** and **gallbladder** are located inferior to the right costal margin.
- **body of sternum**—Palpate area of the sternum inferior to the sternal angle.
- **xiphisternal joint**—Joint between body of the sternum and xiphoid process of the sternum. Palpate the costal margin. Move fingers anteriorly along the costal margin until they reach the superior edge of the costal margin. Medial to this point is the xiphisternal joint. The **heart** rests on the **diaphragm** deep to the xiphisternal joint.
- **xiphoid process of sternum**—Inferior portion of the sternum. Palpate the xiphoid process inferior to the xiphisternal joint.
- **pectoralis major muscle**—Major muscle of the chest. In males the inferior border can be observed as a curved line under the breasts. This line is at the level of the fifth rib. Bend over your lab bench and push yourself up with one limb. Use your opposite hand to palpate the pectoralis major muscle.
- **serratus anterior muscle**—Located on the lateral chest wall, extending from the ribs under the arm to the scapula. Flex your forearm and abduct your elbow. Use your opposite hand to palpate the serratus anterior muscle as it abducts the scapula. If you move too far posteriorly, you will be palpating the latissimus dorsi instead.

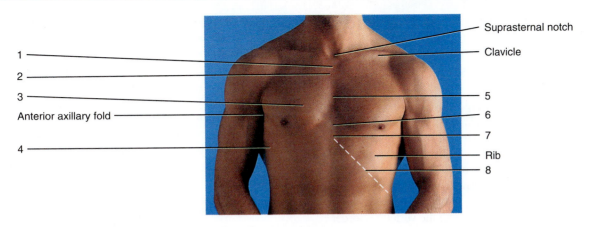

(a) Anterior view of the chest

FIGURE 15.2 Surface anatomy of the trunk.

- body of sternum
- costal margin
- manubrium of sternum
- pectoralis major muscle
- serratus anterior muscle
- sternal angle
- xiphisternal joint
- xiphoid process of sternum

1 _____

2 _____

3 _____

4 _____

5 _____

6 _____

7 _____

8 _____

F. Surface Anatomy of the Back

Structures located in Figure 15.2(b).

- **trapezius muscle**—Large muscle of the posterior neck and middle of back.
- **scapula (spine and vertebral border)**—With left arm at your side, run the fingers of your right hand medially from the point of left shoulder to feel the spine of the scapula. Draw left shoulder back and run the fingers of your right hand just lateral to the midline on the left side to feel the vertebral border of the scapula. Notice that the scapula moves depending on position of the upper limb.
- **infraspinatus muscle**—Located inferior to the spine of the scapula. Palpate area inferior to the spine of the scapula to feel this muscle.

- **teres major muscle**—Located inferior to the infraspinatus muscle.
- **latissimus dorsi muscle**—Broad muscle located between the lumbar region and axillary region. Together with teres major muscle forms the posterior axillary fold.
- **triangle of auscultation**—Triangular region of the back formed by the latissimus dorsi, trapezius, and vertebral border of scapula. This region is not covered by superficial muscles enabling respiratory sounds to be heard with a stethoscope.
- **erector spinae muscle**—Large muscle of lower back. Bend over and palpate this muscle while extending the vertebral column.
- **spinous processes of vertebrae**—Located along the midline of the back. Run your fingers along midline to feel the spinous processes.

(b) Posterior view of the back

FIGURE 15.2 Surface anatomy of the trunk, *continued.*

- erector spinae muscle
- infraspinatus muscle
- latissimus dorsi muscle
- spinous processes of vertebrae
- teres major muscle
- trapezius muscle
- triangle of auscultation
- vertebral border of scapula

1 _____

2 _____

3 _____

4 _____

5 _____

6 _____

7 _____

8 _____

G. Surface Anatomy of the Abdomen

Structures located in Figure 15.2(c).

- **rectus abdominis muscle**—Longitudinal muscles lateral to the linea alba. Lie back in your chair and palpate these muscles as you sit up.
- **external oblique muscle**—Inferior to serratus anterior muscles. Palpate while twisting trunk toward the opposite side.
- **umbilicus**—Most notable feature of the abdomen and a common incision site. The umbilicus is located between L_3 and L_4. The **abdominal aorta** bifurcates at the level of L_4 to form the **common iliac arteries.**
- **linea alba**—Tendinous raphe between the xiphoid process and pubic symphysis forming a vertical groove along midline. This is a common incision site because there is little damage to muscles and little bleeding.
- **tendinous intersection**—Transverse grooves across rectus abdominis muscles that are observable in muscular individuals.

- **linea semilunaris**—Lateral margin of rectus abdominis that is observable in lean people as a groove. The **gallbladder** is located deep to the spot where the linea semilunaris crosses the costal margin of the rib cage.
- **iliac crest**—Marks the inferior border of the abdomen. Palpate the iliac crest by placing your hands on your hips.
- **anterior superior iliac spine**—Located at the anterior end of the iliac crest. Palpate the iliac crest with your fingers and move your fingers medially until you feel the "bump" at the anterior end of the iliac crest.
- **McBurney's point**—The **appendix** is deep to McBurney's point, which is located along the line between the umbilicus and the right anterior superior iliac spine, about 1 to 2 inches away from the latter. During appendicitis, pressure on McBurney's point results in tenderness. This is the most common site of incision for an appendectomy.
- **pubic symphysis** [*not shown in Figure 15.2(c)*] Anterior joint between os coxae (hip bones). Palpate midline of the inferior pelvic area.

(c) Anterior view of abdomen

FIGURE 15.2 Surface anatomy of the trunk, *continued.*

- anterior superior iliac spine
- external oblique muscle
- iliac crest
- linea alba
- linea semilunaris
- McBurney's point
- rectus abdominis muscle
- serratus anterior muscle
- tendinous intersection
- umbilicus

1 _____

2 _____

3 _____

4 _____

5 _____

6 _____

7 _____

8 _____

9 _____

10 _____

ACTIVITY 3 **Surface Anatomy of the Upper Limb**

1 Read the location and description of each structure.

2 Locate and label each structure in Figure 15.3(a), (b), (c), (d), (e), and (f).

3 Palpate the designated structures on your body.

H. Surface Anatomy of the Shoulder

Structures located in Figure 15.3(a).

- **spine of scapula**—The ridge across the posterior surface of the scapula is the spine. Trace this bone, marking with your fingers from its medial border to its lateral border. The spine is easier to palpate on a lean body.

- **acromion of scapula**—The acromion is the flattened lateral end of the spine and is located at the peak of the shoulder. Palpate this prominent projection.

- **clavicle**—Starting at the manubrium of the sternum, trace the clavicle's S-shaped curvature laterally to its acromial end. Palpate the acromioclavicular joint just posterior to the acromial end of the clavicle as you thrust your shoulder joint anteriorly.

- **deltoid muscle**—This large, triangular muscle forms the rounded protrusion of the shoulder. It is used as a site for intramuscular injections. Palpate this muscle as you flex, abduct, and extend the arm to locate the deltoid's anterior, lateral, and posterior portions.

- **greater tubercle of humerus**—This bony landmark can be palpated through the deltoid muscle on the lateral part of the shoulder. As you bend your forearm at the elbow and abduct your arm, move your fingers a little anteriorly to feel the intertubercular groove where the proximal tendon of the long head of the biceps traverses. The greater tubercle is just posterior to this groove.

(a) Right lateral view of the shoulder

FIGURE 15.3 Surface anatomy of the upper limb.

- acromion of scapula
- clavicle
- deltoid muscle
- greater tubercle of humerus
- spine of scapula

1 _____

2 _____

3 _____

4 _____

5 _____

I. Surface Anatomy of the Arm and Elbow

Structures located in Figure 15.3(b).

Arm (from Shoulder to Elbow)

- **biceps brachii muscle**—Flex your forearm and tighten the biceps brachii as you palpate this muscle on the anterior surface of your arm.
- **groove for brachial artery**—Palpate the medial border of the flexed biceps brachii muscle. You can feel your **brachial pulse** if you press into this groove with your fingers.
- **triceps brachii muscle**—This muscle can be palpated on the posterior surface of the arm when the forearm is flexed against resistance. If a person has a muscular arm with definition, all three heads of the triceps can be distinguished.

Elbow

- **tendon of biceps brachii muscle**—The distal tendon of the biceps brachii can best be palpated with the forearm flexed. The tendon is located in the middle of the anterior arm and runs from just proximal of the **cubital fossa,** through the cubital fossa, to insert on the radial tuberosity.
- **olecranons of ulna**—With the forearm flexed, palpate this prominent point at the elbow. Extend and flex the forearm as you feel this bony landmark.
- **medial epicondyle of humerus**—Move your fingers medially from the olecranon to feel the medial epicondyle. The groove between the olecranon and the medial epicondyle has the **ulnar nerve** that you can palpate when the forearm is extended.

1
2
3
4
5

Deltoid muscle

Anterior axillary fold
Axilla
Posterior axillary fold

6

(b) Medial view of the arm and elbow

FIGURE 15.3 Surface anatomy of the upper limb, *continued.*

- biceps brachii muscle
- groove for brachial artery
- medial epicondyle of humerus
- olecranon of ulna
- tendon of biceps brachii muscle
- triceps brachii muscle

1 _____

2 _____

3 _____

4 _____

5 _____

6 _____

J. Lateral Surface of Arm and Elbow

Structures located in Figure 15.3(c).

- **Lateral epicondyle of humerus**—Move your fingers lateral of the olecranon to palpate the lateral epicondyle of the humerus.

K. Surface Anatomy of the Antecubetal Region

Structures located in Figure 15.3(d).

- **cephalic vein**—This superficial vein can be viewed on the lateral side of the anterior forearm and ascends to the arm. This vein is discernible under the skin of lean people.
- **basilic vein**—This superficial vein is seen on the medial side of the anterior forearm and ascends to the arm. This vein is discernible under the skin of lean people.
- **cubital fossa**—This triangular-shaped concavity is on the anterior surface of the elbow joint.
- **median cubital vein**—This superficial vein diagonally crosses the anterior surface of the elbow in the cubital (antecubital) fossa. This vein connects the lateral cephalic vein with the medial basilic vein. The median cubital vein is a typical site for drawing blood or for inserting an intravenous (IV) catheter.
- **brachioradialis muscle**—As its name indicates, this muscle originates in the distal arm (brachium), borders the lateral side of the cubital fossa, and extends to the proximal forearm to insert on the radius. Observe and palpate this muscle as you flex your forearm against resistance.

(c) Right lateral view of the arm and elbow

- biceps brachii muscle
- lateral epicondyle of humerus
- olecranon of ulna
- triceps brachii muscle

1 _____

2 _____

3 _____

4 _____

(d) Anterior view of the right cubital fossa

- basilic vein
- brachioradialis muscle
- cephalic vein (used twice)
- cubital fossa
- median cubital vein

1 _____

2 _____

3 _____

4 _____

5 _____

6 _____

FIGURE 15.3 Surface anatomy of the upper limb, *continued.*

L. Anterior Surface of the Wrist and Hand

The muscles of the forearm are difficult to differentiate from surface anatomy, but four of the muscle tendons are readily discernible in the anterior forearm.

- **flexor carpi radialis tendon**—Make a fist to observe and palpate this laterally located tendon.
- **radial artery**—Note the location of the radial artery just lateral to the flexor carpi radialis tendon. With three fingers, press down on the radial artery and feel your pulse.
- **palmaris longus tendon**—Make a fist to observe and palpate this center-most tendon that is located just medial to the flexor radialis tendon.

- **flexor digitorum superficialis tendon**—Make a fist to observe and palpate this tendon located medial to the palmaris longus tendon.
- **flexor carpi ulnaris tendon**—Make a fist to palpate this most medially located tendon. Tendon is not readily apparent.
- **pisiform bone**—Located on the medial side just distal to the wrist crease. Palpate this small bone that feels like a bump.
- **hypothenar eminence**—Palpate this small hand pad located medially just distal to the pisiform bone.
- **thenar eminence**—Palpate this larger elevation located between the wrist crease and the thumb.

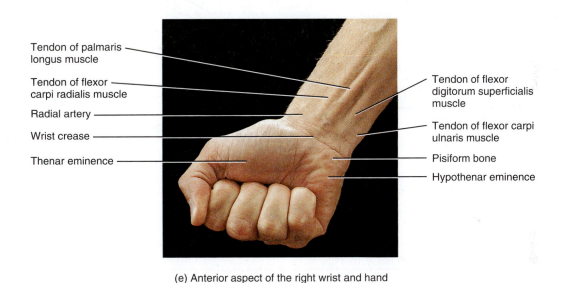

Tendon of palmaris longus muscle

Tendon of flexor carpi radialis muscle

Radial artery

Wrist crease

Thenar eminence

Tendon of flexor digitorum superficialis muscle

Tendon of flexor carpi ulnaris muscle

Pisiform bone

Hypothenar eminence

(e) Anterior aspect of the right wrist and hand

FIGURE 15.3 Surface anatomy of the upper limb, *continued.*

M. Dorsal Surface of the Wrist and Hand

Structures located in Figure 15.3(f).

> **NOTE:** *Remember that the directional terms refer to the anatomical position.*

- **head of ulna**—Palpate this distal end of the ulna on the medial side of the wrist.
- **styloid process of ulna**—Palpate just distal to the head of the ulna.

- **styloid process of radius**—Palpate this distal end of the radius on the lateral side of wrist.
- **cephalic vein** and **dorsal venous arch**—The dorsal venous arch on the dorsum of the hand drains into the cephalic vein. The plexus of veins in the dorsal venous arch is also a site for drawing blood and inserting an IV catheter.
- **anatomical snuffbox**—Palpate this depression located on the dorsum of the hand between the tendon of the extensor pollicis longus muscle and the tendon of the extensor pollicis brevis muscle.
- **tendons of extensor muscles**—Extend the fingers and palpate the tendons of the extensor digitorum muscles.

Head of ulna

Styloid process of ulna

Cephalic vein

Styloid process of radius

"Anatomical snuffbox"

Tendon of extensor pollicis brevis muscle

Tendon of extensor pollicis longus muscle

Dorsal venous arch

Tendons of extensor digitorum muscle

(f) Dorsum of the right wrist and hand

FIGURE 15.3 Surface anatomy of the upper limb, *continued.*

ACTIVITY 4 **Surface Anatomy of the Lower Limb**

1 Read the location and description of each structure.

2 Locate and label each structure in Figure 15.4(a), (b), (c), (d), (e), (f), and (g).

3 Palpate the designated structures on your body.

N. Surface Anatomy of the Buttocks

Structures located in Figure 15.4(a).

- **iliac crest**—Superior boundary of the ileum. The iliac crest can be palpated when you "put your hands on your hips."
- **gluteus maximus muscle**—Most students are familiar with these muscles that make up the largest portion of the buttocks. Palpate this prominent muscle as you extend your thigh.
- **gluteus medius muscle**—These muscles are inferior to the iliac crests in the upper, outer quadrant of the buttocks. Palpate this muscle as you shift your weight onto the palpated leg, causing the gluteus medius to contract. These muscles are the site of intramuscular (IM) injections.
- **greater trochanter**—Palpate this bony landmark on the lateral side of your hip. Flex and extend your thigh to feel this structure move with the joint action.
- **ischial tuberosity**—Sit down to palpate this structure that is near the gluteal folds of the inferior buttocks. This is the bony prominence of the hipbone that you sit upon.
- **gluteal (natal) cleft**—A midline crevice between the buttocks.
- **sacrum**—Superior to the gluteal cleft, palpate the medial sacral crest (fused spinous processes).
- **coccyx**—The tip of the coccyx can be palpated in the superior portion of the gluteal cleft.

(a) Posterior view of the buttocks

FIGURE 15.4 Surface anatomy of the lower limb.

- coccyx
- gluteal cleft
- gluteus maximus muscle
- gluteus medius muscle
- greater trochanter
- iliac crest
- ischial tuberosity
- sacrum

1 _____	5 _____
2 _____	6 _____
3 _____	7 _____
4 _____	8 _____

O. Anterior Surface of the Thigh

Structures located in Figure 15.4(b).

- **sartorius muscle**—Anteriorly located, this muscle diagonally spans the area between the anterior superior iliac spine laterally and the medial region of the knee. This muscle forms the lateral border of the femoral triangle.
- **adductor longus muscle**—Located in the superior part of the thigh in the femoral triangle, just medial to the sartorius.
- **rectus femoris muscle**—Located in the anterior compartment of the thigh, lateral to the sartorius.

This muscle is in the midline and is the most superficial of the quadriceps femoris muscle group.
- **vastus lateralis muscle**—Large muscle of the quadriceps femoris group located on the lateral surface of the thigh.
- **vastus medialis muscle**—Large muscle of the quadriceps femoris group located on the medial surface of the thigh.
- **gracilis muscle**—Located in the inner thigh region, medial to the femoral triangle. This muscle has fibers that run longitudinally from the pubic bone to the knee.
- **femoral triangle**—Triangle of the medial thigh bordered laterally by the sartorius and medially by the gracilis. The **femoral artery, vein,** and **nerve** traverse through this triangle. Palpate your femoral artery to feel your pulse.

Tensor fasciae latae muscle

1
2
3
4
5
6
7

(b) Anterior view of the right thigh

FIGURE 15.4 Surface anatomy of the lower limb, *continued.*

- adductor longus muscle
- femoral triangle
- gracilis muscle
- rectus femoris muscle
- sartorius muscle
- vastus lateralis muscle
- vastus medialis muscle

1 _____

2 _____

3 _____

4 _____

5 _____

6 _____

7 _____

P. Surface Anatomy of the Popliteal Region

Structures located in Figure 15.4(c).

- **popliteal fossa**—A depression located on the posterior of the knee.
- **biceps femoris muscle**—Lateral hamstring muscle. Palpate muscle while flexing the knee. Move your fingers inferiorly until you feel the ropelike **biceps femoris tendon** that crosses the knee joint.

- **semitendinosus and semimembranosus muscles**—Medial hamstring muscles. Palpate these muscles while flexing the knee. Move your fingers inferiorly until you feel the **tendon of semitendinosus muscle** that crosses the knee joint.

1

2

3

4

5

Gastrocnemius muscle (medial and lateral heads)

(c) Posterior view of the left popliteal fossa

FIGURE 15.4 Surface anatomy of the lower limb, *continued.*

- biceps femoris
- biceps femoris tendon
- popliteal fossa
- semitendinosus and semimembranosus muscles
- tendon of semitendinosus

1 _____

2 _____

3 _____

4 _____

5 _____

Q. Anterior Surface of the Knee

Structures located in Figure 15.4(d).

- **patella**—The patella or kneecap is the most anterior bone of the knee. Place your fingers on the patella and move it slightly.
- **tibial tuberosity**—Palpate this bump on the anterior surface of the proximal tibia just inferior to the patella.
- **patellar ligament**—The quadriceps femoris tendon extends beyond the patella as the patellar ligament and inserts on the tibial tuberosity. Press your fingers on the tendon between the patella and the tibial tuberosity to palpate the patellar ligament as you extend and flex the leg at the knee.
- **lateral condyle of femur**—Palpate the bump or projection on the distal femur lateral to the patella.
- **medial condyle of femur**—Palpate the bump or projection on the distal femur medial to the patella.
- **lateral condyle of tibia**—Palpate the bump or projection on the proximal tibia lateral and inferior to the patella.
- **medial condyle of tibia**—Palpate the bump or projection on the proximal tibia medial and inferior to the patella.

(d) Anterior view of the right knee

FIGURE 15.4 Surface anatomy of the lower limb, *continued.*

- lateral condyle of femur
- lateral condyle of tibia
- medial condyle of femur
- medial condyle of tibia
- patella
- patellar ligament
- tibial tuberosity

1 _____

2 _____

3 _____

4 _____

5 _____

6 _____

7 _____

R. Anterior Surface of the Leg and Ankle

Structures located in Figure 15.4(e).

- **anterior border of tibia** (shin)—With your fingers on the tibial tuberosity, slide down the anterior border of the tibia, better known as the shin.
- **tibialis anterior muscle**—Move your fingers laterally from the anterior border of the tibia. Palpate this muscle as you dorsiflex your foot. Note the thick tendon of this muscle as it crosses medially at the ankle to insert on the medial cunieform bone and the first metatarsal.
- **fibularis (peroneus) longus muscle**—This lateral muscle of the leg is superficial to the fibula and can be palpated as you plantar flex the foot. Its tendon inserts on the plantar surface of the foot.
- **lateral malleolus of fibula**—Palpate the projection at the lateral side of the ankle. This bump is formed by the lateral malleolus of the fibula.
- **medial malleolus of tibia**—Palpate the projection on the medial side of the ankle. This bump is formed by the medial malleolus of the tibia.

(e) Anterior view of the leg and ankle

FIGURE 15.4 Surface anatomy of the lower limb, *continued.*

- anterior border of tibia
- lateral malleolus of fibula
- medial malleolus of tibia
- fibularis (peroneus) longus muscle
- tibialis anterior muscle

1 _____

2 _____

3 _____

4 _____

5 _____

S. Posterior Surface of the Leg and Ankle

Structures located in Figure 15.4(f).

- **gastrocnemius muscle** (medial and lateral heads)—Two large bellies of the calf muscles that form most of the superior portion of the calf. To palpate these two heads, extend the knee and place your fingers over the medial and lateral heads of the gastrocnemius. Plantar flex the foot.
- **soleus muscle**—The flatter calf muscle that lies deep to the gastrocnemius. This muscle extends laterally from the gastrocnemius and its tendon is in the mid- to distal portion of the calf. Palpate the soleus as you plantar flex the foot.
- **calcaneal (Achilles) tendon**—The tendons of both the gastrocnemius and the soleus merge to insert on the calcaneus bone of the foot. Palpate this thick tendon as you alternately plantar flex and dorsiflex the foot.
- **calcaneus**—Heel bone of the foot.

(f) Posterior view of the leg and ankle

FIGURE 15.4 Surface anatomy of the lower limb, *continued.*

- calcaneal (Achilles) tendon 1 _____
- calcaneus
- gastrocnemius muscle 2 _____
 (medial and lateral heads)
- soleus muscle 3 _____

 4 _____

T. Dorsal Surface of the Foot

Structures located in Figure 15.4(g).

- **dorsal venous arch**—Palpate the dorsal venous arch on the dorsum of the foot.
- **tendons of the foot**—Palpate the tendons of the **extensor digitorum longus muscle** and the tendon of the **extensor hallucis longus muscle** on the dorsum of the foot as you extend your toes.

Medial malleolus of tibia

Lateral malleolus of fibula

Dorsal venous arch

Tendons of extensor digitorum longus muscle

Tendon of extensor hallucis longus muscle

(g) Dorsum of foot

FIGURE 15.4 Surface anatomy of the lower limb, *continued.*

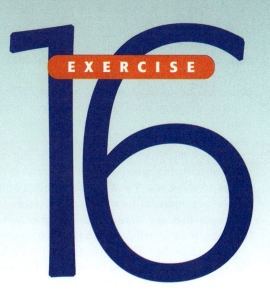

Nervous Tissue

Objectives

**After completing this exercise,
you should be able to:**

- Describe the functions of the nervous system
- Name the two major divisions of the nervous system and their organs
- Explain the difference in function between neurons and neuroglia
- Identify neuron structures and describe their functions
- Describe the reception of a nervous impulse by a neuron
- Describe the generation and transmission of a nervous impulse to other neurons or target organs
- Describe how neurons are classified structurally and functionally
- Identify unipolar, bipolar, and multipolar neurons on photomicrographs or prepared slides

Materials

- models or charts of the nervous system
- models or charts of a neuron and neuroglia
- compound microscope, lens paper, prepared slides of astrocytes, motor neurons, dorsal root ganglia, cerebral cortex, and teased, myelinated nerve fibers
- section of brain (human or animal)
- section of spinal cord (human or animal)

Nervous tissue is found in the organs of the nervous system—the nerves, brain, and spinal cord—and contains cells that enable the nervous system to generate and transmit electrical signals called nerve impulses or action potentials.

A. Overview of Nervous System

The nervous system senses changes in our internal and external environments, coordinates and integrates this data, and initiates and transmits action potentials. It is organized

into two basic components; the central nervous system (CNS), which consists of the brain and spinal cord, and the peripheral nervous system (PNS). The PNS contains an afferent division composed of sensory receptors and sensory neurons, and an efferent division composed of motor neurons.

Sensory receptors detect changes in the environment and transmit this information along sensory or afferent nerves to the CNS. The CNS coordinates and integrates information received from sensory receptors and initiates responses that are transmitted by neurons to effectors (neurons, muscle cells, or glands). Motor nerves transmit impulses from the CNS to effectors in the PNS. The nervous system is stream-lined to send rapid signals from cell to cell to maintain homeostasis and coordinate body organs and functions.

B. Structure of Nervous Tissue

Nervous tissue is located in the brain, spinal cord, ganglia, and nerves, and is composed of two types of cells, neurons and neuroglia. **Neurons** conduct action potentials and are the structural and functional units of nervous tissue. **Neuroglia** (*neuro-* = nerve; *-glia* = glue) are cells that support, protect, and furnish nutrients to neurons, and augment the speed of neuron transmission.

Neuroglia

Neuroglial cells are generally smaller and more abundant than neurons. Although they do not create action potentials, neuroglial cells have important roles in the nervous system. Of the six types of neuroglial cells, four are in the CNS and two are in the PNS. The four neuroglial cells in the CNS are astrocytes, oligodendrocytes, microglia, and ependymal cells. **Astrocytes** (*astro-* = star; *-cyte* = cell) have many processes that make them look star-shaped. Their **perivascular** (*peri* = around; *vascular* = vessel) **feet**

wrap around and cover neurons and blood vessels to keep neurons in place. Astrocytes also guide neurons during development and control the composition of the chemical environment of the neurons by forming a **blood–brain barrier.** This barrier only allows certain substances to enter the nervous tissue at the blood vessel sites. **Oligodendrocytes** (*oligo-* = few; *dendro-* = tree) support the CNS neurons and have processes that form **myelin sheaths** around axons that increase the speed of nerve impulses. **Microglia** (*micro-* = small) are the phagocytes (*phago-* = to eat) of the CNS that engulf debris, necrotic tissue, and invading bacteria or viruses. **Ependymal cells** (*epen-* = above; *dym-* = garment) line all four ventricles (spaces or cavities) of the brain, as well as the central canal of the spinal cord. These cells form cerebrospinal fluid (CSF), and their cilia move the CSF through the ventricles. The two neuroglial cells in the PNS are Schwann cells and satellite cells. **Schwann cells** are flattened cells that wrap around the axons in the PNS. Many Schwann cells form the myelin sheath around one axon. The myelin sheath increases nerve impulse speed and aids in the regeneration of PNS axons. **Satellite cells** have processes that are flattened and surround the neuron cell bodies located in ganglia in the PNS. They give support to these neurons and regulate their chemical environment.

ACTIVITY 1 **Neuroglia**

1 Complete Table 16.1 using the following list of cells. Under *location*, circle CNS or PNS.
 - astrocyte
 - ependymal cells
 - microglia
 - oligodendrocytes
 - satellite cells
 - Schwann cells

2 Label the neuroglial cells in Figure 16.1(a) and (b).

3 Pronounce the terms as you label the figures.

TABLE 16.1 **Neuroglia**

CELL TYPE	LOCATION	FUNCTION
1. _____	CNS or PNS	Entire cell forms myelin sheath around a segment of an axon; helps regeneration of axons.
2. _____	CNS or PNS	Lines four brain ventricles; forms and circulates CSF.
3. _____	CNS or PNS	Engulfs invading microbes; clears debris; migrates to injured nerves.
4. _____	CNS or PNS	Maintains environment around neurons; forms blood–brain barrier.
5. _____	CNS or PNS	Covers sensory neuron cell bodies; maintains neuron environment.
6. _____	CNS or PNS	Processes from cell form myelin sheaths around axons of neurons.

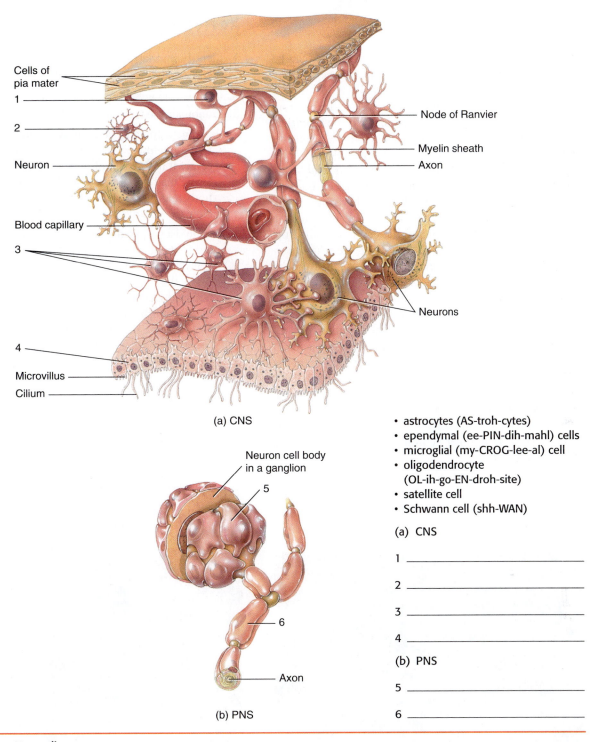

Cells of
pia mater

1 _____

2 _____

Neuron _____

Blood capillary _____

3 _____

4 _____

Microvillus _____

Cilium _____

Node of Ranvier

Myelin sheath

Axon

Neurons

(a) CNS

Neuron cell body
in a ganglion

5

6

Axon

(b) PNS

- astrocytes (AS-troh-cytes)
- ependymal (ee-PIN-dih-mahl) cells
- microglial (my-CROG-lee-al) cell
- oligodendrocyte
 (OL-ih-go-EN-droh-site)
- satellite cell
- Schwann cell (shh-WAN)

(a) CNS

1 _____

2 _____

3 _____

4 _____

(b) PNS

5 _____

6 _____

FIGURE 16.1 Neuroglia

Structure of a Neuron

The longest cells in the body are **neurons,** which can be over 3 feet long. Think about one neuron being long enough to reach from your spinal cord to the tips of your fingers or toes. There are three basic parts to any neuron: dendrites, a cell body, and an axon. Both the dendrites (*dendr-* = little trees) and the single axon are **processes** (extensions) of the neuron cell body. **Dendrites** receive information from receptors or other neurons and send it as a change in membrane potential to the neuron **cell body** or **soma** (*soma* = body). Neuron cell bodies have most of the organelles present in other types of cells. There is usually a triangular or cone-shaped area of the cell body called the **axon hillock** (*hillock* = small hill). The **axon** (*axon* = axis), a longer process than the dendrites, extends from the axon hillock. Changes in membrane potential travel to the axon hillock where they are integrated to determine whether an action potential will be initiated in the axon. The first part of the axon is known as the **trigger area** (initial segment), where the action potential begins. The axon may be a single process, or it may have side branches called **axon collaterals.** Axons and axon collaterals conduct action potentials along their full lengths to end at fine processes called **axon terminals.** Neurotransmitter molecules are released from axon terminals and transmit signals across a synapse to other neurons or to effectors such as muscles or glands.

ACTIVITY 2 Structure of a Neuron

1 Identify and label the structures in Figure 16.2(a) and (b). Note: the magnification of the cells in these figures.

2 Identify and label the structures in Figure 16.3.

3 Pronounce the terms as you label Figures 16.2 and 16.3.

4 Examine a prepared slide of **astrocytes.**
 • Using the low-power objective, find an astrocyte and center it in the field of view.
 • Using the high-power objective, identify the **cell body** and **processes.**
 • Note the size of the astrocytes.

5 Examine a prepared slide of **motor neurons.**
 • Using the low-power objective, find a large motor neuron and center it in the field of view.
 • Using the high-power objective, identify the **cell body, nucleus,** and **dendrites.** The axon is difficult to distinguish.
 • Note the size of the motor neuron and the numerous small, dark-stained **neuroglial cells** near the neuron.

6 With your lab group compare the size of astrocytes and motor neurons.

(a) Astrocyte LM 700×

(b) Motor neuron LM 260×

FIGURE 16.2 Photomicrographs of nervous tissue cells.

(a)
• cell body
• processes

1 _____

2 _____

(b)
• axon
• axon hillock
• cell body
• dendrites
• nucleus

3 _____ 6 _____

4 _____ 7 _____

5 _____

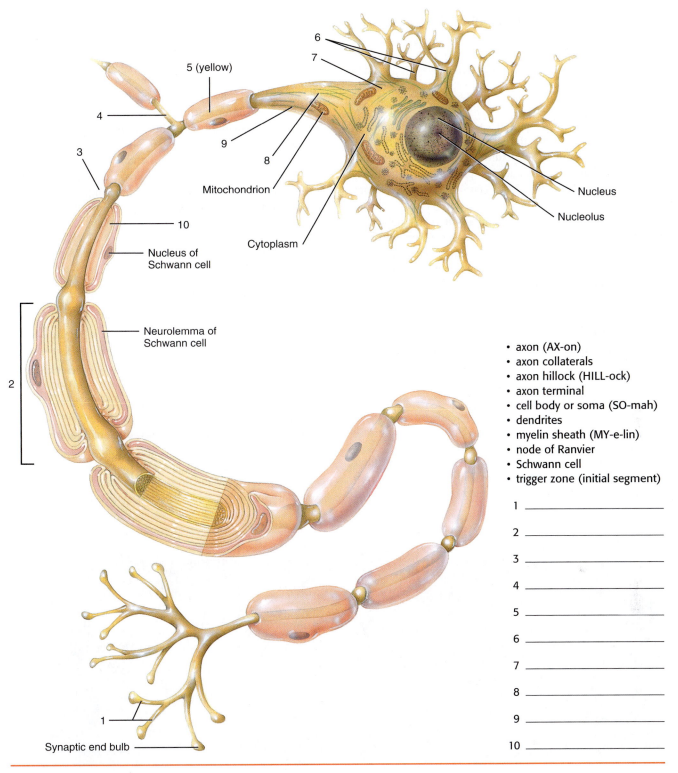

5 (yellow)

4

3

9

8

Mitochondrion

Cytoplasm

6

7

Nucleus

Nucleolus

10

Nucleus of
Schwann cell

Neurolemma of
Schwann cell

2

1

Synaptic end bulb

- axon (AX-on)
- axon collaterals
- axon hillock (HILL-ock)
- axon terminal
- cell body or soma (SO-mah)
- dendrites
- myelin sheath (MY-e-lin)
- node of Ranvier
- Schwann cell
- trigger zone (initial segment)

1 _____

2 _____

3 _____

4 _____

5 _____

6 _____

7 _____

8 _____

9 _____

10 _____

FIGURE 16.3 Parts of a motor neuron.

C. Classification of Neurons

Structural Classification of Neurons

Neurons are classified both structurally and functionally. The number of processes that project from the cell body of the neuron determine its structural classification. The **multipolar neurons** have numerous processes, with many dendrites and one axon. Motor neurons and interneurons (association neurons) are multipolar neurons and comprise most of the CNS neurons. **Bipolar neurons** have two processes—one dendrite and one axon—on either side of the cell body and are found in the special senses like the retina of the eye, the olfactory cells of the nose, and the inner ear. **Unipolar neurons** have only one process, an axon, leading to and from the neuron cell body. The dendrites are small and attach to the axon instead of the neuron cell body. Unipolar neurons are sensory neurons that bring sensory information from the skin, muscles, and organs to the spinal cord.

Functional Classification of Neurons

There are three classifications of neurons based on their functions: sensory, interneuron (association neuron), and motor neuron. Changes in the environment produce a stimulus that is detected by the receptors associated with the dendrites of a **sensory (afferent) neuron.** This neuron changes the stimulation into an action potential or nervous impulse that travels along the axon to the spinal cord. General sensory neurons are structurally unipolar neurons. In the spinal cord, the axon of the sensory neuron synapses with either a motor neuron or an interneuron. The **interneuron (association neuron)** is structurally a multipolar neuron and comprises about 90% of the neurons in the CNS. In the spinal cord, the interneuron can synapse with a chain of interneurons that sends the signal to the brain, and/or it can synapse with a **motor (efferent) neuron** that takes the impulse out of the spinal cord via a spinal nerve to an effector. Motor neurons are structurally multipolar neurons.

ACTIVITY 3 **Structural and Functional Classifications of Neurons**

1 Identify and label the neurons in Figure 16.4(a) and (b).

2 Identify and label the structures listed in Figure 16.5(a) and (b).

3 Examine prepared microscope slide of **dorsal root ganglia.**
 • Using the low-power objective, locate a neuron and place it in the center of the field of view.
 • Identify the **neuron cell body, nucleus,** and **processes.**

4 Examine a prepared microscope slide of **cerebral cortex.** Follow the same steps as above.

(a) Dorsal root ganglia, unipolar neurons **LM** 300×

(b) Cerebral cortex, multipolar neurons **LM** 575×

FIGURE 16.4 Sectional views of the dorsal root ganglion and cerebral cortex.

(a)
• neuron cell body
• nucleus
• process
• satellite cells

1 _____
2 _____
3 _____
4 _____

(b)
• axon
• dendrites
• neuron cell body

5 _____
6 _____
7 _____

1 2 3

(a) Structural classification of neurons

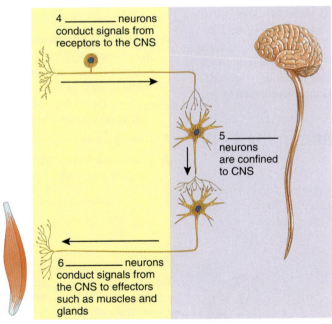

4 _____ neurons conduct signals from receptors to the CNS

5 _____ neurons are confined to CNS

6 _____ neurons conduct signals from the CNS to effectors such as muscles and glands

Peripheral nervous system Central nervous system

(b) Functional classification of neurons

FIGURE 16.5 Structural and functional classifications of neurons.

(a) Structural classifications
• bipolar neuron
• multipolar neuron
• unipolar neuron

(b) Functional classifications
• interneuron (association neuron)
• motor neuron (efferent)
• sensory neuron (afferent)

1 _____

2 _____

3 _____

4 _____

5 _____

6 _____

D. Myelination of Axons

Two neuroglial cells, the **oligodendrocytes** (CNS) and the **Schwann cells** (PNS), form insulated wrappings called **myelin sheaths** around axons. As Schwann cells wrap around axons, their cytoplasm is pushed to the periphery and is called **neurolemma.** The multiple layers of myelin that surround the axons are composed of lipoprotein (about 80% lipids and 20% protein), similar to the makeup of plasma membranes. The high amount of lipid in the myelin sheath gives the axons a whitish appearance. Because babies and children do not have the complete myelin coverings encircling the axons, their need for dietary fat is different from that of an adult and is necessary for proper nervous system development. There is not a continuous myelin sheath around the axon, and gaps do exist between the cells. Gaps in the myelin sheaths are called **nodes of Ranvier** and are more numerous in the PNS than in the CNS. The myelin sheath provides protection and insulation for the axon, and also increases the speed of conductivity of the nervous impulse (action potential). Myelinated axons are also called myelinated fibers. The axons that are not myelinated are called **unmyelinated fibers.** They still have neuroglial cells but possess only a thin coating of the Schwann cell or oligodendrocyte plasma membrane covering the axons. Unmyelinated fibers conduct impulses slower than myelinated fibers.

E. Gray and White Matter in the CNS

Myelin sheaths are white in color, giving nervous tissue with many myelinated axons a white color. Nervous tissue with few myelinated axons appears gray, the color of nerv-ous tissue cells. Groups of myelinated axons in the CNS form **tracts** and are called **white matter.** The outer area or cortex of the brain is comprised of neuron cell bodies, their dendrites, axon terminals, and neuroglia and is called **gray matter** (not myelinated). There are also deeper areas within the brain that have isolated areas of gray matter; these are called **nuclei** of the brain and also contain neuron cell bodies and their dendrites. Unlike the brain, the spinal cord has gray matter in a central H-shaped area made up of neuron cell bodies, dendrites, unmyelinated axons, axon terminals, and neuroglia.

ACTIVITY 4 **Myelination of Axons, Gray Matter, and White Matter**

1 Label the structures on Figure 16.6(a) and (b) and identify (a) and (b) as myelinated or unmyelinated axons.

2 Pronounce the terms as you label the figure.

3 Label the structures in the photomicrograph of a teased myelinated nerve fiber in Figure 16.7.

4 Examine a microscope slide of a teased myelinated nerve fiber.
• Using the low-power objective, locate a nerve fiber and place it in the center of the field of view.
• Using the high-power objective, identify the **nodes of Ranvier, myelin sheath, axon,** and **neurolemma.**

5 Label gray and white matter in Figure 16.8(a) and (b).

6 Examine a section of brain and spinal cord and identify the gray and white matter.

(a) _____
• axon • node of Ranvier
• myelin sheath • Schwann cell cytoplasm

1 _____

2 _____

3 _____

4 _____

(b) _____
• axons • Schwann cell cytoplasm

5 _____

6 _____

FIGURE 16.6 Myelinated and unmyelinated axons.

- axon
- myelin sheath
- neurolemma
 (ner-oh-LEM-ma)
- node of Ranvier

1 _____

2 _____

3 _____

4 _____

FIGURE 16.7 Teased myelinated nerve fiber.

Frontal plane

Transverse plane

(a) Sectional view of brain

(b) Transverse view of spinal cord

FIGURE 16.8 Gray and white matter of brain and spinal cord.

- gray matter in brain and spinal cord
- white matter in brain and spinal cord

1 _____

2 _____

Reviewing Your Knowledge

A. Nervous Tissue

Go to www.wiley.com/college/allen and click on the picture of your lab manual. Then click on Student Companion Site, Anatomy Drill and Practice, and Exercise 16.

B. Organization of Nervous System

Fill in the blanks with the correct term chosen from the following:

afferent motor
effectors peripheral
efferent receptors

If you touch a hot stove with your hand, the sensory (1) _____ in your hand send

a signal of pain to the CNS through the (2) _____ nerve fibers of the

(3) _____ nervous system. When the information

reaches the CNS and is processed, a/an (4) _____ response

is sent through the (5) _____ nerve fibers of the

PNS system to skeletal muscles that are (6) _____.

C. Nervous Tissue Cells

Write the name of the nervous tissue cell described.

_____ 7. ⎫
 ⎬ Supporting cells of PNS.
_____ 8. ⎭

_____ 9. ⎫
 ⎪
_____ 10. ⎪
 ⎬ Supporting cells of CNS.
_____ 11. ⎪
 ⎪
_____ 12. ⎭

_____ 13. ⎫
 ⎬ Form myelin sheaths.
_____ 14. ⎭

_____ 15. ⎫
 ⎬ Regulate chemical
_____ 16. ⎭ environment of neurons.

_____ **17.** Generate and transmit nerve impulses.

_____ **18.** Line cavities of brain and spinal cord; form and move CSF.

_____ **19.** Phagocytes that destroy debris, dead tissue, and pathogens.

D. Structural Classification of Neurons

Identify the neuron type described.

_____ **1.** Many processes associated with the cell body.

_____ **2.** Has two cell processes.

_____ **3.** One short process extends from the cell body and divides.

_____ **4.** Neuron is rare and is the sensory neuron in the eye and nose.

E. Functional Classification of Neurons

Identify the neuron type described.

_____ **1.**
_____ **2.** } Functional neuron types that are structurally multipolar neurons.

_____ **3.**
_____ **4.** } Neuron types whose cell bodies are in the spinal cord (CNS).

_____ **5.** Neuron type that is structurally either a unipolar neuron or bipolar neuron whose cell body is found in the PNS.

_____ **6.** Functional neuron type most prevalent in the CNS.

F. Myelination of Axons and Gray and White Matter

Match the description to the appropriate term.

a. gray matter
b. white matter
c. myelinated fibers
d. unmyelinated fibers
e. nodes of Ranvier

____ **1.** Contains myelinated fibers.

____ **2.** Contains neuron cell bodies and unmyelinated fibers.

____ **3.** Nerve fibers that are white in color and conduct nerve impulses faster.

____ **4.** Nerve fibers that are gray in color.

____ **5.** Gaps in the myelin sheath.

EXERCISE

16

Using Your Knowledge

A. Conduction of a Nervous Impulse

Write the following parts of a multipolar neuron in the correct order (1–8) of receiving and sending the nerve impulse. Start with the dendrites as number one.

____ **1.** axon

____ **2.** axon hillock

____ **3.** axon terminal

____ **4.** cell body

____ **5.** dendrites

____ **6.** second neuron or effector

____ **7.** synapse

____ **8.** trigger zone

B. Nervous Tissue and Diseases

Using your textbook or other references, identify the nervous tissue cell(s) that is(are) involved in the following diseases:

9. Multiple sclerosis

10. Epilepsy

Spinal Cord Structure

The spinal cord and the brain comprise the central nervous system. Being continuous with the brain, the spinal cord begins at the foramen magnum and terminates between vertebrae L_1 and L_2. It is suspended within the vertebral canal, an area formed by the vertebral foramina of the cervical, thoracic, and lumbar vertebrae. The spinal cord has two functions: (1) it carries sensory information to the brain and motor output to nerves, and (2) it mediates spinal reflexes. Spinal reflexes process sensory input from and convey motor output to the spinal nerves.

A. Protective Structures and Spinal Meninges

The **spinal cord** is protected by the bony vertebrae, adipose tissue, spinal meninges, and cerebrospinal fluid. Adipose tissue cushions the spinal cord and is found within the space between the vertebrae and the meninges, the **epidural space.** The three **meninges** (meninx, sing.) or connective tissue membranes cover the spinal cord and are continuous with the cranial meninges that protect the brain. **Dura mater,** the outer meninx, is a tough, single-layered membrane that is deep to the epidural space and superficial to the spider weblike **arachnoid mater.** The inner meninx, the **pia mater** is delicate and hugs the spinal cord. **Denticulate ligaments** are lateral extensions of pia mater that fuse with arachnoid mater and secure the spinal cord. Between the pia and arachnoid mater is the **subarachnoid space** that contains **cerebrospinal fluid,** which also cushions the spinal cord.

FIGURE 17.1 Spinal cord transverse section showing meninges.

- dura mater
- epidural space
- pia mater
- subarachnoid space
- weblike projection of arachnoid mater

1 _____

2 _____

3 _____

4 _____

5 _____

ACTIVITY 1 **Spinal Meninges**

1 Identify and label the structures in Figure 17.1.

2 Identify the spinal meninges on a spinal cord model or chart.

3 Pronounce the terms as you point to them.

B. External Features of the Spinal Cord

The long, cylindrical spinal cord has 31 pairs of spinal nerves attached to it, with each pair of spinal nerves arising from a different segment of the cord. The cord is wider in the cervical and lumbar regions, forming two enlargements. The **cervical enlargement** is located at levels C_3 or C_4 through T_1. This bulge designates the location of nuclei

(collection of neuron cell bodies) for the upper extremities. The inferior **lumbar enlargement** is located at levels T_9 through T_{12} and contains nuclei for the lower extremities. The spinal cord ends inferiorly as the **conus medullaris** between vertebrae levels L_1 and L_2. Nerves arising from the inferior portion of the spinal cord continue inferiorly as a group called the **cauda equina** (*cauda* = tail; *equin-* = horse), or "horse's tail." An extension of the pia mater continues past the conus medullaris as the **filum terminale** (*filum* = filament; *termin-* = terminal) and connects the inferior end of the spinal cord to the coccyx.

ACTIVITY 2 **External Features of the Spinal Cord**

1 Identify and label the structures in Figure 17.2.

2 Identify these structures on a spinal cord model or chart.

3 Pronounce the terms as you point to them.

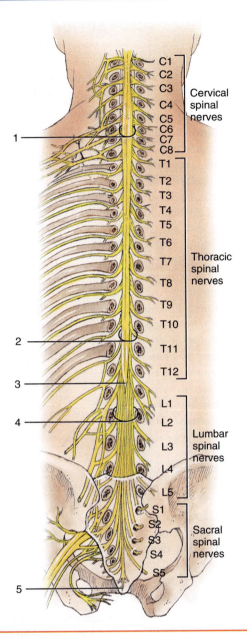

C1
C2
C3
C4 Cervical
C5 spinal
C6 nerves
C7
C8

T1
T2
T3
T4
T5
T6
T7 Thoracic
 spinal
T8 nerves
T9
T10
T11
T12

L1
L2
L3 Lumbar
 spinal
L4 nerves
L5

S1
S2
S3 Sacral
 spinal
S4 nerves
S5

1
2
3
4
5

FIGURE 17.2 Longitudinal view of posterior spinal cord.

- cauda equina (CAU-da ee-QUI-na)
- cervical enlargement
- conus medullaris (CO-nus med-u-LAR-is)
- filum terminale (FI-lum ter-min-A-lee)
- lumbar enlargement

1 _____

2 _____

3 _____

4 _____

5 _____

C. Transverse Section of the Spinal Cord

The most obvious parts of the spinal cord in cross-section are the anterior median fissure, the posterior median sulcus, and the gray and white matter. The **anterior median fissure** is a wide, deep groove on the anterior surface of the spinal cord, and the **posterior median sulcus** is a narrow groove on the posterior surface. The **gray matter** looks like a butterfly or a modified H and is more centrally located than the white matter. The gray matter is divided into the anterior, lateral, and posterior gray horns and consists of nerve cell bodies and dendrites. Somatic motor neuron cell bodies are located in the **anterior (ventral) gray horns,** whereas the **lateral gray horns** (not present in cervical cord segments) contain cell bodies of autonomic motor neurons. The **posterior gray (dorsal) horns** contain neuron cell bodies that receive impulses from sensory neurons. The **gray commissure** is a narrow bridge of gray matter that connects the right and left sides of gray matter in the middle of the spinal cord. The **central canal** is in the center of the gray commissure and contains cerebrospinal fluid. **White matter** surrounds the gray matter and forms the **anterior, lateral,** and **posterior white columns.** These columns or funiculi are made up of white, myelinated fibers (axons) that are either sensory or motor fibers.

A **spinal nerve** is formed from a **posterior (dorsal) root** and an **anterior (ventral) root.** Roots are collections of axons that are going to and leaving the spinal cord. The posterior (dorsal) root carries sensory fibers, whereas the anterior (ventral) root carries motor fibers. Because the sensory and motor roots merge to form the spinal nerve, these nerves are called **mixed nerves.** Near the spinal cord, there is a bulge in the posterior (dorsal) root called the **posterior (dorsal) root ganglion.** The posterior (dorsal) root ganglion consists of somatic sensory neuron cell bodies that synapse onto interneuron and/or motor neuron cell bodies in the spinal gray matter.

ACTIVITY 3 Transverse Section of Spinal Cord

1 Label Figures 17.3 and 17.4.

2 Identify these parts on a transverse section model or chart of the spinal cord.

3 Pronounce these terms as you point to them.

4 Label the photomicrograph of a transverse section of the spinal cord in Figure 17.5.

5 Examine a prepared slide of a transverse section of the spinal cord.
- Using the low-power objective lens, identify the structures listed in Figure 17.5.
- Using the high-power objective, observe myelinated axons in the white matter and unmyelinated processes, neuron cell bodies, and neuroglia in the gray matter.

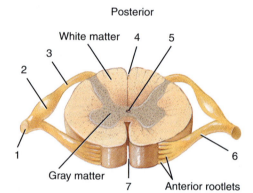

FIGURE 17.3 Transverse section of spinal cord.

- anterior median fissure
- anterior (ventral) root
- central canal
- posterior (dorsal) root
- posterior (dorsal) root ganglion (GANG-li-on)
- posterior median sulcus
- spinal nerve

1 _____

2 _____

3 _____

4 _____

5 _____

6 _____

7 _____

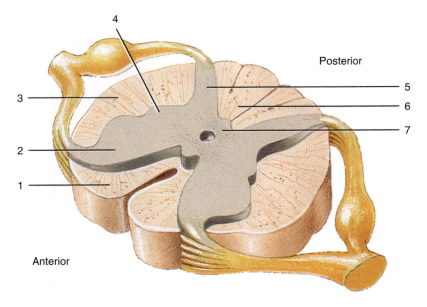

Posterior

Anterior

Transverse section of the thoracic spinal cord

FIGURE 17.4 Transverse section of spinal cord with areas of gray and white matter.

- anterior gray horn
- anterior white column
- gray commissure (COM-mis-sure)
- lateral gray horn
- lateral white column
- posterior gray horn
- posterior white column

1 _____ 5 _____

2 _____ 6 _____

3 _____ 7 _____

4 _____

Posterior

Anterior

FIGURE 17.5 Photomicrograph of spinal cord cross-section with spinal nerve.

- anterior gray horn
- anterior median fissure
- anterior (ventral) root
- anterior white column
- central canal
- gray commissure
- lateral white column
- posterior (dorsal) root
- posterior (dorsal) root ganglion
- posterior gray horn
- posterior white column
- posterior median sulcus

1 _____ 7 _____

2 _____ 8 _____

3 _____ 9 _____

4 _____ 10 _____

5 _____ 11 _____

6 _____ 12 _____

D. Dissection of the Spinal Cord

Use either a preserved cow or sheep spinal cord or a fresh spinal cord from a butcher. Remember: if the specimen is preserved, it will be firmer and look different than a fresh specimen.

> **SAFETY NOTE:** *Use safety glasses and remember to properly dispose of any removed waste tissue as directed by your instructor. Always wash your dissection tools, tray, lab surfaces, and hands before leaving.*

ACTIVITY 4 Spinal Cord Dissection

1 Observe the posterior structures. Refer to Figure 17.6.
 • After putting on your gloves, place the spinal cord in the dissection pan.
 • Use a blunt probe or forceps to separate the **spinal meninges.**
 • Identify the **dura** and **arachnoid mater.**
 • Use a pointed dissection probe or pin to detach the **pia mater** from the spinal cord.
 • Identify the **denticulate ligaments.**
 • Peel back the meninges to uncover the **posterior (dorsal)** and **anterior (ventral) roots** of the spinal nerve. Locate the **rootlets.**
 • Identify the **posterior (dorsal) root ganglion.**

2 Observe transverse section structures. Refer to Figure 17.4.
 • Cut a 1 × 4 inch to 1 × 2 inch section from your spinal cord specimen.
 • Use Figure 17.5 as a reference for the transverse section.
 • Identify the **anterior median fissure, posterior median sulcus, central canal, gray commissure, gray horns,** and **white columns.**

SUPERIOR

Fourth ventricle

Cerebellum of brain (cut)

Occipital bone (cut)
Posterior median sulcus
Vertebral artery

Posterior rootlets

Denticulate ligaments

Dura mater and arachnoid

INFERIOR

Posterior view

FIGURE 17.6 Photograph of posterior view of spinal cord with meninges and spinal nerve roots.

Reviewing Your Knowledge

A. Spinal Cord Anatomy

Go to www.wiley.com/college/allen and click on the picture of your lab manual. Then click on Student Companion Site, Anatomy Drill and Practice, and Exercise 17.

B. Meninges

Write the name of the structure described.

_____ **1.** Middle meninx; weblike.

_____ **2.** Tough, outer meninx.

_____ **3.** Space filled with adipose tissue.

_____ **4.** Thin meninx intimate with spinal cord.

_____ **5.** Contains cerebrospinal fluid.

_____ **6.** Extension of pia mater attaching to dura.

C. Spinal Cord Structures

Write the terms that match the correct description.

_____ **1.** Contains neuron cell bodies that receive impulses from sensory neurons.

_____ **2.** Contains neuron cell bodies and unmyelinated processes.

_____ **3.** Shallow groove on dorsal side.

_____ **4.** Connects right and left halves of gray matter in spinal cord.

_____ **5.** Sensory branch of spinal nerve entering spinal cord.

_____ **6.** Tapered end of spinal cord.

_____ **7.** Motor branch of spinal nerve exiting spinal cord.

_____ **8.** Contains sensory neuron cell bodies.

_____ **9.** Collection of spinal nerves that arise from inferior end of spinal cord.

_____ **10.** Contains myelinated axons.

_____ **11.** Contains somatic motor neuron cell bodies.

_____ **12.** Space in center of spinal cord that contains cerebrospinal fluid.

_____ **13.** Bulge in spinal cord containing cell bodies of motor neurons supplying upper limb.

_____ **14.** Wide, deep groove on ventral side.

_____ **15.** Extension of pia mater that attaches spinal cord to coccyx.

_____ **16.** Bulge in spinal cord at T_9–T_{12}.

Using Your Knowledge

A. Spinal Cord Transverse Section

Label Figure 17.7.

POSTERIOR

Spinous process of
vertebra

1

2

3

Transverse foramen

Body of vertebra

4 (meninx)

Spinal cord

5 (meninx)

6

Spinal nerve

Vertebral artery in
transverse foramen

FIGURE 17.7 Photographic cross-section of spinal cord and cervical vertebra.

1 _____ 4 _____

2 _____ 5 _____

3 _____ 6 _____

B. Spinal Cord Questions

7. The polio virus can cause skeletal muscle paralysis by destroying neuron cell bodies. Identify the area of the spinal cord that is destroyed.

8. Shingles is a condition characterized by pain, discoloration of the skin, and eruption of skin blisters along a sensory nerve. It is caused by the herpes zoster virus, the same virus that causes chicken pox. Using your textbook or another source, identify the spinal cord structure to which herpes virus retreats after chicken pox. The virus remains dormant there until it is activated and causes a shingles outbreak.

9. Identify the structural class(es) of neurons whose cell bodies are present in the spinal cord—unipolar, bipolar, or multipolar.

10. Nerve fibers are classified according to diameter and presence or absence of a myelin sheath. Using your textbook or another reference, name the nerve fiber types (Type A, B, C) present in the:
(a) anterior root
(b) posterior root.

11. When removing cerebrospinal during a spinal tap, the needle is inserted below L_2. Explain why spinal taps are not done above this level.

C. Pathway of Sensory and Motor Impulses

Numbering 1–5, indicate the order of structures through which sensory impulses pass as they enter the spinal cord and travel toward the brain.

____ **12.** posterior (dorsal) root ganglion

____ **13.** posterior (dorsal) root

____ **14.** posterior gray horn

____ **15.** white column

____ **16.** spinal nerve

Numbering 1–4, indicate the order of structures through which motor impulses pass as they descend from the brain and leave the spinal cord.

____ **17.** spinal nerve

____ **18.** white column

____ **19.** anterior (ventral) gray horn

____ **20.** anterior (ventral) root

Spinal Nerves

Objectives

**After completing this exercise,
you should be able to:**

- Describe the connective tissue coverings of the spinal nerves
- Identify the rami that carry impulses to and away from the spinal cord
- Describe the organization and distribution of spinal nerves
- Identify the four spinal plexuses and the major nerves arising from each plexus

Materials

- compound microscope and lens paper
- prepared slide of peripheral nerve cross-section
- prepared slide of peripheral nerve longitudinal section
- model or chart of vertebral column with spinal cord and spinal nerves
- model or chart with major nerves of the upper and lower extremities

Spinal nerves send information from peripheral **sensory receptors** to the spinal cord and information from the spinal cord to **effectors** (muscles and glands). The 31 pairs of spinal nerves emerge from each side of the spinal cord through the **intervertebral foramina** and are named for the vertebral region and level from which they arise. Spinal nerves connect to the spinal cord via a **posterior (dorsal) root** and an **anterior (ventral) root** and are called **mixed nerves** because each nerve contains **sensory** and **motor axons.**

A. Connective Tissue Coverings of Spinal Nerves

Each spinal nerve has three protective connective tissue layers: the **epineurium** (*neuri-* = nerve) that surrounds the whole nerve, the **perineurium** that encases each **fascicle** (*fasciculus* = little bundle), and the **endoneurium** that covers myelinated and unmyelinated axons.

ACTIVITY 1 Connective Tissue Coverings of Spinal Nerves

1 Label Figures 18.1 and 18.2.

2 Pronounce the terms as you label the figures.

3 Examine a prepared slide of a cross-section of a spinal nerve.
- Using the low-power objective lens, identify the **epineurium, fascicles,** and **perineurium.**
- Using the high-power objective lens, identify the **endoneurium, axons,** and the **myelin sheath** of **myelinated neurons.**

B. Rami of the Spinal Nerves

The spinal nerves branch lateral to the intervertebral foramen. These branches or rami (*rami* = branches; *ramus* = branch) are the posterior (dorsal) ramus, the anterior (ventral) ramus, the meningeal branch, and the rami communicantes. The **posterior (dorsal) ramus,** the largest branch, curves around to the dorsal surface and innervates the skin and deep muscles of the back or trunk. The **anterior (ventral) ramus** supplies the muscles and skin of all four limbs, as well as the anterior and lateral parts of the body. The **meningeal** branch serves the vertebrae, vertebral ligaments, blood vessels of the spinal cord, and the meninges. The two **rami communicantes** (*communicans* = communicating) connect to the **sympathetic ganglion** (*sympathein* = to feel with) of the autonomic nervous system.

- axon (AX-on)
- endoneurium (endo-NEUR-i-um)
- epineurium (epi-NEUR-i-um)
- fascicle (FAS-i-cul)
- perineurium (peri-NEUR-i-um)
- spinal nerve

1 _____

2 _____

3 _____

4 _____

5 _____

6 _____

FIGURE 18.1 Transverse section of a spinal nerve with coverings and fascicles.

- axon
- endoneurium
- myelin sheath
- perineurium

1 _____

2 _____

3 _____

4 _____

FIGURE 18.2 Photomicrograph of a cross-section through a fasicle of a spinal nerve.

ACTIVITY 2 Rami of the Spinal Nerves

1 Label Figure 18.3.

2 Identify the structures from Figure 18.3 on a model or chart.

3 Pronounce the terms as you point to them on the model or chart.

• anterior (ventral) ramus (RAY-mus)
• posterior (dorsal) ramus
• rami communicantes
 (RAY-my com-mun-i-CAHN-tayce)
• spinal nerve

1 _____

2 _____

3 _____

4 _____

POSTERIOR

Posterior (dorsal) root

Anterior (ventral) root

3

2

1

Meningeal branch

4

Sympathetic ganglion

ANTERIOR

(a) Transverse section

• anterior (ventral) ramus
• intervertebral foramen
• posterior (dorsal) ramus
• rami communicantes
• spinal nerve
• sympathetic ganglion

5 _____

6 _____

7 _____

8 _____

9 _____

10 _____

6

7

5

8

9

10

(b) Lateral view

FIGURE 18.3 Spinal cord with branches of a spinal nerve.

C. Spinal Nerve Divisions and the Four Spinal Plexuses

Of the 31 pairs of spinal nerves, there are **8 cervical, 12 thoracic, 5 lumbar, 5 sacral,** and **1 coccygeal.** Many spinal nerves join with other spinal nerves to form a braided network or **plexus** (braid) before they innervate body structures. This occurs in four regions of the body where the networks form the **cervical, brachial, lumbar,** and **sacral plexuses.** The **thoracic** (intercostal) spinal nerves (T_2–T_{12}) do not participate in forming plexuses.

ACTIVITY 3 Spinal Nerve Divisions and Spinal Plexuses

1 Label the cervical, thoracic, lumbar, sacral, and coccygeal nerves in Figure 18.4.

2 Label the four spinal plexuses also in Figure 18.4.

3 Identify the spinal nerve divisions and plexuses from Figure 18.4 on a model or chart.

4 Pronounce the terms as you point to them.

- brachial plexus (PLEX-us)
- cervical nerves
- cervical plexus
- coccygeal nerve (cox-sih-GEAL)
- lumbar nerves
- lumbar plexus
- sacral nerves
- sacral plexus
- thoracic nerves

1 _____

2 _____

3 _____

4 _____

5 _____

6 _____

7 _____

8 _____

9 _____

FIGURE 18.4 Posterior view of the four spinal plexuses.

D. Major Nerves from the Cervical and Brachial Plexuses

The **cervical plexus** is formed from the anterior (ventral) rami of C_1–C_5 on both the right and left sides of the spinal cord. An important paired nerve from this plexus is the **phrenic nerve** (*phreni-* = relating to the diaphragm), C_3–C_5, which innervates the diaphragm and is important for breathing. *Remember the saying, "Cervical nerves 3, 4, and 5 keep the diaphragm alive".* Other cervical nerves mainly supply the scalp, neck, shoulders, and chest.

The **brachial plexus** is formed from the ventral rami of C_5–T_1. This plexus serves the shoulders and upper limbs. The main nerves that arise from the brachial plexus are the **axillary, median, musculocutaneous** (*musculo-* = muscle; *cutan-* = skin), **radial,** and **ulnar.**

ACTIVITY 4 **Major Nerves from the Cervical and Brachial Plexuses**

1 Label the phrenic nerve and the five main brachial plexus nerves in Figure 18.5 using the name of the nerve and the muscle(s) it innervates (Table 18.1) to assist in identification.

2 Identify the nerves from Figure 18.5 on a model or chart.

3 Pronounce the terms as you point to them.

4 Test nerve conduction.
 - Flex your forearm to 90°.
 - Palpate a cordlike nerve between your medial epicondyle and olecranon process.
 - Place pressure on the nerve with your fingers until you feel a difference in your forearm, hand, and digits.
 - Describe what sensations you felt and what nerve was compressed.
 - Describe what digits are involved and what produced your symptoms.

Anterior view

1
2
3
4
5 (posterior to bone)
6

FIGURE 18.5 Major nerves from the cervical and brachial plexuses.

- axillary nerve
- median nerve
- musculocutaneous nerve (mus-cu-lo-cu-TAYN-e-ous)
- phrenic nerve (FREN-ic)
- radial nerve
- ulnar nerve

1 _____
2 _____
3 _____
4 _____
5 _____
6 _____

TABLE 18.1 **Motor Function of Major Nerves from Cervical and Brachial Plexuses**

NERVE	MUSCLES INNERVATED BY NERVE
Phrenic	Diaphragm
Axillary	Deltoid and teres minor muscles
Musculocutaneous	Anterior muscles of the arm
Ulnar	Flexor carpi ulnaris, medial-half of flexor digitorum profundis, and most hand muscles
Median	Muscles of anterior forearm (excluding flexor carpi ulnaris and other muscles supplied by ulnar nerve) and some of the muscles of the hand
Radial	Muscles of posterior arm and forearm

E. Major Nerves from the Lumbar and Sacral Plexuses

The **lumbar plexus** is made up of anterior (ventral) rami from L_1–L_4. The lumbar plexus supplies the skin and muscles of the abdominal wall, external genitalia, and part of the lower limbs. The major nerves are the **femoral** and **obturator nerves.**

The **sacral plexus** is formed from the anterior (ventral) rami from L_4–S_4. This plexus supplies the buttocks, perineum, and most of the lower limbs. Its major nerves are the **pudendal** and **sciatic.** The sciatic is composed of the **tibial** and **common fibular (peroneal)** nerves.

ACTIVITY 5 **Major Nerves from the Lumbar and Sacral Plexuses**

1 Label the nerves listed on Figure 18.6 using the name of the nerve, plexus of origin, and muscle innervation. Muscle innervation is described in Table 18.2.

2 Identify the nerves from Figure 18.6 on models or charts.

3 Pronounce the terms as you point to them.

ANTERIOR POSTERIOR

FIGURE 18.6 Major nerves of the lumbar and sacral plexuses.

- femoral nerve (FEM-o-rul)
- obturator nerve (OB-tur-a-tor)
- common fibular (peroneal)
- pudendal (pyoo-DEN-dal)
- sciatic (sci-A-tic)
- tibial

1 _____
2 _____
3 _____
4 _____
5 _____
6 _____

TABLE 18.2 **Motor Function of Nerves from the Lumbar and Sacral Plexuses**

NERVE	MUSCLES INNERVATED BY NERVES
Femoral	Iliacus, quadriceps femoris, sartorius, pectineus
Obturator	Adductor longus, adductor brevis, and part of adductor magnus, gracilis
Common fibular	Fibularis longus muscle, tibialis anterior, extensor digitorum longus
Pudendal	Muscles of perineum
Sciatic	Biceps femoris, semimembranosus, semitendinosus (hamstrings)
Tibial	Gastrocnemius, soleus, tibialis posterior, flexor digitorum longus, plantaris, flexor hallucis longus

Reviewing Your Knowledge

A. Spinal Nerve Review

Go to www.wiley.com/college/allen and click on the picture of your lab manual. Then click on Student Companion Site, Anatomy Drill and Practice, and Exercise 18.

B. Connective Tissue Coverings

Identify the connective tissue covering.

_____ **1.** Covers unmyelinated or myelinated axons.

_____ **2.** Covers fascicles.

_____ **3.** Covers nerves.

C. Rami of Spinal Nerves

Identify the spinal nerve ramus.

_____ **1.** Branch that serves the deep muscles and skin of the back.

_____ **2.** Branches that belong to the sympathetic nervous system.

_____ **3.** Branch that forms nerves serving the limbs.

D. Major Nerves of the Nerve Plexuses

Identify the nerve plexus for each nerve.

_____ **1.** femoral n.

_____ **2.** sciatic n.

_____ **3.** phrenic n.

_____ **4.** ulnar n.

_____ **5.** axillary n.

_____ **6.** tibial n.

_____ **7.** obturator n.

_____ **8.** radial n.

_____ **9.** common fibular n.

_____ **10.** pudendal n.

E. Spinal Nerves

Complete the sentences with the correct term about the spinal nerves.

1. The sciatic nerve is composed of two nerves, the _____

2. and the _____.

3. There are _____ pairs of spinal nerves.

4. The nerve that supplies the posterior thigh and most of the leg and foot is the _____.

5. There is/are _____ pair(s) of coccygeal nerves.

6. If the anterior (ventral) root of a spinal nerve is severed, there is a _____ (motor or sensory) loss.

7. There is/are _____ pair(s) of thoracic nerves.

8. The nerve that supplies the anterior thigh is the _____.

9. There is/are _____ pair(s) of sacral nerves.

10. There is/are _____ pair(s) of lumbar nerves.

11. If the posterior (dorsal) root of a spinal nerve is severed, the functional loss is _____ (motor or sensory).

12. There is/are _____ pair(s) of cervical nerves.

13. The _____ nerve supplies the deltoid and teres minor muscles.

14. The _____ nerve supplies the triceps brachii muscles and the extensor digitorum longus.

15. The _____ nerve supplies the biceps brachii muscles.

16. The _____ nerve supplies the diaphragm.

17. The _____ nerve supplies most hand muscles.

18. The _____ nerve supplies the flexor carpi radialis.

Using Your Knowledge

Identify the correct nerve for each question.

_____ **1.** A broken forearm resulted in an inability to pronate the forearm and loss of finger movement in digits 1–3. Name the nerve that was injured.

_____ **2.** An injection into the shoulder results in an inability to extend the wrist and fingers. Name the nerve that was injured.

_____ **3.** Health care professionals are taught how to properly administer gluteal injections to avoid pain and injury caused by inadvertently striking a major nerve. Name the nerve to avoid.

_____ **4.** Hitting the medial epicondyle results in a tingling sensation in part of the hand. Name the nerve hit and the part of the hand that tingles.

_____ **5.** John Jones injured his spinal cord. He has use of his serratus anterior muscle, biceps brachii, and deltoid. However, he does not have movement in most muscles of his hand and digits 4 and 5. Where is his spinal cord injury?

_____ **6.** Following the birth of her daughter, Mary had trouble adducting her lower limbs. Which nerve was injured during childbirth?

_____ **7.** Charles broke his leg playing softball. The fracture was a compound fracture of the fibula. After the cast was removed, he experienced difficulty dorsiflexing his foot. Which nerve was affected?

Identify the nerve that would carry pain impulses from each of the following injured areas.

_____ **8.** Greenstick fracture of the tibia.

_____ **9.** Sunburn on the skin over deltoid muscles.

_____ **10.** Splinter in digit 5 of the hand.

Somatic Reflexes

Objectives

After completing this exercise, you should be able to:

- Identify and describe the five components of a somatic reflex arc
- Describe how monosynaptic and polysynaptic reflex arcs differ
- Explain the significance of clinical testing of reflexes

Materials

- reflex hammer (with rubber head)
- cross-section spinal cord model

R eflexes are rapid, involuntary motor responses to an environmental stimulus detected by sensory receptors. A nerve impulse travels from the receptor through a neural **reflex arc** pathway to an effector. If the motor response is contraction of skeletal muscle, the reflex is a **somatic reflex.** If the motor response involves cardiac muscle, smooth muscle, or glands, the reflex is an **autonomic (visceral) reflex.** Reflexes mediated by spinal nerves are called **spinal reflexes,** whereas reflexes mediated by cranial nerves are called **cranial reflexes.** Most reflexes help our bodies maintain homeostasis and therefore have a protective function.

A. Reflex Arcs

There are five components of a reflex arc:

1. **Sensory receptor.** If the stimulus to the sensory receptor is strong enough, an action potential is generated in the sensory neuron.

2. **Sensory neuron.** The sensory neuron propagates the action potential and synapses with neurons in the spinal cord or brain stem.

3. **Integrating center.** The integrating center is located within the gray matter of the central nervous system (CNS), and transfers information from the sensory neuron to the motor neuron. The integrating center of a **monosynaptic reflex arc** is a single synapse between a sensory and motor neuron. In a **polysynaptic reflex arc,** the integrating center consists of multiple synapses involving one or more interneurons (association neurons) between a sensory and a motor neuron.

4. **Motor neuron.** The motor neuron carries the action potential initiated by the integrating center to the effector.

5. **Effector.** An effector can be a skeletal muscle (somatic reflex), cardiac muscle, smooth muscle, or glands (autonomic reflex). A **reflex** is the response of the effector to stimulation by the motor neuron of the reflex arc.

Reflex arcs that involve sensory receptors, sensory nerve fibers, motor neurons, and effectors all on the same side of the body are **ipsilateral. Contralateral** reflex arcs involve sensory receptors and neurons on one side of the body and motor neurons and effectors on the opposite side. **Bilateral** or **consensual** reflexes involve both sides of the body simultaneously.

ACTIVITY 1 Reflexes

1 Identify whether the reflexes in Table 19.1 are somatic reflexes or autonomic reflexes, and whether a spinal nerve or a cranial nerve mediates it.

2 Label Figure 19.1 with your lab group.

3 Answer discussion questions about Figure 19.1 with your lab group.

4 Using the cross-section spinal cord model, trace the pathway of the reflex arc illustrated in Figure 19.1.

? DISCUSSION QUESTIONS: REFLEX ARC

1 Is the reflex arc in Figure 19.1 monosynaptic or poly-synaptic?

2 Is the reflex arc in Figure 19.1 ipsilateral or contralateral?

TABLE 19.1 Types of Reflexes

REFLEX ACTION	SOMATIC OR AUTONOMIC REFLEX	SPINAL NERVE REFLEXES OR CRANIAL NERVE REFLEXES
Plantar flexion of foot when Achilles tendon is stretched		
Salivation in response to lemon juice on tongue		
Blinking in response to touching the cornea		
Flexing of arm in response to touching a hot object		

FIGURE 19.1 Reflex arc components.

- effector
- integrating center
- motor neuron cell body
- motor neuron axon
- sensory neuron cell body
- sensory neuron axon
- sensory receptor

1 _____ 5 _____

2 _____ 6 _____

3 _____ 7 _____

4 _____

B. Reflex Testing

A series of reflex tests are used clinically to evaluate the nervous system and to diagnose an abnormality or dysfunction that may cause an inhibition, exaggeration, or absence of reflexes.

Patellar Reflex (Knee Jerk)

The **patellar reflex** is the extension of the knee that occurs when the quadriceps femoris tendon is stretched. A stretched tendon will stretch the muscle and cause it to contract. This reflex helps prevent injury by preventing muscles from overstretching. The stretch reflex helps maintain posture and equilibrium.

The components of the **patellar reflex arc** are:

- Sensory receptors: **Muscle spindles** located in the quadriceps femoris muscle group. Tapping the quadriceps tendon stretches this muscle group and stimulates muscle spindles (stretch receptors), initiating nerve impulses in axons of sensory neurons.

- Sensory neuron: Sensory axons carry nerve impulses to the integrating center (gray matter) in the spinal cord.

- Integrating center: The **monosynaptic integrating center** (two neurons, one synapse) is located in the anterior gray horn of the spinal cord. The sensory axon synapses with and initiates nerve impulses in the motor neuron that innervates the quadriceps femoris muscle group.

- Motor neuron: Axons of the motor neuron travel in the femoral nerve to the quadriceps femoris muscle group.

- Effector: Quadriceps femoris muscle group (agonists) contracts and extends the leg when stimulated.

The sensory neuron associated with the monosynaptic stretch reflex arc is also a component of a **polysynaptic reflex arc** that has three neurons and two synapses. This reflex arc inhibits the motor neuron to the antagonist muscle group and results in its relaxation. This is referred to as **reciprocal innervation**—stimulation of contraction in agonistic muscles with simultaneous inhibition of antagonistic muscles.

ACTIVITY 2 **Testing the Patellar Reflex**

1 Label the patellar reflex arc and reciprocal innervation in Figure 19.2.

2 Test the patellar reflex.
 • Have the subject sit on the edge of a table or a tall lab chair with his/her knee off the table or chair and the leg dangling and relaxed.
 • Palpate the patella and the tibial tuberosity; also palpate the patellar ligament located between these two structures.
 • With the tapered end of a reflex hammer, gently but firmly tap the patellar ligament which includes a portion of fibers of the quadriceps femoris tendon.
 • Although the grading is subjective, grade the extension of the leg: Results: _____

 0 = no response (hyporeflexive)
 +1 = little response (hyporeflexive)
 +2 = normal response
 +3 = above normal response
 +4 = exaggerated reflex (hyperreflexive)
 • Now ask the subject to clasp both hands in front of the chest and isometrically pull in opposite directions. This action leads to an enhancement of spinal reflexes causing a reinforced patellar reflex.
 • While the subject is pulling, strike the patellar tendon again and observe the distance of leg movement this time.
 • Reinforced patellar reflex results: _____

3 Answer discussion questions about the patellar reflex with your lab group.

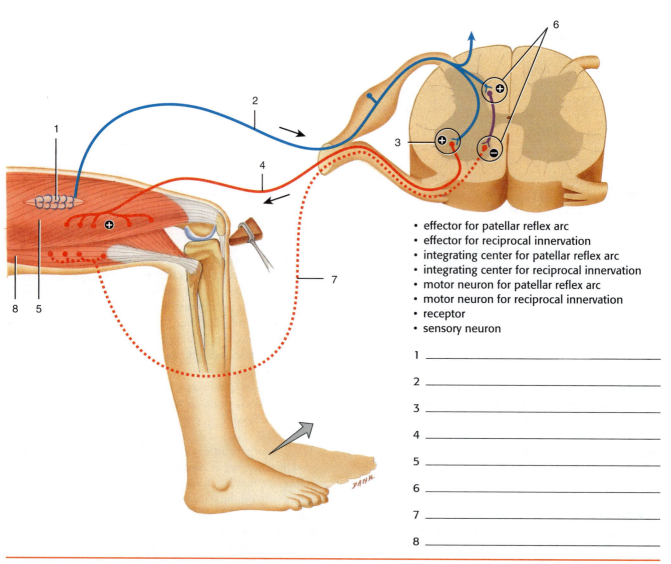

 • effector for patellar reflex arc
 • effector for reciprocal innervation
 • integrating center for patellar reflex arc
 • integrating center for reciprocal innervation
 • motor neuron for patellar reflex arc
 • motor neuron for reciprocal innervation
 • receptor
 • sensory neuron

1 _____
2 _____
3 _____
4 _____
5 _____
6 _____
7 _____
8 _____

FIGURE 19.2 Patellar reflex arc.

? DISCUSSION QUESTIONS: PATELLAR REFLEX

1 Is the patellar reflex ipsilateral or contralateral?

2 During the patellar reflex, are the motor neurons supplying the hamstrings stimulated or inhibited?

3 Describe the effect that clasping and pulling the hands has on the patellar reflex compared with the first patellar reflex.

Biceps Reflex (Biceps Jerk)

The **biceps reflex** the contraction of the biceps brachii muscle that occurs when the biceps tendon is stretched. Although the biceps brachii flexes the forearm, in this test you may only see the contraction or tension in the biceps brachii muscle because the forearm is resting on the table.

ACTIVITY 3 Testing the Biceps Reflex (Biceps Jerk)

1 Test the biceps reflex.
- Have the subject stand with an arm completely relaxed and hanging down at the side.
- Ask the subject to isometrically contract the biceps brachii so you can palpate the tendon in the antecubital fossa.
- After locating the tendon, have the subject relax and place your thumb over the tendon.
- Gently but firmly tap your thumb with the tapered end of the reflex hammer.
- Biceps reflex results: _____
 - 0 = no response
 - +1 = little response
 - +2 = normal response
 - +3 = above normal response
 - +4 = exaggerated reflex
- If the reflex is absent, have the subject clench his/her teeth and try the reflex test again.
- Reinforced biceps reflex results:_____

2 Answer the discussion questions with your lab group.

? DISCUSSION QUESTIONS: BICEPS REFLEX

1 Name the nerve that carries the sensory and motor axons for this reflex arc.

2 Name the antagonistic muscles that are inhibited by reciprocal innervation.

Triceps Reflex (Triceps Jerk)

The triceps reflex the contraction of the triceps brachii muscle that occurs when the triceps brachii tendon is stretched.

ACTIVITY 4 Triceps Reflex (Triceps Jerk)

1 Test the triceps reflex.
- Have the subject stand with an arm completely relaxed and hanging down at the side.
- Palpate the triceps tendon (proximal to olecranon process) as the subject isometrically contracts the triceps muscles.
- Keep your finger on the tendon and have the subject bend the arm across the front of the body.
- Have the subject support the arm by holding it in his/her other hand.
- Use the tapered end of the reflex hammer to tap the triceps tendon.
- Triceps tendon reflex results: _____
 - 0 = no response
 - +1 = little response
 - +2 = normal response
 - +3 = above normal response
 - +4 = exaggerated reflex
- If the reflex is absent, have the subject clench his/her teeth and try the reflex again.
- Reinforced triceps tendon results: _____

2 Answer the discussion questions with your lab group.

? DISCUSSION QUESTIONS: TRICEPS REFLEX

1 Name the nerve that carries the sensory and motor axons for this reflex arc.

2 Name the antagonistic muscles that are inhibited by reciprocal innervation.

? DISCUSSION QUESTIONS: ACHILLES REFLEX

1 Name the nerve that carries the sensory and motor neurons for this reflex arc.

2 Name the antagonistic muscles that are inhibited by reciprocal innervation.

Achilles Reflex (Ankle Jerk)

The response of this reflex is plantar flexion when the Achilles (calcaneous) tendon is tapped with the reflex hammer.

ACTIVITY 5 **Achilles Reflex (Ankle Jerk)**

1 Test the achilles reflex.
- Have the subject keep one foot on the floor and place the knee of the other leg on a chair with the foot dangling over the edge.
- Ask the subject to slightly dorsiflex the foot to stretch the tendon a little, but be relaxed.
- Tap the calcaneal tendon with the tapered end reflex hammer and observe the response of the foot.
- Calcaneal reflex results: _____
 - 0 = no response
 - +1 = little response
 - +2 = normal response
 - +3 = above normal response
 - +4 = exaggerated reflex

2 Answer the discussion questions with your lab group.

Plantar Flexion

Plantar flexion is a superficial cord reflex that is an important neurological test. This test stimulates the cutaneous receptors and involves the brain in addition to the spinal cord. In adults, **plantar flexion** and flexed (curled) toes occurs when the plantar surface of the foot is stroked. A reaction of dorsiflexion with extended flared toes (Babinski's sign) is seen in infants because all the nerve fibers are not myelinated. The Babinski's sign is abnormal in adults.

ACTIVITY 6 **Plantar Flexion**

1 Test the plantar flexion reflex.
- Have the subject seated with a foot propped up and relaxed.
- Using the metal end of the reflex hammer, stroke the plantar surface of the foot starting at the heel, extending up the lateral side of the foot, and crossing over to the great toe area.
- Plantar flexion results: _____
 - 0 = no response
 - +1 = little response
 - +2 = normal response
 - +3 = above normal response
 - +4 = exaggerated reflex

2 Name the nerve tested: _____.

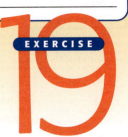

Reviewing Your Knowledge

A. Reflex Arc

1. Name the five components of a reflex arc in order.

 a. _____

 b. _____

 c. _____

 d. _____

 e. _____

2. How many neurons are in a monosynaptic reflex arc? _____ How many synapses are in the integrating center? _____

3. How many neurons are in a polysynaptic reflex arc? _____ How many synapses are in a polysynaptic reflex arc containing two interneurons in the integrating center? _____

4. Which type of neuron does the sensory neuron synapse with in a monosynaptic reflex arc? _____

5. Which type of neuron does the sensory neuron synapse with in a polysynaptic reflex arc? _____

B. Reflex Tests

Name the nerve that is tested in each of the following reflexes.

6. Achilles reflex _____

7. Biceps reflex _____

8. Patellar reflex _____

9. Plantar flexion reflex _____

10. Triceps reflex _____

11. Is a Babinski sign normal in adults? _____

C. Reflexes

1. Define reflex.

2. Describe the difference between a somatic and visceral reflex.

3. Describe the difference between a cranial and spinal reflex.

4. Describe the difference between an ipsilateral and contralateral reflex arc.

5. Define reciprocal innervation.

Using Your Knowledge

A. Flexor and Crossed Extensor Reflex

You are walking along the beach barefoot and step on a sharp object with your right foot. You immediately flex the right leg and balance yourself by extending the left leg. Label the following neurons in Figure 19.3.

1. Sensory neuron

2. Interneuron sending impulses up and down the spinal cord

3. Interneuron synapsing with motor neurons

4. Motor neuron causing flexion

5. Motor neuron causing extension

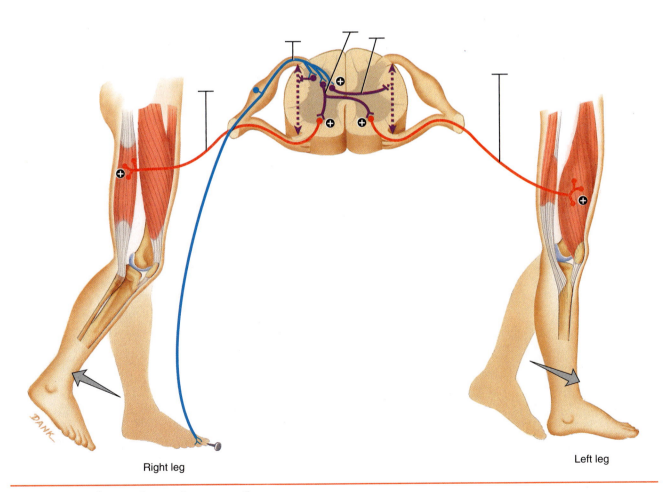

Right leg

Left leg

FIGURE 19.3 Flexor and crossed extensor reflex arc.

For the reflex arc in Figure 19.3, name the:

6. Nerve carrying motor information causing right leg flexion _____

7. Nerve carrying motor information causing left leg extension _____

8. Agonistic muscles for right leg flexion _____

9. Agonistic muscles for left leg extension _____

B. Modification of Reflex Activity

10. Reflex activity can be modified by the cerebral cortex. For example, you are baking a roast in the oven. As you are moving the roast from the oven to the counter, hot juices splash you. Reflex activity would cause you to drop the pan, but you hold on to it. Explain how the cerebral cortex overrides the reflex.

Brain Structure and Function

Objectives

**After completing this exercise,
you should be able to:**

- Identify the major external and internal structures of the brain on a model or chart
- Describe the basic functions of the principal structures of the brain
- Name the three meninges and describe their similarities and differences
- Identify the four ventricles of the brain and describe their function
- Trace the cerebrospinal fluid circulation
- Compare the anatomy of the sheep brain with that of the human brain

Materials

- human brain models and charts
- preserved human brain
- ventricular system model
- sheep brain dissection—preserved sheep brains, sheep brain models, dissecting instruments, dissection trays, disposable gloves, safety glasses

The human brain simultaneously conducts and coordinates a variety of incredible processes with which no computer is presently able to compete. The expansive development of the human brain distinguishes it from that of all other creatures. One of the largest organs in the human body, our brain is responsible for our memory, intellect, ideas, and behavior. The brain is the center for cataloging sensory information, integrating this information with previously recorded information, and producing actions based on the results of the information synthesis.

The brain is well protected, located within the cranium of the skull. During development, the brain and skull are growing simultaneously, so the shape of the cranial cavity mirrors the shape of the brain.

A. Major Brain Regions

The brain is composed of four main regions: the brain stem, cerebellum (*cerebel-* = little brain), diencephalon (*di-* = through; *encephal-* = brain), and cerebrum (*cerebr-* = brain). The **brain stem** is connected to the superior part of the spinal cord, the **cerebellum** is posterior to the brain stem, and the **diencephalon** is superior to the brain stem in the center of the brain. The large **cerebrum** is the dominant brain structure, with many folds and crevices enveloping the diencephalon.

319

2 Locate the structures on human brain models or charts.

3 Pronounce the terms as you point to them.

1 Label the terms in Figure 20.1.

Sagittal plane

CEREBRUM

Corpus callosum

Fornix

DIENCEPHALON:
Thalamus
Hypothalamus
Epithalamus
Pineal gland

BRAIN STEM:
Midbrain
Pons
Medulla oblongata

CEREBELLUM

Infundibulum

Pituitary gland

Spinal cord

POSTERIOR

ANTERIOR

Sagittal section of brain, medial view

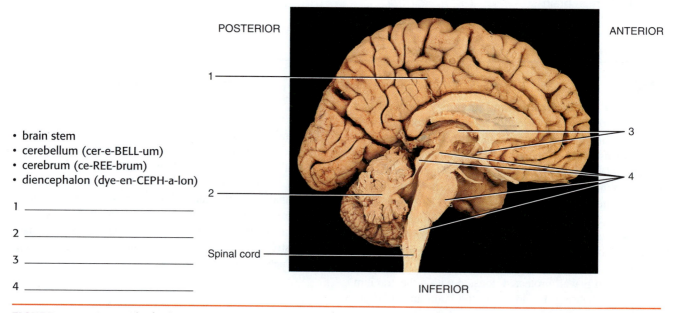

SUPERIOR

POSTERIOR

ANTERIOR

1

3

4

2

Spinal cord

INFERIOR

- brain stem
- cerebellum (cer-e-BELL-um)
- cerebrum (ce-REE-brum)
- diencephalon (dye-en-CEPH-a-lon)

1 _____

2 _____

3 _____

4 _____

FIGURE 20.1 Four major brain regions.

B. The Brain Stem

The brain stem is comprised of three structures: the medulla (*medull-* = marrow) oblongata, the pons, and the midbrain.

The Medulla Oblongata

The first brain structure, the **medulla oblongata,** is immediately superior to the spinal cord and is the most vital part of the brain because it houses the respiratory and cardiovascular control centers. The respiratory center controls the rate and depth of breathing, and the cardiovascular center is responsible for the rate and force of the heartbeat and blood pressure reflexes. Other reflex centers in the medulla are for coughing, vomiting, and sneezing.

The Pons

The **pons** (*pons* = bridge) is an expanded structure located superior to the medulla oblongata and anterior to the cerebellum, and has respiratory centers that assist the medulla oblongata in controlling breathing. The pons also relays information to the diencephalon and the cerebellum.

The Midbrain

The **midbrain** is a smaller area superior to the pons and inferior to the diencephalon, consisting of cerebral peduncles (*ped-* = foot) and the corpora quadrigemina (*corpora* = body; *quad-* = four; *gemin-* = twin; i.e., four twin bodies). The **cerebral peduncles** are white fibers that connect the upper and lower brain areas, and the **corpora quadrigemina** are composed of two superior colliculi (*colliculus* = small mound) and two inferior colliculi. The **superior colliculi** have reflex centers involved in eye, head, and neck movements with visual stimulation, whereas the **inferior colliculi** have reflex centers involved in auditory stimuli that result in head and trunk movements. The midbrain is observed best from a sagittal section, medial view. It spans from the dorsal to the ventral side of the brain.

ACTIVITY 2 **The Brain Stem**

1 Label the structures on Figure 20.2(a) and (b).

2 Identify the structures on a human brain model or chart.

3 Pronounce the terms as you point to them.

SUPERIOR

POSTERIOR

ANTERIOR

Spinal cord

INFERIOR

(a) Sagittal section, medial view

FIGURE 20.2a Brain stem.

- brain stem
- cerebral peduncle
- corpora quadrigemina (cor-POR-a quad-ri-GEM-i-na)
- inferior colliculus (col-LIK-you-lus)
- medulla oblongata
- pons
- superior colliculus

1 _____ 5 _____

2 _____ 6 _____

3 _____ 7 _____

4 _____

Midbrain

Pons

Middle cerebellar peduncle

Medulla oblongata

Brain stem

Spinal cord

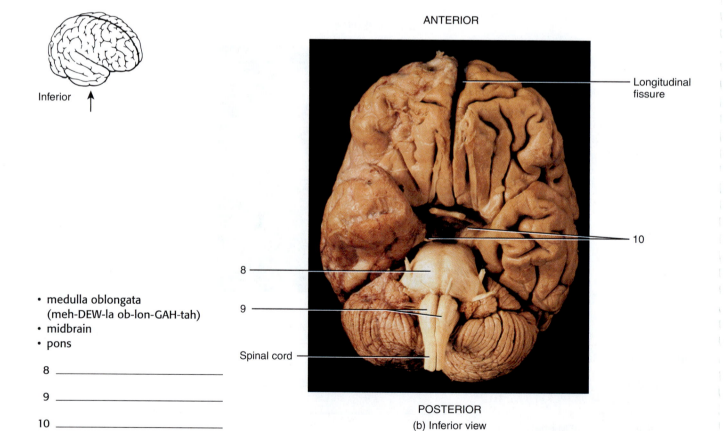

Inferior

ANTERIOR

Longitudinal fissure

10

8

9

Spinal cord

• medulla oblongata
 (meh-DEW-la ob-lon-GAH-tah)
• midbrain
• pons

8 _____

9 _____

10 _____

POSTERIOR

(b) Inferior view

FIGURE 20.2b Brain stem.

C. The Cerebellum

The **cerebel** (*cerebellum* = little brain) is second in size to the cerebrum and is located inferior to it and posterior to the medulla and pons. There are two **cerebellar hemispheres,** with a central area, the **vermis,** connecting them. When cut in sagittal section, gray matter can be observed on the exterior, with deeper white matter called the **arbor vitae** (*arbor* = tree; *vitae* = living) appearing as branches of a tree. The outer layer of gray matter is called the **cerebellar cortex.** Like the cerebral cortex, the cerebellar cortex has folds that increase the surface area for more neuron cell bodies. The cerebellar folds are slender, pleated gyri or **folia** (*folia* = leaves) that look similar to pages in a book.

The cerebellum regulates posture and balance and, in addition, smoothes and coordinates skilled skeletal muscle movements. The cerebellum is connected to the brain stem by the superior, middle, and inferior cerebellar peduncles. Do not confuse them with the cerebral peduncles.

ACTIVITY 3 **Cerebellum Identification**

1 Label the structures in Figure 20.3.

2 Identify these structures on a human brain model, picture, or chart.

3 Pronounce the terms as you point to them on the model.

(a) Posterior view
• cerebellar hemispheres (cer-e-BELL-ar)
• folia (FO-lia)
• vermis (VER-mis)

1 _____

2 _____

3 _____

(b) Sagittal view
• arbor vitae (white matter)
• cerebellar cortex (gray matter)
• cerebral peduncle
• folia
• inferior colliculus
• medulla oblongata
• pons
• superior colliculus

4 _____

5 _____

6 _____

7 _____

8 _____

9 _____

10 _____

11 _____

Anterior

Posterior

(a) Superior view

(b) Sagittal view

FIGURE 20.3 The cerebellum.

D. The Diencephalon

The **diencephalon** (*di-* = two; *-cephalon* = brain) is located in the brain's central area and has three main regions: the thalamus, the hypothalamus, and the epithalamus.

The Thalamus

The **thalamus** (*thala-* = inner chamber) is composed of paired, egg-shaped bodies centrally located in the diencephalon and comprising approximately 80% of this structure. Each cerebral hemisphere contains half of the thalamus, which is connected by a small bridge called the **intermediate mass.** The thalamus is the brain's "grand central relay station" because it is the principal relay station for sensory fibers and some somatic motor fibers. Sensory fibers that synapse in the thalamus are relayed to a particular area of the cerebral cortex to be interpreted, and other fibers relay messages to the somatic motor cortex. The thalamus also filters out unnecessary sensory information and is involved in consciousness, emotions, learning, and memory.

The Hypothalamus

The **hypothalamus** (*hypo-* = below) is located below the thalamus and is a quadrangular-shaped structure. The hypothalamus has important nuclei (a group of nerve cell bodies) that control many body functions and homeostasis. Some of the major functions include integrating and controlling the pituitary gland and hormonal functions, autonomic functions, emotions and behavior, body temperature, eating, and drinking.

The hypothalamus includes the infundibulum and mammillary bodies. The **infundibulum** (*infundibulum* = funnel) is a stalk that connects the pituitary gland to the hypothalamus. The **mammillary bodies** (*mammilla* = nipple)—two small, round masses located just posterior to the infundibulum—are relay stations for smell and taste reflexes.

Other structures that are observed in this area are the optic chiasm and the pituitary gland. The **optic chiasm** (*chi* = χ, Greek letter chi, a crossing over), the area where the optic nerves cross, is anterior to the infundibulum. The **pituitary gland** looks like a large pea and is attached to the end of the infundibulum. The hypothalamus controls the pituitary gland.

The Epithalamus

The **epithalamus** (*epi-* = above) is superior and posterior to the thalamus and includes the **pineal** (*pineas* = pinecone) **gland** or **body,** a small endocrine gland that secretes the hormone melatonin.

ACTIVITY 4 **The Diencephalon**

1 Label the parts of the diencephalon in Figure 20.4(a) and (b). Refer to Figure 20.1 if necessary.

2 Identify the structures listed on the human brain model.

3 Pronounce the terms as you point to them.

SUPERIOR

POSTERIOR ANTERIOR

- diencephalon
- hypothalamus (hypo-THAL-a-mus)
- pineal gland (in epithalamus) (pi-NEE-al)
- thalamus

1 _____

2 _____

3 _____

4 _____

INFERIOR

(a) Sagittal section, medial view

- hypothalamus
- infundibulum (in-fun-DIB-u-lum)
- intermediate mass of thalamus
- mammillary body (MAM-mil-lary)
- pineal gland
- pituitary gland
- thalamus (THAL-a-mus)

5 _____

6 _____

7 _____

8 _____

9 _____

10 _____

11 _____

Optic chiasm

(b) Sagittal section, magnified

FIGURE 20.4 The diencephalon.

E. The Cerebrum

The **cerebrum** is made up of right and left **cerebral hemispheres** and is the largest and most complex division of the brain. The cerebrum is superior to and surrounds the diencephalon and part of the brain stem. The cerebrum is the center of higher mental processes such as intelligence, communication, learning and memory, reasoning, and emotions. In addition, it interprets sensory input and initiates skeletal muscle contraction.

Organization of Cerebral Gray and White Matter

The three main regions of the cerebrum are the **cerebral cortex** (*cortex* = bark), **white matter,** and the deep **basal ganglia.** The **cerebral cortex** (also known as the cortical area) is the superficial gray matter on the exterior of the cerebrum composed of nerve cell bodies and dendrites. The **insula** is part of the cerebrum that lies deeper on the lateral sides of the brain. The cerebral cortex integrates sensory information, initiates motor output, and is also involved in emotions and intellectual processes. **Basal ganglia** (*ganglia* = knot) are areas of gray matter composed of paired nuclei (clusters of neuron cell bodies in the CNS) that are found deep within each cerebral hemisphere. The basal ganglia control automatic skeletal muscle movement and are involved with the limbic system or emotional brain.

White matter lies deep to the outer cortex and is comprised mostly of myelinated axons that give it the white appearance. These axons are organized into three kinds of fiber tracts in the cerebrum that are named according to the direction of the fibers. **Association fibers** transmit nerve impulses within the same hemisphere, whereas **commissural** (*commisura* = connection) **fibers** transmit nerve impulses between the two hemispheres. **Projection fibers** are ascending and descending tracts that project nervous impulses from inferior to superior brain areas or vice versa.

The **corpus callosum** (*corpus* = body; *callosus* = hard), a prominent commissural fiber tract that is readily observable in midsagittal sections of the brain, connects the two cerebral hemispheres. The **fornix** looks like a group of commissural fibers but is actually a tract of arched association fibers. The **internal capsule,** a large group of projection fibers, contains sensory and motor tracts that connect the cerebral cortex to the brain stem and spinal cord.

ACTIVITY 5 **Organization of Gray and White Matter in Cerebrum**

1 Label the structures in Figure 20.5(a) and (b) using Figure 20.1 if needed.

2 Identify these structures on a human coronal section of brain or brain model.

3 Pronounce the terms as you point to them.

Frontal section

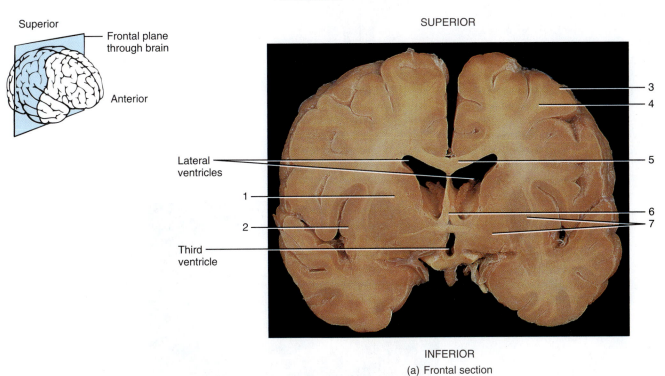

(a) Frontal section

FIGURE 20.5a Gray and white matter in the cerebrum.

- basal nuclei (gray matter)
- cerebral cortex (gray matter)
- corpus callosum (white fibers)
- fornix (white fibers)
- insula (gray matter)
- internal capsule (white fibers)
- white matter

1 _____ 5 _____

2 _____ 6 _____

3 _____ 7 _____

4 _____

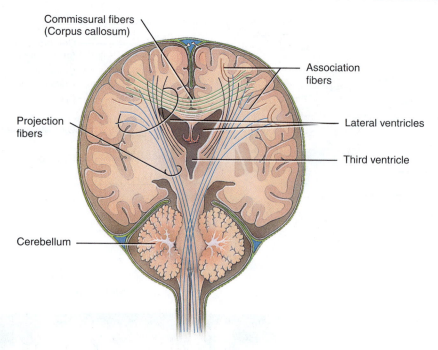

Commissural fibers
(Corpus callosum)

Association fibers

Projection fibers

Lateral ventricles

Third ventricle

Cerebellum

Frontal section

Superior

Midsagittal plane through brain

Anterior

9

SUPERIOR

Cerebral cortex

8

10

11

POSTERIOR

ANTERIOR

INFERIOR

(b) Midsagittal section

FIGURE 20.5b White fiber types.

- association fibers
- association fibers (fornix)
- commissural fibers (corpus callosum)
- projection fibers

8 _____

9 _____

10 _____

11 _____

Surface Features of the Cerebrum

There are four lobes composing the exterior of the cerebrum that are mainly named for the overlying cranial bones. These lobes are the **frontal, parietal, occipital,** and **temporal lobes.** An inner lobe, the **insula,** lies deep to the lateral cerebral fissure and is not visible from the exterior.

Other obvious external anatomical features of the cerebrum include the following:

- **gyrus** (*gyros* = circle; **gyri,** plural)—elevation or fold in the cerebral cortex; increases the surface area for neuron cell bodies
- **sulcus** (*sulcus* = furrow; **sulci,** pl.)—shallow groove between elevations
- **central sulcus**—shallow groove separating frontal lobe from parietal lobe
- **precentral gyrus**—elevation located just anterior to the central sulcus
- **postcentral gyrus**—elevation located just posterior to the central sulcus
- **lateral cerebral sulcus**—shallow groove separating frontal and temporal lobes
- **parieto-occipital sulcus**—shallow groove separating parietal and occipital lobes
- **longitudinal fissure**—deep groove separating the two cerebral hemispheres at the midline
- **transverse fissure**—deep groove separating the cerebrum from the cerebellum in the posterior/inferior part of the brain

ACTIVITY 6 **External Features of the Cerebrum**

1 Label the structures listed for Figure 20.6(a), (b), and (c).
2 Identify these features on a brain model or chart.
3 Pronounce the terms as you point to them on a model.

Postcentral gyrus
Parietal lobe (blue area)
Parieto-occipital sulcus
Occipital lobe
Transverse fissure
Cerebellum
Central sulcus
Precentral gyrus
Frontal lobe
Insula (gold area)
Lateral cerebral sulcus
Temporal lobe

Right lateral view

- central sulcus
- frontal lobe
- lateral cerebral sulcus
- occipital lobe
- parietal lobe
- parietal–occipital sulcus
- postcentral gyrus
- precentral gyrus
- temporal lobe
- transverse fissure

1 _____
2 _____
3 _____
4 _____
5 _____
6 _____
7 _____
8 _____
9 _____
10 _____

(a) Right lateral view

FIGURE 20.6 External features of the cerebrum.

ANTERIOR

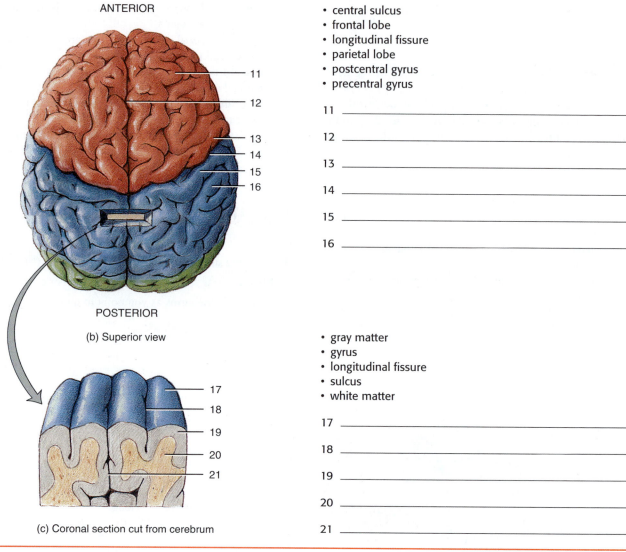

POSTERIOR

(b) Superior view

(c) Coronal section cut from cerebrum

- central sulcus
- frontal lobe
- longitudinal fissure
- parietal lobe
- postcentral gyrus
- precentral gyrus

11 _____

12 _____

13 _____

14 _____

15 _____

16 _____

- gray matter
- gyrus
- longitudinal fissure
- sulcus
- white matter

17 _____

18 _____

19 _____

20 _____

21 _____

FIGURE 20.6 External features of the cerebrum, *continued.*

Functional Areas of the Cerebral Cortex

The cerebral cortex is composed of three types of functional areas: motor, sensory, and association areas. **Sensory areas** receive and interpret impulses from sensory receptors, while **motor areas** initiate impulses to skeletal muscles. **Association areas,** which perform complex integrative functions, receive and send information to multiple areas of the cortex via association fibers. The majority of the cortex is composed of association areas.

Motor Areas

- **primary motor cortex**—located in the precentral gyrus of each frontal lobe; initiates impulses to skeletal muscles
- **Broca's speech area**—located superior to the lateral sulcus and anterior to the primary motor cortex, usually in the left hemisphere; initiates impulses that result in speech

Sensory Areas

- **primary somatosensory cortex**—located in the postcentral gyrus of each parietal lobe; receives nerve impulses for touch, proprioception, pain, and temperature
- **primary auditory area**—located in each temporal lobe across the lateral sulcus from the gustatory area; receives impulses when the auditory receptors of the ear are stimulated
- **primary gustatory area**—in each postcentral gyrus, just superior to the lateral sulcus; receives impulses when the taste buds are stimulated
- **primary olfactory area**—located on the medial side of each temporal lobe; cannot be seen from the lateral view; receives impulses when the olfactory receptors of the nose are stimulated
- **primary visual area**—in the posterior occipital lobe; receives impulses from the thalamus when the retina is stimulated

Selected Association Areas

- **Wernicke's area**—recognizes spoken words, translates words into thoughts, and possibly helps us sound out strange or new words
- **somatosensory, visual, and auditory association areas**—larger areas adjacent to the corresponding sensory cortex; integrate sensory information from the sensory cortex with past experiences allowing us, for example, to identify objects by touch or to identify sound as music or speech

ACTIVITY 7 **Functional Areas of Cerebral Cortex**

1 Label the brain cortical area with its function in Figure 20.7.
2 Pronounce the terms as you point to the area on a human brain model.

- Broca's speech area
- central sulcus
- primary auditory cortex
- primary gustatory cortex (GUS-tah-tory)
- primary motor cortex
- primary somatosensory cortex (so-ma-to-SEN-sory)
- primary visual cortex
- Wernicke's area

1 _____
2 _____
3 _____
4 _____
5 _____
6 _____
7 _____
8 _____

ANTERIOR Right lateral view POSTERIOR

FIGURE 20.7 Functional areas of the cerebral cortex.

F. Protection of the Brain

The brain is protected both physically and chemically by cranial bones, the blood–brain barrier, the cranial meninges, and cerebrospinal fluid.

Cranial Bones

The cranial bones form a vault called the cranium, which surrounds and protects the brain. The floor of the cranium contains depressions—the anterior, middle, and posterior cranial fossae—which support parts of the brain. The **anterior cranial fossae** support the frontal lobes of the cerebrum; the **middle cranial fossae** support portions of the temporal and parietal lobes of the cerebrum and the diencephalon; the **posterior cranial fossae** support portions of the temporal, parietal, and occipital lobes of the cerebrum, the cerebellum, and the brain stem.

Cranial Meninges

There are three cranial meninges (connective tissue membranes): the dura mater, arachnoid mater, and pia mater. The **dura mater** (*dura* = hard; *mater* = mother) is the first meninx (sing.) located deep to the cranial bones. It is comprised of two layers: the **periosteal layer,** which is a tough membrane attached to the cranial bones, and the **meningeal layer,** which is exterior to the arachnoid mater. The two dural layers split to form the **dural sinuses** that eventually drain cranial blood into the jugular veins. The **superior sagittal sinus,** located superior to the longitudinal fissure, is one of the main dural sinuses. The double-layered dura mater extends deep into the longitudinal fissure forming the **falx** (*falx* = sickle-shaped) **cerebri,** into the transverse fissure forming the **tentorium** (tent) **cerebelli,** and between the cerebellar hemispheres forming the **falx cerebelli.**

The **arachnoid** (*arachnoid* = spiderlike) **mater** is the second meninx located deep to the dura mater. The **pia** (*pia* = delicate) **mater** is the thin, inner meninx. It hugs and overlays the cerebral cortex, following each gyrus and sulcus. Between the arachnoid and pia is the **subarachnoid** space.

ACTIVITY 8 Cranial Bones and Cranial Meninges

1 Identify the cranial fossae in the cranial floor.
2 Observe the slight depressions in the anterior cranial fossa. These depressions mirror the gyri of the frontal lobes.
3 Label the structures on Figure 20.8.
4 Pronounce these terms as you label the figure.
5 Using a model, show where the falx cerebri and tentorium cerebelli would be found.

FIGURE 20.8 Cranial meninges.

- arachnoid mater (a-RAK-noid MAH-ter)
- arachnoid villus (VIL-us)
- cerebral cortex (ce-REE-bral)
- falx cerebri (FAL-x ce-REE-bree)
- meningeal layer of dura mater
 (meh-NIN-gee-ul)
- parietal bone
- periosteal layer of dura mater
 (peri-OS-tee-ul)
- pia mater (PEE-ah)
- subarachnoid space (sub-a-RAK-noid)
- superior sagittal sinus (SA-jih-tahl)
- white matter

1 _____ 7 _____

2 _____ 8 _____

3 _____ 9 _____

4 _____ 10 _____

5 _____ 11 _____

6 _____

Cerebrospinal Fluid and Ventricles of the Brain

Cerebrospinal fluid (CSF) constantly bathes the brain and spinal cord with oxygen, nutrients, and vital chemicals. Although different in content, CSF is made from blood plasma that leaks out of specialized, tiny blood vessels (capillaries) called the **choroid plexus** (*choroid* = membranelike; *plexus* = pleated), and passes through ependymal cells into four small brain cavities or ventricles. The ependymal cells have cilia that move the CSF in one direction. There are choroid plexuses in the roof of all four ventricles. A **lateral ventricle** is located in each **cerebral hemisphere** with a thin membrane, the **septum pellucidum** (*septum* = partition; *pellucid* = transparent), separating the two ventricles anteriorly. Each arched lateral ventricle has an **interventricular foramen** that opens medially into the third ventricle. The **third ventricle** is medially located between the paired masses of the thalamus and is narrower and smaller than the other ventricles. Connecting the third ventricle to the fourth ventricle is a thin tube, the **cerebral aqueduct** (*aqua* = water; *duct* = way; i.e., waterway). The **fourth ventricle** is located between the **pons** and the **cerebellum. Lateral** and **median apertures** or openings allow the CSF to flow from the fourth ventricle into the **subarachnoid space** surrounding the brain and the spinal cord. CSF also flows through the central canal of the spinal cord and back out into the subarachnoid space. Just as CSF enters the brain ventricles from the bloodstream, it is returned to the blood by reabsorption through the **arachnoid villi** (tiny projections) located in the **superior sagittal sinus.**

ACTIVITY 9 CSF Circulation

1 Label the structures in Figure 20.9(a) and (b).

2 Refer to Figure 20.5(a) and (b) to observe the lateral and third ventricles in a frontal section of the brain.

3 Identify these structures on a human brain model or chart.

4 Pronounce the terms as you point to them.

5 With your laboratory partner, trace a drop of CSF from its origin in a lateral ventricle through the other ventricles and subarachnoid space, and follow it until it is reabsorbed into the bloodstream.

- central canal of the spinal cord
- cerebral aqueduct (AH-que-duct)
- fourth ventricle
- interventricular foramen (in-ter-ven-TRIK-u-lur)
- lateral ventricles
- third ventricle

1 _____

2 _____

3 _____

4 _____

5 _____

6 _____

(a) Right lateral view (ventricles superimposed)

- arachnoid villus (a-RACH-noid VIL-us)
- central canal
- cerebral aqueduct
- choroid plexus (CHOR-oid PLEX-us)
- fourth ventricle
- lateral ventricle
- subarachnoid space (sub-ah-RAK-noid)
- superior sagittal sinus
- third ventricle

7 _____

8 _____

9 _____

10 _____

11 _____

12 _____

13 _____

14 _____

15 _____

(b) Sagittal section of brain and spinal cord

FIGURE 20.9 Ventricles of the brain and cerebrospinal fluid circulation.

G. Sheep Brain Dissection

Remember that preserved material does not look like a fresh specimen. Usually more detail may be observed in a preserved brain because the tissue is firmer.

ACTIVITY 10 **Dissection of Sheep Brain**

1 Rinse the sheep brain to remove preservative.

2 Observe meninges and main brain regions.
- Examine the brain to see if the tough, outer **dura mater** is present. If present, note the **falx cerebri** and **tentorium cerebelli.** Carefully remove the dura mater.
- Now look for the stringy, weblike **arachnoid mater** beneath and adhering to the dura mater. Deep to this membrane is the very thin **pia mater** that follows the contour of the gyri and sulci.
- Compare the sheep brain with the main external regions of the human brain. Identify the **cerebrum, brain stem (medulla** and **pons),** and **cerebellum.**

3 Identify dorsal structures (Figure 20.10).
- With the dorsal side up, identify the **cerebral hemispheres, gyri, sulci, longitudinal fissure,** and **transverse fissure.**
- Note the four main **lobes** of the brain—**frontal, parietal, occipital,** and **temporal.**
- At the longitudinal fissure, gently separate the two parts and look down between them for the thick band of white fibers, the **corpus callosum.**

4 Identify ventral structures (Figure 20.11).
- Place the sheep brain ventral side up.
- Identify the **olfactory bulb, olfactory tract, optic nerve, optic chiasm,** and **optic tract.** Why are the olfactory tracts not called olfactory nerves?
- Posterior to these structures, locate the *one* large **mammillary body.** How does the human mammillary body look different from the sheep?
- If present, identify the **infundibulum** and **pituitary gland.**
- Look at the three parts of the **brain stem—the midbrain, pons,** and **medulla oblongata.** How do these compare with the human brain?
- The large **trigeminal nerves** (V) are located laterally at the junction of the pons and medulla.
- The small **abducens nerves** (VI) are found medial and slightly posterior to the trigeminal nerves, arising from the pons.

5 Identify dorsal structures (Figure 20.12).
- Carefully pull the cerebellum away from the cerebrum. Identify the **pineal body, superior colliculi,** and **inferior colliculi.**

6 Identify midsagittal section structures (Figure 20.13).
- Using a sharp knife or scalpel, carefully make a midsagittal section.
- Locate the **brain stem** components: the **medulla oblongata,** the **pons,** and the **midbrain.** Compare this with your human brain model.
- Identify the **arbor vitae** in the **cerebellum.** Does this structure look similar to or different from the human brain?
- Note the **cerebral aqueduct** and the **fourth ventricle.**
- Identify the **thalamus, corpus collosum, septum pellucidum, fornix,** and **lateral ventricles.**

7 Observe coronal and transverse sections.
- Observe the coronal section your instructor may have as a demonstration. Note the **gray matter, white matter, longitudinal fissure, corpus callosum, thalamus, lateral ventricles,** and **third ventricle.**
- Observe the transverse or horizontal section your instructor may have as a demonstration. Identify the **gray matter, white matter,** and **ventricles.**

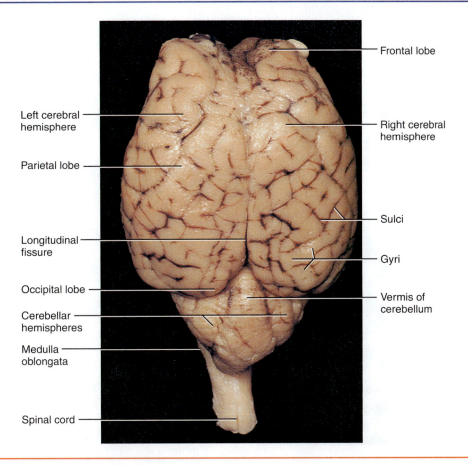

FIGURE 20.10 Dorsal view of the sheep brain.

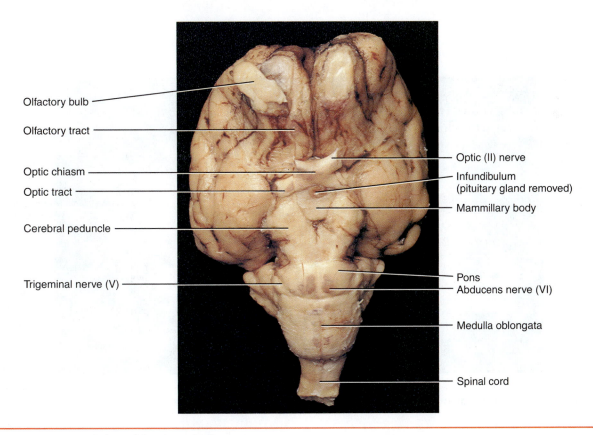

FIGURE 20.11 Ventral view of the sheep brain.

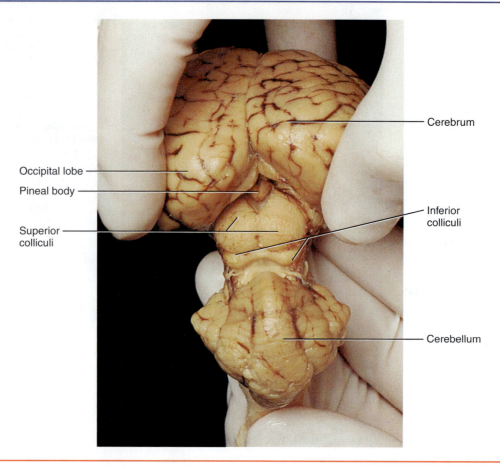

FIGURE 20.12 Posterior view of the sheep brain.

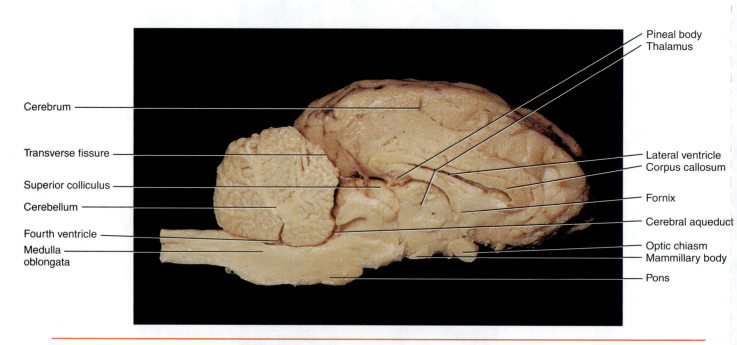

FIGURE 20.13 Midsagittal section of the sheep brain.

Reviewing Your Knowledge

A. Anatomy Drill and Practice

Go to www.wiley.com/college/allen. Click on the picture of your lab manual, then on Student Companion Site, Exercise 20, and Anatomy Drill and Practice.

B. Structure of Brain Regions

Label the structures on Figure 20.14 in the spaces below.

1. _____

2. _____

3. _____

4. _____

5. _____

6. _____

7. _____

8. _____

9. _____

10. _____

11. _____

12. _____

View

ANTERIOR

(Lobe) 1

2

3

(Lobe) 4

5

6

7

8

9

10

11

(Lobe) 12

POSTERIOR

Inferior aspect of brain

FIGURE 20.14 Ventral view of the human brain.

C. Flow of Cerebrospinal Fluid

Fill in the numbered blanks with the name of the structure that corresponds with the number in the following paragraph.

Fluid for the CSF is derived from the bloodstream. The sites of CSF formation are the (1), special tiny capillaries located in the walls of (2), (3), and (4). Cells that line the ventricles have cilia that move the CSF and are called (5). The two lateral ventricles are separated by a thin membrane called the (6). CSF flows by cilia movement from the two lateral ventricles through the interventricular foramen to the (7). From here, the CSF flows through the (8) into the fourth ventricle. The CSF leaves the fourth ventricle through three openings: the median aperture and two lateral apertures. CSF is now located in the (9) space around the brain, and circulates all around the cerebrum and cerebellum. CSF continues to flow into the inner part of the spinal cord by flowing through the tiny (10) of the spinal cord, as well as around the exterior of the spinal cord in the (11) space. Because CSF is continually being made at the rate of about 20 mL/hr, it has to exit back into the bloodstream by being reabsorbed through the (12) that protrude into the dural venous sinuses. The venous sinus that overlies the brain superiorly is called the (13).

1. _____

2. _____

3. _____

4. _____

5. _____

6. _____

7. _____

8. _____

9. _____

10. _____

11. _____

12. _____

13. _____

D. Functions of Brain Regions

Give the brain region for the functions described.

_____ 1. Contains vital centers that regulate heartbeat, breathing, blood pressure, vomiting, coughing.

_____ 2. Smoothes and coordinates skilled skeletal muscle movement; also posture and balance or equilibrium.

_____ 3. Secretes melatonin, that controls the sleep-wake cycle.

_____ 4. Controls and integrates the autonomic nervous system; regulates hormones, emotional behavior, temperature, eating, and drinking behavior.

_____ 5. Interprets sensory input, controls skilled skeletal muscle movements, and is involved in emotional and intellectual processes.

_____ 6. Helps control breathing; conducts impulses to and from the cerebellum, midbrain, and medulla.

_____ 7. Relays all sensory input to the cerebral cortex; involved in skeletal muscle actions and memory processing.

_____ 8. Coordinates visual and auditory reflexes.

_____ 9. Coordinates gross, automatic muscle movements; also involved with limbic system.

_____ 10. White fiber tracts communicating between hemispheres.

E. Location of Functional Areas of the Cerebral Cortex

Match the areas of the cerebral cortex with the location described.

a. Broca's speech area (motor)
b. primary auditory area
c. primary gustatory area
d. primary motor area
e. primary somatosensory area
f. primary visual area
g. Wernicke's area

_____ **1.** precentral gyrus

_____ **2.** postcentral gyrus

_____ **3.** anterior to the primary motor cortex in frontal lobe

_____ **4.** tip of occipital cortex

_____ **5.** inferior end of postcentral gyrus near lateral cerebral sulcus

_____ **6.** parietal lobe superior and posterior to lateral cerebral sulcus

_____ **7.** posterior to gustatory cortex in superior part of temporal lobe; near the lateral cerebral fissure

Label Figure 20.15 with the functional areas of the cerebral cortex.
Color each complete area with a different color pencil.

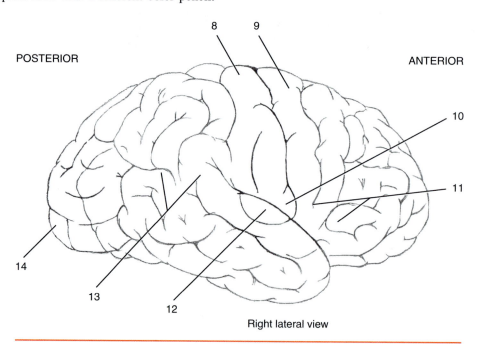

POSTERIOR ANTERIOR

Right lateral view

FIGURE 20.15 Functional areas of the cerebral cortex.

8 _____ 12 _____

9 _____ 13 _____

10 _____ 14 _____

11 _____

Using Your Knowledge

FIGURE 20.16 Midsagittal section of the human brain, CT scan.

A. CT Scan of the Human Brain

Identify the structures in Figure 20.16, midsagittal view.

1. _____

2. _____

3. _____

4. _____

Superior

Transverse plane through brain

Anterior

ANTERIOR

5

6
7

8

POSTERIOR

FIGURE 20.17 Human brain, transverse section.

B. Transverse Section of a Human Brain

Identify the structures in Figure 20.17, a transverse section.

5. _____

6. _____

7. _____

8. _____

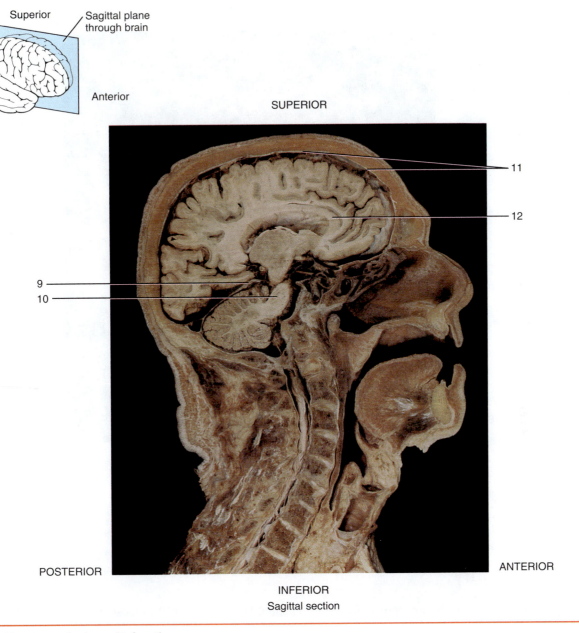

FIGURE 20.18 Human brain, sagittal section.

C. Sagittal Section of Human Brain

Identify the structures in Figure 20.18, sagittal section.

9. _____

10. _____

11. _____

12. _____

SUPERIOR

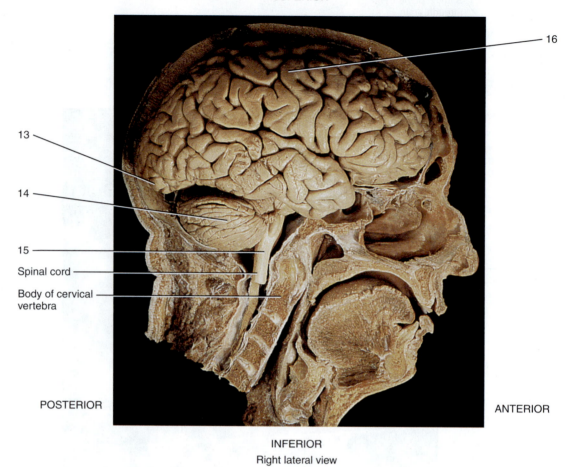

POSTERIOR

ANTERIOR

INFERIOR

Right lateral view

FIGURE 20.19 Human brain, lateral view.

D. Brain Injury

Match the brain injury to the appropriate change in function. Use Figure 20.19 to locate the area injured.

 a. cessation of breathing
 b. loss of equilibrium
 c. loss of use of left arm
 d. loss of vision

_____ **13.** The effect of a blow to the back of the head that damages this area.

_____ **14.** The effect of alcohol on this area.

_____ **15.** The effect of a head injury (i.e. diving into a pool) that forces the dens into this area.

_____ **16.** The effect of a stroke that damages this area.

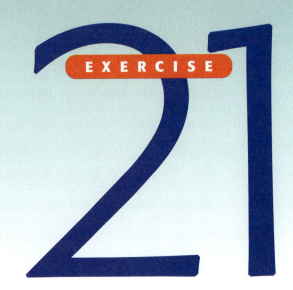
Cranial Nerves

Objectives

**After completing this exercise,
you should be able to:**

- Identify the 12 pairs of cranial nerves by name and Roman numeral on brain models and/or preserved human brains
- State the function of the 12 pairs of cranial nerves
- Test cranial nerve function
- Identify specific cranial nerves on the sheep brain

Materials

- human brain models or charts
- preserved human brain (if available)
- skull with or without cranial nerves
- preserved sheep brain
- dissection equipment and trays
- disposable gloves and safety glasses
- materials for testing cranial nerve function—spice or aromatic oil, penlight, cotton, ice water, sugar, quinine, tuning fork, tongue depressors

A. Name and Location of Cranial Nerves

Cranial nerves are categorized as a part of the peripheral nervous system. The 12 pairs of cranial nerves exit from the brain, are numbered sequentially from anterior to posterior, and pass through foramina to reach their destinations. The nerve's name sometimes indicates the structure it innervates or its function. The names of the cranial nerves I–XII are in order as follows: **olfactory** (*olfactus* = sense of smell), **optic, oculomotor** (*oculo-* = eye; *-motor* = mover), **trochlear** (*trochlea* = pulley), **trigeminal** (*tri* = three; *gemini* = twins), **abducens** (*ab-* = away; *-ducens* = to lead), **facial, vestibulocochlear** (*vestibulo-* = vestibule of ear/equilibrium; *cochlear* = cochlea of ear/hearing), **glossopharyngeal** (*glosso-* = tongue; *-pharyngeal* = throat), **vagus** (*vagus* = wandering), **accessory,** and **hypoglossal** (*hypo-* = below; *-glossal* = tongue). The vagus nerve has the distinction of being different from the other 11 cranial nerves in its distribution. As its word derivative states, it is the "wandering nerve" that branches extensively and innervates the viscera of the thoracic and abdominopelvic cavities.

You may want to use the following mnemonic device to help remember the names of the cranial nerves in order: "**On Old Olympus' Towering Top A Friendly Viking Grew Vines And Hops.**" The bold first letter of each word matches with the first letter in the name of the cranial nerve.

The olfactory nerves are very small and will not be seen in Figure 21.1. The olfactory bulbs and tracts can be observed and are marked instead.

When locating the foramen for the cranial nerves, note that cranial nerves III, IV, VI, and the ophthalmic division of V, all exit the same foramen, the superior orbital fissure. Cranial nerves IX, X, and XI all exit the jugular foramen.

ACTIVITY 1 Name and Location of Cranial Nerves

1 Learn the mnemonic for recalling the names of the cranial nerves.

2 Label the cranial nerves on Figure 21.1.

3 Identify each cranial nerve on a human brain model or chart.

4 Locate the foramen for each cranial nerve on a skull with or without cranial nerves. Refer to Table 21.1.

5 Pronounce the names as you point to them on the brain model or chart.

TABLE 21.1 Foramen of Cranial Nerves

I	Olfactory	Olfactory foramina of cribriform plate
II	Optic	Optic foramen
III	Oculomotor	Superior orbital fissure
IV	Trochlear	Superior orbital fissure
V	Trigeminal	Ophthalmic—Superior orbital fissure Maxillary—Foramen rotundum Mandibular—Foramen ovale
VI	Abducens	Superior orbital fissure
VII	Facial	Stylomastoid foramen
VIII	Vestibulocochlear	Internal acoustic canal (meatus)
IX	Glossopharyngeal	Jugular foramen
X	Vagus	Jugular foramen
XI	Accessory	Jugular foramen
XII	Hypoglossal	Hypoglossal canal

- Olfactory bulb
- Optic nerve
- Trochlear nerve
- Abducens nerve
- Vestibulo-cochlear nerve
- Vagus nerve
- Hypoglossal nerve
- Oculomotor nerve
- Trigeminal nerve
- Facial nerve
- Glossopharyngeal nerve
- Accessory nerve

- abducens (ab-DUE-senz)
- accessory
- facial
- glossopharyngeal (gloss-oh-fah-RIN-jeal)
- hypoglossal (hypo-GLOSS-al)
- oculomotor (ok-u-low-MO-tor)
- olfactory (OHL-fac-tory) tract
- optic
- trigeminal (tri-GEM-i-nal)
- trochlear (TROH-klee-ur)
- vagus (VAY-gus)
- vestibulocochlear (ves-tib-u-lo-COKE-lee-ur)

1 _____

2 _____

3 _____

4 _____

5 _____

6 _____

7 _____

8 _____

9 _____

10 _____

11 _____

12 _____

FIGURE 21.1 Cranial nerves of the human brain.

B. Testing Cranial Nerve Functions

Not all cranial nerves function in the same way. Some cranial nerves are primarily special sensory in function, others are primarily motor in function, and still others have both sensory and motor fibers. A mnemonic device for recalling the function of cranial nerves is: "**S**ome **S**ay **M**arry **M**oney **B**ut **M**y **B**rother **S**ays **B**ad **B**usiness **M**arry **M**oney." The three types are: **S** = sensory; **M** = motor; **B** = both sensory and motor (mixed nerve). Motor nerves also have sensory proprioceptors located in the muscles they innervate, but their main function is motor. Cranial nerve VIII has been shown to have some motor activity, but its main function is sensory. Two of the cranial nerves have branches. The trigeminal nerve (V) has three branches: ophthalmic, maxillary, and mandibular. The accessory nerve (XI) has a cranial portion and spinal portion. In addition to somatic motor fibers, cranial nerves III, VII, IX, and X also have parasympathetic motor fibers.

An easy way to remember the names of the cranial nerves that innervate the eyeball is by using the following formula: LR_6SO_4 = the lateral rectus muscle (LR) is innervated by cranial nerve VI; the superior oblique muscle (SO) is innervated by cranial nerve IV; the other four external muscles of the eyeball are innervated by cranial nerve III.

TABLE 21.2 Cranial Nerve Function, Distribution, and Action

NUMBER AND NAME	DISTRIBUTION	ACTION
I Olfactory	Nasal mucosa	Smell
II Optic	Eye	Vision
III Oculomotor	Levator palpebrae superioris; four extrinsic eye muscles (inferior oblique, superior rectus, medial rectus, inferior rectus); ciliary muscle (intrinsic eye muscle); iris muscles of eye (intrinsic eye muscles)	Movement of eyelid; movement of eyeball; accommodation of lens; pupillary constriction
IV Trochlear	Superior oblique muscle of eyeball	Movement of eyeball
V Trigeminal	Ophthalmic branch Maxillary branch Mandibular branch	Cutaneous sensations from ophthalmic, maxillary, and mandibular areas; chewing
VI Abducens	Lateral rectus muscle of eyeball	Movement of eyeball
VII Facial	Anterior 2/3 of tongue; facial, scalp, and neck muscles; lacrimal glands; salivary glands	Taste; facial expression; secretion of tears; salivation
VIII Vestibulocochlear	Semicircular canals and cochlea of ear	Equilibrium and hearing
IX Glossopharyngeal	Posterior 1/3 of tongue; pharyngeal muscles; parotid gland	Taste; swallowing and speech; secretion of saliva
X Vagus	Pharyngeal muscles and epiglottis; smooth muscles of thorax and GI tract; cardiac muscle; glands of GI tract	Taste and somatic sensation from pharynx and epiglottis; swallowing, coughing, and voice production; smooth muscle contraction of GI tract; slows heart rate; secretion by digestive glands
XI Accessory	Cranial portion—muscles of pharynx, larynx, and soft palate	Swallowing
	Spinal portion—sternocleidomastoid and trapezius muscles	Movement of head and shoulders
XII Hypoglossal	Tongue muscles	Speech and swallowing

ACTIVITY 2 Functions of Cranial Nerves, Their Distributions and Actions

1 Perform the cranial nerve tests to determine if function is normal, consulting Table 21.2 for each nerve action. Record results in Table 21.3 by circling either normal (N) or abnormal (A) for each nerve.

> **SAFETY NOTE:** INSTRUCTIONS FOR TESTING CRANIAL NERVE FUNCTION *Anything that was used in the mouth (tongue depressor or cotton-tipped swab or applicator) needs to be immediately placed in an autoclavable bag after use. Do not place these used items on your lab bench.*

TABLE 21.3 Testing Cranial Nerve Function

NUMBER	CRANIAL NERVE QUICK TEST	RESULTS	
I	Sniff aromatic oil, spice, or coffee grounds (ability to smell is normal)	N	A
II	Read a Snellen eye chart at 20 ft. (ability to read is normal)	N	A
III, IV, VI	• Observe eyelids (eyelid not drooping is normal)	N	A
	• Look up, down, medially, laterally, upper lateral, lower lateral (ability to move eyeball is normal)	N	A
	• Shine penlight on pupils (constriction is normal)	N	A
V	• Lightly touch the cornea of the subject with a wisp of cotton (blinking is normal)	N	A
	• Subject bites down on tongue depressor; try to pull it out (good resistance to pulling is normal)	N	A
VII	• Have subject smile, raise eyebrows, whistle, and close the eyes (ability to do these actions is normal)	N	A
	• Check tip of tongue for taste with sugar (ability to taste is normal)	N	A
VIII	• Use a tuning fork to check hearing in each ear (ability to hear is normal)	N	A
	• Have the subject stand straight with eyes closed (absence of swaying is normal)	N	A
IX, X	• Have subject say, "Ah"; check uvula position with mouth open (midline is normal)	N	A
	• Using a cotton-tipped applicator, *gently* touch the subject's uvula to elicit a gag reflex (gag response is normal)	N	A
	• The posterior 1/3 of tongue can be tested with cotton applicator dipped in quinine (taste is normal)	N	A
XI	• Have subject shrug shoulders against resistance as you are holding them down (both sides having equal strength is normal)	N	A
XII	• Ask your partner to stick out and retract his/her tongue (tongue not deviating to one side is normal)	N	A

Reviewing Your Knowledge

A. Anatomy Drill and Practice

Go to www.wiley.com/college/allen. Click on the picture of your lab manual, then on Student Companion Site, Exercise 21, and Anatomy Drill and Practice.

B. Cranial Nerve Numbers

Give the Roman numeral for the 12 pairs of cranial nerves.

_____ **1.** Abducens

_____ **2.** Accessory

_____ **3.** Facial

_____ **4.** Glossopharyngeal

_____ **5.** Hypoglossal

_____ **6.** Oculomotor

_____ **7.** Olfactory

_____ **8.** Optic

_____ **9.** Trigeminal

_____ **10.** Trochlear

_____ **11.** Vagus

_____ **12.** Vestibulocochlear

C. Cranial Nerve Function

Identify if each cranial nerve is mainly sensory, motor, or both.

S = sensory M = motor B = both sensory and motor

_____ **1.** Olfactory

_____ **2.** Optic

_____ **3.** Oculomotor

_____ **4.** Trochlear

_____ **5.** Trigeminal

_____ **6.** Abducens

_____ **7.** Facial

_____ **8.** Vestibulocochlear

_____ **9.** Glossopharyngeal

_____ **10.** Vagus

_____ **11.** Accessory

_____ **12.** Hypoglossal

Using Your Knowledge

A. Normal Cranial Nerve Function

Give the Roman numeral of the cranial nerve(s) involved with the following functions or activities.

_____ **1.** Hearing the crack of a bat hitting a baseball

_____ **2.** Smelling dinner cooking

_____ **3.** Being dizzy after going on a Tilt-a-Whirl ride at the fair

_____ **4.** Lifting your shoulders while doing warm-up exercises

_____ **5.** Chewing a tender steak

_____ **6.** Swallowing the tender steak

_____ **7.** Salivation and crying

_____ **8.** Reading a book

_____ **9.** Seeing a sunset

_____ **10.** Looking cross-eyed

_____ **11.** Smiling

_____ **12.** Moving the head from side to side

_____ **13.** Feeling a hot curling iron touching the forehead

_____ **14.** Decreased heart rate

_____ **15.** Rolling your eyes

_____ **16.** All of the cranial nerves that are involved with the digestive system in some way, starting with the mouth

_____ **17.** Speaking

B. Cranial Nerve Injury

Write the name of the cranial nerve(s) involved.

18. Injury to this cranial nerve causes Bell's palsy—a loss of taste, decreased salivation, and loss of the ability to close the eyes, even during sleep. Name this nerve.

19. Injury to this cranial nerve causes a loss of taste sensation, decreased salivation, and difficulty in swallowing. Name this nerve.

20. Injury to this cranial nerve causes paralysis of the vocal cords, interferes with swallowing, interrupts sensations from many organs, and causes the heart rate to increase. Name this nerve.

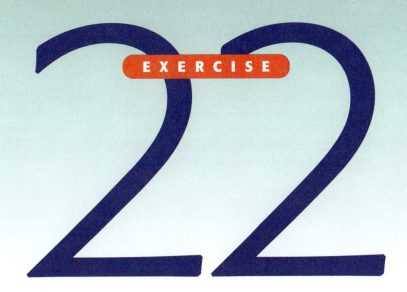

General Senses

Sensory receptors provide information about the environment outside and within our body. General sensory information is either from somatic sensory receptors (skin and skeletal muscle) or visceral sensory receptors (visceral organs). Special sensory information is from special sense organs (eye, inner ear, nasal mucosae, or taste buds).

A. Sensory Receptors

Structural classes of general sensory receptors include: free nerve endings, encapsulated nerve endings, or specialized receptor cells. **Free nerve endings** are dendrites of sensory nerves that convey to the brain the general sensations of pain, temperature, tickle, itch, and some touch sensations. **Encapsulated nerve endings** are dendrites of sensory neurons enclosed by a connective tissue capsule that convey the general sensations of touch and pressure to the brain. Sensory receptors of special sense organs are **receptor cells** that form synapses with sensory neurons.

ACTIVITY 1 General Sensory Receptors

1 Identify the somatic sensory receptors in Figure 22.1(a) and (b). Refer to Table 22.1 that identifies location, structure (free or encapsulated nerve ending), and stimuli of somatic sensory receptors.

2 Point to the sensory receptors identified in Figure 22.1(a) on a model or anatomical chart. Pronounce the names as you point to them.

3 Examine prepared microscope slides of cutaneous sensory receptors.

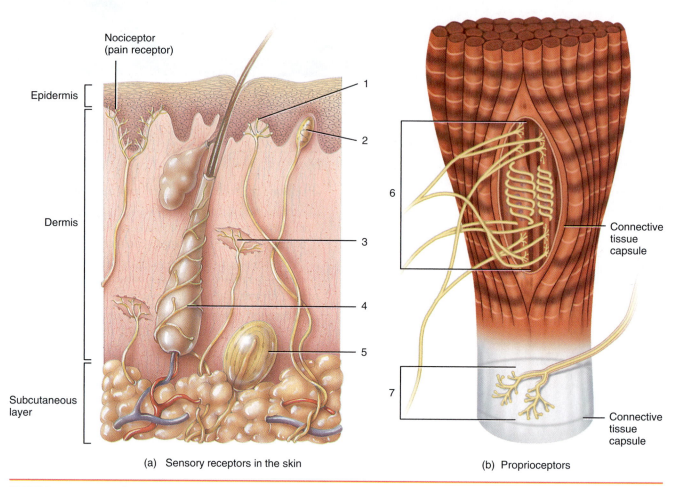

Nociceptor (pain receptor)

Epidermis

Dermis

Subcutaneous layer

1
2
3
4
5

6

7

Connective tissue capsule

Connective tissue capsule

(a) Sensory receptors in the skin

(b) Proprioceptors

FIGURE 22.1 Somatic sensory receptors.

(a)
- corpuscle of touch
- hair root plexus
- lamellated corpuscle
- type I cutaneous mechanoreceptor
- type II cutaneous mechanoreceptor

1 _____
2 _____
3 _____
4 _____
5 _____

(b)
- muscle spindle
- tendon organ

6 _____
7 _____

TABLE 22.1 Somatic Sensory Receptors

RECEPTOR	LOCATION	STRUCTURE	STIMULI
Corpuscles of touch (Meissner's corpuscles)	Dermal papillae of hairless skin	Encapsulated nerve endings	Touch and pressure
Hair root plexuses	Surrounds hair follicles	Free nerve endings	Touching hair
Lamellated (Pacinian corpuscles)	Subcutaneous and submucosal tissue, joints, tendons, and muscles	Encapsulated nerve endings	Touch and pressure
Type I cutaneous mechanoreceptors (Merkel discs)	Associated with Merkel cells in stratum basale layer of epidermis	Free nerve endings	Touch and pressure
Type II cutaneous mechanoreceptors (Ruffini's corpuscles)	Dermis, ligaments, and tendons	Encapsulated nerve endings	Stretching of digits and limbs
Muscle spindles	Found within most skeletal muscles	Encapsulated nerve endings	Respond to changes in muscle length
Tendon organs	Found at junction of tendons and muscles	Encapsulated nerve endings	Respond to changes in joint position and movement
Warm receptors	Dermis	Free nerve endings	Thermoreceptor; responds to temperatures between 32° and 48°C (90°–118°F)
Cold receptors	Stratum basale of epidermis	Free nerve endings	Thermoreceptor; responds to temperatures between 10° and 40°C (50°–105°F)
Nociceptors	Every body tissue	Free nerve endings	Pain receptors; respond to harmful stimuli

B. Sensory Pathways

Sensations that are consciously perceived include touch, pressure, pain, temperature, and some proprioception. These sensations have sensory pathways consisting of a chain of three neurons (first-, second-, and third-order neurons) that terminate in the cerebral cortex. **Sensory neurons (first-order neurons)** located in the peripheral nervous system synapse with an **association neuron** (second-order neurons) in the spinal cord or brain. **Second-order neurons** synapse with third order neurons in the thalamus. **Third-order neurons** transmit information to the primary somatosensory area of the cerebral cortex. These sensory pathways cross to the opposite side of the CNS somewhere during their ascent.

Sensations that do not reach the cerebral cortex are not consciously perceived, and include some proprioception and information from visceral sensory receptors. These pathways do not contain third-order neurons.

ACTIVITY 2 Sensory Pathways

1 Label the sensory pathways in Figure 22.2(a) and (b).

2 Point out the pathway of the first- and second-order neurons in Figure 22.2(a) and (b) on a cross-section spinal cord model or chart.

3 With your laboratory group, answer the discussion questions on sensory pathways.

? DISCUSSION QUESTIONS: SENSORY PATHWAYS

Refer to Exercise 16: Nervous Tissue, if necessary.

1 Are sensory neurons from general sensory receptors unipolar, bipolar, or multipolar neurons?

2 Where are the cell bodies for the sensory neurons (first-order neurons) located?

Where are their axon terminals located?

3 Where are the cell bodies for the second-order neurons located?

Where are their axon terminals located?

4 Where are the cell bodies for the third-order neurons located?

Where are their axon terminals located?

5 Where does the posterior column-medial lemniscus pathway cross the opposite side of the CNS?

6 Where does the anteriolateral (spinothalamic) pathway cross to the opposite side of the CNS?

FIGURE 22.2 Selected sensory pathways.

(a)
- first-order neuron
- second-order neuron
- third-order neuron

(b)
- first-order neuron
- second-order neuron
- third-order neuron

1 _____

2 _____

3 _____

4 _____

5 _____

6 _____

C. Tactile Sensitivity of Different Body Areas

Some areas of the skin have greater **tactile sensitivity** than others. The greater the number of cutaneous receptors in an area (receptor density), the greater the tactile sensitivity of that area. The size of the somatosensory cortex area receiving sensory information from a specific body area is directly proportional to the cutaneous receptor density.

The **two-point discrimination test** is an indirect measure of cutaneous receptor density. A subject is touched by two closely spaced points and asked if he or she can feel both points. The objects are moved farther apart until two points can be felt. An area of skin with a greater density of touch receptors is more sensitive to touch and can discriminate between two points closer together than an area with a lower density of touch receptors.

ACTIVITY 3 **Experiment: Tactile Sensitivity**

1 **Prediction:** List the cutaneous receptor density of the following body areas (cheek, fingertip, palm, forearm, back of leg) from the greatest to the least.

- _____ (greatest)

- _____

- _____

- _____

- _____ (least)

2 **Materials:** Obtain materials for the tactile sensitivity experiment (see Materials list)

3 **Data Collection:** Perform the two-point discrimination test on the different body areas and record your results in Table 22.2.
- Decide who will be the activity coordinator, subject, data collector, and data recorder.
- Have the subject's eyes closed during the experiment.
- Put the two caliper points together.
- Place caliper points on skin area to be tested. Touch both caliper points to the skin at the same time.
- Ask the subject if one or two points of the caliper can be felt.
- Increase the distance between caliper points. For the fingertip and palm, increase the distance 1 mm. For the cheek, forearm, and back of leg, increase the distance by 2 mm.
- Continue to increase the distance between the caliper points until the subject can feel two points. This distance is **the two-point discrimination distance** and is to be measured with a millimeter ruler. Record this value in Table 22.2
- With eyes closed, feel objects of different textures (sandpaper, velvet or synthetic fur, smooth surface such as glass) with fingertips, posterior surface of forearm, and side of leg.
- Clean up as directed by your instructor.

TABLE 22.2 Two-Point Discrimination

AREA OF BODY	TWO-POINT DISCRIMINATION DISTANCE (mm)	RECIPROCAL (mm)	CLASS AVERAGE TWO-POINT DISCRIMINATION DISTANCE (mm)	CLASS AVERAGE RECIPROCAL (mm)
Cheek				
Fingertip				
Palm				
Forearm				
Back of leg				

4 Data Analysis:

- Calculate the reciprocal $\left(\dfrac{1}{\text{two-point distance}}\right)$ of the two-point discrimination distance for each area and record the value in Table 22.2. The reciprocal represents the portion of the somatosensory cortex that receives information from sensory receptors for a given body area. Areas with high sensory receptor density are represented by a corresponding greater area of cerebral cortex.
- Collect the individual two-point discrimination distance and reciprocal values for each body area from each lab group.
- Calculate the average two-point discrimination distance and reciprocal for each body area and record in Table 22.2.
- Create a bar graph with the body areas on the X axis and the class averages of the reciprocals on the Y axis.
- Complete the Experimental Report with your lab group.

EXPERIMENTAL REPORT: TACTILE SENSITIVITY

Results

- State which body area tested has the highest average of cutaneous receptor density and which has the lowest.

- State whether variation in tactile sensitivity per body area was observed among subjects.

- State which body area tested had the least sensitivity to different textures.

Discussion

- Discuss which body area tested was represented by the largest area of cerebral cortex.

- Discuss which body area tested was represented by the smallest area of cerebral cortex.

- Discuss why cutaneous receptor density varied or did not vary among individuals.

Conclusion

- Write a sentence that postulates a correlation between the average two-point discrimination and the number of cutaneous receptors (receptor density) in a body area.

- Write a sentence that states which body area has the greatest and the least density of cutaneous receptors.

D. Adaptation of Sensory Receptors

Adaptation occurs when the continued stimulus of sensory receptors results in a decrease in nerve impulse generation in the sensory neuron(s) associated with the receptors. Receptors may adapt rapidly or slowly to continued stimulation. In this exercise, we will examine adaptation of temperature receptors that are initially rapidly adapting, but maintain some response to stimuli.

ACTIVITY 4 Adaptation of Temperature Receptors

1 Observe adaptation for warm and cold temperature receptors.

PART 1

- Choose a subject, a timer, and a data collector and recorder.
- Obtain 2 bowls or 1,000-mL beakers of water: one warm (45°C) and one room temperature (25°C).
- Have the subject immerse the *left hand* into 45°C water and record the immediate sensations in Table 22.3.
- Keep the left hand immersed for 1 minute then immerse the *right hand* into the same water. Now record the temperature sensation of both hands in Table 22.3.
- In which hand did adaptation occur? _____
- Have the subject wash hands in room temperature water and then wait for 5 minutes before beginning the next procedure.

PART 2

- Obtain 3 bowls or 1,000-mL beakers of water: one room temperature (25°C), one warm (45°C), and one ice water (10°C).
- Have the subject place the *left hand* in 45°C water and the *right hand* in ice water simultaneously. Record the immediate temperature sensation of each hand in Table 22.3.
- Record the sensations of each hand after 1 minute.
- Have the subject simultaneously put both hands into the room temperature bowl. Record the immediate temperature sensation of each hand in Table 22.3.

2 Discuss the results with the class.

3 Answer the discussion questions about adaptation of receptors with your lab group.

TABLE 22.3 Adaptation of Temperature Receptors

BODY AREA	TIME	TEMPERATURE SENSATION
Part 1		
Left hand	Immediate (45°C)	
	1 minute (45°C)	
Right hand	Immediate (45°C)	
Part 2		
Left hand	Immediate (45°C)	
	1 minute (45°C)	
	Immediate (25°C)	
Right hand	Immediate (10°C)	
	1 minute (10°C)	
	Immediate (25°C)	

? **DISCUSSION QUESTIONS: ADAPTATION OF SENSORY RECEPTORS**

1 How long did it take for the cold and warm temperature receptors to adapt for the subject in your group?

2 When sensory receptors adapt, does the cerebral cortex receive an increased or decreased number of sensory nerve impulses?

3 Indicate whether each of the following general sensory receptors adapt quickly, slowly, or not at all. Use your own experience.
 a. Cutaneous touch receptors (stimulated when skin touched by object, for example, when clothes touch the skin)

b. Cutaneous pressure receptors (stimulated when pressure applied to skin, for example, the pressure applied to skin when coins stacked on forearm)

c. Proprioceptors associated with skeletal muscles and tendons that are important for balance (proprioceptors maintain muscle contraction needed for standing erect and keeping head erect)

d. Pain receptors

E. Referred Pain

Pain that is felt in areas that are not injured or stimulated is called **referred pain.** Sometimes this pain originates in visceral organs and is perceived in the skin that is innervated by the same spinal segments. The pain of a heart attack is usually felt in the skin over the heart and down the left arm.

ACTIVITY 5 Referred Pain

1 Observe referred pain in the elbow.
 - Choose a subject, a timer, and a data collector and recorder. This experiment works best on a subject who has thin arms. The subject will be asked to immerse his/her elbow in ice water and describe the sensation (pain, discomfort, or tingling) and where the sensation is located.
 - Obtain a bowl of ice water (10°C).
 - Have the subject immerse an elbow in ice water and immediately describe the sensation and where it is located. Record results in Table 22.4. Have the subject keep the elbow in the water until directions indicate to remove it.
 - After 1 minute, ask the subject to describe the sensation and where it is located. Record results in Table 22.4.
 - After 2 minutes, ask the subject to describe the sensation and where it is located. Have the subject remove elbow from water. Record results in Table 22.4.
2 With your lab group, answer the discussion questions about referred pain.

? DISCUSSION QUESTIONS: REFERRED PAIN

1 The location of the sensation changed over time. Name in order the areas where the sensation was felt.

2 Which nerve supplies the areas where the sensation occurred?

TABLE 22.4 Referred Pain

TIME (MIN)	SENSATION	LOCATION OF SENSATION
Immersion of elbow (0 min)		
1 minute after immersion		
2 minutes after immersion		

Reviewing Your Knowledge

A. Anatomy Drill and Practice

Go to www.wiley.com/college/allen. Click on the picture of your lab manual, then on Student Companion Site, Exercise 22, and Anatomy Drill and Practice.

B. Sensory Receptors

1. Describe the function of sensory receptors.

2. Identify general sensory receptors that belong to each structural class.

 a. Free nerve endings

 b. Encapsulated nerve endings

C. Sensory Pathways

Write the name in the blank that corresponds with the description.

 cerebral cortex
 first-order neuron
 posterior (dorsal) root ganglion
 second-order neuron
 thalamus
 third-order neuron

_____ **1.** Receives input from third-order neurons; sensory information is consciously perceived.

_____ **2.** Cell body located in spinal cord or brain stem.

_____ **3.** Gateway to cerebral cortex; location of synapse of second- and third-order neurons.

_____ **4.** Neuron associated with sensory receptor.

_____ **5.** Relays sensory nerve impulses from thalamus to cerebral cortex.

_____ **6.** Cell body for first-order neuron located here.

D. Tactile Sensitivity

1. Rank body areas 1 to 5 from least to greatest tactile receptor density. 1 = least tactile receptor density; 5 = greatest tactile receptor density.

_____ back of leg

_____ cheek

_____ fingertip

_____ forearm

_____ palm

2. Describe correlation between tactile density of a body area and the size of the cerebral cortex receiving information from those receptors.

E. Adaptation

1. Define adaptation.

2. Do warm receptors adapt quickly?

3. Do cold receptors adapt quickly?

F. Referred Pain

1. Define referred pain.

2. After the subject's elbow was immersed in water, where was the referred pain located?

Using Your Knowledge

Questions 1–3. Refer to the map of the somatosensory cortex in Figure 22.3. Based on the size of the area represented on the cortex and the size of the body area, circle which body area has the greatest density of receptors:

1. The *lips* or the *thigh*.

2. The *tongue* or the *toes*.

3. The *foot* or *hand*.

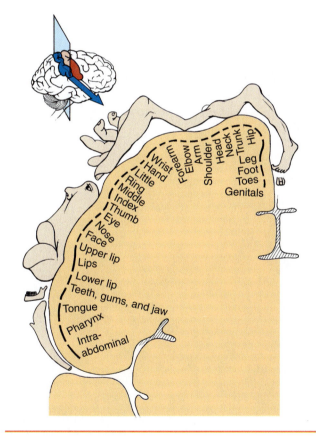

FIGURE 22.3 Somatosensory cortex.

4. Using your knowledge, compare the size of the somatosensory cortex representing the fingertips for a visually impaired person who reads Braille (increased tactile sensitivity) to that of someone with normal vision who has normal tactile sensitivity.

5. Leprosy often results in loss of pain to infected body areas. Describe the hazards of this.

6. When you hit your funny bone, where do you experience referred pain? *Hint:* What nerve would you hit?

Using your textbook or other reference, trace the sensory pathway from proprioceptors located in the quadriceps muscle to the spinocerebellar tract travelling to the cerebellum.

7. Identify the nerve carrying the axon of the first-order neuron.

8. Identify the levels where the first-order neuron enters the spinal cord.

9. Do the cerebellar hemispheres receive proprioceptive nerve impulses ipsilaterally or contralaterally?

10. Using your knowledge, is proprioception consciously perceived by the cerebellum?

Special Senses

Objectives

After completing this exercise, you should be able to:

The Eye and Vision

- Identify the accessory structures of the eye and the structures of the eyeball on models or charts
- Identify the three principal layers of the retina on prepared slides or photomicrographs
- Perform visual tests that are used to determine distance visual acuity, near point of vision, and astigmatism

The Ear, Hearing, and Equilibrium

- Identify the structures of the external, middle, and internal ear on models or charts
- Identify the major structures of the cochlea, vestibule, and semicircular canals on models or charts
- Perform tests that are used in determining the cause of hearing loss
- Perform tests that exhibit the function of the receptors for dynamic and static equilibrium

The Nose and Olfaction

- Identify the location of receptors for olfaction
- Examine the structure of the receptors for olfaction
- Determine time of adaptation for olfactory receptors

The Taste Buds and Gustation

- Identify the location of receptors for taste
- Examine the structure of the receptors for taste
- Observe the influence of smell and texture on taste

Materials

The Eye and Vision

- eye models or charts
- compound microscope, lens paper, prepared slides of an eyeball
- dissection equipment and trays, cow eyes, and gloves
- Snellen eye chart, ophthalmoscope, metric ruler

The Ear, Hearing, and Equilibrium

- ear models or charts
- skull, tuning forks, cotton balls
- compound microscope, lens paper, and prepared slides of the cochlea, macula, and crista
- swivel chair or stool

The Nose and Olfaction

- skull, stopwatch, cotton balls
- clove oil, peppermint oil, or cinnamon oil

Taste Buds and Gustation

- compound microscope, lens paper, prepared slides of taste buds
- mirror, 1/2-inch cubes of apple, banana, cheese, carrot, and raw potato

A. Structure of the Eye and Vision

Accessory Eye Structures

Accessory structures of the eye include the eyebrows, eyelids or **palpabrae, eyelashes,** conjunctiva lacrimal apparatus, and the extrinsic eye muscles. The **conjunctiva** is the thin, protective mucous membrane that covers the anterior eye and folds to cover the inner eyelid. The **palpebral conjunctiva** covers the interior part of the white of the eye, and the **bulbar conjuctiva** covers the anterior part of the white of the eye.

The **lacrimal apparatus** (*lacrim-* = tears) is a group of structures involved in producing and draining tears. The **lacrimal gland** produces and secretes tears onto the eye surface, and the **lacrimal canals** drain the tears from the eyes into the enlarged **lacrimal sac.** The enlarged **nasolacrimal duct** receives tears from the lacrimal sac and drains the tears into the nasal cavity.

The **extrinsic eye muscles** are six skeletal muscles that insert on the exterior of the eyeball to move the eyeball in all directions. The **superior, inferior, medial,** and **lateral** rectus are parallel to the long axis of the eyeball. The **superior** and **inferior oblique** attach to the eyeball at an angle.

ACTIVITY 1 **Accessory Structures of the Eye**

1 Label the parts of the conjunctiva in Figure 23.1(a) and structures of the lacrimal apparatus in Figure 23.1(b).

2 Label the extrinsic eye muscles in Figure 23.2.

(a) Lateral view of eyeball

(b) Lacrimal apparatus

FIGURE 23.1 Accessory structures of the eye.

(a)
- bulbar conjunctiva (con-junk-TIE-va)
- conjunctival fold
- palpebral conjunctiva (PAL-puh-bral)

1 _____

2 _____

3 _____

(b)
- lacrimal canals (LAK-rih-mal)
- lacrimal gland
- lacrimal sac
- nasolacrimal duct

4 _____

5 _____

6 _____

7 _____

3 Identify all the accessory structures of the eye, including all parts of the lacrimal apparatus, on an eye model or chart and pronounce the terms as you point to them.

4 Identify the extrinsic eye muscles on an eye model or chart and pronounce the terms as you point to them.

5 Determine how each muscle moves the eye by observing where it inserts on the eye and how the eye would move if the insertion is moved toward the origin.

6 Write the function of each extrinsic eye muscle in Table 23.1. Choose one of the following eye movements for each muscle: *moves eye superiorly, inferiorly, medially, laterally, superiorly and laterally, or inferiorly and laterally.*

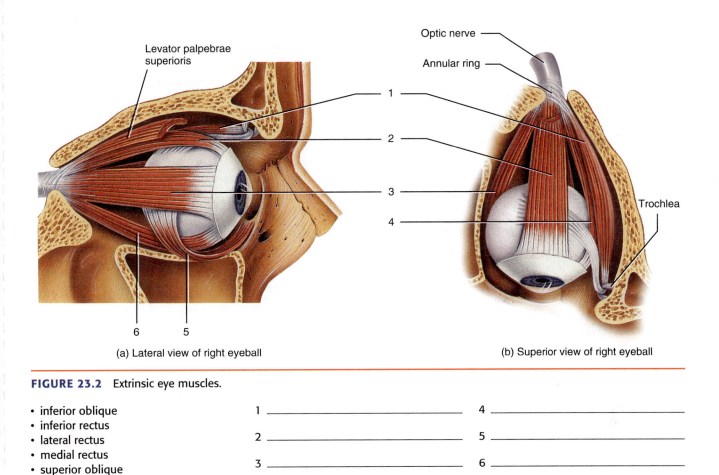

(a) Lateral view of right eyeball

(b) Superior view of right eyeball

FIGURE 23.2 Extrinsic eye muscles.

- inferior oblique
- inferior rectus
- lateral rectus
- medial rectus
- superior oblique
- superior rectus

1 _____ 4 _____

2 _____ 5 _____

3 _____ 6 _____

TABLE 23.1 Function of the Extrinsic Eye Muscles

EXTRINSIC EYE MUSCLE	FUNCTION
Inferior oblique	
Inferior rectus	
Lateral rectus	
Medial rectus	
Superior oblique	
Superior rectus	

Structure of the Eyeball

The wall of the eyeball has three layers: the outer fibrous tunic, the middle vascular tunic, and the inner retina. The **fibrous tunic** is composed of the cornea and sclera. The **cornea** is the transparent anterior portion that covers the iris and pupil, and the **sclera** (*scler-* = hard) is the tough, white part of the eye that forms the majority of the eyeball. The **scleral venous sinus** (canal of Schlemm) is an opening found at the junction of the cornea and the sclera.

The middle **vascular tunic** is composed of the iris, ciliary body, and choroid. The **iris** is the most anterior portion of the vascular tunic and contains pigmented cells. It is made of circular smooth muscle and controls the pupil size. The **pupil** is the opening in the middle of the iris that allows light to enter the eyeball and changes size in response to the intensity of light. The **ciliary body** begins posterior to the iris at the junction of the cornea and sclera and consists of the ciliary muscle and ciliary processes. The **ciliary muscle** is a circular smooth muscle that contracts to control the shape of the lens. **Ciliary processes** are folds that protrude from the ciliary body toward the lens. They contain capillaries that secrete aqueous humor, the fluid in the anterior chamber of the eyeball. **Suspensory ligaments (zonular fibers)** are thin fibers that attach the lens to these processes. The **choroid** is the most posterior part of the vascular tunic that lines most of the interior of the sclera. It contains many blood vessels that nourish the retina.

The **retina** (sensory tunic) is the inner coat that begins at the **ora serrata,** the serrated boundary between the ciliary muscle and the retina. The retina continues posteri- orly, lining the interior of the choroid. The **pigmented layer** of the retina is the outer portion and the **neural layer** is the inner portion that contains photoreceptors and associated neurons.

Interior of the Eyeball

The interior of the eyeball contains the lens, anterior cavity, and vitreous chamber. The **lens** divides the interior of the eyeball into an anterior cavity and a vitreous chamber (posterior cavity). The **anterior cavity** is a space between the cornea and the lens that is filled with watery **aqueous humor** (*aqua* = water; *humor* = moist). This cavity is subdivided into an **anterior chamber** (between the cornea and the iris) and a **posterior chamber** (between the iris and the lens). The **scleral venous sinus** (canal of Schlemm) is an opening found at the junction of the cornea and sclera that drains aqueous humor back into the bloodstream. The **vitreous chamber** is the larger, posterior cavity located between the lens and the retina. This cavity is filled with a gel-like substance called the **vitreous body** (humor) that holds the retina flat against the choroid.

ACTIVITY 2 Structure of the Eyeball

1 Label the eyeball structures in Figure 23.3.
2 Identify these structures on an eye model or chart.
3 Pronounce the terms as you point to them.

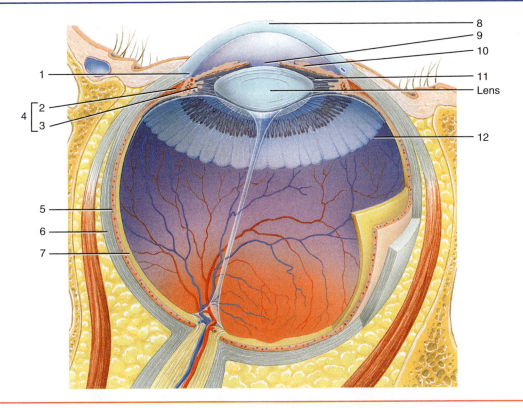

FIGURE 23.3 Structure of the eyeball.

- choroid
- ciliary body (SIL-ee-air-ec)
- ciliary muscle
- ciliary process
- cornea (KOR-nee-ah)
- iris
- ora serrata (ser-RAH-tah)
- pupil
- retina
- sclera (SKLER-ah)
- scleral venous sinus
- suspensory ligament (zonular fibers)

1 _____ 7 _____

2 _____ 8 _____

3 _____ 9 _____

4 _____ 10 _____

5 _____ 11 _____

6 _____ 12 _____

CLINICAL NOTE: GLAUCOMA *is caused by an increase in pressure within the eye called* intraocular pressure. *Blockage of the scleral venous sinus prevents drainage of aqueous humor, increasing the amount in the anterior cavity that causes increased intraocular pressure.*

ACTIVITY 3 **Anterior and Posterior Cavities of the Eye**

1 Label the interior spaces and structures of the eyeball in Figure 23.4.

2 Identify these spaces and structures on a model or chart.

3 Pronounce the terms as you point to them.

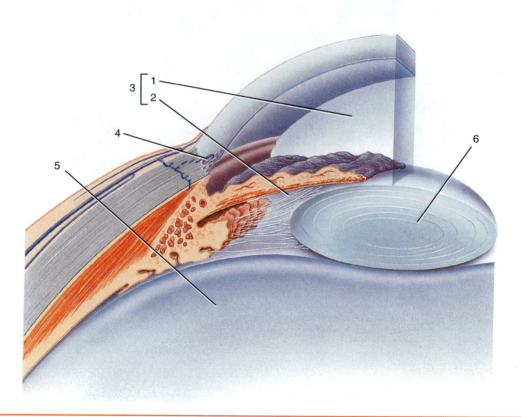

FIGURE 23.4 Anterior and posterior cavities of the eyeball.

- anterior cavity
- anterior chamber
- lens
- posterior chamber
- scleral venous sinus
- vitreous chamber

1 _____ 4 _____

2 _____ 5 _____

3 _____ 6 _____

ACTIVITY 4 Dissection of Cow Eye

> **SAFETY PRECAUTION:**
> *Use safety glasses and remember to properly dispose of any removed waste tissue as directed by your instructor. Always clean your dissection tools, tray, and lab surfaces as directed.*

1 Obtain a laboratory tray, blunt probe, scalpel, scissors, and disposable gloves.

2 Using gloves, obtain a cow eye and rinse it to remove excess preservative.

3 The posterior portion of the eye may be encased in adipose tissue. Carefully remove the adipose tissue protecting the optic nerve.

4 Identify the external eye structures: **cornea, sclera, optic nerve,** and **extrinsic eye muscles.** Refer to Figure 23.5(a). The preservative causes the cornea to change from being transparent and smooth to opaque and wrinkled.

5 Using the point of a scalpel, punch an opening ¼-inch posterior to the cornea through the very tough sclera. Be careful not to squeeze the eyeball too tightly or liquid may squirt out. Use the scissors to cut an incision all the way around the eyeball, separating it into two parts.

6 Carefully separate the anterior and posterior parts of the eyeball so the **vitreous body** remains in the posterior part of the eyeball and the **lens** in the anterior part.

7 Carefully remove the lens, noting the transparent **suspensory ligaments** that are attached to the lens.

8 Using Figure 23.5(b), identify the following structures in the anterior portion of the eyeball: **pupil, iris,** and **ciliary muscle.**

9 Carefully remove the vitreous body from the posterior portion of the eyeball. Refer to Figure 23.5(b) to identify the structures in the posterior portion of the eyeball. The **retina,** a thin beige layer, will probably separate from the choroid. The only point of attachment of the retina to the wall of the eyeball is at the **optic disc.**

10 Pull the retina away from the **choroid,** the dark middle layer of the eyeball that contains the pigment melanin. The choroid of the cow's eye has an iridescent reflecting surface called the **tapetum lucidum** that is not found in humans. The tapetum lucidum reflects light within the eye and enables animals to see in low light conditions.

11 Separate the choroid from the **sclera.** Observe how the three coats or tunics—the **retina, choroid,** and **sclera**—form the wall of the eyeball.

12 Discard the eye as instructed by your instructor.

13 Clean your dissecting tray and equipment and return to the proper place.

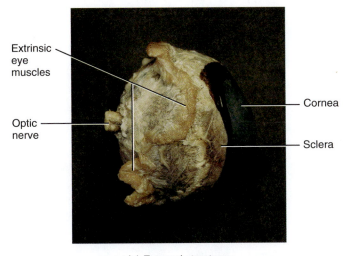

(a) External structures

Extrinsic eye muscles

Optic nerve

Cornea

Sclera

(b) Anterior and posterior sections

Sclera

Retina

Optic disc

Choroid

Ciliary muscle

Pupil

Iris

FIGURE 23.5 Cow eye dissection.

Anatomy of the Retina

The neural portion of the retina is an outgrowth of the brain and contains three layers of neurons: the photoreceptor layer (deepest cell layer), the bipolar cell layer (middle layer), and the ganglion cell layer (the superficial cell layer). The **photoreceptor cell layer** contains the rods and cones. **Rods** are used in night vision and respond to low levels of light, allowing us to perceive shades of gray, black, and white. Visual acuity with rods is low. **Cones** require brighter light for stimulation, but allow us to see color and provide high visual acuity. The rods and cones synapse with the bipolar neurons in the **bipolar cell layer,** which synapses on the ganglion cells in the **ganglion cell layer.** Axons from the ganglion cells extend through the optic disc and leave the eyeball as the optic nerve. The **optic disc** does not contain photoreceptors and forms the **blind spot** of the retina. It is also the site where the central retinal artery and vein enter and leave the retina, and the only place where the retina is secured to the other layers of the eyeball.

The **macula lutea** (*macula* = flat spot; *lutea* = yellow), the site of macular degeneration, is in the center of the neural portion of the retina. In the middle of the macula lutea is the **central fovea** (fovea centralis). This area of the retina has the highest density of cones of any area of the retina and is not covered by ganglion and bipolar cell layers. Therefore, this area has the highest visual acuity (sharpness of vision) of any area of the retina. When we look at an object, the light rays reflected from the object are focused onto the central fovea.

The retina can be viewed with an **ophthalmoscope.** The ophthalmoscope illuminates the interior of the eye, and the retina appears red from the many blood vessels. Blood vessels can be seen branching from the optic disc while the circular macula lutea appears dark because of the absence of blood vessels. The central fovea is in the middle of the macula lutea.

ACTIVITY 5 **Anatomy of the Retina**

1 Label the structures of the retina in Figure 23.6(a) and (b).

2 Label the structures of the retina as observed with an ophthalmoscope in Figure 23.7.

3 Optional activity: View the retina with an ophthalmoscope. Instructions will be provided by your instructor.

4 Observe a longitudinal section of the eyeball in Figure 23.8(a), and label the photomicrograph in Figure 23.8(b).

5 Examine a slide showing a cross-section through an eyeball.
 - Using the low-power objective lens, find the posterior surface of the eyeball and identify the **sclera, choroid,** and **retina.** Place the retina in the center of the field of view and switch to the high-power objective lens.
 - Using the high-power objective lens, identify the **pigmented epithelium of the retina** and the **neural layer of the retina.** Within the neural layer of the retina, identify the **ganglion layer,** the **bipolar layer,** and the **photoreceptor layer.**

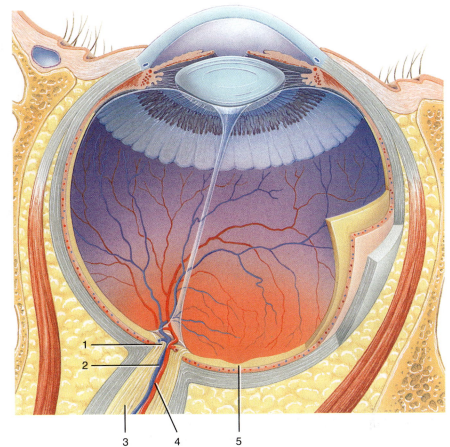

- central fovea
- central retinal artery
- central retinal vein
- optic disc
- optic nerve

1 _____

2 _____

3 _____

4 _____

5 _____

(a) Transverse section of eyeball

- bipolar cell layer
- ganglion cell layer
- neural portion of retina
- optic nerve fibers (axons)
- photoreceptor layer
- pigmented epithelium of retina

6 _____

7 _____

8 _____

9 _____

10 _____

11 _____

(b) Microscopic structure of retina

FIGURE 23.6 Anatomy of the retina.

- blood vessel
- central fovea
- macula lutea
- optic disc

1 _____

2 _____

3 _____

4 _____

FIGURE 23.7 Normal retina viewed with an ophthalmoscope.

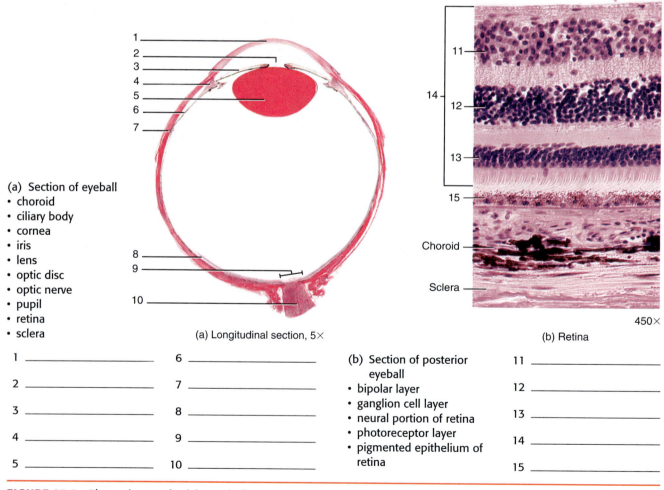

(a) Section of eyeball
- choroid
- ciliary body
- cornea
- iris
- lens
- optic disc
- optic nerve
- pupil
- retina
- sclera

(a) Longitudinal section, 5×

(b) Retina

450×

1 _____ 6 _____

2 _____ 7 _____

3 _____ 8 _____

4 _____ 9 _____

5 _____ 10 _____

(b) Section of posterior eyeball
- bipolar layer
- ganglion cell layer
- neural portion of retina
- photoreceptor layer
- pigmented epithelium of retina

11 _____

12 _____

13 _____

14 _____

15 _____

FIGURE 23.8 Photomicrograph of the eyeball.

ACTIVITY 6 **Locating Blind Spot**

1 Looking at Figure 23.9, hold the lab manual in both hands and extend both arms in front of you. Position the X in line with your right eye and close your left eye. Keeping your right eye focused on the X, *slowly* move the figure toward you. The dot will dissappear when it crosses your blind spot, but if you move the figure too fast, you will miss it.

2 Have a lab partner measure the distance the book is from your face when this happens. Then move the figure closer to your face, and the dot will reappear.

Blind spot distance for right eye _____

3 Repeat the steps for your left eye, only reposition the X in front of the left eye and close your right eye.

Blind spot distance for left eye _____

4 Explain why the dot disappears and reappears.

Visual Tests

Visual acuity tests measure the ability of the lens to focus light reflected from an object on the central fovea of the retina. The lens can **accommodate** or change shape to bend light rays to focus them on the central fovea. At 20 feet, light rays from an object are nearly parallel and do not have to bend as much to focus on the central fovea. At this distance, the lens is flattened and the refractive power (ability to bend light rays) of the lens is lowest. To observe objects closer than 20 feet, the lens must change shape or accommodate to focus the light rays on the central fovea. The lens bulges to increase the refractive power. Individuals who have normal distance vision and near vision are **emmetropic,** individuals who have normal distance vision but blurry near vision are **hyperopic** (farsighted), and individuals who have blurry distance vision but normal near vision are **myopic** (nearsighted).

As we age, the ability of the lens to accommodate diminishes and the ability to focus on very close objects decreases, a condition called **presbyopia** (*presby-* = old). The **near point of vision** is the closest distance that a person can focus on an object. The average near point of vision is 10 cm for a young adult, 20 cm for an adult in their 40s, and 80 cm for someone in their 60s.

Astigmatism is caused by irregularities in the curvature of the cornea or lens. This causes parts of an image to be blurry.

FIGURE 23.9 Blind spot test.

ACTIVITY 7 Visual Tests

1 **Distance visual acuity** is measured using a **Snellen eye chart** (provided by your instructor). *If you wear eyeglasses or contact lenses, remove them to determine visual acuity without correction or wear them to determine visual acuity with correction.* Circle *corrected* or *uncorrected* in Table 23.2.
- Have the subject stand 20 feet from the Snellen eye chart that is placed in a well-lighted area, and cover the left eye with a hand.
- Have the subject read the smallest line of letters without squinting. If the subject can correctly read half of the letters or more, then ask the subject to read the letters on the next, smaller line.
- Record the number of the line with the smallest-sized letters read with half or greater accuracy in Table 23.2.
- Have the subject cover the right eye and repeat the procedure.
- A value of 20/20 indicates that the subject has normal vision. A value of 20/40 indicates that the subject sees at 20 feet what a person who has normal vision sees at 40 feet. This is not as good as normal vision. A value of 20/15 indicates that the subject sees at 20 feet what a person with normal vision sees at 15 feet. This is better than normal vision.

2 **Near visual acuity** is measured using a **Snellen acuity card** (Fig. 23.10).
- Have the subject hold the Snellen acuity card 14 inches from his or her face. The card should be illuminated with the light source either behind or above the subject. Instruct the subject to cover the left eye with his or her hand.
- Repeat the steps used to measure distance visual acuity. Start with the second bulleted step.

3 Measure the **near point of vision** and compare it with the average for the age of your subject.
- Have the subject hold the Snellen visual acuity card (Figure 23.10) 14 inches from his or her face.
- Have the subject cover the left eye and read letters from a line above his or her near visual acuity.
- Instruct the subject to slowly move the chart closer to his or her face until the letters are blurry.
- Measure the distance from the card to the subject's eye in centimeters.
- Record the value on Table 23.2.

TABLE 23.2 Visual Test Results

TEST	RESULTS
Visual acuity tests (corrected or uncorrected)	Distance vision Left eye Right eye Near vision Left eye Right eye
Near point of vision (corrected or uncorrected)	
Astigmatism	Left eye Right eye

FIGURE 23.10 Snellen acuity card.

	distance equivalent
	$\frac{20}{800}$
Point	$\frac{20}{400}$
26	$\frac{20}{200}$
14	$\frac{20}{100}$
10	$\frac{20}{70}$
8	$\frac{20}{50}$
6	$\frac{20}{40}$
5	$\frac{20}{30}$
4	$\frac{20}{25}$
3	$\frac{20}{20}$

are lighter or less distinct than others, you have astigmatism.

• Cover the left eye and repeat the procedure.

5 Answer the discussion questions with your lab group.

CLINICAL NOTE: *To obtain an accurate acuity test, professionals also count the number of letters a patient missed on the line recorded as the best vision. For example, if the best vision was 20/40, and the patient missed two letters on that line, it would be recorded as 20/40$^{(-2)}$. If a patient read two letters on the 20/15 line, it would be recorded as 20/15$^{(+2)}$.*

? DISCUSSION QUESTIONS: VISUAL TESTS

1 Why must the subject stand 20 feet from the Snellen eye chart to test distance vision?

2 Why must the subject hold the Snellen acuity card close to his or her eyes to test near vision?

4 Determine whether you have **astigmatism.**
• Remove corrective lenses.
• Cover the right eye and look at the center of the astigmatism chart in Figure 23.11.
• If all the radiating lines are equally sharp and dark, you do not have astigmatism. However, if some lines

FIGURE 23.11 Astigmatism chart.

B. The Ear, Hearing, and Equilibrium

Anatomy of the Ear

The ear is divided into three regions: the external (outer) ear, the middle ear, and the internal (inner) ear. The **external ear** consists of the auricle, external auditory canal, and tympanic membrane and extends from the auricle to the tympanic membrane. The **auricle,** the flexible external structure that is commonly called the ear, collects sound waves and directs them toward the external auditory canal. The **external auditory canal** conducts sound waves from the auricle to the tympanic membrane. The **tympanic membrane** (*tympan-* = drum) or eardrum converts sound waves to vibrations that are transferred to middle ear structures.

The **middle ear** is an ear-filled cavity within the temporal bone that extends from the tympanic membrane to the oval window. Middle ear structures include: auditory oscicles, oval window, round window, and auditory tube. **Auditory oscicles** are small bones within the cavity that are connected by synovial joints. These bones transfer vibrations from the tympanic membrane to the oval window.

The **malleus** is the outermost bone and is attached to the **tympanic** membrane. The **incus** is the middle bone and connects to the stapes. The innermost bone is the **stapes** which connects to the incus and oval window. The **oval window** is the membrane-covered opening that separates the middle and inner ear and transfers vibrations to the inner ear. The **round window** is a membrane-covered opening between the middle ear and cochlea. The **auditory tube** (pharyngotympanic or Eustachian tube) connects the middle ear to the nasopharynx (part of the throat near the nasal cavity), and equalizes the air pressure of the middle ear with atmospheric air.

ACTIVITY 8 Anatomy of the External and Middle Ear

1 Label the structures of the external and middle ear in Figure 23.12.

2 Identify the structures on a model or chart.

3 Pronounce the terms as you point to them.

4 Identify the area of the temporal bone that houses the middle and internal ear on the skull.

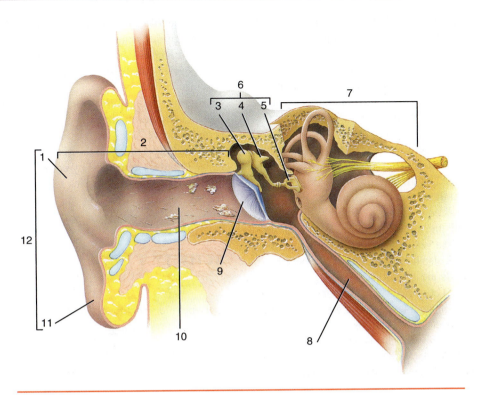

FIGURE 23.12 Anatomy of the external and middle ear.

- auditory tube
- auricle
- external auditory canal
- external ear
- helix (HEE-liks)
- incus (INK-us)
- internal ear
- lobule
- malleus (MAL-ee-us)
- middle ear
- stapes attached to oval window (STAY-peez)
- tympanic membrane (tim-PAN-ik)

1 _____

2 _____

3 _____

4 _____

5 _____

6 _____

7 _____

8 _____

9 _____

10 _____

11 _____

12 _____

The internal ear is housed within the temporal bone. It consists of cavities within the bone called the **bony labyrinth** that encloses a series of connected membranous sacs, the **membranous labyrinth.** The bony labyrinth contains a fluid called **perilymph** that surrounds the membranous labyrinth. **Endolymph** is the fluid within the membranous labyrinth. The bony labyrinth has three main regions: the vestibule, the semicircular canals, and the cochlea.

The **vestibule** is the middle area of the bony labyrinth that encircles two sections of membranous labyrinth, the utricle and the saccule. The **utricle** (*utricle* = little bag) is the posterior section of the membranous labyrinth within the vestibule, and it houses equilibrium receptors. The **saccule** (*saccule* = little sac) is the anterior section of the membranous labyrinth within the vestibule. The saccule is continuous with the utricle and also houses equilibrium receptors.

Semicircular canals are three bony canals posterior to the vestibule that project posteriorly, laterally, and superiorly from the vestibule. Each canal is at right angles to the other two. **Semicircular ducts** are sections of membranous labyrinth within the semicircular canals which contain equilibrium receptors and connect with the utricle. The **ampulla** is the widened end of each semicircular canal and duct.

The **cochlea** is the spiral area of the bony labyrinth anterior to the vestibule. The **cochlear duct** is the section of membranous labyrinth within the cochlea. The cochlear duct contains the hearing receptors and is connected to the saccule.

Hearing and equilibrium receptors initiate nerve impulses which are carried by the **vestibulocochlear nerve** (cranial nerve VIII) to the brain. The vestibulocochlear nerve has two branches: the **vestibular branch** that carries nerve impulses generated by equilibrium receptors and the **cochlear branch** that carries nerve impulses generated by the hearing receptors.

ACTIVITY 9 Anatomy of the Internal (Inner) Ear

1 Label the structures of the internal ear in Figure 23.13.
2 Identify the structures on a model or chart.
3 Pronounce the terms as you point to them.

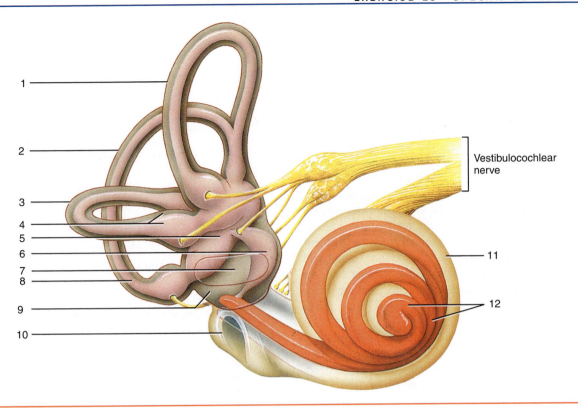

FIGURE 23.13 Anatomy of the internal ear.

- ampulla of semicircular canal and duct
- anterior semicircular canal
- cochlea (COKE-lee-uh)
- cochlear duct
- lateral semicircular canal
- membranous semicircular duct
- oval window
- posterior semicircular canal
- round window
- saccule (SAK-yool)
- utricle (YOU-trih-cul)
- vestibule

1 _____

2 _____

3 _____

4 _____

5 _____

6 _____

7 _____

8 _____

9 _____

10 _____

11 _____

12 _____

Microscopic Anatomy of the Cochlea and the Spiral Organ of Corti

The cochlea is a spiral cavity that resembles the space within a snail shell. The cochlea makes three turns around a bony core. A section through the cochlea shows three channels: the scala vestibule, the cochlear duct (scala media), and the scala tympani. The **scala vestibuli** is part of the cochlea and is superior to the cochlear duct. It is separated from the cochlear duct by the **vestibular membrane.** The **scala tympani** is also part of the cochlea and is posterior to the cochlear duct. It is separated from the cochlear duct by the **basilar membrane.** The scala vestibuli and scala tympani are continuous with one another and are filled with perilymph. The **cochlear duct,** part of the membranous labyrinth, houses the spiral organ of Corti and is filled with endolymph.

The **spiral organ of Corti** sits on the basilar membrane. It contains hair cells (receptors for hearing) and supporting cells. The hair cells have a **hair bundle** composed of stereocilia at their apical end. Superior to and in contact with the stereocilia is the **tectorial membrane** (*tector* = covering). The basal end of the hair cells synapse with sensory and motor neurons from the cochlear branch of the vestibulocochlear nerve.

Movement of the basilar membrane, caused by perilymph movement, forces the hair bundle into the tectorial membrane and bends the stereocilia. This results in a generation of nerve impulses in the sensory neurons.

ACTIVITY 10 **Microscopic Anatomy of the Cochlea and Spiral Organ of Corti**

1 Label the structures of the cochlea and spiral organ of Corti in Figure 23.14.

2 Identify the structures on a model or chart.

3 Pronounce the terms as you point to them.

4 Examine a slide showing a cross-section through the cochlea.
- Using the low-power objective, identify the **basilar membrane, cochlear duct, spiral organ of Corti, scala tympani, scala vestibuli,** and **vestibular membrane.**
- Using the high-power objective, identify the **basilar membrane, hair bundle, hair cells,** and **tectorial membrane.**

20×

(a) Cross-section through cochlea

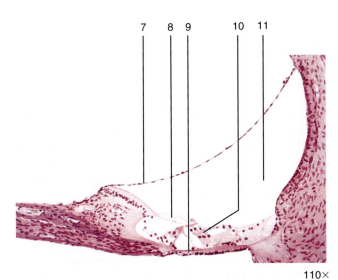

110×

(b) Cross-section through spiral organ of Corti

FIGURE 23.14 Photomicrographs of the cochlea and spiral organ of Corti.

(a)
- basilar membrane (BAY-sih-lur)
- cochlear duct
- scala tympani (SCAY-lah)
- scala vestibuli
- spiral organ of Corti
- vestibular membrane

(b)
- basilar membrane
- cochlear duct
- hair cells
- tectorial membrane
- vestibular membrane

1 _____

2 _____

3 _____

4 _____

5 _____

6 _____

7 _____

8 _____

9 _____

10 _____

11 _____

Microscopic Anatomy of the Equilibrium Receptors of the Internal Ear

There are two types of equilibrium receptors, the maculae (macula, sing.) located in the utricle and saccule, and the cristae (crista, sing.) located in the membranous semicircular ducts within the ampullae. The **maculae** provide information on head position (static equilibrium), as well as linear acceleration and deceleration, a type of dynamic equilibrium. The maculae consist of **hair cells** with **hair bundles** and **supporting cells.** The hair bundles are in contact with a gelatinous membrane, the **otolithic membrane** (*oto* = stone), which contains calcium carbonate crystals called **otoliths.** Movement of the head causes movement of the otoliths and otolithic membrane, which bends the hair bundles. The direction of movement will determine if the hair cells release more or less neurotransmitter to the associated sensory neurons.

The **crista** (ampullaris) detects rotational acceleration and deceleration, a type of dynamic equilibrium. Each crista consists of **hair cells** and **supporting cells.** The hair bundles of the hair cells are covered by a gelatinous structure called the **cupula.** When the head moves, movement of endolymph pushes the cupula causing the hair cells to bend. Bending of the hair bundles results in generation of nerve impulses in the vestibular branch of the vestibulocochlear nerve.

ACTIVITY 11 Microscopic Anatomy of the Macula and Crista

1 Label the structures of the macula and crista in Figure 23.15.

2 Identify the structures on a model or chart.

3 Pronounce the terms as you point to them.

4 Examine the demonstration slide showing a cross-section through the crista with the high-power objective lens. Identify the **cupula** and **hair cells.**

(a) Macula

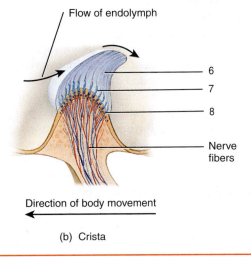

Flow of endolymph

6
7
8

Nerve
fibers

Direction of body movement

(b) Crista

FIGURE 23.15 Microscopic anatomy of the macula and crista.

(a)
- hair bundle
- hair cell
- otoliths (OH-toe-liths)
- otolitic membrane
- vestibular branches of vestibulocochlear nerve

(b)
- cupula (KYOU-pyul-uh)
- hair bundle
- hair cell

1 _____

2 _____

3 _____

4 _____

5 _____

6 _____

7 _____

8 _____

Auditory and Equilibrium Tests

Hearing loss can be described as either conduction deafness or sensorineural deafness. **Conduction deafness** occurs when there is a decreased ability to conduct the energy of sound waves through the external and middle ear to hearing receptors in the inner ear. Ear wax buildup, damage to the tympanic membrane, or fusion of auditory ossicles may cause conduction deafness. **Sensorineural deafness** is caused by damage to hearing receptors, damage to the cochlear branch of the vestibulocochlear nerve, or damage of the neural pathways to the auditory cortex.

Equilibrium receptors provide information that enables the body to maintain balance. There are two types of equilibrium receptors, static and dynamic. **Static equilibrium receptors** provide information about body position relative to the force of gravity (standing upright vs. being upside down). **Dynamic equilibrium receptors** provide information about body position in response to sudden movement such as rotation, acceleration, and deceleration (spinning, going faster, stopping). Inflammation of, or injury to equilibrium receptors results in an inability to maintain body position, vertigo, and/or dizziness. **Vertigo** is the sensation of circular motion either of oneself or external objects, while dizziness is often used to describe faintness, unsteadiness, or lightheadedness. Severe vertigo may be accompanied by **nystagmus**—rapid, involuntary movement of eyeballs.

ACTIVITY 12 Auditory and Equilibrium Tests

1 Test for unilateral (one side only) deafness using the **Weber test.**
 • Choose a subject from your group. Have the subject sit with the head erect and facing forward.
 • Strike a tuning fork (middle C preferably) and place it medially on the subject's forehead (bone conduction).
 • Ask the subject if the sound is equally loud in both ears or louder in one ear. Circle the result in Table 23.3.
 • If the sound is equally loud in both ears, the subject either has normal hearing or bilateral deafness (equal hearing loss in both ears). If the subject hears the sound louder in one ear, then the subject may have unilateral deafness (either conduction deafness in that ear or sensorineural deafness in the opposite ear).
 • If the subject has unilateral deafness, conduct the Rinne test to determine if hearing loss is conduction deafness.
 • *Optional: Mimic unilateral conduction deafness by placing a cotton ball in one external auditory canal and repeat the Weber test.*

TABLE 23.3 **Results of Weber, Rinne, and Barnay Tests**

TEST	RESULTS (CIRCLE YOUR RESULTS)		
Weber test	Equal loudness in both ears = normal hearing or equal hearing loss in both ears *or* Sound is louder in right ear = conduction deafness in right ear or sensorineural deafness in left ear *or* Sound is louder in left ear = conduction deafness in left ear or sensorineural deafness in right ear		
Rinne test	Conduction deafness *or* no conduction deafness		
Balance test	Static equilibrium receptors: functioning *or* not functioning		
		Direction of eye movement	Name semicircular canal stimulated
Barnay test	Head slightly forward:	lateral *or* vertical *or* rotational	lateral *or* anterior *or* posterior
	Head toward shoulder:	lateral *or* vertical *or* rotational	lateral *or* anterior *or* posterior
	Head on chin:	lateral *or* vertical *or* rotational	lateral *or* anterior *or* posterior

2 Test for conduction deafness using the **Rinne test.**
- This test will be conducted on the ear that may have conduction deafness as indicated by the Weber test.
- Have the subject sit with head erect and facing forward.
- Strike a tuning fork and place it on the subject's mastoid process to test hearing by **bone conduction.**
- Ask the subject to tell you when the sound can no longer be heard. Immediately place the still-vibrating tuning fork close to the subject's ear to test hearing by air conduction. If the subject can hear the tuning fork again when it is placed next to his or her ear, the subject does not have conduction deafness in that ear. If the subject cannot hear the tuning fork again, then the subject *may* have conduction deafness in that ear.
- Circle the results in Table 23.3. If the subject does not have conduction deafness, the test is complete.
- To verify conduction deafness, test the same ear again. This time you are testing hearing by **air conduction** first.
- Strike the tuning fork again and place the tuning fork close to the subject's ear.
- Ask the subject to tell you when the sound can no longer be heard, then place the tuning fork on the subject's mastoid process (bone conduction). If the subject hears the sound again, there is conduction deafness in that ear. Record whether the subject has conduction deafness or no conduction deafness in that ear by circling the results in Table 23.3.

3 Conduct a balance test to evaluate **static equilibrium** receptors.
- Have the subject stand in front of a whiteboard or chalkboard with arms at the sides. The subject cannot lean against the wall or support him or herself in any manner.
- Tell the subject to stand perfectly still. Mark the outline of the shoulders to help determine when he or she sways.
- Tell the subject to close his or her eyes. Observe movement in the subject's shoulders. Notice that, although the subject may sway slightly, posture is always corrected. Signals from the static equilibrium receptors are helping the subject to maintain posture. If the static equilibrium receptors are not functioning, the subject will not be able to maintain posture

and will exhibit large swaying movements or will fall. Be prepared to support the subject if necessary.
- Record the results in Table 23.3.

4 Conduct the **Barany test** to evaluate function of semicircular canals and **dynamic equilibrium** receptors.
- Choose a subject from your lab group who does not readily experience dizziness or become nauseated when rotated. If the subject experiences nausea during the demonstration, immediately stop rotation.
- Provide the subject with a chair or stool that can be rotated. Have the subject sit on the chair and hold onto the arms or seat for safety. Decide how the subject will position his or her legs during rotation to ensure safety and prevent interference. Position 3 to 4 students around the chair to prevent the subject from falling off.
- Tell the subject to slightly tilt his or her head forward, focus on a distant object, and keep both eyes open during the rotation.
- Carefully turn the chair or stool clockwise (to the right). Complete 10 turns, one turn per 2 seconds and stop suddenly. **Be prepared to support the subject until vertigo and/or dizziness has passed.** The subject will still experience rotation, indicating that the semicircular canals are functioning. Endolymph continues to move within the membranous semicircular ducts for a short time after rotation has stopped.
- Observe which way the subject's eyeballs are moving immediately after stopping rotation. Record direction in Table 23.3.

 Lateral movement of the eyes indicates stimulation of the dynamic equilibrium receptors in the lateral semicircular canals.
 Vertical movement of the eyes indicates stimulation of the dynamic equilibrium receptors in the anterior semicircular canals.
 Rotational movement of the eyes indicates stimulation of the dynamic equilibrium receptors in the posterior semicircular canals.

- Repeat the demonstration with the subject's head tilted toward one shoulder, and then again with the subject's chin resting on his or her chest.

C. The Nose and Olfaction

The nose contains the receptors for the sense of smell or olfaction (*olfact-* = smell). Olfactory receptors are found within the olfactory epithelium, a specialized area of the epithelium lining the nasal cavity. The **olfactory epithelium** covers the inferior surface of the cribriform plate, the superior nasal concha, and the upper part of the middle nasal concha.

The olfactory epithelium contains olfactory receptor cells, supporting cells, basal stem cells, and ducts of olfactory glands. The **olfactory receptor cells** are bipolar neurons whose dendritic end is imbedded in the mucus layer covering the surface of the olfactory epithelium and whose axons form the **olfactory nerves.** The olfactory receptors are located on **olfactory hairs** that project from the dendrites of the olfactory receptor cells.

Olfactory nerves pass through olfactory foramina in the **cribriform plate** and synapse on neurons in the **olfactory bulb.** Nerve impulses then travel along the **olfactory tract** to the lateral olfactory area of the cerebral cortex.

Olfactory receptors adapt to odors very quickly. This explains why when we are trying to determine the source of an odor, we often lose the smell before we find it. If we leave the area and return, we can smell the odor again.

ACTIVITY 13 **Structure of the Olfactory Epithelium and Olfactory Adaptation**

1 Label the structures of the olfactory epithelium in Figure 23.16.

2 Identify the following areas on a skull: location of **olfactory epithelium** and the **olfactory foramina** in the cribriform plate.

3 Observe olfactory adaptation (see below).

4 Answer discussion questions about olfactory adaptation with your lab partners.

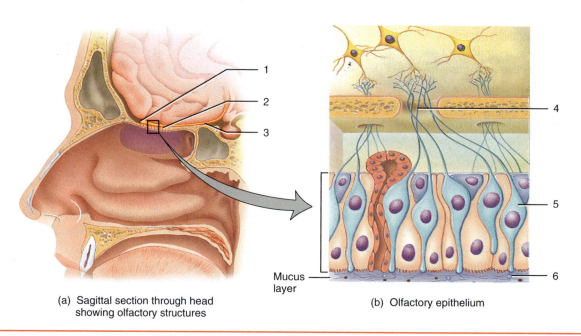

(a) Sagittal section through head showing olfactory structures

Mucus layer

(b) Olfactory epithelium

FIGURE 23.16 Olfactory structures.

(a)
- cribriform plate (CRIB-ri-form)
- olfactory bulb (OHL-fak-tore-ee)
- olfactory tract

(b)
- olfactory hair
- olfactory nerve
- olfactory receptor cell

1 _____

2 _____

3 _____

4 _____

5 _____

6 _____

Olfactory Adaptation

1. Choose a subject, a timer, an experimenter (holds vial under subject's nose), and a recorder.

2. Have the subject plug one nostril with cotton and close both eyes.

3. Noting the time, hold a container of cloves just under the open nostril and ask the subject to inhale through the open nostril and exhale through the mouth.

4. Ask the subject to tell you when the odor has disappeared, note the time, and record in Table 23.4. Instruct the subject to immediately pull out the cotton in the other nostril and inhale (vial still under nose).

5. Ask the subject if he or she can smell cloves. Record the result in Table 23.4.

6. Repeat the experiment with another distinct smell, such as peppermint or cinnamon. Avoid irritating odors.

? DISCUSSION QUESTIONS: OLFACTORY ADAPTATION

1 Is the time of adaptation the same for all odors?

2 After the nostril was unplugged, explain why the subject was able to smell the odor again.

D. Taste Buds and Gustation

Taste buds, which are found on the tongue, soft palate, pharynx (throat), and larynx, are microscopic, onion-shaped structures that contain gustatory cells, gustatory hairs, and supporting cells. Each **gustatory** (*gust-* = taste) **cell** has one **gustatory hair** that projects through an opening, the **taste pore,** on the apical end of the taste bud. **Gustatory receptors** are located on the gustatory hairs. The basal end of gustatory cells synapse onto the dendritic end of sensory neurons. Axons from the sensory neurons contribute fibers to the facial nerve (cranial nerve VII), glossopharyngeal nerve (IX), or vagus nerve (X), depending on the location of the taste bud.

Taste buds on the tongue are located in **papillae,** elevated structures that give the tongue its rough appearance. There are four types of papillae: vallate (circumvallate), fungiform, foliate, and filiform. **Vallate** (circumvallate) **papillae** are the largest papillae and form an inverted V at the posterior of the tongue. **Fungiform papillae** are mushroom-shaped and are scattered over the surface of the tongue. **Foliate papillae** are present mostly in children and are located in lateral margins of the tongue. **Filiform papillae** are slender, pointed projections that cover the surface of the tongue and give the tongue a rough texture. These papillae have tactile receptors but no taste buds. Taste buds are found in vallate, fungiform, and foliate papillae.

TABLE 23.4 **Results of Olfactory Adaptation Experiment**

ODOR	TIME TO ADAPTATION (SEC)	CAN SUBJECT SMELL ODOR AFTER NOSTRIL UNPLUGGED?
Cloves		
Peppermint		
Cinnamon		

There are four primary taste sensations: sweet, bitter, salty, and sour, and a possible fifth, MSG (monosodium glutamate). Gustatory receptors most sensitive to sweet and salty sensations are found on the tip of the tongue, while bitter sensations are in the back and sour sensations are on the sides of the tongue. Other taste sensations are a mixture of these four. Smell, temperature, and texture (tactile sensation) contribute to our sense of taste. A person with a cold often has a loss of taste due to a loss of smell. Cold French fries are not as tasty as hot ones, and mushy apples are not as good as crisp ones.

ACTIVITY 14 Gustatory Structures and Sensations

1 Label the gustatory structures in Figure 23.17(a), (b), and (c).

2 Examine a microscope slide of a vallate papilla.
 • Using the low-power objective lens, identify the taste bud in the wall of the papilla.
 • Using the high-power objective lens, identify the **taste pore, gustatory hairs,** and **gustatory receptor cells** forming the wall of the taste bud.

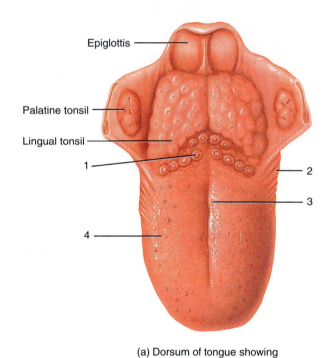

(a) Dorsum of tongue showing location of papillae

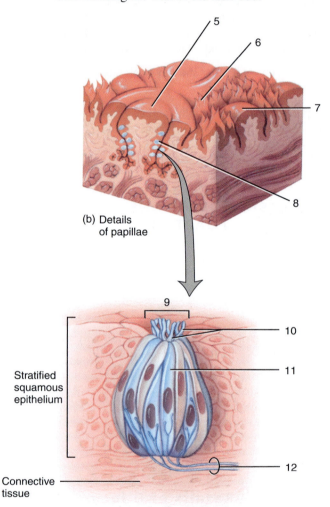

(b) Details of papillae

(c) Structure of a taste bud

FIGURE 23.17 Gustatory structures.

(a)	(b)	(c)
• filiform papilla	• filiform papilla	• gustatory hairs
• foliate papilla	• fungiform papilla	• gustatory receptor cell
• fungiform papilla	• taste bud	• sensory axons
• vallate (circumvallate) papilla	• vallate papilla	• taste pore

1 _____	5 _____	9 _____
2 _____	6 _____	10 _____
3 _____	7 _____	11 _____
4 _____	8 _____	12 _____

3 Use a mirror to identify fungiform and filiform papillae on your tongue. Try to see vallate papillae on the back of the tongue.

4 Examine the contribution of texture and smell to the sense of taste. Since many labs do not allow eating or drinking, this activity may need to be done in another area or at home.

- Choose a subject, an experimenter (who will give the food cubes to the subject), and a recorder. The subject will first try to identify food by texture only (rolling food on surface of tongue), then by taste (chewing food increases the amount of chemicals dissolved in saliva and capable of interacting with taste receptors), and finally with the addition of smell.
- Place ½-inch cubes of carrot, banana, apple, raw potato, and cheese on a plate.
- Have the subject pinch both nostrils, close both eyes, and open his or her mouth. Randomly choose one of the cubes and place it in the subject's mouth.
- Instruct the subject to roll the food around the surface of the tongue and attempt to identify the food. If identification is correct, check the "texture only" column next to the food in Table 23.5.

- Instruct the subject to chew the food and attempt to identify the food. If identification is correct, check "texture and taste" column in Table 23.5.
- Instruct the subject to open both nostrils and attempt to identify the food. If identification is correct, check the "texture, taste, and smell" column in Table 23.5.
- Repeat procedure with the other food cubes.

5 Answer the discussion question.

? **DISCUSSION QUESTION: GUSTATORY STRUCTURES AND SENSATIONS**

1 Compare the number of foods identified by texture, identified by taste, and identified by the sense of smell.

TABLE 23.5 **Results of Experiment: Contribution of Texture and Smell to Taste**

FOOD	TEXTURE ONLY	TEXTURE AND TASTE	TEXTURE, TASTE, AND SMELL
Carrot			
Banana			
Apple			
Raw potato			
Cheese			

EXERCISE

23

Reviewing Your Knowledge

A. Anatomy Drill and Practice

Go to www.wiley.com/college/allen. Click on the picture of your lab manual, then on Student Companion Site, Exercise 23, and Anatomy Drill and Practice.

B. Accessory Eye Structures

Choose the structure that corresponds to each statement.

a. bulbar conjunctiva
b. lacrimal canals
c. lacrimal gland

d. lacrimal sac
e. nasolacrimal duct
f. palpebral conjunctiva

_____ 1. drains tears into nasal cavity

_____ 2. produces tears

_____ 3. membrane that covers the inner surface of eyelid

_____ 4. tears from surface of eye drain into here

C. Extrinsic Eye Muscles

Choose the muscle that applies to each statement.

a. inferior oblique
b. inferior rectus
c. lateral rectus

d. medial rectus
e. superior oblique
f. superior rectus

_____ 1. moves eyeball superiorly

_____ 2. moves eyeball laterally

_____ 3. moves eyeball inferiorly

_____ 4. moves eyeball medially

_____ 5. moves eyeball laterally and inferiorly

_____ 6. moves eyeball laterally and superiorly

D. Eye Structures

Choose the structure that applies to each statement. More than one structure may apply to a statement and a structure may be used more than once.

a. anterior cavity
b. anterior chamber
c. choroid
d. ciliary body
e. ciliary muscle
f. ciliary process
g. cornea
h. iris
i. lens

j. ora serrata
k. posterior chamber
l. pupil
m. retina
n. sclera
o. scleral venous sinus
p. suspensory ligaments
q. vitreous chamber

_____ **1.** produces aqueous humor

_____ **2.** structures that are part of vascular tunic

_____ **3.** contains photoreceptors

_____ **4.** controls the size of the pupil

_____ **5.** drains aqueous humor from the anterior chamber

_____ **6.** structures that are part of the fibrous tunic

_____ **7.** most anterior part of the eyeball

_____ **8.** anterior boundary of retina

_____ **9.** attaches lens to ciliary body

_____ **10.** changes shape to focus light on retina

_____ **11.** location of aqueous humor

_____ **12.** white, tough, outer layer of eyeball

Answer the following questions:

13. Why can the retina pull away from the back of the eyeball?

14. Name the instrument used to view the retina during a physical exam.

15. Name the two layers of the retina.

E. Anatomy of the Retina

Choose the structure that corresponds to each statement.

a. bipolar cell layer
b. central fovea
c. cones
d. ganglion cell layer

e. macula lutea
f. optic disc
g. photoreceptor layer
h. rods

_____ **1.** has the highest density of cones in the retina

_____ **2.** axons form optic nerve

_____ **3.** does not contain photoreceptors; blind spot

_____ **4.** photoreceptor that allows us to see color

_____ **5.** contains rods and cones

_____ **6.** light rays strike this layer of the retina first

F. Visual Acuity Tests

Answer the following questions:

1. Define accommodation.

2. A 10-year-old patient's distance visual acuity was tested and determined to be 20/80. (a) Explain what that means.

(b) Is 20/80 better or worse than 20/20? Explain.

3. A 60-year-old man has to hold his newspaper at arm's length to read it. What condition does he have?

4. A person with _____ has blurry near vision but normal distance vision.

5. A person with _____ does not see everything within a visual field clearly; some parts are blurry.

6. A person with _____ has normal near vision but blurry distance vision.

G. External and Middle Ear Structures

Choose the structure that applies to each statement. More than one structure may apply to a statement, and a structure may be used more than once.

a. auditory tube
b. auricle
c. external auditory canal
d. external ear
e. helix
f. incus
g. lobule

h. malleus
i. middle ear
j. oval window
k. stapes
l. round window
m. tympanic membrane

_____ **1.** auditory ossicle attached to tympanic membrane

_____ **2.** equalizes air pressure in middle ear with external air pressure

_____ **3.** external ear structures

_____ **4.** ear drum

_____ **5.** external feature of ear that contains the helix and lobule

_____ **6.** stapes is attached to this membrane-covered opening

_____ **7.** middle auditory ossicle

_____ **8.** small bones of middle ear that are connected by synovial joints

H. Internal Ear Structures

Choose the structure that applies to each statement. More than one structure may apply to a statement, and a structure may be used more than once.

a. ampulla
b. cochlea
c. cochlear duct
d. endolymph
e. oval window
f. perilymph

g. round window
h. saccule
i. semicircular canals
j. semicircular ducts
k. utricle
l. vestibule

_____ **1.** interconnected components of membranous labyrinth

_____ **2.** fluid found within all bony labyrinth structures

_____ **3.** sections of the membranous labyrinth found within vestibule

_____ **4.** section of the membranous labyrinth that contains hearing receptors

_____ **5.** sections of the membranous labyrinth that contain equilibrium receptors

_____ **6.** interconnected components of bony labyrinth

I. Microscopic Anatomy of the Cochlea, Spiral Organ of Corti, Macula, and Crista

Choose the structure that applies to each statement. More than one structure may apply to a statement, and a structure may be used more than once.

a. basilar membrane	f. hair cells	k. spiral organ of Corti
b. cochlear duct	g. macula	l. supporting cells
c. crista	h. otolithic membrane	m. tectorial membrane
d. cupula	i. scala tympani	n. vestibular membrane
e. hair bundle	j. scala vestibuli	

_____ **1.** receptor for hearing

_____ **2.** receptor(s) that contain(s) hair bundles, hair cells, and supporting cells

_____ **3.** components of macula

_____ **4.** membrane separating the superior chamber of cochlea from cochlear duct

_____ **5.** structure(s) that bends stereocilia of hair cells

_____ **6.** spiral organ of Corti sits on this membrane

_____ **7.** equilibrium receptors

_____ **8.** contains endolymph

_____ **9.** contains perilymph

_____ **10.** equilibrium receptor found within ampullae of semicircular canals

J. Auditory and Equilibrium Tests

Answer the following questions:

1. What causes nystagmus?

2. What causes conduction deafness?

3. What causes sensorineural deafness?

4. If a vibrating tuning fork is placed on the mastoid process, who would "hear" the sound—someone with normal hearing, someone with conduction deafness, or someone with sensorineural deafness? Circle all that apply.

5. Inability to maintain posture while standing would indicate a problem with which equilibrium receptor?

K. Olfaction

Choose the structure that applies to each statement.

 a. cribriform plate
 b. superior portion of nasal cavity
 c. olfactory hair
 d. olfactory nerve
 e. olfactory receptor cells

_____ **1.** bipolar neurons

_____ **2.** part of olfactory receptor cell that contains the olfactory receptors

_____ **3.** formed of bipolar neuron axons

_____ **4.** olfactory nerves pass through this structure before synapsing onto olfactory bulb

_____ **5.** identify the location of the olfactory epithelium

L. Taste

Choose the structure that applies to each statement. More than one structure may apply to a statement, and a structure may be used more than once.

 a. filiform papillae f. gustatory hair
 b. fungiform papillae g. supporting cell
 c. pharynx h. taste buds
 d. soft palate i. taste pore
 e. gustatory cell j. vallate papillae

_____ **1.** contain taste buds

_____ **2.** part of gustatory cell that contains gustatory receptor

_____ **3.** opening in taste bud through which gustatory hair projects

_____ **4.** components of taste bud

_____ **5.** mushroom-shaped projections on tongue

_____ **6.** projections that form inverted V on back of tongue

Answer the following questions:

7. Identify where on the tongue the receptors most sensitive to each of the four taste sensations are located.

8. What other sensations contribute to the sensation of taste?

Using Your Knowledge

A. The Eye and Vision

1. Strabismus, a misalignment of the eyeball caused by an imbalance in the extrinsic eye muscles, is also called lazy eye or wandering eye. Identify the extrinsic eye muscle that would cause lateral strabismus of the right eye.

2. Where in the nasal cavity does the nasolacrimal duct empty?

3–4. Name all the accessory eyeball structures, fluids, and retinal layers that light passes through before it hits the photoreceptors.

5. Explain why the retina can detach from the eyeball with just a blow to the back of the head.

6. Is the retinal area in Figure 23.18 the macula lutea, fovea centralis, or optic disc? Explain your choice.

Retina
Choroid
Sclera

FIGURE 23.18 Retinal area.

(a) Normal

(b) Abnormal

FIGURE 23.19 Visual fields.

7. Explain why a pituitary tumor may affect vision.

8. A 10-year-old boy has 20/20 distance vision, and 20/60 near vision. What refraction abnormality does he have?

9. Using your textbook or another source, research how and why glaucoma and macular degeneration affect vision. Compare Figure 23.19(a) and (b) and identify whether the visual field in (b) is representative of a person with macular degeneration or glaucoma.

10. Explain why the condition represented in Figure 23.19(b) causes loss of vision in the center of the field of vision and blurriness in the periphery.

11. What structure keeps a contact lens on the cornea and prevents it from becoming lodged on the posterior surface of the eyeball?

B. The Ear, Hearing, and Balance

12–13. Sound waves enter the external auditory canal and cause vibrations in structures and fluids of the ear. Identify in order the structures and fluids that vibrate in the pathway from the external auditory canal to the spiral organ of Corti.

14. Diving to the bottom of a deep pool will sometimes cause discomfort in the middle ear. Explain the cause of this discomfort.

15. What do the receptor cells for hearing, static equilibrium, and dynamic equilibrium have in common?

16. Would the receptors for equilibrium work in space at zero gravity?

17. Describe a series of movements that would stimulate all semicircular duct receptors and if repeated may cause vertigo. (*Hint:* Think of a carnival ride that causes vertigo.)

C. Olfaction and Gustation

18. Why do we sniff when we wish to smell something better?

19. When you lick ice cream off an ice cream cone, which papillae are involved in scraping the ice cream off the cone?

20. Licking cold ice cream off an ice cream cone stimulates a variety of sensory receptors associated with the tongue. Name all the sensory receptors involved.

Autonomic Nervous System Structure and Function

Objectives

After completing this exercise, you should be able to:

- Compare and contrast the anatomical components of the sympathetic and parasympathetic divisions of the autonomic nervous system (ANS)
- Name and locate the ganglia of the sympathetic and parasympathetic nervous systems on models or charts

Materials

- spinal cord model or chart with autonomic ganglia
- brain model or chart showing the cranial nerves
- cross-section of the spinal cord model or chart with white and gray rami communicantes
- millimeter rulers, temperature strips (forehead)

The autonomic nervous system (ANS) is the division of the peripheral nervous system that operates without conscious control in most individuals. The ANS has autonomic sensory neurons that carry nerve impulses from sensory receptors in visceral organs and blood vessels to the CNS and autonomic motor neurons that carry nerve impulses from the CNS to smooth muscle, cardiac muscle, and glands (effectors). There are two autonomic motor neurons between the spinal cord and the effectors. The first motor neuron (**preganglionic**) synapses with the second motor neuron (**postganglionic**) in ANS ganglia. The ANS has two different motor divisions, the sympathetic division and the parasympathetic division. Most ANS effectors are innervated by both the sympathetic and parasympathetic divisions, which have opposing actions.

Activation of the **sympathetic division** occurs during exercise, emergencies, excitement, and embarrassment and results in a series of physiological responses that together are called the fight-or-flight response. The fight-or-flight response includes activities that promote mental alertness, vision, and skeletal muscle activity.

- pupillary dilation
- increased heart rate, force of heart contraction and blood pressure
- increased blood flow to skeletal muscles, heart, liver, and adipose tissue
- increased airway diameter resulting in faster and greater air movement into and out of the lungs
- increased glycogenolysis by liver and increased lipolysis by adipose tissue increases blood levels of glucose and fatty acids used to make ATP
- decrease in digestion and urine formation—activities not needed during exercise, fighting, or fleeing.

Activation of the **parasympathetic division** promotes rest, reading, restoration, and repair. These activities include:

- decreased heart rate and force of contraction and decreased blood pressure
- decreased airway diameter
- pupillary constriction and increased tear production
- increased digestive processes
- increased urine formation

A. Anatomy of the Sympathetic Motor Division

Preganglionic and Postganglionic Motor Neurons

Cell bodies of **preganglionic neurons** are located in the gray matter of the lateral horn of spinal cord segments T_1–L_2. Because the nerve impulses from the sympathetic preganglionic neuron originate in the thoracic and lumbar segments of the spinal cord, the sympathetic division is also called the **thoracolumbar division.**

Axons of the first motor neurons of this division, **preganglionic neurons,** exit the spinal cord in the anterior (ventral) root of the spinal nerves and travel with somatic motor neurons in the spinal nerves a short distance. These myelinated preganglionic axons quickly leave the spinal nerve as **white rami communicantes** that connect with sympathetic trunk ganglia paired chains of ganglia that lie parallel to the spinal cord. In these ganglia, the axons either synapse with postganglionic cell bodies, or they *pass through the sympathetic trunk ganglia without synapsing,* emerging to form **splanchnic nerves** (*splanchnic* = viscera) in the abdominopelvic region.

Nerve cell bodies of the second motor neuron, the **postganglionic neuron,** are located in the sympathetic ganglia. Preganglionic axons synapse with postganglionic neuron cell bodies in the ganglia. Some of the axons of postganglionic neurons return to the spinal nerve at the same level via the **gray rami communicantes.** Postganglionic axons are longer than preganglionic axons because they travel a greater distance to innervate the effectors.

Sympathetic Ganglia

The sympathetic division has two groups of ganglia: the sympathetic trunk (chain) ganglia and the prevertebral (collateral) ganglia.

- **sympathetic trunk** (*chain*) **ganglia**—These paired ganglia chains lie parallel to the spinal cord. Postganglionic neurons from these ganglia innervate organs superior to the diaphragm.
- **prevertebral** (*collateral*) **ganglia: celiac, superior mesenteric,** and **inferior mesenteric**—These three single ganglia lie anterior to the abdominal aorta, and their postganglionic neurons innervate organs inferior to the diaphragm. Preganglionic axons traveling in splanchnic nerves synapse with postsynaptic neuron cell bodies in the prevertebral ganglia or in the adrenal medullae.

ACTIVITY 1 **Anatomy of the Sympathetic Motor Division**

1 Label the structures in Figure 24.1.

2 Identify the following structures on a chart showing the sympathetic nervous system: **right** and **left sympathetic trunk ganglia, prevertebral ganglia, splanchnic nerves, preganglionic,** and **postganglionic motor neurons.**

3 Identify the **sympathetic trunk ganglia** on a longitudinal spinal cord model.

4 Identify the rami communicantes and the sympathetic trunk ganglia on a cross-sectional spinal cord model or chart.

Trachea

Arch of aorta

Left primary bronchus

Esophagus

Thoracic aorta

Diaphragm

Left kidney

Abdominal aorta

1

2

3

4

FIGURE 24.1 Location of sympathetic motor ganglia and nerves.

- left sympathetic trunk ganglia
- prevertebral ganglia
- right sympathetic trunk ganglia
- splanchnic nerves

1 _____

2 _____

3 _____

4 _____

B. Anatomy of the Parasympathetic Motor Division

Cell bodies of the preganglionic motor neurons of this division are located in the nuclei of four cranial nerves in the brain stem and in the lateral horns of spinal cord segments S2–S4. Because the outflow is from these areas, this division is also known as the **craniosacral division.**

Axons of the *cranial* **preganglionic neurons** exit the brain in cranial nerves III, VII, IX, and X and synapse with postganglionic neurons in **terminal** (intramural) **ganglia** located near their target organs in the head region. The major terminal ganglia in the head region are: **ciliary, pterygopalatine, submandibular,** and **otic ganglia.**

Axons of the *sacral* preganglionic neurons leave the spinal cord in the anterior (ventral) roots, form nerves called **pelvic splanchnic nerves,** and synapse with postganglionic neurons in **terminal ganglia** that are within the wall of the target organ.

Nerve cell bodies of **postganglionic neurons** (second motor neurons) are in the terminal ganglia. Axons of postganglionic neurons are short because they travel a short distance to their effectors.

ACTIVITY 2 Parasympathetic Anatomy

1 Label the terminal ganglia in Figure 24.2. Use the name of the ganglia to find the location.

2 Identify the following structures on a chart showing the parasympathetic nervous system: **terminal ganglia, pelvic splanchnic nerves, preganglionic** and **postganglionic motor neurons.**

Oculomotor Nerve (III)

2

Facial Nerve (VII)

Vestibulocochlear
Nerve (VIII)

Glossopharyngeal
Nerve (IX)

Vagus
Nerve (X)

Trigeminal
Nerve (V)
(Mandibular
branch)

1

3

Superior cervical
sympathetic ganglion

Cervical
sympathetic trunk

4

FIGURE 24.2 Parasympathetic nerves and ganglia.

- ciliary ganglion
- otic ganglion
- pterygopalatine ganglion
- submandibular ganglion

1 _____

2 _____

3 _____

4 _____

C. Autonomic Reflexes

Autonomic reflexes control many body functions such as blood pressure, respiration, and digestion. The **autonomic reflex arc** contains the same components as the somatic reflex arc: receptor, sensory neuron, integrating center, motor neurons, and effector. Although most autonomic sensory receptors are located in visceral organs, some autonomic reflexes are responses to changes in the external environment. Autonomic integrating centers are **polysynaptic** and most are located in the hypothalamus and brain stem. However, the autonomic reflex arcs that control urination and defecation have integrating centers in the spinal cord. ANS reflex arcs have *two motor neurons,* preganglionic and postganglionic neurons, which synapse in the autonomic ganglia. ANS effectors are cardiac muscle, smooth muscle, and glands.

Most ANS effectors receive motor neurons from both the sympathetic and the parasympathetic branches of the ANS. Since sympathetic and parasympathetic responses usually oppose each other, the hypothalamic integrating center determines which response is appropriate.

ACTIVITY 3 **Autonomic Reflex Arc**

1 Label the parts of the autonomic reflex arc in Figure 24.3.

2 Pronounce the terms as you point to them on the figure.

Pupillary Light Reflex

When photoreceptors in the retina of the eye are stimulated by light, nerve impulses are sent via the optic nerve to integrating centers in the brain. Interneurons from the integrating centers send impulses to motor neurons in the midbrain. Axons from these preganglionic motor neurons travel along cranial nerve III (oculomotor nerve) to the ciliary ganglion where they synapse with the postganglic motor neuron. The postganglionic axons travel a short distance to smooth muscles in the iris of the eye. When these muscles contract, the pupil constricts.

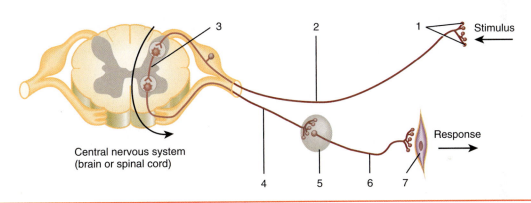

FIGURE 24.3 Autonomic reflex arc.

- autonomic ganglion (sympathetic or parasympathetic)
- interneuron
- postganglionic motor neuron
- preganglionic motor neuron
- sensory neuron (afferent)
- sensory receptor
- visceral effector

1 _____ 5 _____

2 _____ 6 _____

3 _____ 7 _____

4 _____

ACTIVITY 4 Pupillary Light Reflex

1 Test the pupillary light reflex.
- This experiment works best if conducted in a dimly lit room.
- Have the subject cover his or her right eye by holding a hand over the eye.
- Using a penlight, shine a light into the subject's left eye. Observe the pupil and describe what happened. _____
- Uncover the right eye and observe the size of the right pupil. What happened to the right pupil? _____
- Now cover the left eye and shine the light in the right eye. Observe the pupil and describe what happened to the right pupil. _____
- Uncover the left eye and record what happened to the left pupil. _____

2 Answer discussion questions about the pupillary light reflex with your lab group.

? DISCUSSION QUESTIONS

1 Is this reflex a somatic or autonomic reflex?

2 Identify the nerve that is carrying the sensory information from the eye to the brain.

3 Identify the cranial nerve carrying the preganglic parasympathetic axons to the ciliary ganglia.

4 Is this a monosynaptic or polysynaptic reflex?

5 Is the pupillary light reflex ipsilateral, contralateral, or both?

The Effect of Higher Brain Centers on ANS Functions

The cerebral cortex, limbic system (emotional brain), and thalamus can also alter the ANS responses. For example, when some people tell a lie, the limbic system causes the following ANS responses: dilation of the blood vessels in the face and neck area resulting in redness, constriction of the pupils, and an increase in breathing rate, heart rate, and/or blood pressure. However, detection of autonomic changes by observation is a subjective exercise. Also, some people can lie without feeling emotion, avoiding ANS responses.

ACTIVITY 5 Lie Detector Test and the ANS

1 Decide who will be: the subject, the interrogator, the observer, and the pulse taker.

2 Use the questions below or other questions that the group devises and has the instructor approve. Assign each question a number (1–10) indicating the order the questions are to be asked.
How old are you? _____
You were born in what city and state? _____
How many brothers and sisters do you have? _____

When driving a car, do you purposely exceed the speed limit over twice a week? _____
Have you ever flunked a college or university class? _____

Are you single or married? _____
Do you like this anatomy and physiology class? _____
Do you study for this class as much as you should? _____

Are you getting the grade in this class that you want? _____

What is your major? _____
Have you ever cheated on your income taxes? _____
What is your weight? _____
What is your height? _____
Have you ever cheated on an exam? _____

3 Have the subject note privately which three questions are going to be lies. In a real polygraph test, the subject sees all questions before the test commences.

4 Perform the lie detector test.
• Have the observer obtain baseline information on: the subject's redness of face and neck, pupil size, pulse rate over 15 seconds, nervousness of hands or fingers, or any other nervous changes that might indicate lying. You may take the temperature on the forehead with temperature strips, if available. Record the results in Table 24.1.

TABLE 24.1 ANS Responses and Lie Detector Test

PRE-QUESTION OBSERVATIONS	ANS RESPONSES	ANS RESPONSES FOR QUESTION									
		1	2	3	4	5	6	7	8	9	10
Color _____	Color No Change (N) or Redness (R)										
Pupil Size (mm) _____	Pupil Size No Change (N), Increased (I) or Decreased (D)										
Pulse Rate (beats/15 sec) _____	Pulse Rate (beats/15 sec) N, I, or D										
Temperature (°F or °C) _____	Temperature (°F or °C) N, I, or D										
	Nervousness Yes (Y) or No (N)										
	Prediction: Was the answer the Truth or a Lie?										

- Have the interrogator ask the 10 chosen questions.
- Observe ANS changes after each question and record in Table 24.1 before continuing to the next question.

5 Based on the collected data, have the experimenters privately predict which three questions the subject lied about, and then ask the subject to give the answers that were lies.

- Question numbers that experimenters predict the subject is lying about. _____
- Actual question numbers the subject did lie about. _____
- Number of correct predictions chosen by experimenters in your group. _____
- Pool your data with all the lab groups. What is the average number of correct predictions? _____

6 Answer discussion questions with your lab group.

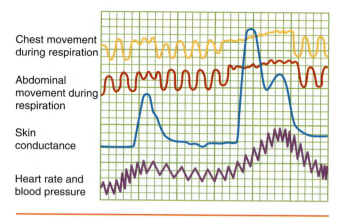

FIGURE 24.4 Polygraph test reactions.

? DISCUSSION QUESTIONS

1 Based on the pooled class data, give reasons why using this type of lie detector test to solve a crime may or may not be advantageous.

ration, abdominal movement during respiration, skin conductance (sweating), and heart rate/blood pressure (Figure 24.4). Note the response (truthful answer) on the left side of the figure, compared with a response (lie) on the right side.

2 A professional **polygraph test** is administered by someone who is trained to perform the test and lasts 2–3 hours with additional time to analyze the results. Four reactions are recorded by four sensors attached to a person's body that record chest movement during respi-

3 Give reasons why the use of a polygraph test given by a professional may or may not reveal if the person is telling the truth.

Reviewing Your Knowledge

A. Anatomy Drill and Practice

Go to www.wiley.com/college/allen. Click on the picture of your lab manual, then on Student Companion Site, Exercise 24, and Anatomy Drill and Practice.

B. Comparison of Somatic and Autonomic Nervous Systems

Answer each of the following questions:

1. Compare the number of motor neurons in a somatic and autonomic reflex arc.

2. Compare the effectors of the somatic and autonomic nervous systems.

C. Comparison of Sympathetic and Parasympathetic Divisions

Match the terms with the appropriate ANS division: P = parasympathetic division; S = sympathetic division. Use one or both choices to answer the question.

_____ **1.** two motor neurons

_____ **2.** short preganglionic fibers

_____ **3.** long preganglionic fibers

_____ **4.** innervates adrenal medullae

_____ **5.** terminal ganglia

_____ **6.** trunk or chain ganglia

_____ **7.** prevertebral ganglia

_____ **8.** craniosacral division

_____ **9.** thoracolumbar division

_____ **10.** activated during exercise, fighting or fleeing

_____ **11.** rami communicantes

_____ **12.** promotes digestion

_____ **13.** short postganglionic fibers

_____ **14.** long postganglionic fibers

D. Autonomic Reflex Arc

Answer each of the following questions:

1. List the components of the autonomic reflex arc. Compare with somatic reflex arc.

2. Where are autonomic integrating centers located?

3. Are the autonomic integrating centers monosynaptic or polysynaptic?

4. Can the autonomic integrating centers be influenced by higher brain area?

Using Your Knowledge

A. Sympathetic Division

Refer to a sympathetic nervous system chart (or textbook) to answer questions 1–6.

Brittany was visibly upset that a class she needed to take was cancelled because of low enrollment. Jim observed that her eyes were dilated. Trace the sympathetic pathway from the lateral gray horn of the spinal cord to the iris of her eye.

1. Identify the spinal cord area (cervical or thoracic) that the preganglionic axons exit in the anterior (ventral) root.

2. Name the sympathetic ganglia in which the preganglionic and postganglionic neurons synapse.

Paul ate a big lunch and then decided to jog a mile, which was not a very great idea. His digestive system was put "on hold" while his sympathetic nervous system was stimulated to run. Trace the sympathetic pathway from the lateral gray horn of the spinal cord to the stomach.

3. Identify the nerve that the preganglionic axons form.

4. Name the effector (the tissue within the organ that is innervated by the postganglionic neuron.)

Burke is in the final 100 yards of a race with an opponent at his heels. His sympathetic nervous system is stimulating his adrenal medullae. Trace the sympathetic pathway from the lateral gray horn of the spinal cord to the adrenal medulla.

5. Name the sympathetic ganglia in which the preganglionic axons synapse.

6. What is different about these postganglionic cells?

B. Parasympathetic Division

Refer to parasympathetic nervous system chart (or textbook) to answer questions 7–10.

Tom had anesthesia for surgery and was unable to void for a short time after awakening. Trace the normal parasympathetic pathway from the lateral gray horn of the spinal cord to the urinary bladder.

7. Name the spinal cord segments that the preganglionic axons exit in the anterior (ventral) root.

8. Name the effector (the tissue within the organ that is innervated by the postganglionic neuron).

Michael just ate lunch and his salivary glands are responding with secretions. Trace the parasympathetic pathway from the medulla oblongata to the salivary glands.

9. Name the two nerves that carry the preganglionic axons to the parasympathetic ganglia.

10. Name the parasympathetic ganglia in which the preganglionic and postganglionic neurons synapse.

C. Functions of the Sympathetic and Parasympathetic Nervous Systems

Choose the correct division of the ANS that generates the following results: S = sympathetic and P = parasympathetic.

_____ **11.** sweaty palms

_____ **12.** blushing

_____ **13.** urination

_____ **14.** stomach churning

_____ **15.** salivary gland secretion

_____ **16.** constriction of pupils

_____ **17.** increased blood pressure

_____ **18.** increased respiration

_____ **19.** decreased heart rate

_____ **20.** digestive enzyme secretions

Structure and Function of the Endocrine System

Objectives

After completing this exercise, you should be able to:

- Identify each major endocrine gland on models or charts
- Name the main hormones secreted by each major endocrine gland
- Distinguish selected endocrine tissue on microscope slides

Materials

- human torso (with a dissectable brain)
- compound microscope, lens paper, prepared slides of selected endocrine glands: hypothalamus, pituitary gland, thyroid gland, parathyroid gland, adrenal gland, pancreas

The endocrine system (*endo-* = within; *krinein* = to secrete) has many glands that secrete hormones into the bloodstream. These chemicals are transported throughout the body in blood and bind to target cells that have specific cell membrane receptors for a specific hormone. Hormones cause changes in activity in these target cells and direct a variety of cellular activities to keep the body in homeostasis.

A. Identification of Major Endocrine Glands

The selected, major endocrine glands that we study in this exercise are shown in Figure 25.1. From the head inferiorly, they are the **pineal gland (body), hypothalamus, pituitary gland, thyroid gland, parathyroid glands, thymus, adrenal glands, pancreas, ovaries,** and **testes.**

ACTIVITY 1 Identification of the Major Endocrine Organs

1 Using Figure 2.1, label the major endocrine glands in Figure 25.1.
2 Identify these 10 major endocrine glands on a human torso with a brain.
3 Pronounce the terms out loud as you point to them.

- adrenal glands (a-DREE-nul)
- hypothalamus (hypo-THAL-a-mus)
- ovaries
- pancreas
- parathyroid glands (para-THY-roid)
- pineal gland (pie-NEE-ul)
- pituitary gland or hypophysis (hy-POF-ih-sis)
- testes (TES-teez)
- thymus (THY-mus)
- thyroid gland

1 _____

2 _____

3 _____

4 _____

5 _____

6 _____

7 _____

8 _____

9 _____

10 _____

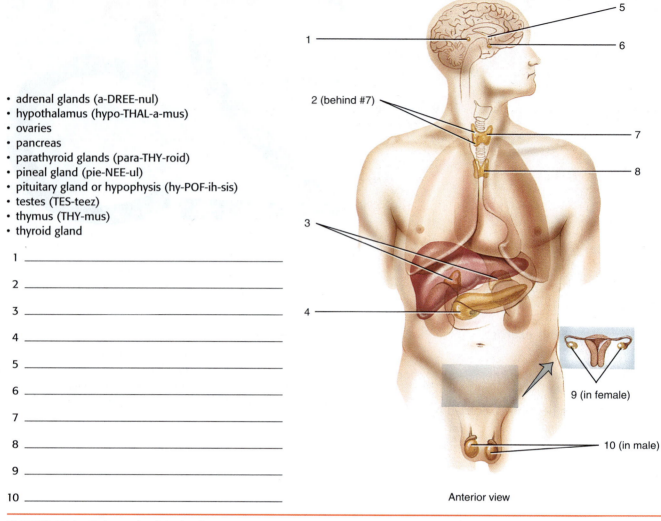

Anterior view

FIGURE 25.1 Major endocrine glands.

B. Anatomy, Histology, and Hormones of the Pituitary Gland

The **pituitary gland** is also called the **hypophysis** (*hypo-* = under; *phyein* = to grow; translated as "to grow under the hypothalamus") and is comprised of the **anterior pituitary** or **adenohypophysis** (*adeno-* = gland) and the **posterior pituitary** or **neurohypophysis** (*neuro-* = nerves). An **intermediate lobe** is present in the fetus but atrophies during fetal development. The **infundibulum** is the stalk that connects the hypothalamus to the pituitary gland.

Although in the past the pituitary gland was called the "master gland," it is now known that the pituitary gland is stimulated by the hypothalamus. Located inferior to the thalamus in the brain, the **hypothalamus** is an important component of both the nervous and endocrine systems, and couples these two regulatory systems together.

The **anterior pituitary** is composed of glandular epithelial tissue that makes and secretes seven hormones: **human growth hormone** (hGH) or somatotropin, **thyroid-stimulating hormone** (TSH), **follicle-stimulating hormone** (FSH), **luteinizing hormone** (LH), **prolactin** (PRL), **adrenocorticotropic hormone** (ACTH), and **melanocyte-stimulating hormone** (MSH). Hormones secreted by the hypothalamus either stimulate or inhibit secretion of anterior pituitary hormones.

The **posterior pituitary** is composed of nervous tissue that stores and releases two hormones into the blood that are not synthesized in the posterior pituitary: **antidiuretic hormone** (ADH) and **oxytocin** (OT). These two hormones are synthesized within cell bodies of hypothalamic neurons, are packaged into vesicles, and travel down axons that pass through the infundibulum to the posterior pituitary gland.

ACTIVITY 2 The Pituitary Gland

1 Examine and label the drawing and photomicrographs of the pituitary gland in Figure 25.2(a), (b), and (c).

2 Pronounce the terms as you point to them on each figure.

3 Observe a slide of the anterior and posterior pituitary with a compound microscope and identify the structures listed with Figure 25.2(c).

- anterior pituitary or adenohypophysis (a-den-oh-hy-POF-ih-sis)
- hypothalamus (hypo-THAL-uh-mus)
- infundibulum (in-fun-DIB-u-lum)
- posterior pituitary or neurohypophysis (neur-oh-hy-POF-ih-sis)

1 _____

2 _____

3 _____

4 _____

(a) Hypothalamus and pituitary gland

10×

(b) Survey photomicrograph of infundibulum and pituitary

- anterior pituitary
- hypothalamus
- infundibulum
- posterior pituitary

5 _____

6 _____

7 _____

8 _____

90×

(c) Photomicrograph of the anterior and posterior pituitary

- anterior pituitary
- posterior pituitary

9 _____

10 _____

FIGURE 25.2 The pituitary gland.

C. Anatomy, Histology, and Hormones of the Thyroid and Parathyroid Glands

The thyroid gland has two lobes and a connecting isthmus and lies on both sides of the trachea near the thyroid cartilage (Adam's apple) of the larynx. The thyroid gland has a striking microscopic structure with large **follicles** that are filled with a protein **colloid** (large insoluble molecules). Simple cuboidal or simple columnar cells called **follicular cells** form the follicular walls. Follicle cells synthesize **thyroglobulin,** a thyroid hormone precursor stored in colloid. Two **thyroid hormones (TH), thyroxine (T$_4$)** or **tetraiodothyronine** (*tetra-* = four; *iodo-* = iodine) and **triiodothyronine (T$_3$),** are produced from thyroglobulin. Located between the follicles are **parafollicular cells** or **C cells,** which synthesize and secrete **calcitonin.**

Embedded in the posterior surface of the thyroid gland are typically four small, round **parathyroid glands.** Two types of epithelial cells are found in the parathyroid glands: the principal cells and the oxyphil cells. The **principal cells,** which produce **parathyroid hormone (PTH),** are more numerous and smaller than the **oxyphil cells** whose function is unknown.

ACTIVITY 3 **Structure and Histology of the Thyroid and Parathyroid Glands**

1 Label the structures of the thyroid and parathyroid glands in Figure 25.3(a) and (b).

2 Pronounce the terms as you point to them.

3 Place your fingers on each side of the trachea, three fingers' breadth superior to the suprasternal (jugular) notch, and feel the thyroid gland move upward as you swallow.

4 Label the photomicrographs of the sections of thyroid and parathyroid glands in Figure 25.4(a) and (b).

5 Observe a slide of the thyroid and parathyroid glands, and identify the structures listed in Figure 25.4(a) and (b).

6 Answer the discussion question with your lab group.

? **DISCUSSION QUESTION: THYROID AND PARATHYROID GLANDS**

1 Examine Figure 25.5. Thyroglobulin is stored in the colloid, and calcitonin and PTH are stored in the parafollicular and principal cells, respectively. Compare the storage space for thyroglobulin (TH precursor) to the storage space for calcitonin and PTH.

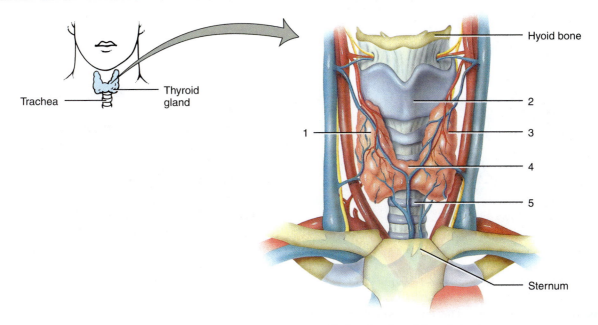

(a) Anterior view of thyroid gland

(b) Posterior view of thyroid gland
with parathyroid glands

FIGURE 25.3 The thyroid and parathyroid glands.

(a)
• isthmus of thyroid gland
• left lobe of thyroid gland
• right lobe of thyroid gland
• thyroid cartilage of larynx
• trachea

1 _____

2 _____

3 _____

4 _____

5 _____

(b)
• left parathyroid glands
• right parathyroid glands
• thyroid gland
• trachea

6 _____

7 _____

8 _____

9 _____

60×

(a) Survey photomicrograph of
thyroid and parathyroid glands

3 4 5 225×

(b) Photomicrograph of
thyroid gland

FIGURE 25.4 Sectional views of the thyroid and parathyroid glands.

(a)
• parathyroid gland
• thyroid gland

1 _____

2 _____

(b)
• colloid-filled follicle
 (COL-oid)
• follicular cell
 (fohl-LIK-u-lar)
• parafollicular cell or C cell
 (para-fohl-LIK-u-lar)

3 _____

4 _____

5 _____

D. Anatomy, Histology, and Hormones of the Adrenal Glands

The **adrenal glands** (*ad-* = addition to; *renal* = kidneys) are also known as the suprarenal glands because of their location on the superior part of the kidneys. Each adrenal gland is surrounded by a **capsule** and is composed of an outer cortex and a central medulla. The cortex and medulla are formed of different tissue types, and therefore produce and secrete different types of hormones.

The **adrenal cortex** is glandular and can be divided histologically into three layers, each layer secreting a different hormone. The **zona glomerulosa** (*zona* = belt; *glomerulosa* = little ball) is the outermost layer whose cells are arranged in little columns. The middle layer, the **zona fasciculata** (*fasciculata* = little bundle), has long cords and is the largest layer. The **zona reticularis** (*reticul-* = network) is the innermost layer and has branching cords.

Hormones secreted by the adrenal cortex are mineralcorticoids, glucocorticoids, and androgens (gonadocorticoids). The zona glomerulosa secretes **mineralcorticoids,** with **aldosterone** being the main hormone secreted.

The zona fasciculata secretes **glucocorticoids,** with **cortisol** being the main hormone secreted. In males and females, the zona reticularis mainly secretes a small amount of the male androgen **DHEA** (dehydroepiandrosterone) which has weak hormonal effects. Androgens (*andros* = man) are hormones that increase male characteristics. Some of the peripheral target tissues of DHEA convert this hormone to a more powerful hormone, testosterone, while other peripheral target tissues convert DHEA into estrogen.

The **adrenal medulla** is comprised of nervous tissue that is stimulated by the sympathetic nervous system to secrete two hormones: **epinephrine** and **norepinephrine (NE).** Like the posterior pituitary gland, the chemicals secreted by this nervous tissue are called hormones because they are secreted into the blood and travel to another part of the body to target cells.

ACTIVITY 4 Adrenal Glands

1 Label the terms in Figure 25.5(a), (b), and (c).

2 Observe a slide of the adrenal cortex and medulla and identify the structures listed with Figure 25.5(c).

3 Pronounce the terms as you point to them.

(a) Anterior view
- kidney
- left adrenal gland
- right adrenal gland

1 _____

2 _____

3 _____

(a) Anterior view

(b) Section through left adrenal gland

(b) Section through left adrenal gland
- adrenal cortex
- adrenal medulla
- capsule

(c) Photomicrograph of the adrenal gland
- adrenal cortex
- adrenal medulla
- capsule
- zona fasciculata (ZOH-na fah-sick-you-LAH-ta)
- zona glomerulosa (glow-mare-you-LOH-sa)
- zona reticularis (reh-tik-you-LAIR-is)

4 _____

5 _____

6 _____

7 _____

8 _____

9 _____

10 _____

11 _____

12 _____

65×

(c) Photomicrograph of the adrenal gland

FIGURE 25.5 The adrenal glands.

E. Anatomy, Histology, and Hormones of the Pancreas

The **pancreas** is composed of a head, body, and tail. Its head is cradled in the curvature of the duodenum, and the body and tail are inferior and posterior to the stomach near the spleen. The pancreas has both endocrine and exocrine (*exo-* = outside; literally "to secrete outside") functions. The exocrine or acinar (*acinus* = grapelike) cells vastly outnumber the endocrine cells. Exocrine enzymes are secreted into the duodenum of the small intestine through a duct called the **pancreatic duct** and function in digestion. Endocrine cells, called **pancreatic islets (islets of Langerhans),** are located as little islands among the exocrine cells. The **alpha** or **A cells** in the pancreatic islets secrete **glucagon,** and the **beta** or **B cells** secrete **insulin.**

- body of pancreas
- head of pancreas
- tail of pancreas

1 _____

2 _____

3 _____

- exocrine or acinar cells (AS-ih-nur)
- pancreatic islet (EYE-let)

4 _____

5 _____

ACTIVITY 5 Pancreas

1 Label the structures on the drawing in Figure 25.6(a).

2 Using the labeled line drawing in Figure 25.6(b), label the structures in Figure 25.6(c).

3 Observe a slide of the pancreas and identify the structures listed in Figure 25.6(c).

4 Pronounce the terms as you point to them.

Spleen (elevated)

1

2

Duodenum

3

(a) Anterior view

Blood capillary

Exocrine acinus

Alpha cell (secretes glucagon)

Beta cell (secretes insulin)

4 5

450×

(b) Pancreatic islet and surrounding acini

(c) Photomicrograph of pancreatic islet and acinar cells

FIGURE 25.6 The pancreas.

F. Hormones of the Ovaries, Testes, Pineal Gland, and Thymus

Ovaries are female gonads that not only produce and house the ova as they mature, but are also endocrine organs that produce two major hormones: **estrogen** and **progesterone.** These two hormones and the microscopic anatomy of the ovary will be studied further in the reproductive exercise. The **testes** are male gonads that not only produce sperm, but also produce and secrete androgens, primarily **testosterone.** These hormones and the microscopic anatomy of the testes will also be studied in the reproductive exercise.

The **pineal gland** (*pinea-* = pinecone) or **pineal body** is a small, cone-shaped gland that is located in the brain posterior to the thalamus and superior to the cerebellum. This gland secretes **melatonin** in darkness and, to a lesser degree in daytime, helps set the body's biological clock.

The **thymus** plays a role in immune function and is located anterior and superior to the heart. This gland is much larger in babies and children and regresses in size as a person ages. The main hormone produced by this gland is **thymosin.** The thymus will be studied further in the immune system exercise.

G. Hormone Functions

Hormones act by changing target cell activity. Changes in target cell activity include:

- Synthesis of molecules within the cell (example: estrogen)
- Changing plasma membrane permeability (example: aldosterone)
- Altering cellular metabolism (example: thyroxin)
- Secretion of cell products (example: thyroid stimulating hormone)
- Contraction of smooth muscle (example: oxytocin)
- Contraction of cardiac muscle (example: epinephrine)

ACTIVITY 6 Hormone Function

1 Using Table 25.1, review the hormone(s) produced by each gland and the target cells and function of each hormone.

2 On a human torso with brain or on an endocrine system chart, point to each endocrine gland and name the hormones it secretes.

3 For each hormone, point to its target cells on a torso model, and describe how the hormone changes target cell activity.

TABLE 25.1 Endocrine Glands, Hormones, Target Cells, and Hormone

GLAND	HORMONE	LOCATION OF TARGET CELLS	HORMONE FUNCTION
Anterior pituitary	1. Human growth hormone (hGH)	Cartilage, bone, skeletal muscle, liver, and other body tissues	Stimulates secretion of hormones that stimulate body growth and metabolism.
	2. Thyroid-stimulating hormone (TSH)	Thyroid gland	Stimulates growth of thyroid gland and secretion of its hormones.
	3. Follicle-stimulating hormone (FSH)	Testes	Stimulates sperm production.
		Ovaries	Stimulates oocyte production and estrogen secretion.
	4. Luteinizing hormone (LH)	Testes	Stimulates secretion of testosterone.
		Ovaries	Triggers ovulation and stimulates secretion of estrogen and progesterone.
	5. Prolactin (PRL)	Mammary gland	Stimulates production and secretion of milk.
	6. Adrenocorticotropic hormone (ACTH)	Adrenal cortex	Stimulates secretion of hormones by adrenal cortex.
	7. Melanocyte-stimulating hormone (MSH)	Skin	Darkens skin pigmentation.
Posterior pituitary	1. Antidiuretic hormone (ADH)	Kidneys	Decreases water lost in urine by returning water to the blood.
	2. Oxytocin (OT)	Uterus and mammary glands	Stimulates uterine contractions and milk ejection during suckling.
Thyroid gland	1. Thyroxine (T_4)	Most body cells	Increases metabolism and basal metabolic rate (BMR).
	2. Triiodothyronine (T_3)	Most body cells	Increases metabolism and BMR.
	3. Calcitonin	Osteoclast cells in bones	Decreases blood calcium levels by inhibiting osteoclasts.
Parathyroid glands	Parathyroid hormone (PTH)	Osteoclast cells in bones	Increases blood calcium levels by stimulating osteoclasts to break down bone matrix.
Adrenal cortex	1. Aldosterone	Kidneys	Decreases sodium and water loss in urine by returning sodium and water to the blood.
	2. Cortisol	Liver, muscle, and cells involved in body defenses	Increases resistance to stress, increases blood glucose levels, and decreases inflammation.
	3. Androgens	Uterus, mammary glands, and other body cells involved in secondary sex characteristics	Insignificant in males; increases sex drive in females.
Adrenal medulla	1. Epinephrine	Body cells involved in fight-or-flight response	Promotes fight-or-flight response.
	2. Norepinephrine (NE)	Body cells involved in fight-or-flight response	Promotes fight-or-flight response.
Pancreas	1. Insulin	Most body cells	Decreases blood glucose levels by transporting glucose into body cells.
	2. Glucagon	Liver	Increases blood glucose levels by stimulating liver to break down glycogen into glucose.

TABLE 25.1 *(Continued)*

GLAND	HORMONE	LOCATION OF TARGET CELLS	HORMONE FUNCTION
Ovaries	1. Estrogen	Uterus, mammary glands, and other body cells involved in female sexual characteristics	Stimulates development of female sex characteristics; helps regulate menstrual cycle.
	2. Progesterone	Uterus, mammary glands and other body cells involved in female sexual characteristics	Stimulates development of female sex characteristics; helps regulate menstrual cycle.
Testes	Testosterone	Testes, muscle, and other body cells involved in male sexual characteristics	Stimulates development of male sex characteristics; stimulates male sex drive; regulates sperm production.
Pineal Gland	Melatonin	Brain	Helps to set biological clock.
Thymus	Thymosin	T cells (type of white blood cell involved in immune response)	Promotes the maturation of T cells for the immune response.

Reviewing Your Knowledge

A. Anatomy Drill and Practice

Go to www.wiley.com/college/allen. Click on the picture of your lab manual, then on Student Companion Site, Exercise 25, and Anatomy Drill and Practice.

B. Hormone Abbreviations

Write the name of the hormone next to the abbreviation.

1. ACTH Adrenocorticotropic hormone
2. ADH AntiDiuretic Hormone
3. DHEA dehydroepiandrosterone
4. FSH Follicle-Stimulating Hormone
5. hGH human growth hormone
6. LH Lutenizing
7. NE Norepinephrine
8. OT Oxytocin
9. PRL Prolactin
10. PTH Parathyroid Hormone
11. T_3 Triiodothyronine
12. T_4 Thyroxine
13. TSH Thyroid-Stimulating hormone
14. MSH Melanocyte-Stimulating hormone
15. TH Thyroid hormones

C. Main Endocrine Organs and Their Hormones

Write the name of the endocrine gland that secretes the following hormones.

Hormone	**Endocrine Gland**
1. ACTH	Anterior Pituitary
2. ADH	Posterior pituitary
3. aldosterone	Adrenal cortex
4. androgens (DHEA)	Adrenal Cortex
5. calcitonin	Thyroid gland
6. cortisol; cortisone	Adrenal cortex
7. epinephrine/NE	Adrenal medulla
8. estrogen; progesterone	Ovaries
9. FSH	Anterior Pituitary
10. glucagon	Pancreas
11. hGH	Anterior pituitary
12. insulin	Pancreas
13. LH	Anterior pituitary
14. melatonin	Pineal gland
15. OT	Posterior pituitary
16. PRL	Anterior pituitary
17. PTH	Parathyroid glands
18. T_3 and T_4	Thyroid gland
19. testosterone	Testes
20. TSH	Anterior pituitary
21. thymosin	Thymus
22. MSH	Anterior pituitary

D. Hormone Function

Write the name of the hormone that matches its function.

ACTH	glucagon	progesterone
ADH	hGH	PTH
aldosterone	insulin	T$_3$
androgens	LH	T$_4$
calcitonin	melatonin	testosterone
cortisol	MSH	thymosin
epinephrine	NE	TSH
estrogen	OT	
FSH	PRL	

Hormone Function

_____ **1.** Stimulates uterine contractions and milk ejection during suckling.

_____ **2.** Stimulates secretion of hormones by the adrenal cortex.

_____ **3.** Decreases water loss by increasing reabsorption of water into blood and decreasing urine production.

_____ **4.** Increases sex drive in females.

_____ **5.**⎫
 ⎬ Increases metabolism and BMR.
_____ **6.**⎭

_____ **7.** Triggers ovulation and stimulates secretion of estrogen and progesterone.

_____ **8.** Increases blood calcium levels by stimulating osteoclast activity.

_____ **9.**⎫
 ⎬ Promotes fight-or-flight response.
_____ **10.**⎭

_____ **11.** Stimulates production and secretion of milk.

_____ **12.** Increases resistance to stress, increases blood glucose levels, and decreases inflammation.

_____ **13.** Helps to set the biological clock.

_____ **14.** Darkens skin pigmentation.

_____ **15.** Stimulates secretion of hormones that stimulate body growth and metabolism.

_____ **16.** Decreases blood calcium levels by inhibiting osteoclasts.

_____ **17.** Promotes the maturation of T cells for the immune response.

_____ **18.** Stimulates oocyte production and estrogen secretion.

_____ **19.** Decreases blood glucose levels by transporting glucose into body cells.

_____ **20.** Increases blood glucose levels by stimulating the liver to break down glycogen into glucose.

_____ **21.**⎫
_____ **22.**⎭ Stimulates development of female sex characteristics and helps regulate menstrual cycle.

_____ **23.** Regulates sperm development, stimulates development of male sex characteristics, and stimulates male sex drive.

_____ **24.** Stimulates secretion of testosterone.

_____ **25.** Stimulates secretion of thyroid hormones.

_____ **26.** Stimulates sperm production.

_____ **27.** Increases reabsorption of sodium and water into blood and decreases urine output.

Using Your Knowledge

A. Endocrine Organs of the Thorax

Label the structures in Figure 25.7.

1. _____ Thyroid gland _____
2. _____ Parathyroid gland _____
3. _____ Thymus _____

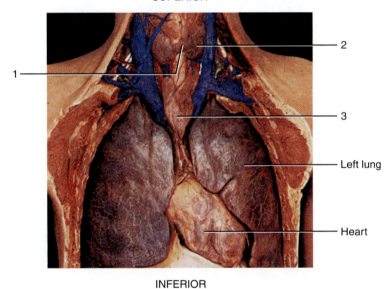

SUPERIOR

1 ——

2

3

Left lung

Heart

INFERIOR

FIGURE 25.7 Neck and thorax region, anterior view.

B. Hormones

Answer the following questions.

4. Name two hormones that are also neurotransmitters.

 epinephrine & Norepinephrine

5. Explain why hormones are only able to affect certain cells and not other cells.

 Because they have target cells that must pair up or match the receptor cells.

437

C. Hormone Imbalance

Using your textbook or another reference book, identify the gland and hormone(s) affected by each of the following operations or conditions.

6. Oophorectomy

7. Orchiectomy

8. Graves' disease Thyroid Gland

9. Acromegaly Gland
Anterior Pituitary , HGH Hormone

10. Diabetes insipidus Gland
Pancreas , Hormone
Insulin

Blood Components and Blood Tests

Objectives

After completing this exercise, you should be able to:

- Describe the functions of blood
- Name the components of blood
- Describe the structure, characteristics, and function of red blood cells (RBCs), each type of white blood cell (WBC), and platelets
- Identify RBCs, each type of WBC, and platelets on prepared blood smear slides
- Describe the importance of and perform the following blood tests: differential WBC count, ABO and Rh blood typing, hematocrit, hemoglobin, and coagulation time.

Materials

- prepared slides of normal human blood (with Wright's stain)
- prepared slides of abnormal blood (allergies or infections)
- compound microscopes, lens paper, hand counter
- Blood Typing: blood typing kit or antisera (anti-A; anti-B; anti-D), new test cards or glass slides (two per student), toothpicks, wax marking pencil, Rh warming tray
- materials needed if using human or animal blood: biohazardous waste container; sharps container; safety glasses; disposable gloves; 10% bleach solution in spray bottle; sterile, disposable lancets or Autolets; alcohol swabs; cotton balls
- Hematocrit: human or animal blood, heparinized capillary tubes, clay or Seal-ease, microcentrifuge, mm rulers (or hematocrit reader)
- Hemoglobin: human or animal blood, Tallquist paper or hemoglobinometers
- Coagulation Time: human or animal blood, nonheparinized capillary tubes, small metal file

A. Components of Blood

The adult cardiovascular system contains approximately 5.5 liters of blood. When centrifuged, **blood** separates visually into two main components, plasma and formed elements. The clear, straw-colored liquid is called **plasma** and the dark-red and buff-colored portions are the **formed elements** (see Figure 26.1). The formed elements include **red blood cells** (RBCs), **white blood cells** (WBCs), and **platelets** (cell fragments). RBCs are also called **erythrocytes** (*erythro-* = red; *-cytes* = cells), WBCs are known as **leukocytes** (*leuko-* = white), and platelets are known as **thrombocytes** (*thrombo-* = clot). The formed elements constitute approximately 45% of whole blood volume, and plasma composes about 55%. Plasma is about 91.5% water and 8.5% solutes. The solutes are mostly plasma proteins, but also include nutrients (glucose, amino acids, and lipids), blood gases (oxygen and carbon dioxide), electrolytes, hormones, enzymes, and waste materials.

A **complete blood count (CBC),** a vital diagnostic tool, screens for abnormalities in the number or structure of formed elements. The CBC includes: a total RBC count, total WBC count, platelet count, differential WBC count, hematocrit, and hemoglobin concentration. The CBC is used along with a battery of blood chemistry tests to put together a comprehensive profile of a person's general level of health based on blood values. Consult Exhibit 26.1 to observe the complexity of a CBC.

ACTIVITY 1 Components of Blood

1 Label the components of blood in Figure 26.1.

1 (55%)

3

2 (45%)

Appearance of centrifuged blood

FUNCTIONS OF BLOOD

1. Transport of oxygen, carbon dioxide, nutrients, hormones, heat, and wastes.

2. Regulation of pH, body temperature, and water content of cells.

3. Protection against blood loss through clotting, and against disease through phagocytic white blood cells and antibodies.

FIGURE 26.1 Components of blood in a normal adult.

- plasma
- red blood cells (packed)
- white blood cells and platelets (buffy coat)

1 _____

2 _____

3 _____

EXHIBIT 26.1 **Complete Blood Count Form**

PATIENT:					DOCTOR:	
ACCOUNT NO:					REPORT FOR LOCATION:	
MEDICAL RECORD NO.:						

HEMATOLOGY

DAY:	−1					
DATE:	AUG 20 04					
TIME:	1000				REFERENCE	UNITS
WBC	8.0				4.5–11.0	×10³/UL
RBC	4.12	L			4.70–6.10	×10⁶/UL
HGB	14.4				14.0–18.0	G/DL
HCT	41.6	L			42.0–52.0	%
MCV	101	H			83–99	FL
MCH	34.8	H			28–34	PG
MCHC	34.5				31.0–37.0	%
RDW	13.1				11.0–15.5	%
PLT	265				150–450	×10³/UL
MPV	8.2				7.4–10.4	FL
NEUT%	47.7	L			49.0–90.0	%
LYMPH%	41.0				22.0–51.1	%
MONO%	6.5				0.0–13.0	%
EO%	4.3				0.0–7.0	%
BASO%	0.5				0.0–3.0	%
NEUT #	3.90				1.50–7.06	×10³/UL
LYMPH #	3.30	H			0.70–3.00	×10³/UL
MONO #	0.5				0.0–1.4	×10³/UL
EO #	0.3				0.0–0.8	×10³/UL
BASO #	0.0				0.0–0.3	×10³/UL

COAGULATION

DAY:	−1			
DATE:	AUG 20 04			
TIME:	1000		REFERENCE	UNITS
PROTIME:	11.4		10.9–13.3	sec
PTT:	28.3		25.9–35.3	sec

WBC = white blood cell; RBC = red blood cell; HGB = hemoglobin; HCT = hematocrit; MCV = mean corpuscular volume; MCH = mean corpuscular hemoglobin; MCHC = mean corpuscular hemoglobin concentration; RDW = red blood cell distribution index; PLT = platelet; MPV = mean platelet volume; NEUT% = neutrophil percent; LYMPH% = lymphocyte percent; MONO% = monocyte percent; EO% = eosinophil percent; BASO% = basophil percent; NEUT# = neutrophil number; LYMPH# = lymphocyte number; MONO# = monocyte number; EO# = eosinophil number; BASO# = basophil number; L = low; H = high; PTT = partial thromboplastin time.

B. The Structure and Function of Red Blood Cells, White Blood Cells, and Platelets

Red Blood Cells (Erythrocytes or RBCs)

Erythrocytes are small, anucleate (without a nucleus) cells that contain **hemoglobin,** a large molecule used to transport oxygen and carbon dioxide in the blood. About 33% of the RBCs' total weight is composed of hemoglobin. Hemoglobin contains a red pigment called heme that gives blood its red color. Blood is bright red when oxygen-rich and darker red when oxygen-poor. There are approximately 4.5 to 5 million RBCs per microliter of blood, with lower amounts being in females and higher amounts in males. An abnormally high number of RBCs is called **polycythemia** (*poly-* = many; *cyto-* = cells; *-emia* = blood), and an unusually low number of RBCs is one type of **anemia.**

White Blood Cells (Leukocytes or WBCs)

WBCs, nucleated cells that are typically larger than RBCs, attack pathogens and other foreign substances in the body. There are five kinds of leukocytes that are divided into two categories, granular and agranular. **Granular leukocytes** have discernible vesicles (granules) in the cytoplasm that can be seen after staining, and **agranular leukocytes** also have granules, but they cannot be observed with the light microscope. These WBCs were named agranular because the microscopes that were used at that time were not powerful enough to distinguish the granules. Granular leukocytes include **neutrophils, eosinophils,** and **basophils.** Agranular leukocytes include **lymphocytes** and **monocytes.** There are normally 5,000 to 10,000 WBCs per microliter of blood. An abnormally high number of WBCs is called **leukocytosis** (*-osis* = an increase in a pathological condition), and a decrease in the number of WBCs is called **leukopenia** (*-penia* = deficiency).

Platelets (Thrombocytes; *thrombo-* = clot)

Platelets are formed in red bone marrow from large, multinuclear cells called **megakaryocytes** (*mega-* = large; *karyo-* = nucleus). Large megakaryocytes break into tiny cytoplasmic fragments called platelets that do not have nuclei and are not considered to be cells. These special cell fragments protect the body by forming a platelet plug to stop bleeding when blood vessels rupture and by secreting chemicals that aid in blood clotting. **Thrombocytopenia** is a deficiency in the number of circulating platelets.

C. Microscopic Examination of the Formed Elements

Blood cells are difficult to see at low magnifications because the cells are so small. The erythrocytes are the most numerous and are small, pinkish cells. The WBCs are not very numerous but are conspicuous because their nuclei stain dark blue to dark purple with Wright's stain. The majority of WBCs are larger than the RBCs. RBCs are used as a standard to compare the size of WBCs. The tiny platelets stain dark purple, are much smaller than RBCs, and may look to you like extra stain on the slide. Keep in mind that blood cells and their nuclei have a three-dimensional spherical shape. You will not always find the nucleus appearing as described or as seen in photomicrographs because of the various views that are possible. Figure 26.2 depicts the three-dimensional structure of RBCs, WBCs, and platelets.

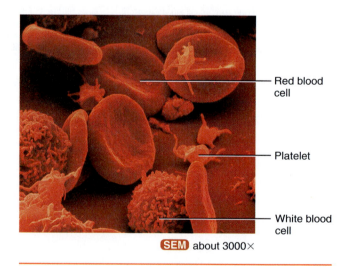

Red blood cell

Platelet

White blood cell

SEM about 3000×

FIGURE 26.2 Scanning electron micrograph of the formed elements of blood.

ACTIVITY 2 Identification of RBCs, WBCs, and Platelets

1 Examine a prepared blood slide using the low-power lens to locate RBCs and WBCs.
 • Using the high-power or oil-immersion lens, locate WBCs and platelets.

2 To identify the WBC type, refer to Table 26.1 and Figure 26.3. Look at the size of the cell compared with RBCs, the presence and color of granules, the shape and number of lobes to the nucleus, and the color of the cytoplasm.

TABLE 26.1 Characteristics of RBCs, WBCs, and Platelets (Wright's Stain)

	RELATIVE NUMBER	SIZE	NUCLEUS	GRANULES	COLOR OF CYTOPLASM
RBCs	700:1 ratio of RBCs to WBCs	7–8 micrometer (μm) diameter	No nucleus	No granules	Light pinkish-red color
Neutrophils (*neutr-* = neutral; *-phil* = loving)	60–70% of all WBCs	About 10–12 μm diameter	Multilobed nucleus that gives them the nickname **polymorphonuclear** leukocytes, or **PMNs** (*poly-* = many; *morpho-* = shapes) with 2–5 or more lobes connected by threads. Also clinically called **"segs"** because of the segmented look of the nucleus	Small, nondistinct pale lilac to neutral-staining granules	Cytoplasm is usually pink but, depending on the stain used, sometimes has a reddish tinge that makes students mistake these cells for eosinophils
Eosinophils (*eosin-* = red, acidic dye)	2–4% of all WBCs	About 10–12 μm in diameter	Bi-lobed nucleus (occasionally three lobes)	Many medium, red-orange granules (some stains make them look very dark brownish-red-orange)	Cytoplasm is pink
Basophil (*baso-* = dark blue, basic dye)	0.5–1% of all WBCs	8–10 μm diameter	Nucleus large, varied in shape; generally obscured by large granules	Large, dark blue-purple granules	Cytoplasm appears purple
Lymphocyte (lymph cell)	20–25% of all WBCs	6–9 μm (small lymphocyte); 10–14 micrometers (large lymphocyte)	Large, round, or slightly indented nucleus; stains very dark purple	Granules not obvious with light microscope	Light sky-blue cytoplasm; small cells have only a rim of cytoplasm; more cytoplasm in larger lymphocytes
Monocyte (*mono-* = one; pertains to having only one nucleus)	3–8% of all WBCs	Very large cell; 12–20 μm	Large kidney bean or horseshoe-shaped lacy nucleus; sometimes oval and indented	Granules not obvious with light microscope	Light blue-gray cytoplasm; sometimes irregular extensions of the cytoplasm
Platelets	20:1 ratio of RBCs to platelets	2–4 μm diameter; small cell fragments	No nucleus	Dark purple granules	Difficult to see because of dark purple granules

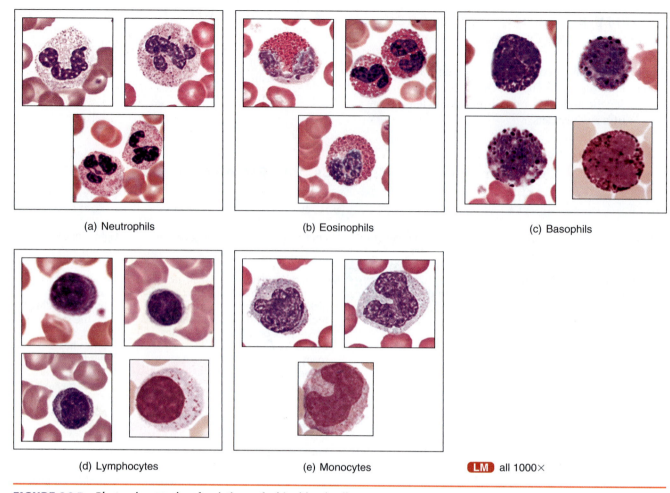

(a) Neutrophils

(b) Eosinophils

(c) Basophils

(d) Lymphocytes

(e) Monocytes

LM all 1000×

FIGURE 26.3 Photomicrographs of variations of white blood cells.

D. Blood Tests

Differential White Blood Cell Count

A **differential WBC count** is performed to determine the percentage of each of the five types of WBCs in a blood sample. When attempting to learn which WBCs are normally present from the greatest to least percentages, it is helpful to remember the following mnemonic: "Never Let Monkeys Eat Bananas" (N = neutrophils; L = lymphocytes; M = monocytes; E = eosinophils; and B = basophils). Because the normal percentage of each WBC type is known, any significant abnormality in these percentages, elevated or depressed, can be indicative of particular disorders. Numerical values for the normal range of WBCs may vary depending on the reference source.

ACTIVITY 3 Differential WBC Count

1 Complete a differential WBC count using a prepared slide of normal blood.

- Complete this activity individually or with a lab partner as your instructor directs. If done together, one person will identify each WBC with the microscope and call out the type to the lab partner. The other person will mark the type of cells encountered by using tally marks (℟ ll) and also keep track of the total number of cells.

- Using a prepared, normal blood slide, focus using the low-power objective lens and then switch to a higher power objective lens as directed by your instructor.

- To avoid recounting any WBCs, begin at one end of the slide that has a good separation of cells and slowly scan systematically, moving the slide down

and over, then up and over, and repeat as seen in Figure 26.4.

- If you encounter a cell that you cannot identify or are not sure of, do not count it and continue on.
- Continue this pattern until you have found and recorded 100 WBCs in Table 26.2.
- Because you are counting 100 cells, the total number of each type of WBC counted is its percentage in the blood sample (i.e., 26 lymphocytes means 26% of the WBCs are lymphocytes).

- Write your results on the board to pool class data.
2 Complete a differential WBC count using a prepared slide of abnormal blood, if available.
- Follow the instructions for a differential WBC using normal blood (listed above).
- Record your tally counts and totals in Table 26.2.
- Determine the disorder by consulting Table 26.3 and record in Table 26.2.
- Write your results on the board to pool class data for comparison.

FIGURE 26.4 Procedure for scanning WBC differential count.

TABLE 26.2 **Results of Differential WBC Count**

A. Normal Differential WBC Count

TYPE OF WBC	OBSERVATION TALLY	TOTAL CELLS YOU COUNTED/ PERCENTAGE	CLASS AVERAGE OF COUNTED CELLS/PERCENTAGE	NORMAL PERCENTAGE FOR WBCs
Neutrophil				60–70%
Lymphocyte				20–25%
Monocyte				3–8%
Eosinophil				2–4%
Basophil				0.5–1%
	Total WBCs			

B. Pathological Differential WBC Count

TYPE OF WBC	OBSERVATION TALLY	TOTAL CELLS YOU COUNTED/ PERCENTAGE	CLASS AVERAGE OF COUNTED CELLS/PERCENTAGE	NORMAL PERCENTAGE FOR WBCs
Neutrophil				60–70%
Lymphocyte				20–25%
Monocyte				3–8%
Eosinophil				2–4%
Basophil				0.5–1%
	Total WBCs			

Type of Disorder

TABLE 26.3 **Significance of Elevated and Depressed WBC Counts**

WBC TYPE	HIGH COUNT MAY INDICATE	LOW COUNT MAY INDICATE
Neutrophils	Bacterial infection, burns, stress, inflammation	Radiation exposure, drug toxicity, vitamin B_{12} deficiency, systemic lupus erythematosus
Lymphocytes	Viral infections, some leukemias	Prolonged illness, immunosuppression, treatment with cortisol
Monocytes	Viral or fungal infections, tuberculosis, some leukemias, other chronic diseases	Bone marrow depression, treatment with cortisol
Eosinophils	Allergic reactions, parasitic infections, autoimmune diseases	Drug toxicity, stress
Basophils	Allergic reactions, leukemias, cancers, hypothyroidism	Pregnancy, ovulation, stress, hyperthyroidism

ABO and Rh Blood Typing

Blood typing is critical for blood transfusions, transplantations, and maternal-fetal compatibility, but is also used in genetic studies, forensic studies, legal medicine, and anthropology. Although there are many different systems for classifying human blood, we will be studying the **ABO** and **Rh systems,** because they are most commonly used. Blood typing is based on the **antigenic** (agglutinogens) molecules that are on the surface of the RBC membranes. An **antigen** is a substance that is able to produce an immune response and will react with a specific antibody. **Antibodies** are plasma proteins that combine with a specific antigen to inhibit or destroy it. In the **ABO system,** there are two types of antigens (A and B) that can be present as surface membrane molecules on RBCs. If the plasma membrane of your RBCs have only the A antigen present, you have **type A** blood; correspondingly, if you have only B antigens present, you have **type B** blood. If you have both antigens A and B present you have **type AB** blood, and if you do not have either A or B antigen present, you have **type O** blood (see Figure 26.5).

ABO antibodies appear in babies' blood a few months after birth. If you have type A blood, you do not have its corresponding anti-A antibody. If the two are mixed, they will form a detrimental antigen-antibody complex and cause clumping. People with type A blood have anti-B antibodies that will become cross-linked and **agglutinate** (clump) if type B blood is given to them (see Figure 26.6). Agglutination is followed by the activation of another plasma protein that attaches to the recipient's RBCs and hemolyzes or bursts them, releasing hemoglobin that can cause kidney damage. ABO antibodies do not cross the placenta because of their large size.

The **Rh blood system** is different from the ABO system, but has some similarities. If you have the Rh antigen as a surface membrane molecule on your RBCs, you are **Rh+.** If you do not have the Rh antigen, you are **Rh−.** An Rh− person is not born with the anti-Rh antibody (as in the ABO system) and does not obtain this antibody until the person is exposed to the Rh antigen from Rh+ blood. This can happen through a blood transfusion, by sharing hypodermic needles, or by an Rh− mother carrying an Rh+ child. During delivery, the baby's blood can leak from the placenta into the mother's bloodstream, causing the mother's body to make Rh antibodies. The first baby would not be affected, but subsequent pregnancies with Rh+ fetuses can result in the small Rh antibodies crossing the placenta causing hemolysis in the fetuses' blood. This condition is called **hemolytic disease of the newborn.** Rh− mothers are typically given RhoGAM so they will not make Rh antibodies.

The following activity will use **antisera** (plural of antiserum), which can be artificial serum or serum from an animal or human containing antibodies against A, B, or D (Rh) antigens. **Serum** is blood plasma without clotting proteins. If anti-A serum clumps a particular blood sample, the blood is type A because of an antibody-antigen complex formed with the A antigens on the RBCs.

ACTIVITY 4 **ABO and Rh Blood Typing**

1 With your lab partner(s), complete Table 26.4 by writing in the antibody present in plasma for each blood type. Also, determine each compatible and incompatible donor for each blood type.

2 Follow the safety precautions below to familiarize yourself with blood handling safety procedures.
- When using human or animal blood samples, you must protect yourself from any blood-borne infectious disease (hepatitis, HIV, etc.).
- Wear gloves and safety goggles while performing blood tests.

- Place all blood slides and toothpicks in a biohazardous container according to your instructor's directions.
- Wash tabletops thoroughly with 10% bleach solution at the end of the lab.
- Wash your hands with soap and water before leaving the lab.
- If a blood spill occurs, cover the spill with paper towels soaked in 10% bleach solution and immediately inform your instructor.

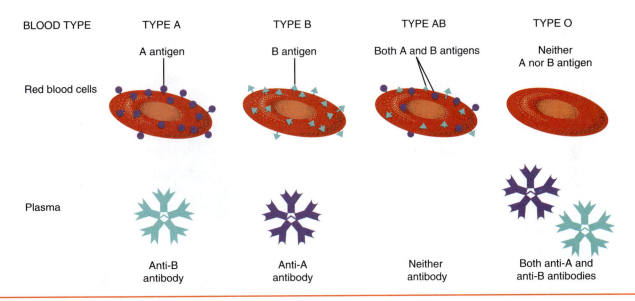

FIGURE 26.5 Antigens and antibodies of the ABO blood types.

TABLE 26.4 **Summary of ABO Blood Group Interactions**

BLOOD TYPE	A	B	AB	O
Antigen on RBCs				
Antibody in plasma				
Compatible donor blood types (no hemolysis)				
Incompatible donor blood types (hemolysis)				

3 Perform ABO blood typing according to the procedure below using Figure 26.6 as a guide.
- Obtain a clean glass slide (test cards if using a typing kit), two new toothpicks, a wax marking pencil, and anti-A and anti-B sera.
- Your instructor will tell you whether you will be using sterile blood, your own blood, or simulated blood.
- Divide the glass slide in half with a wax pencil. Mark A on the left side and B on the right side. Place one drop of anti-A serum on the left side and one drop of anti-B serum on the right side.
- Place one drop of blood on sides A and B of slide. If using your own blood, follow the procedure in #5.
- Using *separate* new toothpicks, mix each sample of blood with the corresponding antiserum.
- Results may take up to 2 minutes. Observe each sample for agglutination or the appearance of granulation (see Figure 26.6); interpret the results.
- Record your results in Table 26.5.
- Write your results on the board to pool class data for comparison.
- Record percentage of ABO blood types for your class in Table 26.5.

4 Perform Rh blood typing according to the procedure below.
- Obtain a clean glass slide and antiserum D (Rh).
- Place one drop of anti-D on the glass slide and add a drop of blood. If using your own blood, follow the procedure in #5.
- Mix the two liquids with a new toothpick.
- Place slide on the warm Rh typing box and rock gently to mix. Rh typing requires a higher temperature than ABO typing does.
- Results may take up to 2 minutes. Observe your sample for clumping or the appearance of granulation (see Figure 26.6); interpret the results.
- Sometimes, it is more difficult to obtain positive results with the anti-Rh serum, depending on the supply and delivery situations.
- Dispose of blood slides and materials according to your instructor's directions.
- Record your results in Table 26.5 for each sample you tested.
- Write your results on the board to pool class data for comparison.
- Record percentage of Rh blood types for your class in Table 26.5.

5 Procedure for obtaining blood sample.
- Thoroughly wash hands with soap and dry with a clean paper towel.
- Thoroughly clean the tip of index finger with an alcohol swab.
- Open a new, sterile blood lancet exposing the sharp tip only (or use an Autolet). Lance just to the side of the finger pad with the new lancet. *Never* reuse a lancet, even your own.
- Deposit the used lancet in the sharps container for biohazardous materials only.
- Wipe away the first drop of blood with cotton ball and dispose in biohazard waste container. Gently squeeze one drop of blood on each side of a clean prepared slide.

FIGURE 26.6 Serum agglutination results.

TABLE 26.5 Blood Typing Results

SERUM AGGLUTINATION RESULTS	YES	NO
Clumping with anti-A		
Clumping with anti-B		
Clumping with anti-D (Rh)		

ABO blood type _____
Rh type _____

BLOOD TYPES	CLASS BLOOD TYPES (PERCENTAGE)
A	
B	
AB	
O	
RH+	
RH−	

Hematocrit (Packed Red Cell Volume)

A **hematocrit** determines the volume of RBCs described as the percentage of RBCs in a whole blood sample. A capillary tube of blood is centrifuged to pack the red cells at the outer end of the tube, separating them from the WBCs, platelets, and plasma. After measuring the length of the RBC column and the total length of the blood column, the percentage of RBCs can be calculated. The normal hematocrit range for females is approximately 38 to 46% and for males it is 40 to 54%. An abnormally high hematocrit (generally 65% or above) is indicative of **polycythemia,** and a hematocrit with RBCs below the normal level indicates a type of **anemia.**

NOTE: *If materials are not available, this activity can be completed without blood by reading the instructions and using the photographs of centrifuged capillary tubes for measurements to calculate the hematocrits.*

SAFETY PRECAUTION:
If using blood, follow all of the safety tips given in Activity 4: Blood Typing.

ACTIVITY 5 Hematocrit

1 Calculate the hematocrit for blood in the capillary tubes in Figure 26.7.
 - Use a millimeter ruler to measure (in mm) the length of the whole column (the length of packed RBCs, WBCs, and plasma) and record it in Table 26.6.
 - Measure the length (in mm) of the RBCs and record in Table 26.6.
 - Use the following formula to calculate the percentage of RBCs (the hematocrit) in whole blood.

$$\frac{\text{Length of RBCs in mm}}{\text{Length of the whole column in mm}} \times 100.$$

 - Record the hematocrit in Table 26.6.

2 Calculate the hematocrit on a blood sample if equipment is available.
 - After reviewing safety tips on using blood, obtain two heparinized capillary tubes, sealing clay, a lancet or Autolet, alcohol swabs, and two cotton balls.
 - After a finger is cleaned with an alcohol wipe, lance it with a new, sterile lancet, and discard the lancet in the proper sharps container for biohazardous material.
 - Touch the red-lined end of a heparinized (prevents blood from coagulating) capillary tube to the blood and hold the tube at a downward angle. Taking care that no air is allowed to enter while the tube fills by capillary action, fill the capillary tube about three-fourths full.

 - Hold a finger over the end of the tube so blood cannot drain out. Place the blood end of the tube into sealing clay or Seal-ease to plug the end.
 - Place the sealed tube in a microhematocrit centrifuge with the sealed end on the rubber gasket pointing outward and away from the center.
 - Prepare a second tube in the same manner.
 - Place the second sealed tube in the centrifuge opposite the first tube for balance. *Always keep the centrifuge balanced.*
 - If others are using the centrifuge, make a note of the groove numbers of your tubes.
 - Secure the inside and outside covers of the centrifuge and set the timer for 4 to 5 minutes.
 - After the centrifuge has stopped, remove your tubes. Note that the RBCs are packed at the bottom near the clay, the WBCs (and platelets) are buff-colored next to them, and the clear plasma is on the top.
 - Use a microhematocrit reader or a millimeter ruler to measure the length of the whole column, the length of the packed RBCs, WBCs, and plasma in millimeters. (Figure 26.7).
 - Record the whole column length (in mm) and the RBC length (in mm) in Table 26.6.
 - Use the following formula to calculate the percentage of RBCs (the hematocrit) in whole blood.

$$\frac{\text{Length of RBCs in mm}}{\text{Length of whole column in mm}} \times 100.$$

 - Record the hematocrit in Table 26.6.

3 Answer discussion questions with your group.

FIGURE 26.7 Capillary tubes for calculating a hematocrit.

TABLE 26.6 Hematocrit Results

HEMATOCRIT CAPILLARY TUBES	RBC COLUMN (mm)	WHOLE COLUMN (mm)	HEMATOCRIT (PERCENTAGE OF RBCs/TOTAL BLOOD)
Hematocrit a			
Hematocrit b			
Hematocrit (your blood)			
MALE CLASS AVERAGE	**FEMALE CLASS AVERAGE**	**NATIONAL AVERAGE FOR MALES** 40–54%	**NATIONAL AVERAGE FOR FEMALES** 38–46%

? DISCUSSION QUESTIONS: HEMATOCRIT

1 For the capillary tubes in Figure 26.7, is the hematocrit normal, low, or high? If abnormal, do values indicate anemia or polycythemia?

2 Is the class average hematocrit for males higher than females?

3 How does your class compare with the national averages?

Determining Hemoglobin Content of Blood

Hemoglobin (Hb) is a protein that carries oxygen in the RBCs. Therefore, hemoglobin concentration in blood determines the oxygen-carrying capacity of the blood. It is possible for you to have a normal hematocrit or RBC volume and still be anemic due to low hemoglobin concentration. Normal values are 12 to 15 g hemoglobin/100 mL blood in females and 13 to 16 g/100 mL in males. In severe anemia, the hemoglobin content can be less than 7 g/100 mL. Clinically, the relationship of the hematocrit (%) is compared with the hemoglobin reading (g/100 mL) and is generally a ratio of 3:1.

Two of the several techniques for determining the hemoglobin content of blood are discussed here. The Tallquist measurement is the simpler, but older method that does not always produce accurate results. Hemoglobinometers are accurately calibrated instruments but are more expensive.

ACTIVITY 6 Determining Hemoglobin Content of Blood

1 Follow the Tallquist procedure for determining hemoglobin content, or if using a hemoglobinometer, follow the procedure given in its instruction booklet.

Procedure for Tallquist Method

- After reviewing safety tips on using blood, obtain a blood sample provided by the instructor or your own blood using methods described previously.
- Place one drop of blood on the absorbent Tallquist paper. Be sure the blood spot is large enough to be seen in all the chart holes simultaneously.
- Allow blood to dry until it is no longer shiny, but not so dry as to turn brownish. Match its color with the color scale provided.
- Record the results as grams of hemoglobin/100 mL of blood in Table 26.7.

2 Answer the discussion questions with your group.

TABLE 26.7 Hemoglobin Content

HEMOGLOBIN CONTENT
Tallquist method: _____ g/100 mL
Hemoglobinometer: _____ g/100 mL

? DISCUSSION QUESTIONS: HEMOGLOBIN CONTENT OF BLOOD

1 Compare your hematocrit results with your grams of hemoglobin/100 mL. Calculate the ratio. _____ Is this close to the normal ratio? (yes/no)

2 To measure the hemoglobin content, RBCs are hemolyzed. How would the results be different if you did not completely hemolyze the RBCs? Would the hemoglobin concentration be higher or lower?

3 Within your class, are there any students who have the same hematocrit, but different hemoglobin concentrations? _____ Explain why this can occur.

Coagulation Time

The process of blood clotting or **coagulation** prevents excessive blood loss. There are many blood clotting factors present in the plasma. Other factors are released by injured tissues and platelets to initiate the chemical chain reaction that forms a blood clot. Fibrin is a long, insoluble thread-like protein strand that forms a mesh to trap platelets. Fibrin is formed from soluble firinogen molecules during clotting.

The coagulation time is the time it takes for clotting to take place when blood is removed from the body. Normal clotting time is within 2 to 6 minutes.

ACTIVITY 7 Coagulation Time

1 Observe the demonstration of clotting blood.
- Your instructor will place animal or sterile human blood into two test tubes, one with and one without an anticoagulant.
- Observe the clot form and settle to the bottom of the tube without anticoagulant. Observe the straw-colored serum above the clot.
- Compare the tube with the blood clot to the blood in the test tube with an anticoagulant.

2 Determine coagulation time.
- After reviewing safety tips for using blood, obtain a blood sample from your instructor, alcohol swabs, a nonheparinized capillary tube, a lancet, cotton balls, and a small metal file.
- If using your own blood, use an alcohol swab on a clean finger and prick the finger with a clean, sterile lancet. Have a free flow of blood.
- Hold the nonheparinized capillary tube at an angle to collect the blood and put the upper end in the drop of blood.
- Fill the tube a minimum of two-thirds full and record the time started. Time: _____
- Lay the tube on a paper towel.
- After 30 seconds, scratch the glass with the metal file close to an end of the tube. Holding the tube close to each side of the scratch, gently break the tube away from you.
- Gently pull the tube apart and observe the blood to see if a fibrin thread is present.
- Repeat this process every 30 seconds until fibrin occurs. Record the amount of time required for coagulation in Table 26.8.
- Write results on the board to pool class data. Calculate average coagulation time for your class and record in Table 26.8.

2 Answer the discussion questions with your group.

? DISCUSSION QUESTIONS: COAGULATION TIME

1 Would a person with hemophilia have higher or lower than normal clotting times?

2 How would coagulation time be affected if a heparinized capillary tube was used?

TABLE 26.8 Coagulation Time

YOUR COAGULATION TIME	CLASS AVERAGE OF COAGULATION TIME

Reviewing Your Knowledge

A. Anatomy Drill and Practice

Go to www.wiley.com/college/allen. Click on the picture of your lab manual, then on Student Companion Site, Exercise 26, and Anatomy Drill and Practice.

B. Characteristics of the Formed Elements and Blood Abnormalities

Fill in the blank with the term that fits the description.

_____ **1.** The oxygen and carbon dioxide carrying cell

_____ **2.** Help the body fight infections and foreign substances

_____ **3.** Form a clot to help the body stop bleeding

_____ **4.** Another name for red blood cells

_____ **5.** Another name for platelets

_____ **6.** Another name for white blood cells

_____ **7.** Large cells that develop into platelets

_____ **8.** A deficiency in number of RBCs or decreased hemoglobin content of blood

_____ **9.** An abnormal increase in RBCs

_____ **10.** An abnormal increase in WBCs

_____ **11.** A deficiency in WBCs

_____ **12.** A deficiency in platelets

C. White Blood Cell Structure and Characteristics

Fill in the blank with the term that fits the description. Terms may be used more than once.

_____ **1.** 60–70% of all WBCs

_____ **2.** 2–4% of all WBCs

_____ **3.** 0.5–1% of all WBCs

_____ **4.** 20–25% of all WBCs

_____ **5.** 3–8% of all WBCs

_____ **6.** 10–12 μm; nucleus with 2–5 connected lobes; pale lilac granules

_____ **7.** 10–12 μm; nucleus with 2 or 3 lobes; red-orange granules

_____ **8.** 8–10 μm; nucleus difficult to see; large deep blue-purple granules

_____ **9.** 6–9 μm; round nucleus that is dark purple; sky blue cytoplasm, no visible granules

_____ **10.** 12–20 μm; kidney-shaped nucleus; blue-gray cytoplasm, no visible granules

_____ **11.** Abbreviation for polymorphonuclear leukocytes

_____ **12.** General name for all of the WBCs

_____ **13.**
⎫
⎬ Nicknames for neutrophils
⎭
_____ **14.**

D. White Blood Cells

Write the name of the white blood cell that matches the description. Terms may be used more than once.

_____ **1.**
⎫
⎬ granulocytes
⎭
_____ **2.**

_____ **3.**

_____ **4.**
⎫
⎬ agranulocytes
⎭
_____ **5.**

_____ **6.** most numerous leukocyte

_____ **7.** least numerous leukocyte

E. ABO and Rh Blood Typing

Name the antigens present on the RBCs and the antibodies present in plasma of each blood type.

	Antigens on RBCs	**Antibodies in Plasma**
1. O+	_____	_____
2. A−	_____	_____
3. B−	_____	_____
4. AB+	_____	_____

5. Based on what you know about antigens and antibodies, what blood type is the universal donor? _____ Explain.

6. What blood type is the universal recipient? _____ Explain.

F. Hematocrit

1. Define hematocrit.

2. What is the anticoagulant used in this type of blood test?

3. Ron has a hematocrit of 47%. Is this within the normal range? (yes/no)

4. Janey has a hematocrit of 58%. Is this within the normal range? (yes/no)

G. Hemoglobin Content and Coagulation Time

1. Do the hematocrit and hemoglobin content of blood measure the same thing? _____ Explain.

2. What is the importance of coagulation time?

3. Would a hemophiliac have an above or below normal coagulation time?

Using Your Knowledge

A. White Blood Cell Functions

1. Ken has a bacterial infection with inflammation. What type of WBC increases in number to help him fight the infection?

B. Differential White Blood Cell Count

2. Lindsey had a differential WBC count that showed 60% neutrophils, 10% eosinophils, 1% basophils, 22% lymphocytes, and 7% monocytes. What could have caused this?

C. Blood Typing

3. Blake's doctor found out that, although Blake is Rh−, he received Rh+ blood for the first time. Will he have a transfusion reaction?

Explain.

4. Jim has type AB+ blood. What type(s) of blood can he receive in a transfusion?

5. If Brittany is blood type O− and her husband is blood type AB−, what effect can the Rh factor have on their second child?

6. Michael is type O−. He is donating blood today and wants to know what blood types can accept his blood. What would you tell him?

7. Staci has antibody A in her blood with no Rh antibody. What blood type does she have?

8. Leslie has AB+ blood and had a blood transfusion with AB− blood. Could this be a life-threatening situation?

 Why or why not?

D. Hematocrit

9. Michaela is an athlete who has been training for several years only in Florida. She had an annual physical, and the doctor read her hematocrit as 59%. What could this reading potentially tell the doctor?

E. Hemoglobin

10. Using your textbook or another reference, determine what type of anemia could cause a reduced hemoglobin level, but the person would still have a normal hematocrit.

Heart Structure

Objectives

After completing this exercise, you should be able to:

- Describe the location of the heart using anatomical landmarks
- Name and describe the two layers of the pericardium and the three layers of the heart wall
- Identify the major heart structures on models or charts
- Explain the flow of a drop of blood through the pulmonary and systemic circulations, listing the vessels, chambers, and valves
- Explain how the heart functions as a double pump
- Explain the differences in structure and function of the two types of heart valves
- Describe the changes that take place in the heart after birth
- Identify the major vessels involved in coronary circulation on models or charts
- Identify the selected heart structures on a dissected sheep heart
- Describe the microscopic structure of cardiac muscle

Materials

- articulated skeleton
- human heart model
- human torso or chart showing the heart/pulmonary/systemic circulations
- model or chart of the coronary circulation
- red, blue, and purple pencils
- preserved sheep heart (or other mammal heart), dissecting instruments, trays, disposable gloves, 5-inch blade knife

Your heart beats without external stimulation and never takes a very long rest. The heart is a small double pump that simultaneously pumps blood to and from body cells through the systemic circulation and to and from the lungs through the pulmonary circulation. The blood vessels that carry blood from the heart are called arteries and the blood vessels that carry blood to the heart are called veins.

A. Location of the Heart

The **heart** is about the size of a fist and lies in the thoracic cavity within the **mediastinum** (*mediastinus* = midway), an area bounded by the lungs laterally, the sternum anteriorly, and the thoracic vertebrae posteriorly. The **base** of the heart is the wide superior portion of the heart from which the great vessels emerge, and the **apex** of the heart is the inferior pointed end.

The location of the heart in the thorax can be further defined by finding the superior and inferior right and left points of the heart. The **superior right point** of the heart is located at the superior border of the 3rd right costal cartilage at its attachment to the rib. The **superior left point** is located at the inferior border of the 2nd left costal cartilage at its attachment to the rib. The **inferior left point** is located in the 5th intercostal space, inferior to the fifth costal cartilage at its attachment to the rib. The **inferior right point** is at the inferior border of the 5th right costal cartilage, a little to the right of the border of the sternum. The heart is tilted at an angle so that its inferior surface lies against the diaphragm, with two-thirds of the heart to the left side of the sternum.

FIGURE 27.1 Location of the heart.

- inferior border of the 2nd left costal cartilage
- superior border of the 3rd right costal cartilage
- 5th left intercostal space (midclavicular)
- 5th right costal cartilage

ACTIVITY 1 Location of the Heart

1 In Figure 27.1, identify the superior and inferior right and left points of the heart. Mark dots at these points and connect these four dots to outline the heart location.

2 Identify the superior and inferior right and left points for the heart location on an articulated skeleton and/or a torso model.

3 Palpate the superior and inferior right and left points on yourself.

B. Major Heart Structures

The Pericardium and the Layers of the Heart Wall

In the mediastinum, the heart is surrounded and protected by the pericardium (*peri-* = around). The pericardium consists of an outer, tough fibrous pericardium and an inner, delicate serous pericardium. The **fibrous pericardium** attaches to the diaphragm and also to the great vessels of the heart. Like all serous membranes, the **serous pericardium** is a double membrane composed of an outer parietal layer and an inner visceral layer. Between these two layers is the **pericardial cavity** filled with **serous fluid.** The outer **parietal** (*paries* = wall) **pericardium** is attached to the fibrous pericardium, and the **inner visceral** (*viscera* = internal organs) **pericardium** covers the cardiac muscle.

The wall of the heart has three layers: the outer epicardium (*epi-* = on, upon; *cardia* = heart), the middle myocardium (*myo-* = muscle), and the inner endocardium (*endo-* = within, inward). The **epicardium** is the *visceral layer of the pericardium*. The majority of the heart is the **myocardium** or cardiac muscle tissue. The **endocardium** is a thin layer of endothelium deep to the myocardium that lines the inside chambers of the heart and the valves.

ACTIVITY 2 **The Pericardium and Layers of the Heart Wall**

1 Label the structures in Figure 27.2.
2 Pronounce each term as you write the answer in the blank.

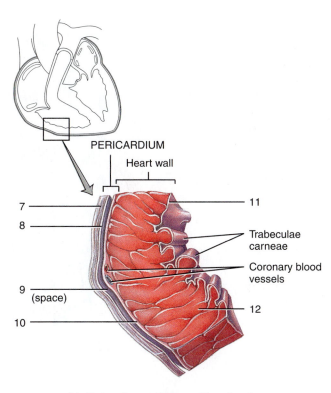

(a) Heart in the mediastinum, anterior view

(b) Pericardium and layers of heart wall

FIGURE 27.2 External heart and pericardium.

(a)
• apex of heart
• base of heart
• diaphragm
• fibrous pericardium (peri-CAR-dee-um)
• left lung
• right lung

1 _____
2 _____
3 _____
4 _____
5 _____
6 _____

(b)
• endocardium
• fibrous pericardium
• myocardium
• parietal layer of serous pericardium
• pericardial cavity
• visceral layer of serous pericardium (epicardium)

7 _____
8 _____
9 _____
10 _____
11 _____
12 _____

Surface Features of the Heart

Like all mammalian hearts, the human heart has four chambers and is divided into right and left sides. Each side has an upper chamber called an **atrium** and a lower chamber called a **ventricle.** The two atria form the base of the heart, and the tip of the left ventricle forms the apex. **Auricles** (*auricle* = little ear) are pouchlike extensions of the atria. From the exterior, the auricles look like flaps with wrinkled edges.

Coronary blood vessels and adipose tissue are found in the **sulci** or grooves that externally mark the boundaries between the four heart chambers. Although a considerable amount of external adipose tissue is present on the heart surface for protection and padding, most heart models do not show this. The **coronary sulcus** is a deep sulcus that externally shows the separation of the atria and the ventricles. The **anterior interventricular sulcus** and the **posterior interventricular sulcus** are shallow grooves that depict the surface boundaries between the two ventricles.

ACTIVITY 3 Surface Features of the Heart

1 Label the structures in Figure 27.3.
2 Identify each term on a model or chart.
3 Pronounce each term as you point to it.

(a) Anterior view

FIGURE 27.3 Surface features of the heart.

- anterior interventricular sulcus
- auricle of left atrium
- auricle of right atrium
- coronary sulcus
- left ventricle
- right ventricle

1 _____

2 _____

3 _____

4 _____

5 _____

6 _____

(b) Posterior view

FIGURE 27.3 Surface features of the heart, *continued.*

- adipose tissue
- coronary sulcus
- left auricle
- left ventricle
- posterior interventricular sulcus
- right auricle
- right ventricle

7 _____

8 _____

9 _____

10 _____

11 _____

12 _____

13 _____

Great Vessels of the Heart

The great vessels of the heart either return blood to the atria or carry blood away from the ventricles. The superior vena cava, inferior vena cava, and coronary sinus return oxygen-poor blood to the right atrium. The **superior vena cava** returns blood from the head, neck, and arms; the **inferior vena cava** returns blood from the body inferior to the heart; and the **coronary sinus** is a smaller vein that returns blood from the coronary circulation. Blood leaves the right atrium to enter the right ventricle. From here, blood passes out the **pulmonary trunk,** the only vessel that removes blood from the right ventricle. This large artery divides into the **right** and **left pulmonary arteries** that carry blood to the lungs where it is oxygenated. Oxygen-rich blood returns to the left atrium through two **right** and two **left pulmonary veins.** The blood then passes into the left ventricle which pumps blood into the large **aorta.** The aorta distributes blood to the systemic circulation. The aorta begins as a short **ascending aorta,** curves to the left to form the **aortic arch,** descends posteriorly, and continues as the **descending aorta.**

The fetal heart contains a short, temporary vascular channel, the **ductus arteriosus** (*ductus* = duct; *arteriosus* = artery), which connects the pulmonary trunk and the aorta. This right heart to left heart shunt re-routes some of the blood destined for the lungs to the systemic circulation via the aorta. In fetal life, oxygen is obtained through the placenta from the mother and not from the lungs. Therefore, it is not detrimental to the baby's health for blood to bypass the lungs. The ductus arteriosus changes into a ligament after birth and remains as the **ligamentum arteriosum.**

ACTIVITY 4 Great Vessels of the Heart

1 Label the great vessels of the heart in Figure 27.4. Blood vessels carrying oxygen-rich blood are red, and those carrying oxygen-poor blood are blue.

2 Identify each great vessel on a model or chart.

3 Pronounce each term as you point to it.

(a) Anterior view

FIGURE 27.4 Great vessels of the heart.

- aortic arch
- ascending aorta
- descending aorta
- inferior vena cava
- left pulmonary artery
- left pulmonary veins
- ligamentum arteriosum
- pulmonary trunk
- right pulmonary artery
- right pulmonary veins
- superior vena cava

1 _____

2 _____

3 _____

4 _____

5 _____

6 _____

7 _____

8 _____

9 _____

10 _____

11 _____

(b) Posterior view

FIGURE 27.4 Great vessels of the heart, *continued.*

- aortic arch
- ascending aorta
- coronary sinus
- inferior vena cava
- left pulmonary artery

- left pulmonary veins
- ligamentum arteriosum
- right pulmonary artery
- right pulmonary veins
- superior vena cava

12 _____

13 _____

14 _____

15 _____

16 _____

17 _____

18 _____

19 _____

20 _____

21 _____

Internal Features of the Heart

The myocardium of the anterior wall of the right atrium has a honeycombed appearance, and these myocardial ridges, called **pectinate muscles** (*pecten* = comblike), continue into the auricles. The walls of the right and left atria are separated by the thin **interatrial septum.** In the fetus, there is a hole in the interatrial septum called the foramen ovale. The **foramen ovale** allows blood to bypass the lungs and go from the right atrium to the left atrium, forming another right heart to left heart shunt. The **fossa ovalis,** a connective tissue membrane, forms over and closes the fetal foramen ovale after birth.

The ventricles have ridges of muscles called **trabeculae carneae** (*trabecula* = little beam; *carnea* = flesh). The larger of these muscles, the **papillary muscles,** have string-like cords attached to them called the **chordae tendineae** (tendinous strands). The opposite ends of these cords are attached to the AV valves. The **interventricular septum** is a thin wall that separates the right and left ventricles.

The heart has four valves that control the one-way flow of blood through the heart: two **atrioventricular (AV) valves** and two **semilunar valves** (*semi-* = half; *lunar* = moon). Blood passing between the right atrium and the right ventricle goes through the right AV valve, the **tricuspid valve** (*tri-* = three; *cusp* = flap). The left AV valve, the **bicuspid valve,** is between the left atrium and the left ventricle. This valve clinically is called the **mitral valve** (*miter* = tall, liturgical headdress) because the open valve resembles a bishop's headdress. The two AV valves are structurally similar, except the tricuspid valve has three cusps or flaps and the bicuspid valve has two cusps or flaps that prevent blood from flowing back into the atria when the valves are closed.

Blood in the right ventricle goes through the **pulmonary (semilunar) valve** to enter the pulmonary trunk (artery). The **aortic (semilunar) valve** is located between the left ventricle and the aorta. These two semilunar valves are identical, with each having three pockets that fill with blood, preventing blood from flowing back into the ventricles when the valves are closed.

The thinner-walled atria receive the blood returning to the heart from veins. The pressure of blood in the atria opens the AV valves, and most of the blood flows into the ventricles. Both atria then contract simultaneously to pump the remaining blood into the ventricles. The larger, thick ventricular walls are double pumps that also contract simultaneously. When the ventricles contract, the papillary muscles contract and close the AV valves. The chordae tendinae prevent the AV valve flaps from opening into the atria. Blood from the right ventricle flows into the pulmonary circulation as blood from the left ventricle simultaneously flows into the systemic circulation. The wall of the left ventricle is thicker than the wall of the right ventricle because the left side requires more force to pump blood throughout the systemic circulation.

ACTIVITY 5 **Internal Features of the Heart**

1 Label the structures in Figure 27.5.

2 Identify each term on a model or chart.

3 Pronounce each term as you point to it.

Frontal plane

1

2

Fossa ovalis

3

4

5

6

7

8

9

10

11

12

13

14

15

16

FIGURE 27.5 Internal features of the heart, frontal section.

- aortic (semilunar) valve
- bicuspid valve (mitral)
- chordae tendineae (CHOR-dee ten-DIN-ee)
- coronary sinus opening
- inferior vena cava opening
- interventricular septum (inter-ven-TRIC-u-lar)
- left atrium
- left ventricle
- papillary muscle (PAP-ih-lary)
- pulmonary (semilunar) valve
- pulmonary vein openings
- right atrium
- right ventricle
- superior vena cava opening
- trabeculae carneae (tra-BEC-u-lee CAR-nee)
- tricuspid valve

1 _____ 9 _____

2 _____ 10 _____

3 _____ 11 _____

4 _____ 12 _____

5 _____ 13 _____

6 _____ 14 _____

7 _____ 15 _____

8 _____ 16 _____

C. Systemic and Pulmonary Circulations

As you trace a drop of blood through the heart to the lungs and then to the rest of the body, you will be examining the pulmonary and systemic circulations. The **pulmonary circulation** takes blood from the right ventricle to the lungs and back to the left atrium. The **systemic circulation** takes blood from the left ventricle to the body tissues and back to the right atrium. Note that each circulation begins and ends at the heart, and each circulation is composed of arteries, capillaries, and veins.

ACTIVITY 6 Systemic and Pulmonary Circulations

1 In Figure 27.6, color the vessels that are carrying oxygen-poor blood blue and the vessels carrying oxygen-rich blood red, being careful to note the color switch in the pulmonary vessels. Color the four capillary beds purple.

2 Trace the pathway of blood in Figure 27.6 through the pulmonary circulation with one color of arrows and the systemic circulation with another color of arrows, starting and ending with the right atrium.

3 Indicate whether the following blood vessels contain oxygen-poor or oxygen-rich blood:
- aorta
- pulmonary arteries
- pulmonary trunk
- pulmonary veins
- venae cavae

Blood vessels with oxygen-poor blood

1. _____

2. _____

3. _____

Blood vessels with oxygen-rich blood

4. _____

5. _____

4 In the blanks below, trace a drop of blood through the heart and lung by listing *in order* all vessels, heart chambers, and valves through which the blood passes, starting with the right atrium.
- aorta
- aortic (semilunar) valve
- bicuspid valve (mitral)
- left atrium
- left ventricle
- pulmonary arteries
- pulmonary capillaries
- pulmonary (semilunar) valve
- pulmonary trunk
- pulmonary veins
- right atrium
- right ventricle
- systemic arteries
- systemic capillaries
- systemic veins
- tricuspid valve
- venae cavae

1. _____

2. _____

3. _____

4. _____

5. _____

6. _____

7. _____

8. _____

9. _____

10. _____

11. _____

12. _____

13. _____

14. _____

15. _____

16. _____

17. _____

FIGURE 27.6 Systemic and pulmonary circulations.

D. Coronary Circulation

The walls of the heart have their own blood supply and circulation, the **coronary** (*corona* = crown) **circulation.** These vessels encompass the heart similar to a crown. The endothelium lining the heart chambers is too thick for blood in the chambers to supply nutrients to cardiac muscle tissue. Coronary blood vessels supply blood to cardiac muscle tissue.

On the anterior surface of the heart, the **right** and **left coronary arteries** branch off the base of the ascending aorta just superior to the aortic semilunar valve. These small arteries are supplied with blood when the ventricles are resting. When the ventricles contract, the cusps of the aortic valve open to cover the openings to the coronary arteries. If the coronary arteries were not covered, they would not be able to withstand the blood pressure and would burst like an overinflated balloon.

As the **left coronary artery** passes the base of the left auricle, it branches into the:

- anterior interventricular branch (left anterior descending branch, or LAD)
- circumflex branch

The **anterior interventricular branch** (LAD) supplies both ventricles with oxygen-rich blood and lies within the anterior interventricular sulcus, a shallow depression between the two ventricles. The anterior interventricular branch is commonly occluded and can result in a myocardial infarct and at times death. The **circumflex branch** continues around the left side of the heart, lying within the coronary sulcus (atrioventricular sulcus), and supplies blood to the left ventricle and left atrium. The circumflex branch forms anastomoses (connections) with the posterior interventricular branch near the posterior interventricular sulcus.

The **right coronary artery** passes the right auricle, supplying it with blood, and continues inferiorly to the auricle in the coronary sulcus. The right coronary artery branches into the:

- marginal branch
- posterior interventricular branch

The **marginal branch** supplies the anterior right side of the right ventricle. The **posterior interventricular branch** lies in the posterior interventricular sulcus on the posterior surface of the heart, supplying oxygen-rich blood to both ventricles.

Arteries branch into smaller vessels, called arterioles, which penetrate the heart muscle and divide into narrower vessels, called capillaries, which deliver oxygen to the cardiac muscle. Capillaries drain into venules which exit the heart muscle and connect to veins that receive oxygen-poor blood, returning it to the heart. The **great cardiac vein** is the principal vein of the coronary circulation, draining the left anterior portion of the heart. It lies near the anterior interventricular branch in the interventricular sulcus. The **small cardiac vein** drains the right anterior portion of the heart. The **middle cardiac vein** that lies next to the posterior interventricular branch in the posterior interventricular sulcus drains the posterior portion of the heart. Both of these veins empty into a large, thin-walled venous sinus called the **coronary sinus** which is located on the posterior surface of the heart. The coronary sinus empties its oxygen-poor blood into the right atrium.

ACTIVITY 7 Coronary Circulation

1 Label the structures in Figure 27.7.
2 Identify each vessel on a model or chart.
3 Pronounce each term as you point to it.

(a) Coronary arteries, anterior view (b) Coronary veins, anterior view

FIGURE 27.7 Coronary circulation.

- anterior interventricular branch (LAD)
- circumflex branch
- coronary sinus
- great cardiac vein
- left coronary artery
- marginal branch
- middle cardiac vein
- posterior interventricular branch
- right coronary artery
- small cardiac vein

1 _____

2 _____

3 _____

4 _____

5 _____

6 _____

7 _____

8 _____

9 _____

10 _____

E. Dissection of a Sheep Heart

The sheep heart is similar to the human heart in both structure and size. It provides students the opportunity to observe the flexibility of the valves and tissues.

ACTIVITY 8 Dissection of a Sheep Heart

1 Obtain a dissecting tray, tools, disposable gloves, and a sheep heart.

2 Examine the **anterior surface** of the heart. Great vessels are often cut close to the base of the heart and may be difficult to find. Refer to Figure 27.8(a) and a heart model to identify the following structures:
 • pericardium (if present)
 • epicardium
 • base

 • apex
 • right auricle
 • left auricle
 • right ventricle
 • left ventricle
 • pulmonary trunk

3 Examine the **posterior surface** of the heart using Figure 27.8(b), and identify the following structures:
 • coronary sulcus
 • left auricle
 • left ventricle
 • right auricle
 • right ventricle

4 Insert a blunt probe into the collapsed superior vena cava and into the right atrium. Maneuver the probe to find the interior opening of the inferior vena cava in the right atrium and push the probe out into this vessel.

ANTERIOR POSTERIOR

FIGURE 27.8 Sheep heart.

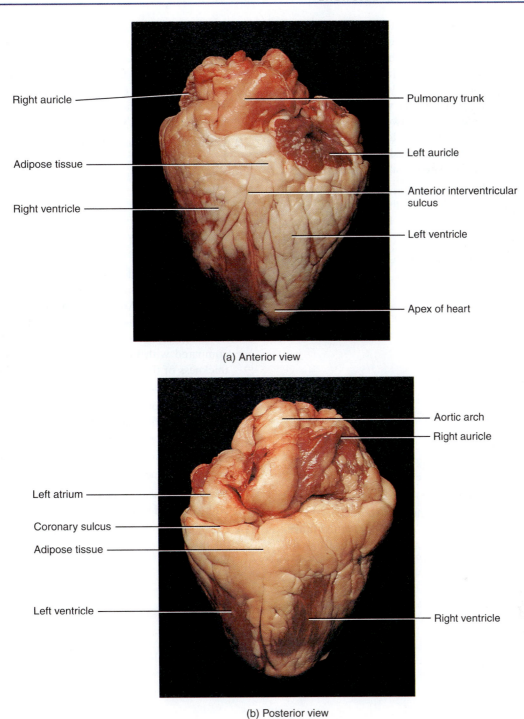

Right auricle

Adipose tissue

Right ventricle

Pulmonary trunk

Left auricle

Anterior interventricular sulcus

Left ventricle

Apex of heart

(a) Anterior view

Aortic arch

Right auricle

Left atrium

Coronary sulcus

Adipose tissue

Left ventricle

Right ventricle

(b) Posterior view

FIGURE 27.8 Sheep heart, *continued*.

5 Examine the **interior of the heart** by making a coronal section of the heart. Using a knife with about a 5-inch blade, make a coronal cut of a sheep heart starting at the apex and cutting toward the base (see Figure 27.8(c)). Cut through both auricles (to ensure cutting through both atria) but not all the way through the base and the great vessels, so the two halves do not get separated. This cut allows you to observe both atria and both ventricles simultaneously (similar to a heart model), and easily compare the size of the walls of the right and left ventricles.

6 Using Figure 27.8(c), identify the following interior structures on the right side of the heart:
- myocardium
- endocardium
- right atrium
- right auricle
- pectinate muscle
- opening of superior and inferior vena cava
- opening of the coronary sinus
- tricuspid valve
- right ventricle
- chordae tendinae
- papillary muscles
- moderator band (cord between the two walls of the right ventricle)
- interventricular septum
- pulmonary trunk
- pulmonary semilunar valve

7 In the right atrium, insert a blunt probe in the small opening of the coronary sinus that is inferior to the opening of the inferior vena cava. Observe the movement of the probe in the coronary sinus from the posterior view of the heart.

8 In the right atrium, insert a blunt probe into the opening of the **pulmonary trunk** and push it through to the superior end of the vessel. Remove the probe and take a scalpel to cut the wall of the pulmonary trunk longitudinally to expose the **pulmonary semilunar valve.** Count the three cusps. How does this valve differ from the tricuspid valve?

9 Continue identifying structures on the left side of the heart.
- left atrium
- left auricle
- bicuspid valve
- left ventricle
- aortic semilunar valve
- aorta

10 How many cusps does the bicuspid valve have? _____ Does this valve look similar otherwise to the other AV valve? _____ Are there **chordae tendineae** and **papillary muscles?** _____ Does the left ventricle have a greater or smaller number of papillary muscles compared with the right side? _____ Compare the thickness of the right and left ventricles. Which one is thicker?_____ Why? _____

11 Look just above the cusps of the aortic valve for the openings to the **right** and **left coronary arteries.** Use the blunt probe to push into these small vessels.

12 Dispose of any removed dissection material in the proper container (NOT the sink!).

13 Wash the dissection pan, instruments, and your hands with soap and water when finished.

14 Clean up your lab space and wash the countertops with disinfectant.

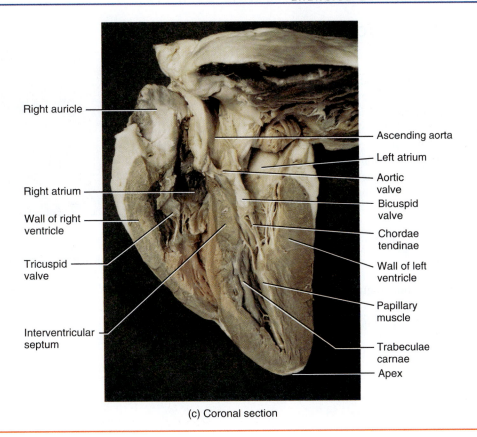

Right auricle

Right atrium

Wall of right
ventricle

Tricuspid
valve

Interventricular
septum

Ascending aorta

Left atrium

Aortic
valve

Bicuspid
valve

Chordae
tendinae

Wall of left
ventricle

Papillary
muscle

Trabeculae
carnae

Apex

(c) Coronal section

FIGURE 27.8 Sheep heart, *continued*.

F. Histology of Cardiac Muscle

The heart is the only structure in the body composed of **cardiac muscle tissue,** which is striated and involuntary. Cardiac fibers have one or two nuclei and demonstrate branching. Cardiac muscle has **intercalated discs** that contain gap junctions and desmosomes that hold cardiac fibers together. **Gap junctions** enable action potentials to spread quickly from cell to cell, allowing the atrial fibers to contract as a unit, as do the ventricular fibers.

ACTIVITY 9 Histology of Cardiac Muscle

1 As a review, label the photomicrograph of cardiac tissue in Figure 27.9.
2 Pronounce the terms as you label the figure.

LM 700×

FIGURE 27.9 Photomicrograph of cardiac muscle fibers.

- branching cardiac fibers
- cardiac muscle fiber
- intercalated discs (in-TER-cal-ated)
- nucleus

1 _____

2 _____

3 _____

4 _____

Reviewing Your Knowledge

A. Anatomy Drill and Practice

Go to www.wiley.com/college/allen. Click on the picture of your lab manual, then on Student Companion Site, Exercise 27, and Anatomy Drill and Practice.

B. Location of the Heart

1. Describe the location of the heart using the lungs, rib cartilages, and intercostal spaces as landmarks.

C. Major Heart Structures

Fill in the blank with the term that fits the description.

_____ **1.** Arteries that supply blood to cardiac muscle

_____ **2.** Layer of heart wall containing cardiac muscle

_____ **3.** Extensions of the atria

_____ **4.** Heart is located here (area between the lungs)

_____ **5.** Lines the heart chambers

_____ **6.** Pointed inferior part of the heart

_____ **7.** Two heart pumps; lower heart chambers

_____ **8.** Superior heart chambers

_____ **9.** Another name for visceral pericardium

_____ **10.** Wide superior part of the heart

_____ **11.** Blood pumped by right ventricle (oxygen-rich or oxygen-poor)

_____ **12.** Blood pumped by left ventricle (oxygen-rich or oxygen-poor)

_____ **13.** Enlarged muscles in ventricles attached to chordae tendinae

_____ **14.** Muscle ridges in ventricles

_____ **15.** Ridges in anterior wall right atrium

_____ **16.** Strings attached to AV cusps

D. Coronary Circulation—Blood Vessels

After reviewing the coronary circulation in Figure 27.7, fill in the blank with the word that fits the description.

_____ **1.** Anterior branch of the left coronary artery

_____ **2.** Posterior branch of the right coronary artery

_____ **3.** Coronary artery that lies in anterior coronary sulcus

_____ **4.** Branch of the left coronary artery that curves around left side

_____ **5.** Branch of left coronary artery that supplies the anterior ventricles

_____ **6.** Shorter coronary artery that is hidden anteriorly by pulmonary trunk

_____ **7.** Vein that drains coronary circulation into right atrium

_____ **8.** Vein that drains most of anterior ventricles

_____ **9.** Vein that drains the posterior ventricles

_____ **10.** Vein that drains the right anterior side

E. The Heart and Pulmonary Circulation

Place the following structures _in order,_ tracing the blood flow from the superior vena cava to the heart, to the pulmonary circulation, and out of the heart to the systemic circulation.

- aorta
- aortic valve
- bicuspid valve
- left atrium
- left ventricle
- pulmonary capillaries
- right atrium
- right ventricle
- pulmonary arteries
- pulmonary trunk
- pulmonary valve
- pulmonary veins
- superior vena cava
- tricuspid valve

1. _____

2. _____

3. _____

4. _____

5. _____

6. _____

7. _____

8. _____

9. _____

10. _____

11. _____

12. _____

13. _____

14. _____

Using Your Knowledge

A. The Heart

1. What is the function of the pericardial fluid? _____

2. If a person has mitral valve prolapse, which circulation is affected the most: pulmonary, systemic, or cardiac

circulation? _____

3. A blood clot (thrombus) in which artery of the coronary circulation would be more likely to cause sudden death?

 a. right coronary artery

 b. posterior interventricular branches

 c. circumflex branch

 d. anterior interventricular branch

4. Explain your choice in question 3: _____

5. The pulmonary trunk and pulmonary arteries are colored blue in all of the figures. Why are they colored blue like

most veins? _____

6. If the ductus arteriosus does not fully close after birth, what will happen? _____

7. If a person has endocarditis, myocarditis, or pericarditis, what areas (*in order*) are infected?

 a. _____ **b.** _____ **c.** _____

8. Athletes have thicker heart walls than nonathletes. Which layer of the heart enlarges? _____

Explain: _____

9. Atrioventricular valves and semilunar valves have different structures. Explain if high pressure from *above*

or *below* each type of valve opens the valves. Atrioventricular valves: _____

Semilunar valves: _____

10. Which ventricular wall is thicker? _____ What is the significance of this? _____

11. What is the significance of the difference in thickness between the aorta and the pulmonary trunk?

12. Using your knowledge, explain why mitral valve prolapse occurs more often than tricuspid prolapse. _____

13. How would heart function be affected if the AV valves did not completely close during ventricular contraction?

B. Coronary Circulation—Tracing Blood Flow

Trace a drop of blood from the ascending aorta to the coronary circulation of the anterior surface of the left ventricle and back to the heart again.

- anterior interventricular branch
- ascending aorta
- coronary sinus
- great cardiac vein

- left coronary artery
- left ventricular capillaries
- right atrium

1. _____

2. _____

3. _____

4. _____

5. _____

6. _____

7. _____

Heart Function

Objectives

After completing this exercise, you should be able to:

- Describe the relationship of auscultated heart sounds, pulse rate, and heart rate
- Calculate changes in length of cardiac cycle with exercise and discuss significance
- Identify the parts of the cardiac conduction system on models or charts
- Describe the pathway of action potentials through the cardiac conduction system and explain its association with contractions of the atria and ventricles
- Describe the association of electrocardiogram (ECG) tracings with electrical events occurring in the heart
- Identify the components of a normal ECG
- Identify normal sinus rhythm, tachycardia, and bradycardia on an ECG

Materials

- Heart sounds: stethoscope, alcohol swabs
- Cardiac cycle length: stethoscope, alcohol swabs, stopwatch or clock with second hand, calculators
- model or chart of the heart electrical conduction system
- ECG recording equipment (if available), electrodes, cream or jelly

A. Heart Sounds

Auscultation (*auscult-* = listening) means listening to body sounds, typically using a stethoscope. Two particular **heart sounds** can be detected during one heartbeat. These sounds occur after the heart valves quietly close and blood strikes against the closed valve, causing turbulence that we can hear with a stethoscope. Although there are four sounds generated during one heartbeat, only the first and second sound can easily be heard without additional ampli-

fication. The first sound, **lubb,** is a little longer and louder than **dupp** (or **dubb**), the second sound that occurs shortly after the first. The first sound (S1) occurs with blood turbulence from the closure of the two atrioventricular (AV) valves at ventricular **systole** when the ventricles are contracting. The second sound (S2) is at ventricular **diastole,** when the ventricles relax and the two semilunar valves close. The sounds that you will hear are lubb-dupp, pause… lubb-dupp, pause. Remember that these two heart sounds, lubb-dupp, equal one heartbeat.

CLINICAL NOTE: *Heart murmurs are abnormal heart sounds caused by the restriction of blood flow or backflow of blood that typically indicate an abnormality in valve structure or function. It usually takes special training and a very quiet room to be able to detect heart murmurs.*

ACTIVITY 1 **Heart Sounds**

1 Preparation for auscultation of heart sounds.
 • Obtain a stethoscope and alcohol swabs.
 • Clean the earpieces with an alcohol swab and let air dry.
 • Earpieces should be pointed forward when placed in the external ear canal to facilitate hearing the heart sounds.
 • Put the earpieces in place and gently tap the bell to check that you can hear sounds.

2 Auscultate the mitral valve.
 • You may auscultate the heart sounds on your own chest or that of your partner while either is in a sitting position.
 • To clearly hear the first heart sound (lubb), you will auscultate the mitral valve. Place the large bell of the stethoscope at the apex of the heart. This is located in the 5th intercostal space just left of the nipple (Figure 28.1).
 • If the heart sound is difficult to hear, have the subject bend forward to tip the heart against the anterior thoracic wall and try the auscultation again.

3 Auscultate the aortic valve.
 • Palpate the suprasternal (jugular) notch and then the sternal angle (marks the insertion of the 2nd rib). Now drop down another 1 inch on the sternum from the sternal angle and then move right 1 inch.
 • You should be in the 2nd intercostal space just to the right of the sternum where you can clearly hear dupp, or the second heart sound (Figure 28.1).

❓ DISCUSSION QUESTION: HEART SOUNDS

1 At rest, which heart is louder, lubb or dupp? Explain.

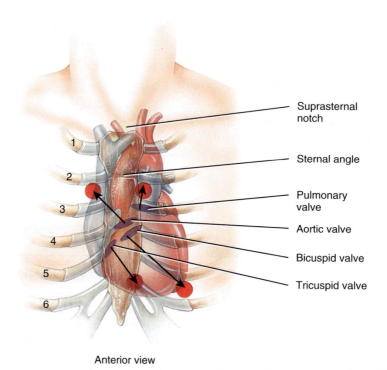

Anterior view

FIGURE 28.1 Locations for auscultation of the heart.

B. Changes in Length of Cardiac Cycle with Exercise

Heart rate is the number of heartbeats per minute. Each heartbeat represents one **systole** (atrial and ventricular contraction and one **diastole** (atrial and ventricular relaxation) or one cardiac cycle.

The heart rate can be measured by counting the number of heartbeats per minute or by the number of pulses per minute. A **pulse** is the blood pressure wave that travels through the arteries when the ventricles contract. Pulses are commonly felt in the radial and carotid arteries. The number of heartbeats per minute (heart rate) will be very close to, but not necessarily equal to, the number of pulses per minute.

One **cardiac cycle** consists of the cardiac events that occur from the beginning of one period of systole and the beginning of the next period of systole. The length of one cardiac cycle in seconds can be calculated by dividing 60 seconds by the heart rate (heartbeats/min). Although the length of the cardiac cycle varies with heart rate, the length of systole remains constant (0.4 sec). Only the length of diastole changes.

ACTIVITY 2 **Experiment: Changes in Length of Cardiac Cycle with Exercise**

1 **Prediction:** Exercise causes the length of the cardiac cycle (seconds) to *increase* or *decrease*. Circle the correct choice in italics.

2 **Materials:** Obtain materials for length of cardiac cycle (see Materials list).

3 Locate the radial and carotid pulses. Decide which pulse will be used for this experiment.
 • Radial pulse: Locate the groove in your partner's wrist between the radius and the ligament just medial to the radius. Palpate the radial artery by pressing down with your index and middle fingers. Because the thumb has a noticeable pulse of its own, it is not used.
 • Carotid pulse: Locate your partner's Adam's apple (thyroid cartilage of larynx). Using your index and middle fingers again, palpate the carotid artery on either side of the larynx.

4 **Data Collection:** Measure heart rate and pulse rate at rest and after exercise and record your results in Table 28.1.
 • Decide who will be the subject, who will count heartbeats and pulses, and who will be the recorder and the timer. The subject should not have heart problems and should be in good health.

Count heart beats and pulses at rest

 • Listen for the subject's heartbeat and count the number of heartbeats in 15 seconds. Record your data in Table 28.1.
 • Count the number of pulses in 15 seconds and record your data in Table 28.1.

Count heart beats and pulses after exercise

 • Have the subject run in place for 2 minutes.
 • Immediately count the number of heartbeats and pulses in 15 seconds and record your data in Table 28.1.
 • Repeat the pulse count every minute until the pulse rate returns to the initial resting rate. Record the time for recovery: _____ minutes.

TABLE 28.1 **Heart Rate and Pulse Rate Before and After Exercise**

ACTIVITY	BEATS IN 15 SEC	BEATS/MIN
Heart rate at rest		
Heart rate after exercise		
Pulse rate at rest		
Pulse rate after exercise		

5 Data Analysis
- Calculate heart rate by multiplying the number of heartbeats in 15 seconds by 4 to calculate the heart rate per minute. Calculate the resting and exercising heart rates (beats/min) and record them in Table 28.1.
- Calculate pulse rate by multiplying the number of pulses in 15 seconds by 4 to calculate the pulse rate per minute. Calculate the resting and exercising pulse rates (pulses/min) and record them in Table 28.1.
- Calculate the length of the cardiac cycle (seconds) by dividing 60 by the heart rate. Calculate resting and exercising cardiac cycle lengths and record them in Table 28.2.

$$\frac{60 \text{ sec}}{\# \text{ of heartbeats/min}} = \text{length of 1 cardiac cycle in sec}$$

- Calculate the length of diastole by subtracting the length of systole (0.4 sec) from the length of the cardiac cycle. Calculate the length of diastole at rest and after exercise and record in Table 28.2.
- Complete the Experimental Report and answer the discussion questions with your lab partners.

EXPERIMENTAL REPORT: LENGTH OF CARDIAC CYCLE

Results

- State whether heart rate and pulse rate increase or decrease immediately after exercise.

- State whether heart rate and pulse rate at rest are the same and whether exercising heart rate and pulse rates are the same.

- State the time for recovery after exercise.

- State whether the length of the cardiac cycle increases or decreases immediately after exercise.

Discussion

- Discuss why heart rate and pulse rate change with exercise.

- Discuss why heart rate equals pulse rate at rest and after exercise.

- Discuss why the length of diastole changes with exercise and discuss the implications of this.

- Discuss factors that affect pulse rate recovery time.

Conclusion

- Write a statement that describes the correlation between heart rate, length of the cardiac cycle, and length of diastole.

TABLE 28.2 **Changes in Length of Cardiac Cycle with Exercise**

AT REST	LENGTH IN SECONDS
Cardiac cycle	
Systole	
Diastole	
AFTER EXERCISE	
Cardiac cycle	
Systole	
Diastole	

C. Electrical Conduction System of the Heart

Your heart beats without any stimulation from nerves and if removed from your body, the heart would still beat. The internal stimulation that makes the heart beat by itself is called **intrinsic** (*intrinsic* = inside) **stimulation** and is caused by specialized, noncontractile cells called **autorhythmic cells** (causing a rhythm by themselves). These cells belong to the intrinsic **conduction system** (nodal system) that (1) initiates the action potential that causes contraction of cardiac muscle fibers, and (2) provides a pathway for conducting the action potential to all cardiac muscle fibers. The autonomic nervous system and hormones, **extrinsic** (*extrinsic* = on the outside) **stimulation,** only increase or decrease the intrinsic pace.

The parts of the conduction system are:

1. **sinoatrial (SA) node**—called the pacemaker because it initiates action potentials first; located in the wall of the right atrium just inferior to the opening of the superior vena cava; stimulates the atria to contract.

2. **atrioventricular (AV) node**—receives action potentials from the atrial muscle fibers; located in the lower interatrial septum anterior to the opening of the coronary sinus; sends the action potentials to the AV bundle (bundle of His).

3. **AV bundle (bundle of His)**—located in a membranous septum between the atria and ventricles superior to the interventricular septum; this is the electrical connection between the atria and ventricles; sends action potentials to the bundle branches.

4. **right and left bundle branches**—located in the interventricular septum; sends action potentials to the Purkinje (conduction) fibers.

5. **Purkinje (conduction) fibers**—located in the apex of the myocardium, as well as in the lateral walls of the right and left ventricles; sends action potentials to the ventricular cardiac muscle fibers and papillary muscles and stimulates them to contract.

ACTIVITY 3 **Cardiac Conduction System**

1 Label the structures of the conduction system on Figure 28.2.

2 Pronounce each term as you point to it on the figure.

FIGURE 28.2 Cardiac conduction system.

- atrioventricular (AV) node
- AV bundle (bundle of His)
- left bundle branch
- Purkinje (conduction) fibers in left ventricle
- Purkinje (conduction) fibers in right ventricle
- right bundle branch
- sinoatrial (SA) node

1 _____

2 _____

3 _____

4 _____

5 _____

6 _____

7 _____

D. Electrocardiography

The conduction of action potentials throughout the heart can be detected as electrical currents. An instrument called an **electrocardiograph** is used to record the electrical changes in the heart. The chart recording of the electrical events that occur before each heartbeat is called an **electrocardiogram** (ECG; originally called an EKG). The electrical events that are recorded are from the entire heart muscle (i.e., the atria and the ventricles) and not just the conduction system activity. The ECG records only voltage changes over time, and not the force of contraction. Cardiac muscle will contract after the action potential has occurred.

ECGs are usually recorded by *indirect leads,* with electrodes placed on the skin of the subject's arms and legs rather than on the heart itself. The electrical impulses generated by the depolarization and repolarization of the atria and ventricles are detected on the body surface due to the conductivity of the ions in extracellular fluid. Clinically, 12 standard leads are used to record an ECG for diagnosing abnormalities of the conduction system, myocardial infarctions, and other clinical situations. However, in anatomy and physiology classes, usually a three-standard lead ECG is examined.

Parts of an ECG

The main parts of a typical three-lead ECG are:

- **P wave**—first wave; small, curved upward deflection; represents **atrial depolarization** that spreads from the sinoatrial (SA) node just before the atria contract.
- **QRS complex**—short downward deflection (Q); tall upward deflection (R); medium downward deflection (S); represents **ventricular depolarization** that spreads from the AV node, AV bundle, right and left bundle branches, and to the Purkinje (conduction) fibers just before the ventricles contract.

> **NOTE:** *Atrial repolarization takes place during the QRS complex, but gets overshadowed by the more prominent ventricular depolarization.*

- **T wave**—medium, curved upward deflection; represents **ventricular repolarization** and occurs just before the ventricles relax.

In addition to these three waves, an ECG also contains intervals or segments between the waves.

- **P-Q interval**—the interval between the beginning of the P wave until the beginning of the Q (in the QRS wave); some texts call this the P-R interval; it represents the time it takes for the electrical conduction (excitation) to travel through the atria and AV node to the Purkinje (conduction) fibers.
- **S-T segment**—the segment from the end of the S (in QRS wave) to the beginning of the T wave; it represents the time the ventricular fibers are fully depolarized.
- **Q-T interval**—the interval that begins at the Q (in the QRS wave) to the end of the T wave; it represents the time from the beginning of ventricular depolarization until the end of ventricular repolarization.

FIGURE 28.3 Three-standard lead positions.

ACTIVITY 4 Electrocardiography

1 Label the terms for an ECG in Figure 28.4, a normal ECG.

2 Record an ECG at rest and after exercise if equipment is available.

- Do this activity with a lab partner. Choose who will be the subject and who will do the recording.
- Swab the skin where the electrodes will be placed with alcohol, and then apply electrode paste, jelly, or cream only to the same area (see Figure 28.3).
- Attach the electrodes to the anterior surface of each wrist and to the inside area of each ankle.
- Connect the patient cables to the electrodes and also to the recording instrument.
- Follow your instructor's directions for recording the ECG with your equipment.
- Label the P wave, QRS complex, T wave, P-Q interval, S-T segment, and Q-T interval on the recorded ECG.

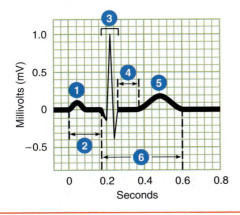

FIGURE 28.4 Section of a normal ECG, Lead II.

- P wave
- P-Q interval
- QRS complex
- Q-T interval
- S-T segment
- T wave

1 _____

2 _____

3 _____

4 _____

5 _____

6 _____

Determining Heart Rate Using an ECG

In an adult, a heart rate of 60 to 100 beats/min is a **normal sinus rhythm** (NSR). A heart rate above 100 beats/min is called **tachycardia,** but in young children, this rate would be considered normal. Heart rates below 60 beats/min are normal for highly conditioned individuals, but in other adults, heart rates below 60 beats/min are called **brady-cardia.** Neither condition is considered to be pathological. Prolonged tachycardia can develop into **ventricular fibrillation,** rapid uncoordinated heart contractions that do not pump blood.

Heart rate can be easily calculated from an ECG. Standard ECGs are printed on paper moving at a paper speed of 25 mm/sec. Therefore, the distance of 1 mm (1 small square on standard ECG paper) is equivalent to 0.04 sec. To **calculate heart rate:**

1. Measure the distance between the start of one P wave to the start of the next P wave by counting the number of small squares between them. (For ease of measurement, the distance between R waves can be measured instead.)

2. Multiply the number of squares by 0.04 sec to give the time in seconds for one beat, or sec/beat. This is the length of one **cardiac cycle.**

3. Since there are 60 seconds in one minute, divide 60 by your answer in step 2 to calculate beats/min.

Example Step 1—20 small squares (20 mm) are counted between two P waves.

Step 2—0.04 sec × 20 mm = 0.8 sec/beat

Step 3—$\dfrac{60 \text{ sec/min}}{0.8 \text{ sec/beat}} = 75$ beats/min

NOTE: *The heart rate calculated using the distance between R waves is the ventricular rate, while the heart rate calculated using the distance between P waves is the atrial rate. In NSR the atrial rate equals the ventricular rate.*

ACTIVITY 5 Heart Rate Calculations

1 Calculate the heart rate for the ECGs in Figure 28.5(a), (b), and (c).
- Count the number of small squares (mm) between two R waves.
- Multiply the number of squares by 0.04 seconds to determine the length of one heartbeat in seconds.
- Divide 60 by your answer to get beats/min. This is the heart rate.

 Figure 28.5(a) = _____ beats/min
 Figure 28.5(b) = _____ beats/min
 Figure 28.5(c) = _____ beats/min

2 Determine which ECG in Figure 28.5 illustrates the following heart rates and write the term next to the figure letter.
- bradycardia Figure 28.5(a) _____
- normal sinus rhythm Figure 28.5(b) _____
- tachycardia Figure 28.5(c) _____

3 Calculate the heart rate before and after exercise on the ECGs your lab group recorded in activity 3.
- Heart rate before exercise = _____ beats/min
- Heart rate after exercise = _____ beats/min

(a)

(b)

(c)

FIGURE 28.5 Tracings of ECGs, Lead II.

Calculating Lengths of ECG Segments

ECGs are important for diagnosing and following the course of abnormal electrical conductions or cardiac rhythms. The normal intervals of an ECG in seconds are given below along with the potential disorder if the intervals are abnormal.

- *P-Q intervals:* Normal P-Q intervals (the time between atrial and ventricular depolarization) are between 0.12 and 0.2 seconds, averaging about 0.16 seconds. P-Q intervals longer than 0.2 seconds could be indicative of a heart block. A **heart block** could be produced by cardiac damage to the AV node or AV bundle, reducing electrical conduction from the atria to the ventricles. A **complete heart block** results in the ventricles depolarizing independently from the atria.
- *QRS complex:* A normal QRS complex is 0.06 to 0.10 seconds. If this complex is longer than 0.12 seconds, a **right** or **left bundle branch block** may be considered, and the two ventricles do not contract simultaneously.
- *Q-T interval:* A normal resting Q-T interval is 0.33 to 0.44 seconds. The interval becomes longer as the heart rate decreases, and the interval becomes shorter as the heart rate increases. Myocardial ischemia (decreased blood flow) or myocardial damage lengthens the normal resting interval.

ACTIVITY 6 Calculate Lengths of the ECG Segments

1 Calculate the lengths of the ECG segments on Figure 28.6. Count the squares between the beginning to the end of that segment. Multiply this number by 0.04 sec to determine the length of the segment in seconds.
- P-Q interval = _____ sec
- QRS complex = _____ sec
- Q-T interval = _____ sec
- Relaxation period (end of T wave to P wave) = _____ sec
- Heart rate (count the squares between the beginning of two R waves) = _____ sec/beat. Convert to beats/min = _____

2 Answer the discussion questions with your lab partner.

? DISCUSSION QUESTIONS

1 Which ECG segments in Figure 28.6 is(are) abnormal?

2 Explain how the relaxation period changes when heart rate increases, and when heart rate decreases. Refer to Figure 28.5.

FIGURE 28.6 ECG tracing, Lead II.

Reviewing Your Knowledge

A. Heart Sounds

Fill in the blank with the correct term.

1. The first heart sound heard is _____ that is due to blood hitting against the _____ valves.

2. The second heart sound heard is _____ that is due to blood hitting against the _____ valve.

3. What is a heart murmur? _____

4. Which heart sound is the loudest sound when auscultated, the lubb or dupp? _____ Why? _____

5. Valerie's radial pulse rate is 70/min. Assuming her heart and arteries are in good health, what is her approximate heart rate/min? _____

6. Jaxton's heart rate is 68/min. What is the length of his cardiac cycle/sec? _____

B. Electrical Conduction System of the Heart

Circle True or False for the following questions.

1. The heart beats without extrinsic stimulation from the autonomic nervous system. True or False

2. Autorhythmic cells are located only in the interventricular septum. True or False

3. Each heart chamber contracts separately, first the right atrium, then the right ventricle, left atrium, and left ventricle. True or False

4. The ECG records the electrical stimulation of cardiac muscle by the conduction system and not the contraction of the muscle itself. True or False

5. The normal pacemaker of the heart is the AV node. True or False

6. Number the following structures of the cardiac conduction system in the normal order of depolarization (1 to 5).

_____ AV bundle (of His)
_____ AV node
_____ Purkinje (conduction) fibers
_____ right and left bundle branches
_____ SA node

491

C. Electrocardiography

Match the following descriptions with the correct terms. A term may be used more than once.

> bradycardia
> fibrillation
> P wave
> QRS complex
> T wave
> tachycardia

_____ **1.** depolarization of the atria

_____ **2.** depolarization of the ventricles

_____ **3.** fast heartbeat above 100 beats/min

_____ **4.** repolarization of the atria

_____ **5.** repolarization of the ventricles

_____ **6.** slow heartbeat below 60 beats/min

_____ **7.** uncontrolled, rapid heart contractions

Measurements Using the ECG

Using the ECG tracing in Figure 28.5(b), calculate the:

8. QRS complex length: _____ sec

9. P-Q interval: _____ sec

10. Q-T interval: _____ sec

11. Heart rate: _____ beats/min

Using Your Knowledge

A. Auscultations of Heart Sounds

1. Greg counted his lab partner's heartbeats as 13 beats in 15 seconds. What is the student's heart rate? What other information would you like to have about the student to decide if this heart rate is abnormal?

B. Electrical Conduction System

2. Larry had damage to his SA node. What will happen now that the "pacemaker" is not functional?

3. In a normal cardiac cycle, what two chambers of the heart contract last?

4. What two structures directly receive an electrical impluse from the SA node?

5. Valerie's heart rate is 70 beats/min. What is the length of one cardiac cycle?

C. Electrocardiography

6. Fibrillation is an asynchronous contraction of cardiac muscle fibers. Is atrial fibrillation or ventricular fibrillation more serious? Explain.

7. If a myocardial infarction damaged the left bundle branch, how would this affect the spread of the action potential?

8. How would damage to the left bundle branch affect blood flow in the systemic circulation?

An abnormal ECG tracing is shown in Figure 28.7.

9. Does the atrial rate equal the ventricular rate in this ECG tracing? Explain.

10. Would you be able to feel a pulse during atrial systole only?

FIGURE 28.7 Tracing of abnormal ECG, Lead II.

Blood Vessel Structure and Function

Objectives

After completing this exercise, you should be able to:

- Compare and contrast the structure of arteries, capillaries, and veins
- Identify layers or tunics present in a transverse section of an artery and vein
- Measure systolic and diastolic blood pressure at rest and after exercise
- Discuss how exercise and body position affect blood pressure

Materials

- compound microscope, lens paper, prepared microscope slide of transverse section through artery and vein
- Blood pressure measurements: sphygmomanometer, stethoscope, alcohol wipes, meterstick, exercise mat
- stopwatch or watch with second hand
- Regulation of blood pressure: sphygmomanometer, stethoscope, alcohol wipes, stopwatch or watch with second hand, graph paper

Arteries carry blood away from the heart and divide into smaller vessels called **arterioles** that branch into the tiniest vessels called **capillaries.** An exchange of nutrients, wastes, and gases occurs between blood and interstitial fluid (fluid surrounding tissue cells) at the capillary level. Capillaries join to form small **venules** that will merge to form larger **veins** that carry blood back to the heart. The structure of these vessels reflects their functions.

There are two main circulatory routes within the body: the systemic circulation and pulmonary circulation. In the **systemic circulation,** arteries carry oxygen-rich blood to the body tissues and veins return oxygen-poor blood to the heart where it will then enter the pulmonary circulation. In the **pulmonary circulation,** arteries carry oxygen-poor blood from the right ventricle to the lungs where gas exchange occurs and veins return oxygen-rich blood back to the left atrium.

A. Blood Vessel Structure

Arteries

Arterial walls have three layers or tunics: the tunica externa, tunica media, and tunica interna. The outer **tunica externa** is composed mainly of elastic and collagen fibers (proteins) that provide support and protection. The **tunica media** is the middle and thickest layer that contains elastic fibers and smooth muscle fibers (cells) encircling the diameter of the vessel. The sympathetic nervous system regulates the diameter of the blood vessel by innervating smooth muscle fibers. Contraction of smooth muscle fibers causes a decrease in lumen diameter or **vasoconstriction,** whereas relaxation causes **vasodilation,** an increase in lumen diameter. Elastic fibers in the tunica media allow the vessel to be stretched and return to its original shape. The **tunica interna (intima)** contains simple squamous

epithelium **(endothelium)**, a basement membrane, and elastic tissue **(internal elastic lamina)** that is adjacent to the tunica media. The endothelium lines the lumen of all blood vessels and forms a smooth, slick barrier that promotes blood flow. Damage to the endothelium causes platelets to stick to the blood vessel wall, forming plaques that narrow the lumen and impede blood flow.

The thickness and exact composition of each tunic varies according to vessel type. **Elastic arteries,** such as the aorta, are large diameter arteries that have more elastic fibers in their tunica media than other arteries. Elastic fibers stretch under greater pressure when blood is pumped into them during systole. During diastole, the recoiling of elastic fibers in elastic arteries continues blood flow. **Muscular arteries,** smaller in diameter than elastic arteries, have more smooth muscle and fewer elastic fibers in their tunica media than elastic arteries. However, they have a layer of elastic tissue, the **external elastic lamina,** between the tunica externa and the tunica media. Muscular arteries distribute blood to different areas of the body (e.g., brachial artery distributes blood to arm) and control blood flow to these areas by vasoconstriction or vasodilation. Muscular arteries branch into smaller and smaller arteries

that eventually form arterioles, small blood vessels from which capillaries branch. **Arterioles** have a tunica media containing mainly smooth muscle and few elastic fibers. These vessels control blood flow into capillaries and play a large role in controlling blood pressure.

Veins

Blood flows from capillaries into **venules** (*venule* = little vein) that drain into veins. Venule walls contain a tunica interna of endothelium only and a tunica media with a few smooth muscle fibers.

The walls of **veins** contain a tunica interna, tunica media, and a tunica externa. Compared with arterial vessels, the tunica interna and tunica media are thinner, and the tunica media contains few smooth muscle fibers and elastic fibers. Veins lack an internal and external elastic lamina. The tunica externa is the thickest of the three layers and like arterial vessels is composed of elastic and collagen fibers. The lumen of veins is larger than arteries and often appears collapsed in tissue sections.

(a) (b)

FIGURE 29.1 Transverse sections through a frog arteriole.

The blood pressure gradient in the veins is very small and often is not enough to overcome gravity. Muscular activity squeezes the veins and pushes blood toward the heart while venous **valves,** found in many veins, prevent the backflow of blood, especially in the limbs. Valves are tunica interna folds whose flaplike cusps point toward the heart. Faulty or incompetent valves allow backflow of blood and result in pooling of blood in veins, a condition called varicose veins.

ACTIVITY 1 Structure of Blood Vessels

1 In Figure 29.1 compare the lumen diameter, thickness of tunics, and shape of tunica interna in two sections of a frog arteriole. Norepinephrine (NE) was applied to one area of the arteriole to induce vasoconstriction. Identify which cross-section is vasoconstricted, (a) or (b).

2 Label the components of a muscular artery wall in Figure 29.2.

3 Label the components of a vein in Figure 29.3.

FIGURE 29.2 Structure of a muscular artery.

• basement membrane
• endothelium
• external elastic lamina
• internal elastic lamina
• smooth muscle
• tunica externa
• tunica interna
• tunica media

1 _____

2 _____

3 _____

4 _____

5 _____

6 _____

7 _____

8 _____

FIGURE 29.3 Structure of a vein.

• basement membrane
• endothelium
• smooth muscle
• tunica externa
• tunica interna
• tunica media
• valve

1 _____

2 _____

3 _____

4 _____

5 _____

6 _____

7 _____

ACTIVITY 2 Arteries and Veins

1 Identify and label the tunics of the blood vessels in Figure 29.4. Items in the list may be used more than once.

2 Examine a prepared microscope slide containing a transverse section of a small artery and vein.
- Using the low-power objective lens, focus and center one blood vessel in the field of view and switch to high power.
- Using the high-power objective lens, focus with fine adjustment knob only and examine the three tunics in the vessel. Note the thickness of the tunica media.
- Repeat procedure with other blood vessel.
- Decide which vessel is the artery and which is the vein.

FIGURE 29.4 Transverse section through small artery and vein.

- lumen of artery
- lumen of vein
- tunica externa
- tunica interna
- tunica media

1 _____

2 _____

3 _____

4 _____

5 _____

6 _____

7 _____

8 _____

Capillaries

Capillaries have the smallest diameter and thinnest walls of any blood vessels. Their lumen is so small that red blood cells can only pass through one at a time. Capillary walls are composed of a single layer of endothelial cells (simple squamous endothelium) supported by a basement membrane. The endothelial cells of capillaries can exhibit:

- **tight junctions**—fusion of plasma membranes of adjacent cells
- **intercellular clefts**—spaces between cells
- **fenestrations** (*fenestra* = window)—holes in plasma membrane covered by a diaphragm (thin membrane)
- **vesicles**

Only substances that are lipid soluble or have membrane carriers can cross capillary walls with tight junctions between cells. However, many other substances can cross capillary walls containing fenestrations, intercellular clefts, and vesicles. The size of the fenestrations and intercellular clefts between endothelial cells determines which molecules can cross. Molecules can also be transported across capillary walls by vesicles, a process called **transcytosis** (*trans* = across; *cyto* = cell).

Hydrostatic pressure (blood pressure) at the arterial end of the capillaries causes plasma to be filtered across the capillary wall into the interstitial fluid. The filtered plasma containing nutrients, hormones, and other substances is called interstitial fluid once it leaves the capillaries and enters the space around tissue cells. Blood osmotic pressure draws interstitial fluid into the venous end of the capillary (reabsorption) bringing fluid, metabolic wastes, hormones, or other substances into the blood. The combination of these opposing pressures causes some filtered plasma to remain in the interstitial fluid to bathe the tissue cells.

ACTIVITY 3 Capillaries

1 Examine the blood vessels in Figure 29.5. Identify the arteriole and a capillary. *Hint:* look at the thickness of the wall and the blood cells in the lumen.

2 Observe the capillaries in Figure 29.6(a) and (b) and label the electron micrographs as exhibiting a tight junction, fenestrations, an intercellular cleft, or vesicles.

250×

FIGURE 29.5 Photomicrograph capillary network.

- arteriole
- capillary

1 _____

2 _____

(a)

35,000×

(b)

19,000×

FIGURE 29.6 Transmission electron micrographs showing capillary walls.

- fenestration
- intercellular cleft
- vesicles involved in transcytosis

1 _____

2 _____

3 _____

B. Blood Pressure

Blood pressure, the pressure exerted by blood against blood vessel walls, is highest in the aorta and large elastic arteries, and decreases as the arteries branch and blood travels further from the heart. Blood pressure drops significantly in the arterioles and steadily decreases through capillaries, venules, and veins, and drops to zero in the right atrium. The difference in blood pressure between two areas of the circulatory system is the **blood pressure gradient.** The blood pressure gradient between the aorta and right atrium causes blood to flow through the systemic circulation, and the blood pressure gradient between the pulmonary trunk and the left atrium causes blood to flow through the pulmonary circulation.

With each ventricular contraction, blood pressure fluctuates in the large arteries (i.e., the aorta, pulmonary trunk, and other elastic and large muscular arteries). Blood pressure during ventricular systole, **systolic blood pressure,** is higher than blood pressure during ventricular diastole, or **diastolic blood pressure.** These fluctuations diminish within the arterioles, and little or no fluctuations in blood pressure between systole and diastole are observed in the capillaries and venous vessels.

The blood pressure gradient between veins and the right atrium is low, making it difficult for venous blood flow to overcome the force of gravity. Therefore, when standing with arms hanging down, blood may pool in the large veins of the limbs.

Blood Pressure Measurements

Arterial blood pressure is reported in millimeters of mercury (mm Hg) and can be measured directly by inserting a pressure transducer into a large artery or indirectly with a sphygmomanometer. A **sphygmomanometer** (SFIG-moe-mah-NOH-meh-ter), or blood pressure cuff, can be used to measure systolic and diastolic blood pressure in any large artery. Clinically, the brachial artery is used most often. The sphygmomanometer contains a pressure gauge attached to an inflatable rubber cuff that is connected by a rubber tube to a hand pump (rubber bulb) or automatic pump. Clinically, it is important to use the right size cuff—pediatric, standard, or large arm—to obtain an accurate reading. The pump is used to inflate the rubber cuff to a pressure greater than the systolic pressure. This puts pressure on the artery, flattens it, and stops blood flow in the artery. A stethoscope is placed over the brachial artery in the antecubital area, and the pressure is slowly released by opening a valve. When blood pressure is greater than the pressure in the cuff, the artery opens and blood flow

returns. The examiner listens for the sound caused by the turbulent flow of blood that can be heard until blood flow returns to normal. These sounds are called the **Korotkoff** (koh-ROT-koff) **sounds.** Although there is a range of values considered normal, the average normal systolic pressure (the first sound heard) is 120 mm Hg, and normal diastolic pressure (the last, faint sounds heard) is 80 mm Hg, usually written as 120/80.

Venous blood pressure can also be measured directly with a pressure transducer inserted into a venous vessel or indirectly. Indirect measurement of venous pressure described in Activity 4 are estimates at best, because blood pressure in veins is very low. The average venous blood pressure is 16 mm Hg.

ACTIVITY 4 **Blood Pressure Measurements**

1 Measure resting systemic arterial blood pressure.
 • Work in teams of two, alternating who will be the subject and who will measure blood pressure.
 • The subject should sit and rest for 5 minutes before blood pressure is measured.
 • Wipe the earpieces of the stethoscope with alcohol and completely deflate the blood pressure cuff.
 • Place the cuff of the sphygmomanometer around the arm as shown in Figure 29.7. If the cuff does not fit the arm, the blood pressure measurements will probably be inaccurate. The inflatable portion of the cuff should be over the anterior surface of the arm. If there is an arrow on the cuff, place it over the brachial artery. The brachial artery is located in the anterior arm, immediately superior to the antecubital fossa and lateral to biceps brachii. The bottom of the cuff should be approximately 1 inch above the elbow.
 • Close the valve on the rubber bulb.
 • Insert earpieces, place the large bell of the stethoscope over the brachial artery inferior to the cuff, and listen for the brachial pulse.
 • Inflate the cuff to 160 to 180 mm Hg.
 • Immediately use the valve on the hand pump to slowly release air to deflate the cuff, and listen for the first sound.
 • The first sound, the systolic pressure, is the return of blood flow through the partially occluded brachial artery. Watch the pressure gauge and continue to listen; the sound will increase, then muffle, and stop. The diastolic pressure is the pressure when the last, faint sound is heard.
 • Record blood pressure measurements in Table 29.1.
 • Deflate cuff and have subject rest for 2 minutes before repeating the blood pressure measurement. Record second value in Table 29.1.

FIGURE 29.7 Measuring arterial blood pressure with a sphygmomanometer.

TABLE 29.1 Resting Blood Pressures

SUBJECT	SYSTOLIC PRESSURE	DIASTOLIC PRESSURE	PULSE PRESSURE	MAP	VENOUS PRESSURE
Subject 1	1.	1.	1.	1.	1.
	2.	2.	2.	2.	2.
	Avg.	Avg.	Avg.	Avg.	Avg.
Subject 2	1.	1.	1.	1.	1.
	2.	2.	2.	2.	2.
	Avg.	Avg.	Avg.	Avg.	Avg.

2 Calculate pulse pressure and mean arterial pressure.
- The **pulse pressure** is the difference between the systolic and diastolic blood pressure. The larger the pulse pressure, the greater the volume of blood ejected from the ventricle. **Mean arterial pressure (MAP or MABP)** is the average blood pressure (BP) over the course of the cardiac cycle.

- Pulse pressure = systolic pressure − diastolic pressure. Record values in Table 29.1.
- Mean arterial pressure (MAP) = $$\text{diastolic BP} + \frac{\text{(pulse pressure)}}{3}.$$
- Record values in Table 29.1

3 Estimate systemic venous pressure.
- Decide who will be the subject, the observer, and the recorder.
- Have the subject stand with arms hanging at the side. Observe the blood pooling in the veins on the dorsal surface of the hands.
- Have the subject abduct one arm and observe the veins become smaller in size and flatten (collapse) as the arm is raised above the level of the right atrium. When the force of gravity exceeds venous pressure, the blood in the veins drains away quickly and does not exert enough pressure on the walls of the veins to keep them open.
- Lower the arm to pool blood in the veins of the hand again.
- Have the subject stand in front of a white (or black) board. With hands at the sides, locate the level of the right atrium (approximately 2 inches above the nipple). Mark on the board where the hand touches the board.
- Have the subject abduct the arm and hand until the veins on the surface of the hand flatten (collapse). Mark where the hand touches the board. Using a meterstick, measure the vertical distance above the right atrium that the hand has traveled in millimeters (1 cm = 10 mm): _____ mm. Each 12.88 mm elevation of the hand above heart level represents an approximately 1 mm rise in Hg.
- Venous pressure (mm Hg) =
$$\frac{\text{mm measured}}{\text{12.88 mm elevation/mm Hg}} = \underline{\hspace{1cm}} \text{ mm Hg}$$
- Repeat measurement and record in Table 29.1.

4 Observe the effect of body position on blood pressure.
- Decide who will be the subject, who will do the measuring, and who will record.
- Have subject lay on mat (supine position) for 2 minutes. Measure BP and heart rate (HR). Record in Table 29.2.
- Have subject stand and *immediately* measure BP and HR. Record in Table 29.2.
- Take the BP and HR again after subject has been standing for 2 minutes. Record in Table 29.2.
- After standing for 2 minutes, measure subject's BP and HR. Record in Table 29.2.
- Pool class data and calculate averages. Record in Table 29.2.

5 Answer discussion questions with your lab partners.

? DISCUSSION QUESTIONS

1 Did the subject's BP change when altering body position?

2 Did the subject's HR change when altering body position?

3 Explain what happens in regard to BP when a person changes from a supine to a standing position. Why does this occur?

Regulation of Blood Pressure

The body maintains blood pressure (BP) to ensure adequate blood flow to body tissues (tissue perfusion). If blood pressure is too low, tissue perfusion may not be sufficient to provide oxygen and nutrients to cells, especially brain neurons, or to remove metabolic wastes. If blood pressure is too high, it may cause damage to capillaries and the endothelial lining of blood vessels. Eye exams can often detect early hypertension. Damaged retinal vessels can be observed with an ophthalmoscope.

TABLE 29.2 Effect of Body Position on Blood Pressure

SUBJECT	SYSTOLIC/DIASTOLIC BP (mm Hg)	HR
Supine		
Immediately after standing		
After standing for 2 minutes		

Mean arterial blood pressure increases when cardiac output increases and when resistance to blood flow increases. Blood viscosity, total blood vessel length, and blood vessel diameter affect resistance to blood flow. However, blood vessel diameter is the only variable that can be changed to make immediate changes in blood pressure. Increasing blood vessel diameter (vasodilation) decreases resistance and lowers BP, while decreasing blood vessel diameter (vasoconstriction) increases resistance and BP rises.

ACTIVITY 5 Experiment: Effect of Exercise on Blood Pressure

1 **Prediction:** Circle the correct choice in each statement.
 - Exercise will cause systolic and diastolic BP to *increase* or *decrease*.
 - Exercise will cause pulse pressure to *increase* or *decrease*.
 - Exercise will cause MAP to *increase* or *decrease*.

2 **Materials:** Obtains materials for the regulation of blood pressure experiment from Materials list.

3 **Data Collection:** Measure the effect of exercise on BP and record the results in Table 29.3.
 - Decide who will be the subject, who will measure BP and heart rate (HR), who will be the recorder and timer. The subject should not have heart problems and should be in good health.

BP and HR at rest

- Have the subject stand at rest for 2 to 3 minutes.
- Measure systolic and diastolic BP and record the results in Table 29.3.
- Measure the number of heartbeats (or pulses) in 15 seconds and multiply by 4 to obtain HR. Record the results in Table 29.3.

BP and HR immediately after exercise

- Have the subject run in place for 5 minutes.
- Immediately measure systolic and diastolic BP and record the results in Table 29.3.

- Immediately count the number of heartbeats (or pulses) in 15 seconds and multiply by 4 to obtain HR. Record the results in Table 29.3.

4 **Data Analysis:**
 - Calculate resting and post exercise pulse pressures and MAP. Record in Table 29.3.
 - Pool class data, calculate averages, and record in Table 29.3.
 - If graph paper is available, graph your data.

5 Complete the **Experimental Report** with your lab group.

EXPERIMENTAL REPORT

Results

- State whether systolic and diastolic BP increased or decreased after exercise.
- State whether pulse pressure and MAP increased or decreased after exercise.
- State whether HR increased or decreased after exercise.

Discussion

- Why was the resting blood pressure taken standing rather than sitting?
- Discuss how body changes BP during exercise.
- Discuss why the body changes BP during exercise.
- Discuss importance of change in pulse pressure and MAP with exercise.

Conclusion

- Write a statement that postulates how increasing levels of exercise affect BP, pulse pressure, and MAP.

TABLE 29.3 Effect of Exercise on Blood Pressure

	SYSTOLIC BP (mm Hg)	DIASTOLIC BP (mm Hg)	HR	PULSE PRESSURE (mm Hg)	MAP (mm Hg)
Subject 1					
At Rest					
Post Exercise					
Class Average					
At Rest					
Post Exercise					

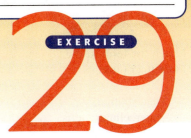

Reviewing Your Knowledge

A. Anatomy Drill and Practice

Go to www.wiley.com/college/allen and click on the picture of your lab manual. Then click on Student Companion Site, Anatomy Drill and Practice, and Exercise 29.

B. Structure of Arteries and Veins

Name the components of each blood vessel tunic. Terms may be used more than once.

smooth muscle
endothelium
elastic fibers
collagen fibers

_____ **1.** Tunica media of arteries

_____ **2.** Tunica media of veins

_____ **3.** Tunica interna of arteries

_____ **4.** Tunica interna of veins

_____ **5.** Tunica externa of arteries

_____ **6.** Tunica externa of veins

Circle True or False for the following questions. If false, underline and change word(s) that are incorrect to make statement true.

7. The tunica media of veins is thicker than the tunica media of arteries. True or False

8. The tunica media of elastic arteries contains more elastic fibers than muscular arteries. True or False

9. Venous valves are folds of the tunica externa. True or False

10. Venous valves prevent backflow of blood. True or False

11. Walls of veins are thicker than the walls of arteries of the same size. True or False

12. Lumens of veins are larger than lumens of arteries of the same size. True or False

C. Structure of Capillaries

Match the term to the correct description. More than one term may apply to each description. Terms may be used more than once.

fenestrations
intercellular clefts
tight junctions
transcytosis

_____ **1.** Holes in plasma membrane through which molecules pass across capillary walls

_____ **2.** Fusion of plasma membranes of adjacent endothelial cells; forms very selective barrier

_____ **3.** Vesicles transport substances across capillary wall

_____ **4.** Spaces between cells through which substances pass

Circle True or False for the following questions. If false, underline and change word(s) that are incorrect to make statement true.

5. Capillary walls are composed of an endothelium and a basement membrane only. True or False

6. Hydrostatic pressure forces plasma across capillary walls at venous end of capillary, and interstitial fluid enters arterial end of capillary by osmotic pressure. True or False

D. Blood Pressure

Fill in the blank with the correct term.

_____ **1.** Term used for arterial pressure during ventricular systole.

_____ **2.** Term used for arterial pressure during ventricular diastole.

_____ **3.** Device used to measure arterial blood pressure in the brachial artery.

_____ **4.** Average normal adult arterial blood pressure.

_____ **5.** Average venous blood pressure.

_____ **6.** Sounds of turbulent blood flow that occur when blood flow resumes in an artery that has been occluded.

Circle True or False for the following questions. If false, underline and change word(s) to make statement true.

7. The blood pressure gradient from the aorta to the capillaries is greater than the blood pressure gradient from the venules to the right atrium. True or False

8. Blood pressure gradient from the aorta to the capillaries is less than the blood pressure gradient from the arterial end of the capillary to the venous end of the capillary. True or False

9. Kate's systolic blood pressure is 115 and diastolic is 72. Her pulse pressure is 47. True or False

10. Scott's blood pressure is 126/83 and his MAP is 97. True or False

E. Regulation of Blood Pressure and Blood Flow

Circle True or False for the following questions. If false, underline and change word(s) that are incorrect to make statement true.

1. Increasing heart rate increases blood pressure. True or False

2. Systemic vasoconstriction decreases blood pressure. True or False

3. Increasing arterial blood pressure increases blood flow. True or False

4. Recoil of muscular arteries maintains blood flow during ventricular diastole. True or False

5. Muscular arteries control the blood flow to different body areas. True or False

6. Vasoconstriction of the renal arteries (arteries supplying blood to kidneys) would decrease blood flow to the kidneys. True or False

7. Blood pressure is higher in the supine position than in the standing position. True or False

8. Blood pressure decreases when going from a supine to standing position. True or False

9. The greater the pulse pressure, the lower the pressure gradient driving blood from the aorta through the systemic circulation. True or False

10. Exercise increases MAP. True or False

Using Your Knowledge

A. Blood Vessel Structure and Function

1. In coronary bypass surgery, a section of vein is used to replace (bypass) occluded coronary arteries. Over time, the vein wall becomes more like an arterial wall. Describe the changes that would occur in the vein wall.

2. The smooth muscle fibers within the wall of an artery do not receive their nutrients from the blood within the lumen of the artery. Why not?

3. Blood flow within the capillaries is slower than arterial or venous vessels. Explain why this is important.

4. Blood flow to the skeletal muscles increases with exercise. Explain how the autonomic nervous system mediates the increase in blood flow to skeletal muscles.

5. Explain how blood flow to the kidneys is decreased with exercise.

6. Explain why the differences in the amount of elastic fibers and smooth muscle fibers in the tunica media of elastic arteries and muscular arteries is important to their function.

7. Why does the tunica interna of the vasoconstricted section of the arteriole in Figure 29.1 exhibit folds?

8. Capillaries within the brain have tight junctions, whereas capillaries within the anterior pituitary and other endocrine glands contain fenestrations, intercellular clefts, and vesicles for transcytosis. Explain how these structural differences are important to the function of the brain and the anterior pituitary.

9. Compare the strength of arterial and capillary walls.

10. Marfan's syndrome is an inherited genetic disorder that results in malformed elastic fibers that are weak. What effect would this have on the aorta?

Blood Vessel Identification

30

Objectives

**After completing this exercise,
you should be able to:**

- Identify the major arteries and veins of the systemic circulation on models or charts
- Identify the major vessels of the cerebral arterial circle (circle of Willis), pulmonary circulation, hepatic portal system, and fetal circulation on models or charts
- Trace blood flow through the cardiovascular system
- Identify selected blood vessels on the cat or fetal pig

Materials

- models or charts showing human arteries, veins, and circulatory pathways
- preserved cats or fetal pigs, disposable gloves, dissecting equipment, and safety glasses, accompanying cat or fetal pig dissection manual

Arteries carry blood from the heart to capillaries where gas and nutrient exchange occurs, while veins carry blood back to the heart. There are two circulatory routes within the body: the pulmonary circulation and the systemic circulation. In the **pulmonary circulation,** the arteries carry oxygen-poor blood from the right ventricle to the lungs where gas exchange occurs, and the veins carry oxygen-rich blood back to the left atrium. In the **systemic circulation,** the arteries carry oxygen-rich blood to body tissues, and the veins return oxygen-poor blood to the heart where it enters the pulmonary circulation.

Anastomoses provide alternative routes or detours for blood flow in case the main route is blocked or damaged. Arterial anastomoses are blood vessels that connect arteries, while venous anastomoses connect veins.

A. Systemic Arteries

All major arteries of the systemic circulation branch off the aorta. These arteries transport blood to different regions of the body where blood is distributed to smaller arteries and finally to arterioles that connect to capillaries.

Major Arteries of the Ascending Aorta and Aortic Arch

Blood is ejected from the left ventricle and into the **ascending aorta.** The **right** and **left coronary arteries** branch off the ascending aorta near its origin and travel to the heart. The ascending aorta curves superiorly toward the left and becomes the **aortic arch.** Three major arteries branch off the aortic arch in the following order from right to left: the **brachiocephalic trunk,** the **left common carotid artery,** and the **left subclavian artery.** The brachiocephalic trunk divides to form the **right common carotid** and **right subclavian arteries.**

ACTIVITY 1 Major Arteries of the Ascending Aorta and Aortic Arch

1 Label the arteries in Figure 30.1 using the location described in Table 30.1.

2 Point to the arteries on a model or chart.

3 Pronounce the names as you point to them.

TABLE 30.1 Major Arteries of the Ascending Aorta and Aortic Arch

ARTERY	LOCATION
Ascending aorta	Begins at left (aortic) semilunar valve and becomes the aortic arch at the level of the sternal angle
Right and left coronary arteries	Branch off ascending aorta just superior to the left aortic valve and end when they divide into their respective arterial branches
Aortic arch	Begins at the level of the sternal angle and becomes the thoracic aorta at the level of the intervertebral disc between the fourth and fifth vertebrae
Brachiocephalic trunk	First branch off aortic arch, divides into right subclavian artery and right common carotid artery at level of sternoclavicular joint
Right common carotid artery	Formed from division of brachiocephalic trunk and divides into the right external and right internal carotid arteries at the superior border of the larynx
Right subclavian artery	Formed from division of brachiocephalic trunk and is renamed axillary artery within axillary region
Left common carotid artery	Second branch off aortic arch, divides into the left external and left internal carotid arteries at the superior border of the larynx
Left subclavian artery	Third branch off aortic arch, renamed axillary artery within the axillary region

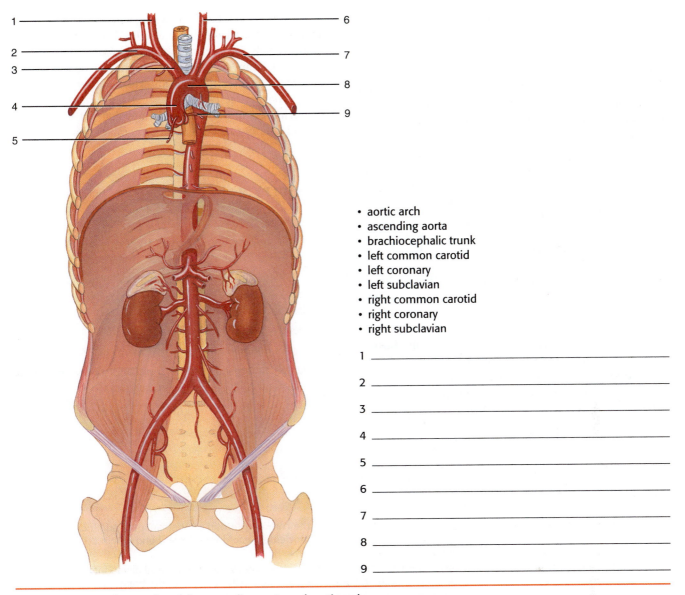

1 _____

2 _____

3 _____

4 _____

5 _____

- aortic arch
- ascending aorta
- brachiocephalic trunk
- left common carotid
- left coronary
- left subclavian
- right common carotid
- right coronary
- right subclavian

1 _____

2 _____

3 _____

4 _____

5 _____

6 _____

7 _____

8 _____

9 _____

FIGURE 30.1 Major arteries of the ascending aorta and aortic arch.

Major Arteries Supplying the Head (Including Circle of Willis)

The common carotid arteries branch to form the internal and external carotid arteries. The **internal carotid arteries** travel through the neck and enter the skull through the carotid canal within each temporal bone and supply structures within the skull. Each internal carotid artery branches to form an **anterior cerebral artery** and a **middle cerebral artery** that supply blood to these areas of the brain. The **external carotid arteries** ascend along the lateral surface of the neck and terminate as two arteries near the temporomandibular joint that supply structures external to the skull.

There are two **vertebral arteries,** one branching off each subclavian artery. The vertebral arteries travel superiorly through the transverse foramen of the cervical vertebrae and then through the foramen magnum to reach the base of the brain where they merge to form the **basilar artery.** The **posterior cerebral artery** is a branch of the basilar artery supplying the posterior part of the brain.

Blood supply to the brain is vital, and the configuration of the blood vessels supplying the brain ensures continuous blood flow even when blockage occurs in a blood vessel. The **cerebral arterial circle** (circle of Willis) is a ring of blood vessels formed by three anastomoses—the anterior communicating artery and the right and left posterior communicating arteries. The **anterior communicating arteries** connect the two anterior cerebral arteries, and the **posterior communicating arteries** connect the internal carotid arteries to the posterior cerebral arteries.

ACTIVITY 2 **Major Arteries Supplying the Head**

1 Label the arteries in Figure 30.2(a) using the location described in Table 30.2.

2 Label the arteries forming the cerebral arterial circle (circle of Willis) in Figure 30.2(b) using the location described in Table 30.2.

3 Identify the arteries on a model or chart.

4 Pronounce the names as you point to them.

5 Palpate the right common carotid artery pulse by placing your fingers just lateral to the larynx.

TABLE 30.2 Major Arteries Supplying the Head

ARTERY	LOCATION
External carotid arteries	Begin as terminal branches off the common carotid arteries near the superior border of the larynx, anterior to the sternocleidomastoid muscle, and end within the parotid gland
Internal carotid arteries	Begin as terminal branches off the common carotid arteries and divide into the anterior and middle cerebral arteries below the inferior surface of the brain
Anterior cerebral arteries	Branch off internal carotid arteries that pass anterior to the optic chiasma and run along the midline of the base of the brain
Middle cerebral arteries	Branch off internal carotid arteries that run toward the lateral surface of the base of the brain
Vertebral arteries	First major branch off subclavian artery that ends near inferior surface of brain where it forms the basilar artery
Basilar artery	Formed when right and left vertebral arteries merge along the inferior surface of the medulla oblongata and divides to form the right and left posterior cerebral arteries near the superior border of pons along the base of the brain
Posterior cerebral arteries	Formed when basilar artery divides and travels along the base of the brain toward the posterior surface of the cerebrum
Anterior communicating artery	Connects right and left anterior cerebral arteries anterior to the optic chiasma
Right and left posterior communicating arteries	Connect posterior cerebral arteries and internal carotid arteries

- anterior cerebral
- basilar
- brachiocephalic trunk
- common carotid
- external carotid
- internal carotid
- middle cerebral
- posterior cerebral
- posterior communicating
- vertebral

Right internal carotid

1 _____

2 _____

3 _____

4 _____

5 _____

6 _____

7 _____

8 _____

9 _____

10 _____

(a) Right lateral view of major arteries of the head and neck

- anterior cerebral
- anterior communicating
- basilar
- internal carotid
- middle cerebral
- posterior cerebral
- posterior communicating
- vertebral

11 _____

12 _____

13 _____

14 _____

15 _____

16 _____

17 _____

18 _____

(b) Inferior view of brain showing cerebral arterial circle

FIGURE 30.2 Arteries of the head and neck.

Major Arteries of the Upper Extremities

The subclavian artery is renamed the **axillary artery** within the axilla and continues as the **brachial artery** as it enters the arm. As the brachial artery enters the forearm, it divides to form the **radial artery** and the **ulnar artery**. The **superficial palmar arch** and the **deep palmar arch**, both of which supply blood to the fingers and palms, connect the radial and ulnar arteries at their distal ends and are examples of arterial anastomoses.

ACTIVITY 3 · Arteries of the Upper Extremities

1 Label the arteries in Figure 30.3 using the location described in Table 30.3.

2 Identify the arteries on a model or chart.

3 Pronounce the names as you point to them.

4 Palpate the radial artery by placing the 2nd and 3rd fingers over the distal end of the radius.

TABLE 30.3 · Arteries Supplying the Upper Extremities

ARTERY	LOCATION
Subclavian arteries	See Table 30.1.
Axillary arteries	Continuation of subclavian artery that begins at the outer border of 1st rib and is renamed the brachial artery as it enters the arm
Brachial arteries	Continuation of axillary artery that begins near the tendon of the teres major muscle and ends just distal to the elbow where it divides into the radial and ulnar arteries
Radial arteries	Terminal branch of brachial artery that runs along lateral surface of forearm into the wrist and hand
Ulnar arteries	Terminal branch of brachial artery that runs along medial surface of forearm into the wrist and hand
Deep palmar arch	Anastomoses that connect radial and ulnar arteries; located deep to the tendons of flexor muscles and proximal to superficial palmar arch
Superficial palmar arch	Anastomoses that connect radial and ulnar arteries; located superficial to flexor tendons and distal to deep palmar arch

1

2

3

4

5

Deep palmar arch

Superficial palmar arch

FIGURE 30.3 Arteries supplying the upper extremities.

- axillary
- brachial
- radial
- subclavian
- ulnar

1 _____

2 _____

3 _____

4 _____

5 _____

Major Arterial Branches of the Descending Aorta

The diaphragm divides the aorta into thoracic and abdominal sections. The **thoracic aorta** lies to the left of the midline, just anterior to the vertebral column, and has branches that supply thoracic structures. Nine pairs of **posterior intercostal arteries** branch off the thoracic aorta to follow the ribs. The abdominal aorta is also anterior to the vertebral column but closer to the midline. It supplies the abdomen, pelvis, and lower extremities.

The major branches of the descending aorta are the **celiac trunk** (which divides into the left gastric artery, splenic artery, and common hepatic artery), the **superior mesenteric artery,** the paired **suprarenal arteries,** the paired **renal arteries,** the paired **gonadal arteries** (**ovarian arteries** in females and **testicular arteries** in males), the **inferior mesenteric artery,** and four pairs of **lumbar arteries.** Within the pelvis, the abdominal aorta divides into the **common iliac arteries** that are the terminal branches of the abdominal aorta.

ACTIVITY 4 Arterial Branches of the Descending Aorta

1 Label the arteries in Figures 30.4 and 30.5 using the location described in Table 30.4.

2 Identify the arteries on a model or chart.

3 Pronounce the names as you point to them.

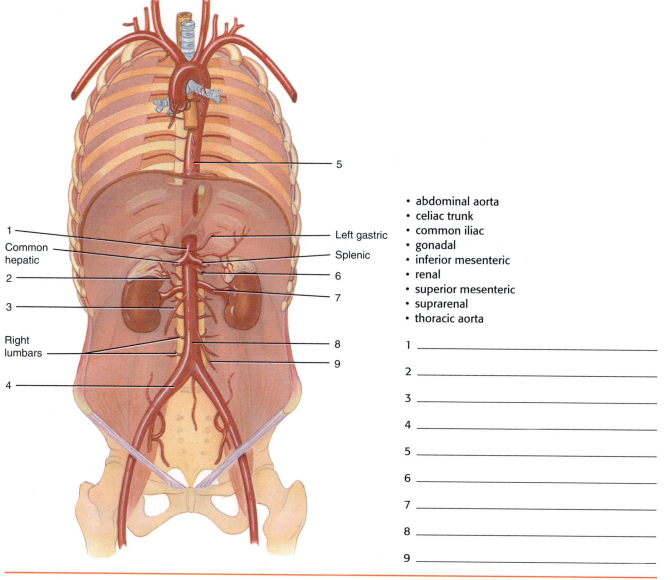

- abdominal aorta
- celiac trunk
- common iliac
- gonadal
- inferior mesenteric
- renal
- superior mesenteric
- suprarenal
- thoracic aorta

1 _____

2 _____

3 _____

4 _____

5 _____

6 _____

7 _____

8 _____

9 _____

FIGURE 30.4 Descending aorta and major arterial branches.

- celiac trunk
- common hepatic
- left gastric
- splenic

1 _____

2 _____

3 _____

4 _____

FIGURE 30.5 Anterior view of celiac trunk and its arterial branches.

TABLE 30.4 Major Arterial Branches of the Descending Aorta

ARTERY	LOCATION
Thoracic aorta	Continuation of aortic arch that starts at the level of the intervertebral disc between the 4th and 5th thoracic vertebrae and becomes the abdominal aorta at the diaphragm
Abdominal aorta	Continuation of the thoracic aorta that begins at the diaphragm and ends at the level of the 4th lumbar vertebra where it divides into the right and left common iliac arteries
Celiac trunk	Short branch off abdominal aorta just inferior to the diaphragm that immediately divides into three arteries: the left gastric, splenic, and common hepatic arteries
Left gastric artery	Smallest of the three celiac trunk branches, it comes off the left anterior side and courses to the left toward the esophagus and stomach
Splenic artery	Largest of the three celiac trunk branches, it comes off the left anterior side and courses to the left toward the spleen and branches into three arteries: the pancreatic artery, the left gastroepiploic artery, and the short gastric artery
Common hepatic artery	Intermediate of three celiac trunk branches, it comes off the right anterior side and courses toward the right and branches into three arteries
Superior mesenteric artery	Arises off the anterior surface of the abdominal aorta just inferior to the celiac trunk at the level of the first lumbar vertebra; it has many anastomoses and branches into five arteries
Right and left suprarenal arteries	Branch off the lateral surface of the abdominal aorta at the level of the first lumbar vertebra and course toward the adrenal glands
Right and left renal arteries	Branch off the lateral surface of the abdominal aorta at the level of the second lumbar vertebra and course toward the kidneys
Right and left gonadal arteries	Branch off the anterior surface of the abdominal aorta inferior to the renal arteries; ovarian arteries course toward the ovaries and testicular arteries pass through the inguinal canal toward the testes
Inferior mesenteric artery	Branches off the abdominal aorta at the level of the third lumbar vertebra; it has many anastomoses and branches into three arteries
Right and left common iliac arteries	Begin at division of abdominal aorta at level of 4th lumbar vertebrae and terminate into internal and external iliac arteries

Major Arteries of the Pelvis and Lower Extremities

The **common iliac arteries** divide into the internal iliac and external iliac arteries. The **internal iliac arteries** are medial terminal branches off the common iliac arteries just anterior to the lumbosacral joint. They course posteriorly to the pelvis. The **external iliac artery,** the larger of the two arteries, becomes the femoral artery as it enters the thigh. The **femoral artery** descends along the middle of the anterior two-thirds of the thigh and travels to the posterior thigh, becoming the popliteal artery as it enters the posterior knee area. The **popliteal artery** divides into the **anterior** and **posterior tibial arteries.** The anterior tibial artery travels to the anterior surface of the leg and descends to the ankles where it becomes the **dorsalis pedis artery.** The **posterior tibial artery** descends along the posterior aspect of the leg and terminates at the ankle where it divides into the **medial** and **lateral plantar arteries.** The **fibular (peroneal) artery** branches off the posterior tibial artery and travels down the lateral side of the leg to the foot.

ACTIVITY 5 **Major Arteries of the Pelvis and Lower Extremities**

1 Label the arteries in Figure 30.6 using the location described in Table 30.5.

2 Identify the arteries on a model or chart.

3 Pronounce the names as you point to them.

4 Palpate the femoral artery pulse by placing your fingertips over the inguinal area.

5 Palpate the popliteal artery pulse by placing your fingertips on the posterior surface of the knee.

6 Palpate the pulse of the dorsal artery of the foot (dorsalis pedis artery) by placing your fingertips over the dorsal surface of the medial side of foot near the ankle.

TABLE 30.5 **Major Arteries of the Pelvis and the Lower Extremities**

ARTERY	LOCATION
Internal iliac arteries	Terminal branches of common iliac arteries that begin at the level of the intervertebral disc between vertebrae L_5 and S_1 and travel posteriorly until they divide in the pelvis
External iliac arteries	Terminal branches of common iliac arteries that run along the psoas major muscles and become femoral arteries posterior to the inguinal ligaments
Femoral arteries	Begin posterior to inguinal ligaments; descend along the midline of the anterior two-thirds of the thighs; travel to the posterior thigh and become the popliteal arteries in the popliteal region of the lower extremity
Popliteal arteries	Continuation of femoral arteries that begin in the popliteal region and end in the upper leg where they divide into the anterior and posterior tibial arteries
Anterior tibial arteries	Terminal branches of popliteal arteries that begin in the upper posterior leg and travel to the anterior leg. They end in the ankles where they become the dorsal arteries of the foot (dorsalis pedis arteries).
Dorsal arteries of the foot (dorsalis pedis arteries)	Continuation of anterior tibial arteries that begin in the ankles and form branches within the foot
Posterior tibial arteries	Terminal branches of popliteal arteries that begin in the upper leg and terminate at the medial malleolus of the tibia where they divide into the medial and lateral plantar arteries
Fibular (peroneal) arteries	Branch off the posterior tibial artery slightly distal to the point where the popliteal artery divides into the anterior and posterior tibial arteries; descends down the lateral surface of the leg, and terminates within the foot

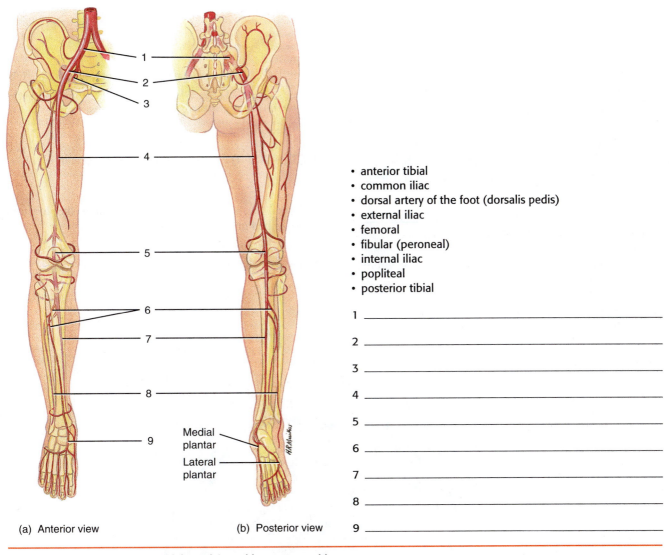

(a) Anterior view (b) Posterior view

- anterior tibial
- common iliac
- dorsal artery of the foot (dorsalis pedis)
- external iliac
- femoral
- fibular (peroneal)
- internal iliac
- popliteal
- posterior tibial

1 _____

2 _____

3 _____

4 _____

5 _____

6 _____

7 _____

8 _____

9 _____

FIGURE 30.6 Major arteries of the pelvis and lower extremities.

B. Systemic Veins

Veins carry blood from the capillaries back toward the heart. Veins can be found under the skin (superficial veins) or deep within the body, typically near an artery. Veins that accompany arteries usually have the same name as the artery. Veins carrying blood from the lungs to the left side of the heart carry oxygen-rich blood and are part of the pulmonary circulation, whereas veins carrying blood from all other body tissues carry oxygen-poor blood and are part of the systemic circulation.

> **CLINICAL NOTE:** *A thrombus in a deep vein can cause a pulmonary embolus if not treated with anticoagulant, whereas a thrombus in a superficial vein does not tend to form an embolus.*

Veins Carrying Blood to the Right Atrium of the Heart

Three large veins carry oxygen-poor blood into the right atrium—the **superior vena cava**, the **inferior vena cava,** and the **coronary sinus.**

ACTIVITY 6 Veins Carrying Blood to the Heart

1 Label the veins in Figure 30.7 using the location described in Table 30.6.

2 Identify the veins on a model or chart.

3 Pronounce the names as you point to them.

FIGURE 30.7 Systemic veins carrying blood to the heart.

- coronary sinus
- inferior vena cava
- superior vena cava

1 _____

2 _____

3 _____

TABLE 30.6 Veins Carrying Blood to the Heart

VEIN	LOCATION
Superior vena cava	Formed from the merger of the right and left brachiocephalic veins at the level of the first costal cartilage and ends where it enters the right atrium at the level of the third costal cartilage
Coronary sinus	Formed from the merger of the great, middle, and small cardiac veins, located in the coronary sulcus, and empties into the right atrium
Inferior vena cava	Formed from the merger of the common iliac veins at the level of the fifth lumbar vertebra and ascends behind the peritoneum just right of the midline and empties into the right atrium

Major Veins Draining the Head and Neck

Three large, paired veins—the internal jugular veins, the external jugular veins, and the vertebral veins—drain blood from the head and neck. The **internal jugular veins** run lateral to the internal carotid and common carotid arteries, the **external jugular veins** are superficial veins descending along the lateral surface of the neck, and the **vertebral veins** descend through the transverse foramina of the vertebral column with the vertebral arteries.

ACTIVITY 7 Major Veins Draining the Head and Neck

1 Label the veins in Figure 30.8 using the location described in Table 30.7.

2 Identify the veins on a model or chart.

3 Pronounce the names as you point to them.

- brachiocephalic
- external jugular
- internal jugular
- subclavian
- vertebral

1 _____

2 _____

3 _____

4 _____

5 _____

FIGURE 30.8 Major veins of the head and neck.

TABLE 30.7 Major Veins of the Head and Neck

VEIN	LOCATION
Internal jugular veins	Begin at the jugular foramina; run lateral to the internal carotid and common carotid arteries; merge with the subclavian veins to form the brachiocephalic veins
External jugular veins	Begin in the occipital region and the parotid gland; descend through the neck along the sternocleidomastoid muscle, and end near the middle of the clavicle where they drain into the subclavian veins
Vertebral veins	Originate inferior to the occipital condyles; travel through the transverse foramina of the 1st through 6th cervical vertebrae, and drain into the brachiocephalic veins

Major Veins Draining the Upper Extremities

Superficial veins are visible running beneath the skin of the arm. As we age, these veins become more visible. The major superficial veins of the upper extremities are the basilic and the cephalic veins. Each arm has a **cephalic vein** that travels along the anteriolateral surface of the entire limb and merges with the axillary vein inferior to the clavicle. Each **basilic vein** travels along the posteromedial surface of the forearm and the anteromedial surface of the arm. Each **median cubital vein,** a common site for obtaining blood samples, is an anastomosis anterior to the elbow that connects the basilic and cephalic veins.

The *major deep veins* of the arm and forearm run along with arteries and are named for the arteries they accom-

pany. The **radial** and **ulnar veins,** found in each forearm, merge to form the **brachial veins.** The brachial veins ascend the arm and join with the basilic vein to form the **axillary vein.** The axillary vein becomes the **subclavian vein** as it leaves the axillary region.

ACTIVITY 8 Major Veins Draining the Upper Extremities

1 Label the veins in Figure 30.9 using the description supplied for superficial veins in Table 30.8 and for deep veins in Table 30.9.

2 Identify the veins on models or charts.

3 Pronounce the names as you point to them.

- axillary
- basilic
- brachial
- cephalic
- median cubital
- radial veins
- subclavian
- ulnar veins

1 _____

2 _____

3 _____

4 _____

5 _____

6 _____

7 _____

8 _____

FIGURE 30.9 Major veins of the right upper extremity.

TABLE 30.8 **Superficial Veins Draining the Upper Extremities**

VEIN	LOCATION
Cephalic veins	Begin at lateral edge of the dorsal venous arches (superficial veins of the hand); travel laterally along the anterior surface of the entire limb, and end inferior to the clavicle where they drain into the axillary vein
Basilic veins	Begin at the medial edge of the dorsal venous arches and travel medially along the posterior surface of the forearm and the anterior surface of the arm; in the middle of the arm, this vein travels deep and eventually merges with the brachial vein in the axillary region to form the axillary vein
Median cubital veins	Travel laterally across the anterior surface of the elbow to connect the basilic and cephalic veins

TABLE 30.9 **Deep Veins Draining the Upper Extremities**

VEIN	LOCATION
Radial veins	From venous palmar arches, travel along lateral forearm with the radial arteries, and end at brachial veins
Ulnar veins	From superficial palmar venous arches, travel along medial forearm with ulnar arteries, and end at brachial veins
Brachial veins	From merger of radial and ulnar veins and run superiorly until merge with basilic veins to form axillary veins
Axillary veins	From merger of basilic and brachial veins and run with axillary arteries until they become subclavian veins at the border of first ribs
Subclavian veins	Continuation of axillary veins that begin at the border of first ribs; end at the sternal end of the clavicle where they join with the internal jugular veins to form the brachiocephalic veins

Major Veins Draining the Thorax, Abdomen, and Pelvis (Including Hepatic Portal Circulation)

The **azygos system** of veins includes the **azygos vein,** the **hemiazygos vein,** and the **accessory hemiazygos vein.** These veins drain most thoracic structures and the abdominal wall. They also form anastomoses with the large veins draining the lower limbs and abdomen. Therefore, the azygous system can serve as a bypass if the inferior vena cava becomes obstructed.

The **internal** and **external iliac veins** in the pelvis merge to form the **common iliac** veins that unite to form the **inferior vena cava.** The **lumbar veins, gonadal veins, renal veins, suprarenal veins,** and **hepatic veins** drain directly into the inferior vena cava.

The veins draining the stomach, intestines, spleen, pancreas, and gallbladder do not drain directly into the inferior vena cava but enter the hepatic portal circulation first. The major veins of the hepatic circulation are the inferior mesenteric vein, the splenic vein, the superior mesenteric vein, and the hepatic portal vein. The **inferior mesenteric vein,** which drains blood from the large intestine, joins the **splenic vein,** which carries blood from the stomach, pancreas, and spleen.

The **superior mesenteric vein** drains blood from the small intestine and merges with the splenic vein to form the **hepatic portal vein,** which carries nutrient-rich blood to the liver for processing. The liver is drained by the **hepatic veins** that empty into the inferior vena cava.

ACTIVITY 9 **Major Veins Draining the Thorax, Abdomen, and Pelvis**

1 Label the veins in Figure 30.10 using the description supplied in Tables 30.10 and 30.11.

2 Label the veins in Figure 30.11 using the location described in Table 30.11.

3 Identify the veins on models or charts.

4 Pronounce the names as you point to them.

- accessory hemiazygos
- azygos
- hemiazygos
- hepatic veins
- inferior vena cava
- left brachiocephalic
- left common iliac
- left external iliac
- left gonadal
- left internal iliac
- left renal
- left suprarenal
- right brachiocephalic
- right common iliac
- right external iliac
- right gonadal
- right internal iliac
- right renal
- right suprarenal
- superior vena cava

1 _____

2 _____

3 _____

4 _____

5 _____

6 _____

7 _____

8 _____

9 _____

10 _____

11 _____

12 _____

13 _____ 17 _____

14 _____ 18 _____

15 _____ 19 _____

16 _____ 20 _____

FIGURE 30.10 Major veins draining the thorax, abdomen, and pelvis.

TABLE 30.10 Veins Draining the Thorax

VEIN	LOCATION
Right and left brachio-cephalic veins (*brachi-* = arm; *cephalic* = head)	Formed by the merger of the subclavian and internal jugular veins and unite to form the superior vena cava
Azygos vein	Forms near the diaphragm; runs anterior and to the right of the vertebral column, and ends in the superior vena cava
Hemiazygos vein	Often begins just below the diaphragm; runs anterior and to the left of the vertebral column, and ends at the level of the 9th thoracic vertebra where it joins the azygous vein
Accessory hemiazygos vein	Superior to the hemiazygous vein; runs anterior and to the left of the vertebral column from the level of the 5th thoracic vertebra to the 8th, joining the azygous vein

TABLE 30.11 Veins Draining the Abdomen and Pelvis

VEIN	LOCATION
Common iliac veins	Formed by the merger of the internal and external iliac veins near the sacroiliac joint and end when both common iliac veins form the inferior vena cava
Internal iliac veins	Begin near the greater sciatic notch and drain into the common iliac veins
External iliac veins	Continuations of femoral veins that are renamed within the inguinal region; drain into the common iliac veins
Gonadal veins	Run adjacent to the gonadal arteries in the posterior abdomen; left gonadal veins drain into the left renal vein; right gonadal veins drain into the inferior vena cava
Renal veins	Run anterior to renal arteries and drain into the inferior vena cava
Suprarenal veins	Run adjacent to the suprarenal arteries and drain into the inferior vena cava
HEPATIC PORTAL CIRCULATION	
Inferior mesenteric vein	Formed from the fanlike network of veins from the inferior mesentery; drains into splenic vein and joins the splenic vein near the pyloric region of the stomach to form the hepatic portal vein
Splenic vein	Drains the spleen and joins the superior mesenteric vein near the inferior region of the stomach to form the hepatic portal vein
Superior mesenteric vein	Formed from the fanlike network of veins of the superior mesentery; joins the splenic vein near the inferior region of the stomach to form the hepatic portal vein
Hepatic portal vein	Begins at merger of splenic vein and superior mesenteric vein and ends in the liver
Hepatic veins	Multiple hepatic veins form in the liver and drain into the inferior vena cava slightly inferior to the diaphragm

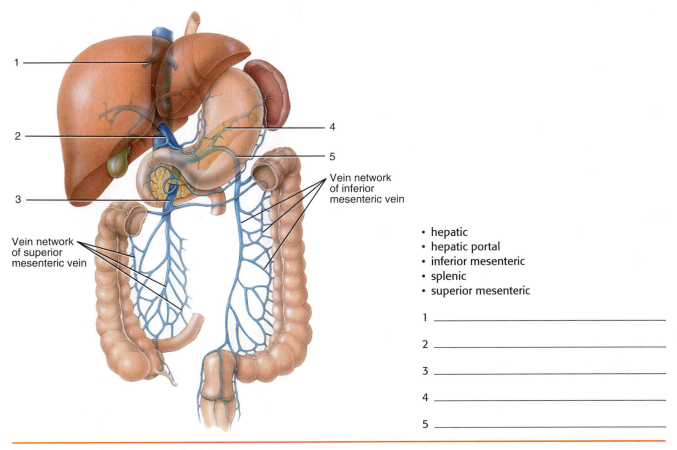

Vein network of inferior mesenteric vein

Vein network of superior mesenteric vein

• hepatic
• hepatic portal
• inferior mesenteric
• splenic
• superior mesenteric

1 _____

2 _____

3 _____

4 _____

5 _____

FIGURE 30.11 Hepatic portal circulation.

Major Veins Draining the Lower Extremities

The great saphenous vein and the small (short) saphenous vein are the major *superficial veins* of the leg. The **great saphenous vein** is the longest vein in the body, running along the medial surface of the leg and thigh, and contains between 10 and 20 valves to prevent backflow of blood. The **small saphenous vein** ascends along the lateral posterior surface of the leg. Varicose veins are common in both of these veins.

Deep veins of the leg ascend in the leg adjacent to the arteries of the same name. The **anterior** and paired **posterior tibial veins** ascend in the anterior and posterior leg, and unite inferior to the popliteal fossa to form the popliteal vein. The paired **fibular (peroneal) veins** travel superiorly along the lateral leg and join the posterior tibial veins. The **popliteal vein** ascends along the posterior surface of the knee and becomes the **femoral vein,** which travels up the posterior thigh and becomes the **external iliac vein** in the pelvis.

ACTIVITY 10 Major Veins Draining the Lower Extremities

1 Label the veins in Figure 30.12 using the description of veins in Table 30.12.
2 Identify the veins on models or charts.
3 Pronounce the names as you point to them.

• anterior tibial
• external iliac
• femoral
• great saphenous
• posterior tibial
• small saphenous

1 _____

2 _____

3 _____

4 _____

5 _____

6 _____

• femoral
• fibular (peroneal)
• great saphenous
• popliteal
• posterior tibial
• small saphenous

7 _____

8 _____

9 _____

10 _____

11 _____

12 _____

(a) Anterior view (b) Posterior view

FIGURE 30.12 Veins of the right lower limb.

TABLE 30.12 Veins of the Lower Extremities

VEIN	LOCATION
Great saphenous veins	Superficial veins that form at the medial end of the dorsal venous arches (venous networks on dorsal surface of foot); run anterior to the medial malleolus; ascend along the medial leg and thigh, and drain into the femoral veins in the groin area
Small saphenous veins	Superficial veins that form at the lateral end of the dorsal venous arches; run posterior to the lateral malleolus; ascend along the lateral posterior leg, and empty into the popliteal veins
Posterior tibial veins	Paired deep veins formed from plantar veins draining the foot posterior to the medial malleolus; ascend deep within the posterior leg accompanying the posterior tibial arteries, and merge with the anterior tibial veins just inferior to the popliteal fossa to form the popliteal veins
Fibular (peroneal) veins	Paired deep veins that run lateral to the posterior tibial veins; drain into the posterior tibial veins two-thirds of the way up the leg
Anterior tibial veins	Form from the dorsal venous arch (venous network on dorsal surface of foot); travel up the anterior surface of the leg accompanying the anterior tibial arteries; merge with the posterior tibial veins to form the popliteal veins just inferior to the popliteal fossa
Popliteal veins	Form from the merger of the anterior and posterior tibial veins just inferior to the popliteal fossa; become the femoral veins just superior to the knee
Femoral veins	Continuation of popliteal veins that begin just superior to the knee; accompany the femoral arteries up the posterior surface of the thigh; become the external iliac veins as they enter the pelvis

C. Pulmonary Circulation

The pulmonary circulation carries oxygen-poor blood from the right ventricle to the capillaries of the lung where oxygen is added and carbon dioxide is removed. The **pulmonary trunk** carries oxygen-poor blood from the right ventricle and divides to form the **right** and **left pulmonary arteries,** which carry blood to the lungs. Oxygen-rich blood leaves the lungs via the **pulmonary veins,** which drain into the left atrium.

ACTIVITY 11 Major Blood Vessels of the Pulmonary Circulation

1 Label the blood vessels in Figure 30.13.
2 Identify the blood vessels on models or charts.
3 Pronounce the names as you point to them.

• left pulmonary artery
• left pulmonary vein
• pulmonary trunk
• right pulmonary artery
• right pulmonary vein

1 _____

2 _____

3 _____

4 _____

5 _____

FIGURE 30.13 Blood vessels of the pulmonary circulation.

D. Fetal Circulation

Because the fetus cannot breathe or ingest food, the mother's circulatory system provides the fetus with oxygen and nutrients, and eliminates carbon dioxide and other wastes from fetal blood. This exchange occurs across the **placenta,** which forms within the uterus early during pregnancy. Uterine blood vessels enter the placenta, and substances are exchanged by diffusion between maternal and placental capillaries without direct mixing of maternal and fetal blood. Two **umbilical arteries** carry oxygen-poor fetal blood to the placenta, and one **umbilical vein** carries the oxygen-rich blood back to the fetus. The **ductus venosus** allows blood to enter the inferior vena cava. Several circulatory modifications in the fetus allow blood to bypass the nonfunctioning fetal lungs and gastrointestinal tract (Table 30.13). The **foramen ovale** and the **ductus arteriosus** allow blood to bypass the fetal lungs, and after birth, vascular changes occur to allow blood to enter the lungs.

ACTIVITY 12 **Fetal Circulation**

1 Label the structures in Figure 30.14 using the description in Table 30.13.
2 Identify the structures on models or charts.
3 Pronounce the names as you point to them.

TABLE 30.13 **Fetal Circulation**

STRUCTURE	LOCATION AND FUNCTION	STRUCTURAL MODIFICATION AFTER BIRTH
Umbilical arteries	Branch off internal iliac arteries; carry fetal blood to the placenta	Close and fill with connective tissue and become the medial umbilical ligaments
Umbilical vein	Formed at the placenta, and travels to the liver where it forms two branches, the ductus venosus and another vein that drains into the hepatic portal vein	Vessel collapses and renamed ligamentum teres (round ligament)
Ductus venosus	Major branch of umbilical vein that carries most of fetal blood from the placenta to the inferior vena cava	Vessel collapses and renamed ligamentum venosum
Foramen ovale	Opening in atrial septum that allows blood to pass from right atrium to left atrium	Closes and becomes the fossa ovalis
Ductus arteriosus	Arterial vessel that connects pulmonary trunk with the aorta	Closes and becomes the ligamentum arteriosum

(a) Fetal circulation

(b) Modifications at birth

FIGURE 30.14 Fetal circulation and modifications after birth.

(a)
- ductus arteriosus
- ductus venosus
- foramen ovale
- placenta
- umbilical arteries
- umbilical vein

1 _____

2 _____

3 _____

4 _____

5 _____

6 _____

(b)
- fossa ovalis
- ligamentum arteriosum
- ligamentum teres
- ligamentum venosum
- medial umbilical ligaments

7 _____

8 _____

9 _____

10 _____

11 _____

E. Animal Dissection of Blood Vessels

Refer to the appropriate accompanying cat or fetal pig dissection manual for dissection instructions and figures.

Reviewing Your Knowledge

A. Anatomy Drill and Practice

Go to www.wiley.com/college/allen. Click on the picture of your lab manual, then on Student Companion Site, Anatomy Drill and Practice, and Exercise 30.

B. Arterial Blood Distribution

Complete the schematic diagrams of arterial blood distribution in Figure 30.15.

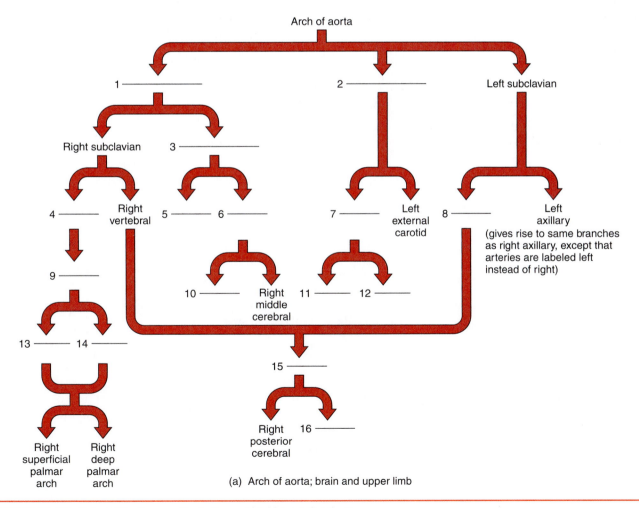

(a) Arch of aorta; brain and upper limb

FIGURE 30.15 Schematic diagram illustrating arterial blood distribution.

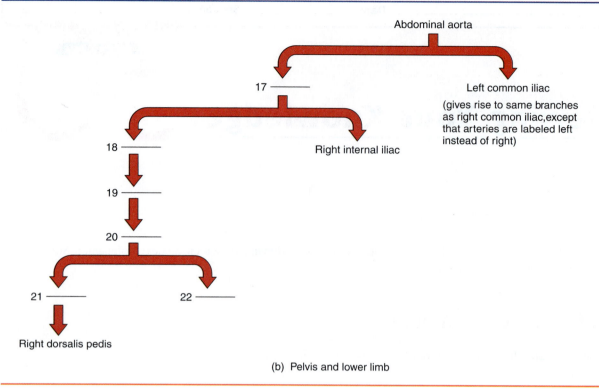

(b) Pelvis and lower limb

FIGURE 30.15 Schematic diagram illustrating arterial blood distribution, *continued.*

Using Your Knowledge

A. Blood Vessel Identification

1. Balloon catheters are inserted into the femoral artery to unblock coronary vessels. Name all the arterial structures, *in order*, that the balloon catheter passes through from the right femoral artery to the right coronary artery.

2. Trace a drop of blood from the aorta to the *small intestine* and back to the right atrium. Include, *in order*, the major arteries and veins.

3. Trace a drop of blood from the aorta to the *skin of the face* and back to the right atrium. Include, *in order*, the major arteries and veins.

4. A blockage in the right internal carotid does not necessarily cause a loss of blood supply to the brain.
 (a) Identify the area of the brain supplied by the right internal carotid, and name the vessels that would supply blood to this area if there is a blockage in the right internal carotid artery.

 (b) Is a thrombus in the external carotid life-threatening? Why or why not?

Identify the blood vessel that supplies blood to each of the following structures in questions 5–7:

5. Radius (bone) and radial nerve

6. Skin covering anterior thigh

7. Rectus femoris muscle

Use the angiogram in Figure 30.16 to answer questions 8–10.

FIGURE 30.16 Angiogram.

8. Identify the blood vessel marked in the angiogram. (*Note:* Only one branch of the aorta can be seen. The other two cannot be observed because of the angle of the angiogram.)

9. Identify the blood vessel marked in the angiogram.

10. Identify the blood vessel marked in the angiogram.

Lymphatic System Structure and Immune System Function

Objectives

After completing this exercise, you should be able to:

- List the functions of the lymphatic system
- Name the organs and tissues of the lymphatic system and list their functions
- Describe lymph formation
- Compare and contrast blood plasma, interstitial fluid, and lymph
- Describe the lymphatic vessels and the flow of lymph through the body
- Describe the gross anatomy and histology of the lymph nodes and spleen
- Identify the main lymphatic organs in the cat or fetal pig

Materials

- human torso or chart with lymphatic organs
- compound microscope, lens paper, prepared slides of the lymph nodes and spleen
- preserved cats or fetal pigs, disposable gloves, dissection equipment, safety glasses, accompanying cat or fetal pig dissection manual

The components of the lymphatic system are **lymphatic organs** and **tissues, lymphatic vessels,** and **lymph.** The **lymphatic system** is the structural location where much of the immune response takes place. The lymphatic system also collects excess interstitial fluid and returns this fluid to the bloodstream. In addition, it delivers dietary lipids and lipid-soluble vitamins (A, D, E, and K) absorbed from the small intestine to the bloodstream.

A. Lymphatic Organs and Tissues

Lymphatic organs and tissues are located throughout the body. **Lymphatic organs** are surrounded by a capsule and include the thymus, lymph nodes, and spleen. **Lymphatic tissues** are not surrounded by a capsule and are called **lymphatic nodules.** Because their functions differ, lymphatic

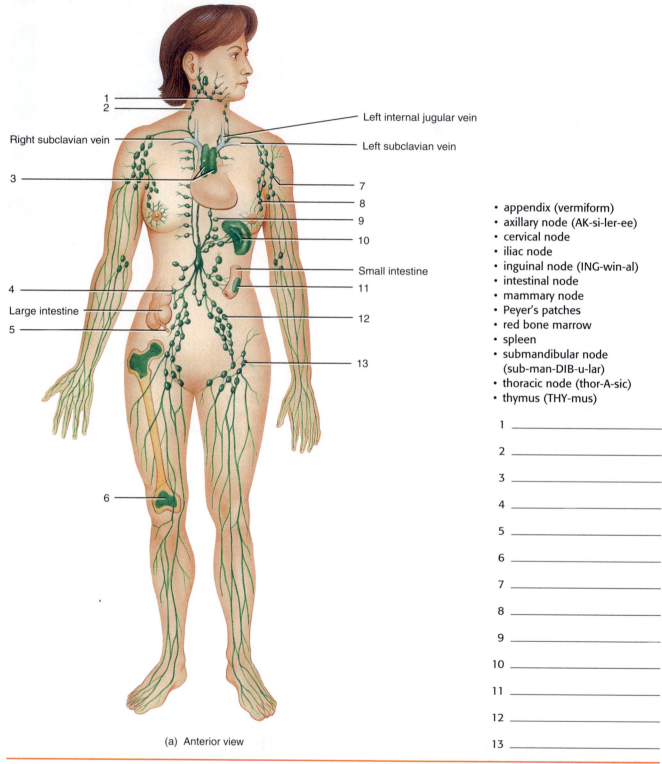

(a) Anterior view

Labels on figure:
- 1
- 2
- Right subclavian vein
- 3
- 4
- Large intestine
- 5
- 6
- Left internal jugular vein
- Left subclavian vein
- 7
- 8
- 9
- 10
- Small intestine
- 11
- 12
- 13

Word list:
- appendix (vermiform)
- axillary node (AK-si-ler-ee)
- cervical node
- iliac node
- inguinal node (ING-win-al)
- intestinal node
- mammary node
- Peyer's patches
- red bone marrow
- spleen
- submandibular node (sub-man-DIB-u-lar)
- thoracic node (thor-A-sic)
- thymus (THY-mus)

1 _____

2 _____

3 _____

4 _____

5 _____

6 _____

7 _____

8 _____

9 _____

10 _____

11 _____

12 _____

13 _____

FIGURE 31.1 Lymphatic organs and tissues.

organs and tissues are divided into two categories: primary lymphatic organs, and secondary lymphatic organs and tissues.

Primary lymphatic organs contain stem cells that produce lymphocytes and are the site where these lymphocytes become immunocompetent (recognize and mount immune system response). The two primary lymphatic organs are **red bone marrow** and the **thymus.** The thymus, which is located in the mediastinum just superior to the heart, is much larger in children and decreases considerably in size as they get older.

Secondary lymphatic organs and **tissues** are sites for defense against invading agents and cancer cells and include the **lymph nodes, spleen,** and **lymphatic nodules** (follicles). Single lymphatic nodules located in the connective tissue of mucous membranes of the gastrointestinal, respiratory, urinary, and reproductive tracts are called **mucosa-associated lymphoid tissue (MALT).** Aggregations of large nodules (follicles) include **Peyer's patches** in the small intestine, lymphoid follicles in the **appendix** (*appendix* = appendage), five tonsils (one **pharyngeal tonsil** and paired **palatine** and **lingual tonsils**), and **bronchial nodules** located in the walls of the respiratory tract. The pharyngeal tonsil is located in the posterior nasopharynx, while the palatine tonsils are in the posterior portion of the oral cavity. The lingual tonsils are located under the tongue. All of these nodules protect the body from foreign substances that are ingested or inhaled, or otherwise enter body openings. *Note:* Lymph nodes are named for their location in the body.

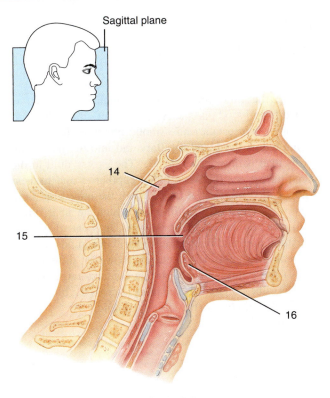

(b) Sagittal view

FIGURE 31.1 Lymphatic organs and tissues, *continued.*

- lingual tonsil (LING-wal) 14 _____
- palatine tonsil (PAL-ih-tine)
- pharyngeal tonsil 15 _____
 (fuh-RIN-gee-al)
 16 _____

ACTIVITY 1 Lymphatic Organs and Tissues

1 Label the structures in Figure 31.1(a) and (b).

2 Identify these structures on a human torso or chart.

3 Pronounce each term as you point to the structure.

B. Lymph and Lymphatic Vessels

Lymph, the fluid found within lymphatic vessels, is formed from blood plasma. As blood flows through blood capillaries, hydrostatic and osmotic pressures force more plasma out of the blood capillary and into the interstitial spaces than is drawn back inside. The excess fluid in the interstitial spaces (between blood capillaries and tissues) is now called **interstitial fluid.** When the interstitial fluid enters the **lymphatic capillaries,** which lie near blood capillaries and are closed at one end, it is called **lymph.** Interstitial fluid and lymph are identical in composition but different in location. Blood plasma, however, differs from both of these fluids in that plasma contains large plasma proteins (albumin) that are too large to cross the blood capillary endothelium into the interstitial fluid.

Lymph flows from **lymphatic capillaries** into many **lymphatic vessels** that take lymph toward the neck region. Lymphatic vessels follow the pathway of veins in the body and also have a similar structure, but they are thinner-walled and have more valves than veins. Just as a body massage is helpful for moving blood through the veins and toward the heart, a light massage is also helpful for moving lymph through the lymphatic vessels and nodes to cleanse the lymph.

Lymphatic vessels flow into a series of lymph nodes where lymph is filtered. The last lymph nodes in each chain of nodes have vessels that merge to form **lymph trunks.** Lymph trunks merge to form two main ducts, the **thoracic duct** (left lymphatic duct) and the **right lymphatic duct.** In the abdominal area, lymph trunks merge to form a saclike reservoir called the **cysterna chyli.** The long thoracic duct begins at the cysterna chyli and continues superiorly to drain the lymph from the legs, abdomen, left arm, and left side of the thorax, neck, and head into the left

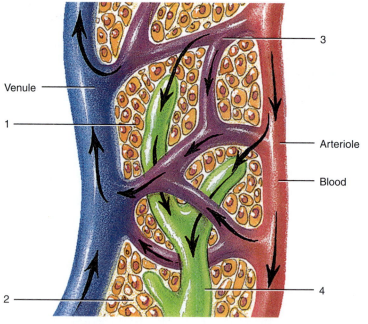

Venule

1

2

3

Arteriole

Blood

4

(a) Lymphatic and blood capillaries

5

Opening into lymphatic capillary

6

Anchoring filament

Endothelium of lymphatic capillary

7

(b) Lymphatic capillary

FIGURE 31.2 Lymphatic capillaries.

- blood capillary
- interstitial fluid (inter-STIH-she-al)
- lymphatic capillary
- tissue cell

1 _____

2 _____

3 _____

4 _____

- interstitial fluid
- lymph in lymphatic capillary
- tissue cell

5 _____

6 _____

7 _____

subclavian vein. The **right lymphatic duct** is the lesser and very short duct that drains lymph from the right arm via the right jugular trunk and right side of the thorax, neck, and head via the right subclavian trunk. Each duct enters the subclavian vein at its juncture with an internal jugular vein to drain lymph into venous blood.

ACTIVITY 2 **Lymphatic Vessels**

1 Label the structures in Figures 31.2 and 31.3.

2 Examine the lymphatic vessel with valve in Figure 31.3(d).

3 Pronounce each term as you point to it on the figure.

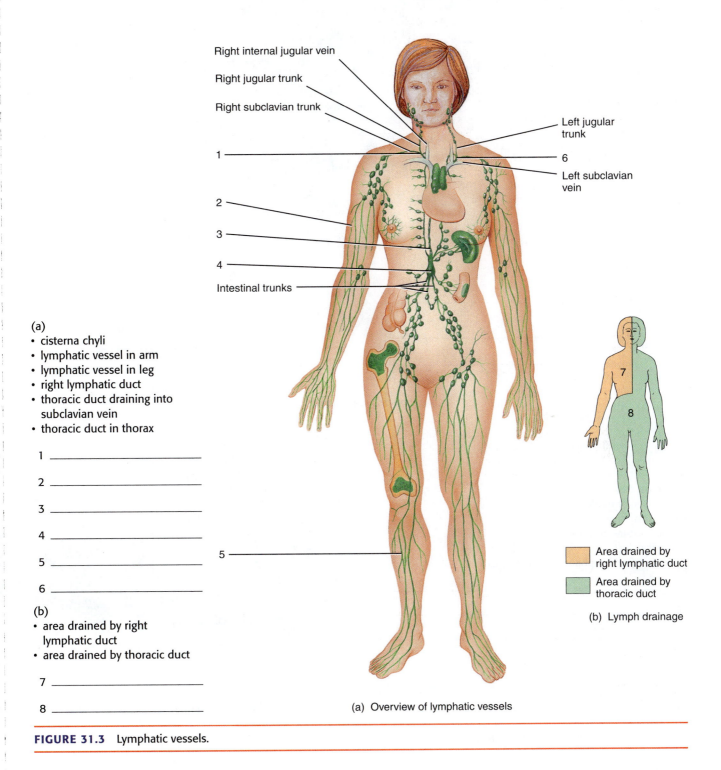

Right internal jugular vein

Right jugular trunk

Right subclavian trunk

Left jugular trunk

1

6

Left subclavian vein

2

3

4

Intestinal trunks

(a)
- cisterna chyli
- lymphatic vessel in arm
- lymphatic vessel in leg
- right lymphatic duct
- thoracic duct draining into subclavian vein
- thoracic duct in thorax

1 _____

2 _____

3 _____

4 _____

5 _____

6 _____

(b)
- area drained by right lymphatic duct
- area drained by thoracic duct

7 _____

8 _____

5

7

8

| | Area drained by right lymphatic duct |
| | Area drained by thoracic duct |

(b) Lymph drainage

(a) Overview of lymphatic vessels

FIGURE 31.3 Lymphatic vessels.

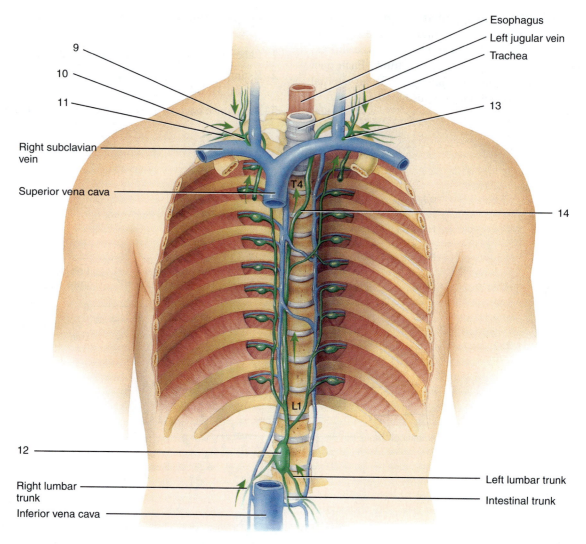

(c) Thoracic area, anterior view

(c)
- cisterna chyli
- right jugular trunk
- right lymphatic duct
- right subclavian trunk
- thoracic duct
- thoracic (left lymphatic) duct in thorax

9 _____

10 _____

11 _____

12 _____

13 _____

14 _____

(d) Photomicrograph of a lymphatic vessel and valve

FIGURE 31.3 Lymphatic vessels, *continued.*

C. Structure of the Lymph Node

Lymph nodes are bean-shaped structures that filter lymph and are usually found in groups. Clusters of these nodes are generally found in different body regions (i.e., the cervical, axillary, and inguinal areas) and are named for their location. Each node has a **capsule** with extensions called **trabeculae** (*trabecula* = little beam) that compartmentalize the node.

The nodes are composed of **reticular** (*reticulum* = process) **tissue,** with reticular fibers forming the netlike support for **reticulocytes** (specialized **fibroblasts**). Two main regions of lymph nodes are the outer **cortex** (*cortex* = bark) and inner **medulla** (*medulla* = marrow). There are many lighter colored oval **lymphatic nodules** in the outer cortex that contain many B lymphocytes in the **germinal center.**

Lymph enters each node via many **afferent vessels** and travels through the **subcapsular sinus, trabecular sinuses,** and **medullary sinuses,** and exits the node via fewer **efferent vessels.** The **hilus** (*hilus* = depression or pit) is the indented area of the node where the efferent vessels leave the node. This design causes the lymph to move slowly through the lymph nodes, allowing time for immune system cells (lymphocytes and macrophages) to attack pathogens, cancer cells, and foreign molecules. Enlarged lymph nodes could be caused by, but not limited to, infection, cancer, or scarring from a chronic lesion.

ACTIVITY 3 Lymph Node Structure

1 Label the structures in Figure 31.4(a), (b), and (c).

2 Pronounce the terms as you write the name or point to them in the figure.

3 Examine a prepared slide of a lymph node.
 • Using low power, identify the structures listed in Figure 31.4(b).
 • Using high power, identify the **reticulocytes** and **reticular fibers** of the lymph node (Figure 31.4c).

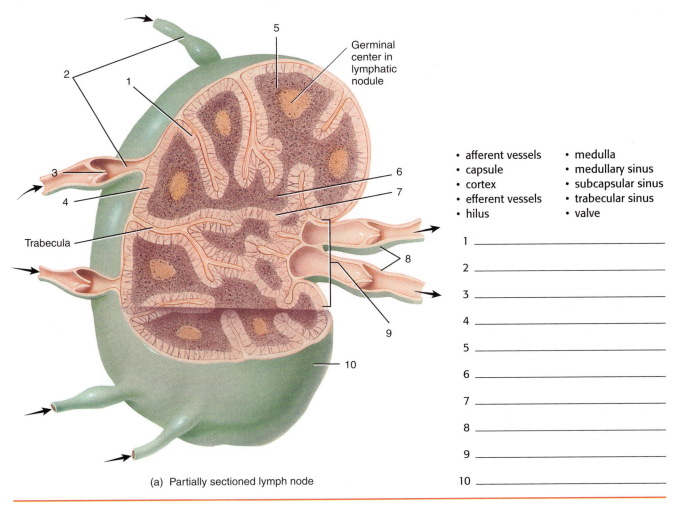

 • afferent vessels • medulla
 • capsule • medullary sinus
 • cortex • subcapsular sinus
 • efferent vessels • trabecular sinus
 • hilus • valve

1 _____

2 _____

3 _____

4 _____

5 _____

6 _____

7 _____

8 _____

9 _____

(a) Partially sectioned lymph node 10 _____

FIGURE 31.4 Lymph node.

(b) Photomicrograph of a portion of a lymph node

(c) Photomicrograph of reticular tissue within a lymph node

FIGURE 31.4 Lymph node, *continued*.

(b)
- capsule
- cortex
- germinal center in lymphatic nodule
- medulla
- trabecula

11 _____

12 _____

13 _____

14 _____

15 _____

(c)
- reticular fiber
- reticulocyte

16 _____

17 _____

D. Structure of the Spleen

The largest lymphatic organ, the **spleen,** filters blood of foreign organisms and particles, eliminates aged erythrocytes, and is a blood reservoir. The spleen is the main organ that filters blood, similar to the lymph nodes being the main organs that filter the lymph. This highly vascular organ lies posterior and lateral to the stomach in the LUQ. The **splenic artery** enters the spleen and the **splenic vein** and **efferent lymphatic vessels** exit the spleen at the **hilus.** The spleen is an encapsulated organ with extensions of the capsule forming **trabeculae.**

Histologically, the spleen has regions of white pulp and red pulp that are named for their appearance in fresh specimens. As students, you will see stained specimens with different colors on your prepared slides. **White pulp** resembles nodules with many lymphocytes and macrophages and appears dark purple when stained. Lymphocytes carry out immune function here, and macrophages engulf any aged blood cells, foreign material, or debris. **Red pulp** stores platelets and contains reticular fibers that cannot be seen at this magnification. It stains a lighter color because its many red blood cells are anucleate. Many macrophages can also be found here to phagocytize foreign substances in the blood.

ACTIVITY 4 Structure of the Spleen

1 Label the structures in Figure 31.5.

2 Pronounce each term as you write it.

3 Examine a prepared slide of the spleen.
 • Using low power, identify the capsule, trabecula, red pulp, and white pulp.
 • Using high power, identify reticular fibers, reticulocytes, and lymphocytes in white pulp and the reticular fibers, reticulocytes, and RBCs in the red pulp.

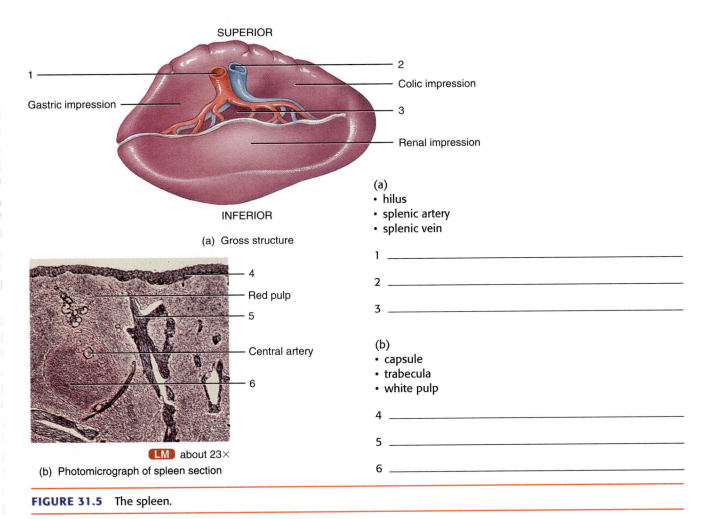

SUPERIOR

1 —————

Gastric impression ————

2
Colic impression

3

Renal impression

INFERIOR

(a) Gross structure

4
Red pulp
5

Central artery

6

LM about 23×

(b) Photomicrograph of spleen section

(a)
• hilus
• splenic artery
• splenic vein

1 _____

2 _____

3 _____

(b)
• capsule
• trabecula
• white pulp

4 _____

5 _____

6 _____

FIGURE 31.5 The spleen.

E. Immune Cells Involved in Body Defense

The cells involved in body defense are the **T cells** (lymphocytes), **B cells** (lymphocytes), and **macrophages** (*macro-* = big; *phago-* = to eat). T and B cells are originally produced from pluripotent stem cells in the red bone marrow. B cells complete their maturation and become immunocompetent in the red bone marrow. T cells migrate to the thymus through the bloodstream to complete maturization and become immunocompetent there.

After being instructed and programmed for their work, both T and B cells are released into the bloodstream and travel to secondary lymphatic organs and tissue to be activated and cloned in the immune response. Secondary lymphatic organs include the lymph nodes and the spleen. Lymphatic tissue (nodules) includes MALT, Peyer's patches, appendix, tonsils, and bronchial nodules.

ACTIVITY 5 **Cells Involved in Body Defenses**

1 Identify where T and B cells are produced and programmed, and where active T and B cells are located in Figure 31.6(a).

2 Observe a macrophage phagocytizing bacteria in Figure 31.6(b).

F. Animal Dissection

You may have already looked at the lymphatic organs in your previous dissections. As you dissected the blood vessels, you may have noticed small, 1/8- to 1/4-inch **lymph nodes** in the axillary or inguinal areas. Because the nodes are small, they are easy to miss if you do not know their structure or location.

Consult the appropriate cat or fetal pig dissection manual for your dissection instructions.

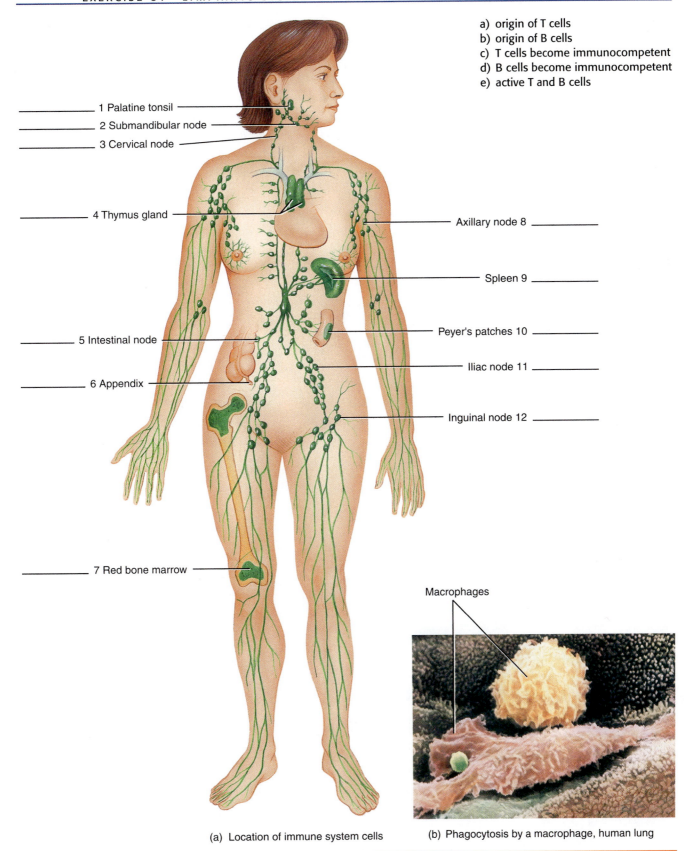

a) origin of T cells
b) origin of B cells
c) T cells become immunocompetent
d) B cells become immunocompetent
e) active T and B cells

1 Palatine tonsil
2 Submandibular node
3 Cervical node

4 Thymus gland

Axillary node 8

Spleen 9

5 Intestinal node

Peyer's patches 10

Iliac node 11

6 Appendix

Inguinal node 12

7 Red bone marrow

Macrophages

(a) Location of immune system cells

(b) Phagocytosis by a macrophage, human lung

FIGURE 31.6 Cells of the lymphatic and immune system.

Reviewing Your Knowledge

A. Anatomy Drill and Practice

Go to www.wiley.com/college/allen. Click on the picture of your lab manual, then on Student Companion Site, Anatomy Drill and Practice, and Exercise 31.

B. Structure of Lymphatic Organs and Tissues

Write the name of the organ or tissue that matches the description. Terms may be used more than once.

_____ **1.** Regresses considerably with age

_____ **2.** Largest lymphatic organ; red and white pulp

_____ **3.** Bean-shaped; has efferent and afferent vessels

_____ **4.** ⎫
 ⎬ Primary lymphatic organs
_____ **5.** ⎭

_____ **6.** ⎫
 ⎬ Secondary lymphatic organs
_____ **7.** ⎭

_____ **8.** Secondary lymphatic nodules associated with the nasal and oral cavities

_____ **9.** Secondary lymphatic nodules located in connective tissue of mucous membranes

_____ **10.** Secondary lymphatic nodules located in the small intestine

C. Physiology of Lymphatic Tissues and Organs

Write the name of the organ or tissue that matches the description. Terms may be used more than once.

> Peyer's patches
> appendix (vermiform)
> lymph nodes
> mucosa-associated lymphoid tissue (MALT)
> red bone marrow
> spleen
> thymus
> tonsils

_____ **1.** Filters lymph

_____ **2.**

_____ **3.**

_____ **4.**

Sites for the immune response

_____ **5.**

_____ **6.**

_____ **7.**

_____ **8.** Filters blood and contains red and white pulp

_____ **9.** Programs T cells for immunocompetence

_____ **10.** Programs B cells for immunocompetence

D. Formation of Lymph

Write the name of the term that matches the phrase.

> blood
> interstitial fluid
> plasma

_____ **1.** Fluid that becomes interstitial fluid

_____ **2.** Fluid that becomes lymph

_____ **3.** Interstitial fluid plus blood cells and large proteins

E. Lymphatic Vessels

Write the name of the vessel that matches the description. Terms may be used more than once.

cisterna chyli
lymphatic vessels
right lymphatic duct
thoracic duct
lymphatic trunks

_____ **1.** Saclike vessel

_____ **2.** Left lymphatic duct

_____ **3.** Merge to form lymphatic trunks

_____ **4.** Merge to form lymphatic ducts

_____ **5.** The thoracic duct begins here

_____ **6.** Drains the legs, abdominal area, and left side of the body

_____ **7.** Drains the right side of thorax, arm, head, and neck

_____ **8.** The longer lymphatic duct

F. Immune Cells Involved in Body Defense

Write the name of the immune cells that matches the description. Terms may be used more than once.

_____ **1.** Maturation and immunocompetency in red bone marrow

_____ **2.** Maturation and immunocompetency in the thymus gland

_____ **3.**
⎱ Located in secondary organs and tissues
_____ **4.**

Using Your Knowledge

A. Lymphatic and Immune Systems

1. How would the lymph that is leaving the cisterna chyli differ from lymph draining into the right lymphatic duct?

2. If you look into a child's mouth, there are "golf balls" puffing from each side of the oral cavity. What are these structures?

3. How do lymphatic capillaries differ from blood capillaries?

4. If you dissected a fetal pig or a young cat, what difference would you expect in the thymus compared to the adult?

5. Why are the walls of lymphatic vessels thin, like veins?

6. Bacteria and viruses in infected tissues easily enter lymph vessels. Explain why.

7. If a woman had a radical mastectomy (removal of a cancerous breast, surrounding tissues with axillary lymph nodes, and anterior thoracic muscles), would you expect the arm on that side to be edematous (have edema)? Explain.

8. Would you describe the flow of lymph through lymph nodes as being fast or slow, compared with blood capillaries? Explain.

9. What would be missing in lymph leaving a lymph node, compared with lymph entering the node?

10. A patient has enlarged right inguinal nodes that are very tender to the touch. Examination of the feet reveals a small cut between the right 3rd and 4th toes which are warm and red. What would your initial impression be and why are the nodes enlarged?

Respiratory System Structure

Objectives

After completing this exercise, you should be able to:

- Locate and identify the organs of the respiratory system on models or charts
- Describe the structure and function of the respiratory system organs
- Trace the path of air from the nose to the alveoli
- Identify the main respiratory organs in the cat or fetal pig

Materials

- models or charts with respiratory organs
- sagittal section model or chart of the human head, neck, and thorax; larynx model
- compound microscope, lens paper, prepared slides of trachea and lung
- preserved cats or fetal pigs, disposable gloves, dissection equipment, and safety glasses, accompanying cat or fetal pig dissection manual

A. Gross Anatomy of the Respiratory Organs

The respiratory system provides an airway for movement of air into and out of the body. It is also the site where atmospheric oxygen diffuses into the bloodstream to be delivered to all body cells, and carbon dioxide produced by these cells diffuses out of the bloodstream to be exhaled into the atmosphere. This gas exchange takes a coordinated effort of the respiratory and cardiovascular systems. The main organs of the respiratory system are the nose, pharynx, larynx, trachea, bronchi, and lungs. Clinically, the nose, pharynx, and larynx are organs of the upper airways, and the trachea, bronchi, and lungs comprise the lower airways.

ACTIVITY 1 **Structures of the Respiratory System**

1 Label the structures of the respiratory system in Figure 32.1.

2 Identify the structures on a model or an anatomical chart with respiratory organs.

3 Pronounce the terms as you point to them on the model.

Nose

The framework of the **nose** is composed of bone and cartilage. Bones comprise the base of the nose, and hyaline cartilage the anterior portion. Two **external nares** (singular = *naris*) are the openings for air to enter a space called the **nasal cavity,** which is lined with a mucous membrane. The **nasal septum** separates this cavity into right and left portions. The nasal septum is highly vascularized and is the site of nosebleeds. A deviated septum may result from trauma, such as a blow to the nose, and can cause difficulty in breathing. On either side of the nose, three curved bony structures extend into the nasal cavity: the **superior,**

Anterior view

FIGURE 32.1 Organs of the respiratory system.

- larynx (LAIR-inks)
- lungs
- nasal cavity
- nose
- pharynx (FAIR-inks)
- right primary bronchus (BRON-kus)
- trachea (TRAY-key-a)

1 _____

2 _____

3 _____

4 _____

5 _____

6 _____

7 _____

middle, and **inferior nasal conchae** (*concha* = shell-shaped), or **turbinates** (*turbinum* = scroll-shaped). Beneath each nasal concha is a concavity called a **meatus** (*meatus* = passage) which increases surface area in the nose. When incoming air passes over the vascularized mucous membranes of the meatuses, the air is warmed and humidified. This is the reason that the pharynx gets dry if a person's nose is stopped up and breathing is done through the mouth. The bony **hard palate** forms the floor of the nasal cavity. At the junction of the hard and soft palates are the **internal nares,** two openings that lead into the tubular nasal pharynx posteriorly. When the skull was discussed in the axial skeleton exercise, it was mentioned that the nasal cavity is surrounded by four paranasal sinuses. The frontal, maxillary, ethmoidal, and sphenoidal bones have mucous membrane-lined sinuses or cavities that also warm and moisten the air. These paranasal cavities have ducts that drain into the nasal cavity.

ACTIVITY 2 Nasal Structures

1 Label the parts of the nose in Figure 32.2(a) and (b).

2 Identify the structures on a sagittal section model or an anatomical chart of the human head.

3 Pronounce the terms as you point to them on the model.

(a) Sagittal section

(b) Coronal section

FIGURE 32.2 Nasal structures within nasal cavity.

(a)
- external naris (NEH-ris)
- hard palate
- inferior nasal concha or turbinate (CON-cha; TUR-bin-it)
- inferior nasal meatus (mee-AY-tus)
- internal naris
- middle nasal concha or turbinate
- middle nasal meatus
- superior nasal concha or turbinate
- superior nasal meatus

1 _____
2 _____
3 _____
4 _____
5 _____
6 _____
7 _____
8 _____
9 _____

(b)
- inferior nasal concha
- inferior nasal meatus
- middle nasal concha
- middle nasal meatus
- nasal septum

10 _____
11 _____
12 _____
13 _____
14 _____

Pharynx

The **pharynx** or throat is divided into three regions: the nasopharynx (*naso-* = nose), oropharynx (*oro-* = mouth), and laryngopharynx (*laryngo-* = larynx), corresponding to the anatomical structures nearby. The **nasopharynx** begins at the internal nares and ends at the soft palate. The **oropharynx** begins at the soft palate and extends to the level of the hyoid bone, and the **laryngopharynx** extends from the hyoid bone to the beginning of the esophagus. The **soft palate** is found in the pharynx as a posterior extension of the hard palate. The oval-shaped **uvula** dangles inferiorly as an extension of the soft palate. During swallowing, the soft palate pushes superiorly to close off the nasopharynx and to direct food toward the laryngopharynx. Two **auditory tubes** (Eustachian tubes) have openings from the middle ear into the nasopharynx, and the single **pharyngeal tonsil** (adenoid) is located in its posterior wall.

The oropharynx has two pairs of tonsils, the **palatine** (palate) and **lingual** (tongue) tonsils. Note that the mucous membranes of the nose, paranasal sinuses, pharynx, and middle ear are connected by ducts, which explains why infections can spread to any adjacent areas.

ACTIVITY 3 The Pharynx

1 Label the three divisions of the pharynx in Figure 32.3(a).

2 Label the structures in Figure 32.3(b).

3 Identify the structures on a sagittal section model or an anatomical chart of the pharynx.

4 Pronounce the terms as you point to them on the model.

Sagittal plane

(a) Divisions of the pharynx

(b) Pharyngeal structure

FIGURE 32.3 The pharynx.

(a)
- laryngopharynx
 (la-rin-go-FAIR-inks)
- nasopharynx
- oropharynx (oro-FAIR-inks)

1 _____

2 _____

3 _____

(b)
- internal naris
- laryngopharynx
- lingual (LING-gwal)
 tonsils (2)
- nasopharynx
- orifice of the auditory tube
 (Eustachian)
- oropharynx
- palatine tonsils (2)
- pharyngeal
 (fair-IN-gee-al) tonsil
- soft palate
- uvula (YOU-view-la)

4 _____

5 _____

6 _____

7 _____

8 _____

9 _____

10 _____

11 _____

12 _____

13 _____

Larynx

The **larynx** or voice box connects the laryngopharynx with the trachea. This organ consists of nine hyaline cartilages and houses the vocal cords. The larynx contains three paired cartilages—the arytenoid, cuneiform, and corniculate cartilages—which are small cartilages located in the posterior wall of the larynx; and three single cartilages—the thyroid cartilage, cricoid cartilage, and epiglottis—which comprise the main body of the larynx. The largest cartilage, the **thyroid cartilage** (*thyroid* = shield-shaped), is seen anteriorly and has a prominence called the Adam's apple, which is made of hyaline cartilage. The **cricoid cartilage** (*cricoid* = ringlike), which is inferior to the thyroid cartilage, is larger on the posterior side than on the anterior side and consists of hyaline cartilage. As the term *cricoid* suggests, this is the only cartilage that is a complete ring in the larynx or trachea. The **glottis** is the opening that allows air into the larynx. The oval-shaped **epiglottis** (*epi-* = over; *-glottis* = tongue) is composed of elastic cartilage and closes over the glottis during swallowing. If you have food or drink in your mouth and start laughing, the glottis opens and you could aspirate these substances into the larynx with

the air. Coughing forces the food or liquid back into the laryngopharynx. Lateral to the glottis are two pairs of folds. The superior folds are the **ventricular** (little belly) **folds** (vestibular folds), or false vocal cords, while the **vocal folds** or true vocal cords are inferior and medial. The vocal folds attach via small muscles to the **arytenoid** (arytenoid = ladle-like) **cartilage.** The vocal folds, which contain elastic tissue, vibrate with phonation or speaking. If these folds become inflamed or have tumors, they cannot vibrate and hoarseness (laryngitis) results.

ACTIVITY 4 The Larynx

1 Label the structures listed in Figure 32.4(a), (b), (c), and (d).

2 Locate these structures on a larynx model or an anatomical chart.

3 Pronounce the terms as you point to them.

4 Palpate your thyroid cartilage. Swallow and feel the thyroid cartilage move upward.

(a) Anterior view (b) Posterior view

FIGURE 32.4 The larynx and associated structures.

(a)
- cricoid cartilage (CRY-koid)
- epiglottis
- thyroid cartilage

1 _____

2 _____

3 _____

(b)
- cricoid cartilage
- epiglottis
- thyroid cartilage

4 _____

5 _____

6 _____

Sagittal plane

Posterior Anterior

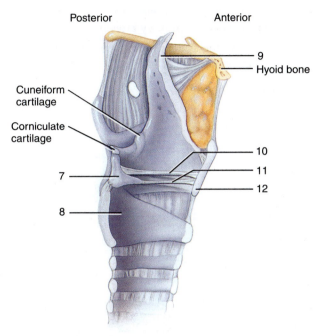

9
Hyoid bone

Cuneiform
cartilage

Corniculate
cartilage

7

8

10
11
12

(c) Midsagittal

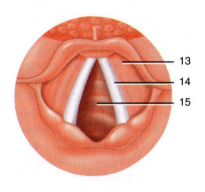

13
14
15

(d) Superior

FIGURE 32.4 The larynx and associated structures, *continued*.

- arytenoid cartilage
 (ar-ih-TEE-noid)
- cricoid cartilage
- epiglottis
- thyroid cartilage
- ventricular folds
- vocal folds

7 _____

8 _____

9 _____

10 _____

11 _____

12 _____

- glottis
- ventricular folds
- vocal folds

13 _____

14 _____

15 _____

Bronchial Tree

Just as a tree has a trunk, branches, twigs, and fruit, the lower airways have a trachea, bronchi, bronchioles, and alveoli. If the trachea and bronchi were inverted, they would resemble the branches of a tree, hence the name **bronchial tree.** The **trachea** (*trachea* = sturdy) or windpipe is a tubelike conduit that conducts air from the larynx to the bronchi. It is located anterior to the esophagus and can be palpated on the anterior surface of the neck. When palpating the trachea, horizontal tracheal cartilages that keep the airway open can be felt as bumps one on top of the other. **Tracheal cartilages** are C-shaped, and a smooth muscle, the **trachealis muscle,** connects the open sides of the C-shaped cartilage. The trachealis muscle is on the posterior surface of the trachea and allows expansion of the esophagus into the trachea during swallowing. The trachea ends at the **carina** (*carina* = keel of a boat) inferiorly and branches into a shorter right and longer left primary bronchus, each serving the corresponding lung. Each **primary bronchus** branches into **secondary (lobar) bronchi** to supply a **lobe** of each lung. The right primary bronchus branches into three secondary bronchi, and the left primary bronchus branches into two secondary bronchi. The secondary bronchi further subdivide into **tertiary (segmental) bronchi** that enter **bronchopulmonary segments** within a lobe. Tertiary bronchi divide into **bronchioles,** each serving small compartments called **lobules.** Bronchioles further subdivide into **terminal bronchioles,** which branch into **respiratory bronchioles** that begin the respiratory zone of the lung. Respiratory bronchioles further divide into **alveolar ducts** that lead into clusters of alveoli called **alveolar sacs. Alveoli** bud off alveolar sacs like individual grapes in a grape cluster. Simple squamous alveolar cells and simple squamous blood capillary cells form the **respiratory membrane,** where gas exchange occurs. Several changes occur as the bronchi turn into bronchioles. The diameter of the airway diminishes, the epithelium changes, cartilage disappears early in the tertiary bronchi, and the bronchioles have only smooth muscle spiraled around them. Because smooth muscle in the bronchioles causes constriction, asthmatic inhalers with bronchial dilators cause the smooth muscle to relax during asthmatic attacks.

ACTIVITY 5 Bronchial Tree

1 Label the structures in Figure 32.5(a) and (b) and Figure 32.6.

2 Locate the structures on a lung model.

3 Pronounce each term as you point to it on the model.

4 Palpate the sternal angle to locate where the trachea divides into the primary bronchi.

5 Observe the sheep pluck demonstration with the bronchial tree dissected.

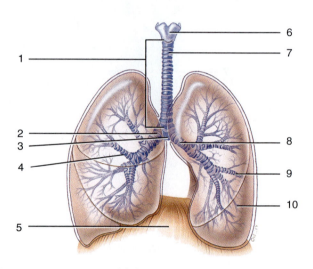

(a) Anterior view

FIGURE 32.5 Bronchial tree.

- bronchiole (BRON-key-ol)
- carina (ka-RYE-na)
- diaphragm
- larynx
- primary bronchus, left
- primary bronchus, right
- secondary bronchus, right
- tertiary bronchus, left
- trachea
- tracheal cartilage

1 _____

2 _____

3 _____

4 _____

5 _____

6 _____

7 _____

8 _____

9 _____

10 _____

CLINICAL NOTE: *The airways can be clinically divided into three areas according to size:*

Large airways—trachea and bronchi

Medium airways—bronchioles and terminal bronchioles

Small airways—respiratory bronchioles, alveolar ducts, alveolar sacs, and alveoli

Respiratory medicine uses these terms to identify the location of diseases. Bronchitis is a large airway disease, asthma is a middle airway disease, and emphysema is a small airway disease.

- alveolar sac (al-VEE-oh-lur)
- primary bronchus
- respiratory bronchiole
- secondary bronchus
- terminal bronchiole
- tertiary bronchus

11 _____

12 _____

13 _____

14 _____

15 _____

16 _____

(b) Bronchial branching

FIGURE 32.5 Bronchial tree, *continued.*

- alveolar ducts
- alveolar sac
- alveoli
- respiratory bronchiole

1 _____

2 _____

3 _____

4 _____

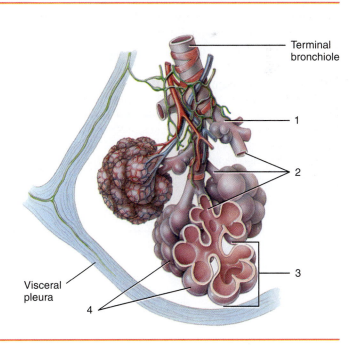

FIGURE 32.6 Diagram of portion of lung lobule.

Lungs

The two **lungs** are divided into *three lobes* on the right side and *two lobes* on the left. The three right lobes are the **superior, middle,** and **inferior lobes,** and the left lobes are called the **superior** and **inferior lobes.** The rounded superior part of the lung is the **apex,** and the broader inferior part is the **base,** which rests on the diaphragm. The left lobe has a concave surface called the **cardiac notch,** which has the apex of the heart projecting into it. Each lung has a **hilus,** an area surrounded with pleura, where the bronchi, blood and lymphatic vessels, and nerves enter or exit the medial side of the lung. The lungs are in the thoracic cavity and are separated from each other by the heart and the mediastinum. **Parietal pleura** lines the thoracic cavity wall, and **visceral pleura** covers the surface of each lung. The **pleural cavity** is the space between the two pleural layers that contains pleural fluid.

ACTIVITY 6 Lungs

1 Label the parts of the lung in Figure 32.7(a), (b), (c), and (d).
2 Identify these structures on a lung model or an anatomical chart.
3 Pronounce the terms as you point to them.
4 Label the parts of the pleura in Figure 32.8(a) and (b).

(a) Anterior right lung

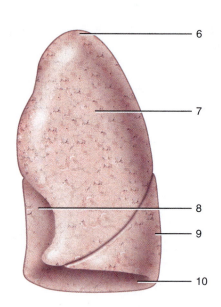

(b) Anterior left lung

FIGURE 32.7 Lung structures.

(a)
• apex
• base
• inferior lobe
• middle lobe
• superior lobe

1 _____
2 _____
3 _____
4 _____
5 _____

(b)
• apex
• base
• cardiac notch
• inferior lobe
• superior lobe

6 _____
7 _____
8 _____
9 _____
10 _____

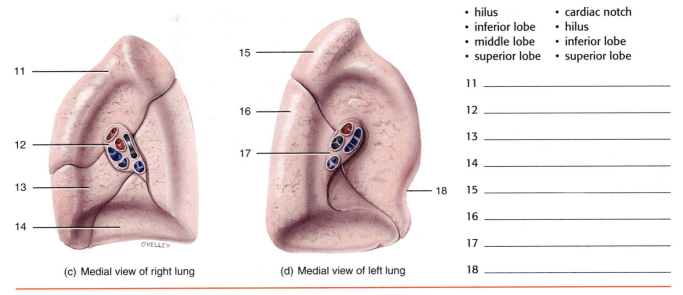

- hilus • cardiac notch
- inferior lobe • hilus
- middle lobe • inferior lobe
- superior lobe • superior lobe

11 _____

12 _____

13 _____

14 _____

15 _____

16 _____

17 _____

18 _____

(c) Medial view of right lung (d) Medial view of left lung

FIGURE 32.7 Lung structures, *continued.*

View
Transverse plane

(a) Anterior view

(b) Cross-section through thorax
(inferior view)

FIGURE 32.8 Pleura.

(a)
- diaphragm
- parietal pleura
- pleural cavity
- visceral pleura

1 _____

2 _____

3 _____

4 _____

(b)
- left lung
- parietal pleura
- pericardium
- pleural cavity
- right lung
- visceral pleura

5 _____

6 _____

7 _____

8 _____

9 _____

10 _____

B. Microscopic Anatomy of the Respiratory System

There are four basic types of epithelium in the respiratory system, with areolar connective tissue as the underlying tissue just beneath the epithelia. As can be seen in Table 32.1, the mucous membranes of the upper airways are lined with **pseudostratified ciliated columnar epithelium.** This epithelium has **goblet cells** which secrete viscous mucus that traps dust and other particles. There are also **cilia** which beat to move mucus toward the pharynx to be swallowed. The oropharynx and laryngopharynx are common areas for both the digestive and respiratory systems. Since abrasive food particles will travel through these two structures, they are lined with nonkeratinized **stratified squamous epithelium.** The multilayers of this type of epithelium protect the underlying tissue. This epithelium continues into the upper part of the larynx. In the lower larynx, pseudostratified ciliated columnar epithelium resumes and extends through the trachea and primary bronchi. Deeper into the bronchial tree, the epithelium gradually changes from high **pseudostratified columnar epithelium** to **simple columnar** and finally to **simple cuboidal,** with an accompanying gradual loss of cilia. The respiratory zone structures (respiratory bronchioles to the alveoli) all have **simple squamous epithelium** that allows the highly vascularized alveoli to facilitate diffusion of respiratory gases, oxygen, and carbon dioxide.

ACTIVITY 7 Microscopic Examination of Respiratory Tract Tissues

1 Label the structures of the trachea and esophagus in Figure 32.9(a).

2 Examine a prepared slide of the trachea and esophagus.
- Using the low-power objective lens, identify the structures listed in Figure 32.9(a).
- Using the high-power objective lens, identify the pseudostratified ciliated columnar epithelium lining the tracheal lumen and the hyaline cartilage within the tracheal wall. Identify a goblet cell within the epithelium.
- Refer to Figures HA-42 and HA-43 in Appendix-A Histology Atlas.

3 Label the lung structures in Figure 32.9(b).

4 Examine a prepared slide of lung tissue.
- Using the low-power objective lens, identify alveoli and, if present, the other structures listed in Figure 32.9(b).
- Using the high-power objective lens, identify the simple squamous epithelium forming the alveolar walls.
- Refer to Figures HA-44 through HA-46 in Appendix-A Histology Atlas.

5 Answer the discussion questions with your lab partner(s).

TABLE 32.1 Respiratory Epithelia

RESPIRATORY STRUCTURE	EPITHELIA
Nasal cavity	Pseudostratified ciliated columnar
Paranasal sinuses	Pseudostratified ciliated columnar
Nasopharynx	Pseudostratified ciliated columnar
Oropharynx	Stratified squamous
Laryngopharynx	Stratified squamous
Larynx (vocal folds and above)	Stratified squamous
Larynx (below vocal folds)	Pseudostratified ciliated columnar
Trachea	Pseudostratified ciliated columnar
Primary bronchi	Pseudostratified ciliated columnar
Secondary bronchi	Simple columnar with fewer cilia
Tertiary bronchi	Simple columnar; no cilia
End of tertiary bronchi	Simple cuboidal; no cilia
Bronchioles, including terminal bronchioles	Simple cuboidal; no cilia
Respiratory bronchioles	Begins as simple cuboidal; ends as simple squamous
Alveolar ducts, alveolar sacs, alveoli	Simple squamous

POSTERIOR

ANTERIOR LM (2.6×)

(a) Transverse section of trachea and esophagus

Terminal bronchiole

Blood vessel

Visceral pleura

(b) Sectional view of lung

FIGURE 32.9 Sectional views of the trachea and lung.

- epithelial lining of trachea
- lumen of esophagus
- lumen of trachea
- tracheal cartilage
- trachealis muscle

1 _____

2 _____

3 _____

4 _____

5 _____

- alveolar ducts
- alveolar sac
- alveoli
- respiratory bronchiole
- simple squamous epithelium

6 _____

7 _____

8 _____

9 _____

10 _____

DISCUSSION QUESTIONS: MICROSCOPIC ANATOMY OF RESPIRATORY SYSTEM

1 Which connective tissue type allows the lungs to expand and regain their original size and shape (recoil) during inhalation and exhalation?

2 Which connective tissue type in the bronchial tree provides the support to maintain an open airway?

3 Discuss the significance of the changing epithlelia along the respiratory tree.

C. Animal Dissection

If you are dissecting a cat or fetal pig to observe respiratory system structures, refer to the appropriate accompanying dissection manual.

The respiratory system organs of a cat or fetal pig are similar to those of the human. This dissection will illustrate the structure of the larynx and trachea, the relationship of respiratory organs to other organs in the mediastinum, and the connective tissues surrounding these organs.

EXERCISE

32

Reviewing Your Knowledge

A. Anatomy Drill and Practice

Go to www.wiley.com/college/allen. Click on the picture of your lab manual, then on Student Companion Site, Anatomy Drill and Practice, and Exercise 32.

B. Air Flow

Trace the air flow through the respiratory system starting with the *external nares*. Number the structures 1 through 17.

_____ alveolar duct _____ nasopharynx

_____ alveolar sac _____ oropharynx

_____ alveolus _____ primary bronchus

_____ bronchiole _____ respiratory bronchiole

_____ external nares _____ secondary bronchus

_____ internal nares _____ terminal bronchiole

_____ laryngopharynx _____ tertiary bronchus

_____ larynx _____ trachea

_____ nasal cavity

C. Structural Changes in the Respiratory Tree

Name three structural changes that occur in the bronchi as they branch into bronchioles.

1. _____

2. _____

3. _____

D. Structure and Function of the Respiratory System

Write the name of the structure described.

_____ **1.** Tubular airways that begin the respiratory zone

_____ **2.** Connects the laryngopharynx with the trachea

_____ **3.** Tubelike structure that conducts air from the larynx to the bronchi

_____ **4.** Closes over the glottis during swallowing

_____ **5.** Keep the trachea from collapsing

_____ **6.** Division of the bronchi that enter bronchopulmonary segments

_____ **7.** Last division of the conducting zone

_____ **8.** Conducts air from the nasopharynx to the laryngopharynx

_____ **9.** Small, round sacs where gas exchange occurs

_____ **10.** Small conduction airway that serves a lobule

E. Location and Function of Epithelial Tissue

Match the type of epithelium with its location in the respiratory system and its function.

Epithelial Tissue

 a. pseudostratified ciliated columnar d. simple squamous
 b. simple columnar e. stratified squamous
 c. simple cuboidal

Location

_____ **1.** Nasal cavity through nasopharynx

_____ **2.** Oropharynx through larynx above vocal cords

_____ **3.** Larynx below vocal cords through primary bronchi

_____ **4.** Secondary bronchi through tertiary bronchi

_____ **5.** Bronchioles through beginning of respiratory bronchiole

_____ **6.** End of respiratory bronchiole through alveoli

Function

_____ **7.** Secretes mucus to trap and remove dust and debris

_____ **8.** Diffusion of respiratory gases

_____ **9.** Protects underlying tissues

EXERCISE

32

Using Your Knowledge

A. Anatomy of the Respiratory System

Answer the following questions.

1. Women in the late nineteenth and early twentieth centuries wore whalebone corsets that severely restricted the respiratory and digestive systems. These corsets were worn to have a wasplike waist. It was quite common for these women to faint—in fact, they had "fainting couches." Explain the physiological reason for their fainting problems.

2. Explain how an inflection in the nasopharynx can also result in an inflection in the paranasal sinuses and/or the middle ear.

3. Asthma attacks are caused by smooth muscle spasms in the bronchial tree. These spasms can close the airway. What airway structures are closed? Why are these areas closed but not the rest of the bronchial tree?

4. Emphysema is characterized by destructive changes of alveolar walls resulting in large air-filled spaces due to over-inflation. There is a loss of lung elasticity, decreased gas exchange, and large air-filled spaces.

 Pneumonia is an acute inflammation of the lungs with the alveoli and bronchioles becoming plugged with an exudate (fluid), and the alveolar cell membranes thicken.

 In Figure 32.10, identify which alveolar section came from a person with no lung disease (normal alveoli), with emphysema, or with pneumonia.

(a) _____ (b) _____ (c) _____

FIGURE 32.10 Normal and diseased alveoliar sections.

Explain how the changes in alveolar structure cause a decrease in blood oxygen levels in:

5. Emphysema

6. Pneumonia

Identify the structures in Figures 32.11 and 32.12.

FIGURE 32.11 Thoracic cavity, cross-section.

7 _____ (membrane)

8 _____ (include right or left)

FIGURE 32.12 Bronchoscopic view of the larynx.

9 _____

10 _____

Respiratory System Function

33

Objectives

After completing this exercise, you should be able to:

- Explain how changes in thoracic and lung volumes and lung pressures result in pulmonary ventilation
- Define and measure or calculate lung volumes and capacities
- Discuss the relationship between blood carbon dioxide levels and pulmonary ventilation

Materials

- models or charts with respiratory muscles
- bell jar model of lungs
- Effect of Exercise on Lung Volumes and Capacities: handheld dry spirometer, disposable mouthpieces, 70% alcohol and cotton, or alcohol wipes
- Formation of Bicarbonate and Hydrogen Ions: paper bag (lunch size), straws, 50-mL beakers, 100-mL beakers, 25-mL graduated cylinders, pH meter, squeeze bottle of deionized water, stopwatch or watch with second hand

The respiratory system supplies oxygen needed by body cells to produce adenosine triphosphate (ATP) for metabolism and removes carbon dioxide produced by metabolic reactions. Respiration involves three steps and requires the cardiovascular system to transport oxygen and carbon dioxide throughout the body.

1. **Pulmonary ventilation,** or breathing, is the movement of air between the atmosphere and the lungs that occurs when we inhale (inhalation) and exhale (exhalation).

2. **External respiration** is the movement of oxygen from the alveoli into pulmonary capillaries and carbon dioxide from pulmonary capillaries to the alveoli.

3. **Internal respiration** is the movement of oxygen from capillaries into body cells and carbon dioxide from body cells into capillaries.

A. Pulmonary Ventilation

During pulmonary ventilation, air moves from an area of higher pressure to an area of lower air pressure. Changes in air pressure in the lung (alveolar pressure) occur when lung volume changes. The relationship between pressure and volume is described by Boyle's Law, which states that the pressure of a gas in a closed container is inversely proportional to the volume of the container. Therefore, when lung volume increases, the pressure of the air inside decreases and air flows into the lungs. When lung volume decreases, the pressure of the air inside increases and air flows out of the lungs.

Changing Thoracic and Lung Volumes

When the diaphragm and other respiratory muscles contract or relax, they change the size of the thorax which in turn changes lung volume. Normal inhalation is caused mainly by contraction of the diaphragm. The **diaphragm** is dome-shaped when relaxed and flattens when contracted. When the diaphragm flattens, the length of the thoracic cavity and its volume increase. During forced inhalation, contraction of the **external intercostals** further increases the width of the thoracic cage by raising the ribs. Contraction of the **sternocleidomastoid, scalenes,** and **pectoralis minor muscles** cause a greater increase in thoracic volume by elevating the rib cage and sternum, resulting in a greater volume of air inhaled.

Normal exhalation is a passive process involving relaxation of the diaphragm and elastic recoil of the chest wall and lungs. The diaphragm becomes dome-shaped and decreases the length of the thoracic cavity and thoracic volume. In forced exhalation, contraction of the **internal intercostals** depresses the rib cage. Contraction of the **abdominal muscles (external oblique, internal oblique, transverse abdominis,** and **rectus abdominis)** pushes the diaphragm superiorly, further decreasing thoracic volume and resulting in a greater volume of exhaled air.

ACTIVITY 1 Respiratory Muscles and Volume Changes

1 Label the respiratory muscles in Figure 33.1. Refer to Exercise 14: Skeletal Muscles or your textbook.

2 Identify the respiratory muscles on a model or anatomical chart.

3 Pronounce the muscle names as you point to them.

4 Complete Table 33.1 and circle the correct choice in the column "Effect on Thoracic Dimensions."

5 Examine the radiographs in Figure 33.2. Decide whether a radiograph was taken after inhalation or exhalation, and put your answer in the appropriate blank.

TABLE 33.1 Respiratory Muscle Functions

MUSCLE	FUNCTION	EFFECT ON THORACIC DIMENSIONS (circle correct choice)
External intercostals	Elevate ribs	Increase or decrease diameter
Internal intercostals	Depress ribs	Increase or decrease diameter
Sternocleidomastoids	Elevate the sternum	Increase or decrease diameter
Scalenes	Elevate 1st and 2nd ribs	Increase or decrease diameter
Pectoralis minor	Elevate 3rd, 4th, and 5th ribs	Increase or decrease diameter
Diaphragm	Flattens when contracted	Increase or decrease length
Abdominal muscles	Compress abdominal contents, increase abdominal pressure, and force diaphragm superiorly	Increase or decrease length

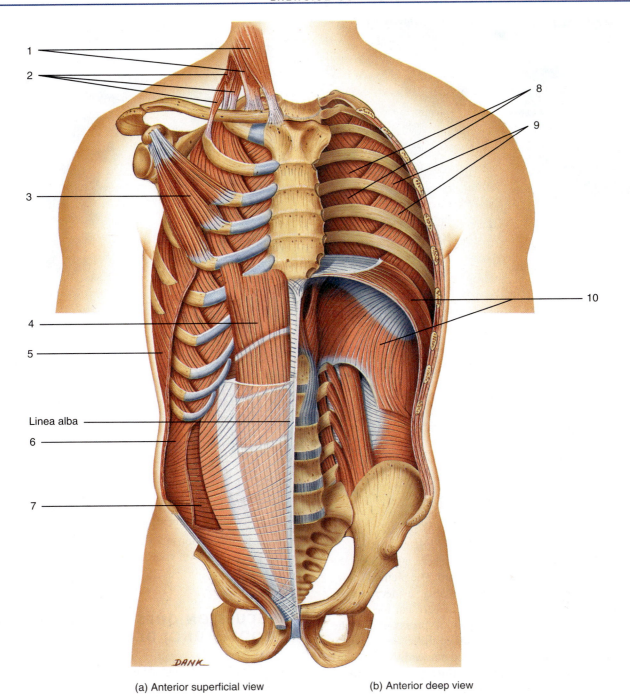

(a) Anterior superficial view (b) Anterior deep view

FIGURE 33.1 Respiratory muscles.

- diaphragm
- external intercostal
- external oblique
- internal intercostals
- internal oblique
- pectoralis minor
 (pek-tor-A-lis)

- rectus abdominis
- scalenes (SKAY-leens)
- sternocleidomastoid
 (ster-no-kli-doh-MAS-toid)
- transverse abdominis

1 _____ 6 _____

2 _____ 7 _____

3 _____ 8 _____

4 _____ 9 _____

5 _____ 10 _____

(a)

(b)

FIGURE 33.2 Radiographs of the thorax taken either after inhalation or exhalation.

Indicate whether the view shows inhalation or exhalation.

a _____

b _____

Pressure Changes During Pulmonary Ventilation

As the thorax expands, the parietal pleura attached to the internal thoracic wall is pulled outward. The **pleural cavity** increases slightly in volume causing a decrease in **intrapleural pressure,** the pressure between the pleural layers. The decrease in intrapleural pressure and the surface tension of pleural fluid cause the visceral pleura, and therefore the lungs, to be pulled outward. As the lungs increase in volume, the **alveolar (intrapulmonic) pressure** decreases to below atmospheric pressure, and air enters the lungs. When the thorax decreases in size the lungs recoil, causing intrapleural pressure and alveolar pressure to increase, and air leaves the lungs.

ACTIVITY 2 **Pressure Changes During Pulmonary Ventilation**

1 Using the model lung (bell jar model), observe what happens when the rubber diaphragm at the base of the model is domed or flattened.

2 Label Figure 33.3.

3 Complete Table 33.2 by circling the correct choice in the column "Movement of Diaphragm."

4 Answer discussion questions concerning model lung demonstration with your lab partner(s).

? **DISCUSSION QUESTIONS: BELL JAR (MODEL LUNG) DEMONSTRATION**

1 (a) In the model lung demonstration, the glass jar is filled with air. What is in the pleural cavity in the thorax? (b) Compare the relative size of the pleural space in the model lung to that in the human lung.

2 (a) In the model lung, is the air in the glass jar at atmospheric pressure or subatmospheric pressure? (b) Is the pressure in the pleural cavity atmospheric or subatmospheric?

3 (a) If the chest wall (including parietal pleura) is punctured, would the pressure in the pleural cavity increase or decrease? (b) Would lung size increase or decrease? (c) What is this condition called?

4 (a) In humans, is there one pleural cavity surrounding both lungs or two separate pleural cavities, one surrounding each lung? (b) Does this differ from the model lung?

5 In the model lung demonstration, air flows into the balloons through a rigid tube that divides once. Compare the rigidity and the number of branches of the conducting structures in the model to the rigidity and number of branches of the conducting structures in the human respiratory system.

6 In the model lung demonstration, the glass bell jar does not change in size. Describe the changes in the corresponding human respiratory system structures during inhalation and exhalation.

FIGURE 33.3 Model lung (bell jar model).

- bronchus
- diaphragm
- lungs covered with visceral pleura
- parietal pleura and thoracic wall
- pleural cavity
- trachea

1 _____
2 _____
3 _____
4 _____
5 _____
6 _____

TABLE 33.2 Bell Jar (Model Lung) Demonstration

OBSERVATIONS	MOVEMENT OF DIAPHRAGM	
	As Diaphragm Domes (circle correct choice)	*As Diaphragm Flattens* (circle correct choice)
Volume of air in bell jar	1. Increased or decreased	5. Increased or decreased
Pressure of air in bell jar	2. Increased or decreased	6. Increased or decreased
Balloon size	3. Inflated or deflated	7. Inflated or deflated
Direction of air flow	4. Into or out of balloons	8. Into or out of balloons

B. Lung Volumes and Capacities

Lung volumes and capacities are defined in Table 33.3. Lung capacities are combinations of lung volumes. Both lung volume and lung capacity values vary according to gender, age, and height. Lung volumes also vary with exercise, but lung capacities do not.

Measuring and Calculating Lung Volumes and Capacities

A **spirometer** is an instrument used to measure lung volumes and capacities. There are different types of spirometers depending on how they measure air volumes and capacities. Some spirometers measure only expiratory volumes and capacities while others measure both inspiratory and expiratory volumes and capacities. In the following activity, directions are given for using a dry, handheld spirometer. *Exhale only into this type of spirometer.* It does not measure inspiratory volumes. Spirometers that measure inspiratory air volumes must be cleaned or supplied with new filters when used by a different person to prevent spread of infection. Your instructor will supply you with instructions if you are using a different type of spirometer instead of the dry, handheld spirometer.

Dry, handheld spirometers cannot be used to measure inspiratory volumes and capacities. However, IRV and IC can be calculated using values for TV, ERV, and VC.

Clinically, FRC is measured indirectly with special equipment and RV is calculated using values for FRC and ERV. In this activity, RV will be calculated with equations that were developed using RV values from many individuals.

ACTIVITY 3 Experiment: Effect of Exercise on Lung Volumes and Capacities

1 **Prediction:** With your lab group, predict how exercise will affect TV, IRV, ERV, VC, and TLC. Circle your answer (in italics) below.
 - During exercise TV will *increase, decrease,* or *stay the same.*
 - During exercise IRV will *increase, decrease,* or *stay the same.*
 - During exercise ERV will *increase, decrease,* or *stay the same.*
 - During exercise VC *will increase, decrease,* or *stay the same.*
 - During exercise TLC *will increase, decrease,* or *stay the same.*

2 **Materials:** Obtain the materials for Effect of Exercise on Lung Volumes and Capacities given in the Materials list.

3 **Data Collection:** Measure lung volumes and capacities before and after exercise and record the values in Table 33.4.
 - Decide who will be the subject, who will prepare the spirometer, who will prepare and coach the subject, and who will be the timer and recorder. The subject should not have heart problems and should be in good health.

TABLE 33.3 Lung Volumes and Capacities

VOLUME OR CAPACITY	ABBREVIATION	DEFINITION	AVERAGE VALUES FOR HEALTHY ADULT
Tidal volume	TV	The amount of air inhaled and exhaled during one normal breath at rest	500 mL
Inspiratory reserve volume	IRV	Maximum amount of air that can be inhaled after a normal inhalation	3,100 mL
Expiratory reserve volume	ERV	Maximum amount of air that can be exhaled after a normal exhalation	1,200 mL
Residual volume	RV	Amount of air that remains in the lungs after a maximal exhalation	1,200 mL
Inspiratory capacity	IC	IC = TV + IRV	3,600 mL
Functional residual capacity	FRC	FRC = RV + ERV	2,400 mL
Vital capacity	VC	VC = IRV + TV + ERV	4,800 mL
Total lung capacity	TLC	TLC = IRV + TV + ERV + RV	6,000 mL

TABLE 33.4 Measured Lung Volumes and Capacities

AT REST	VALUE 1	VALUE 2	VALUE 3	AVERAGE VALUE
TV				
ERV				
VC				
AFTER EXERCISE	**VALUE 1**	**VALUE 2**	**VALUE 3**	**AVERAGE VALUE**
TV				
ERV				
VC				

Preparing the spirometer

- Wipe the nozzle of the spirometer with 70% alcohol.
- Place a clean, disposable mouthpiece on the nozzle.
- With use, water will condense inside the dial. Follow instructions included with your spirometer for removing condensation.

Preparing and coaching the subject

- Use a noseclip or your fingers to close your nostrils to prevent air leaking out of your nose.
- When blowing into the spirometer, hold it in a horizontal position with the face of the dial upward.
- Have subject practice breathing to measure TV, ERV, and VC. When **measuring TV,** the subject should breathe normally. When **measuring ERV,** the subject should take two or three breaths, then inhale normally and forcefully exhale. Verbally encourage the subject to exhale as much air as possible. When **measuring VC,** the subject should take two or three normal breaths, then completely fill the lungs with air and forcefully exhale as much air as possible as fast as possible.
- Remind subject to *exhale only* into a handheld spirometer.

Measuring TV, ERV, and VC at rest

- Measure tidal volume (TV)
 Reset the dial on the spirometer to 0.
 Take two or three normal breaths and then inhale normally, place the spirometer to your mouth and exhale normally into the spirometer. Repeat this process *two more times,* and **do not reset** the dial before each measurement. Divide the volume by 3 to obtain the average TV. Record the average TV in Table 33.4.

- Measure expiratory reserve volume (ERV)
 Reset the dial to 0.
 Take two or three normal breaths, and then inhale normally and exhale normally. Without taking another breath, place the spirometer to your mouth and exhale as much air as you can. Record the value for ERV in Table 33.4. Repeat this process two more times, *resetting the dial to 0 before each measurement.*

- Measure vital capacity (VC)
 Reset the dial to 0.
 Take two or three normal breaths, then inhale as much air as possible. Place the spirometer to your mouth and forcefully exhale as much air as you can as fast as you can. Record the value for VC in Table 33.4. Repeat this process two more times, *resetting the dial to 0 before each measurement.*

Measuring TV, ERV, and VC after exercise

- The subject should not have heart problems and should be in good health.
- Have the subject run in place for 5 minutes.
- Measure tidal volume (TV)
 Reset the dial on the spirometer to 0.
 Take two or three breaths, and then inhale normally. Place the spirometer to your mouth and exhale normally into the spirometer. Record the value for TV in Table 33.4. Repeat this process two more times, but this time **reset the dial before each measurement.**
- Measure expiratory reserve volume (ERV) as before.
- Measure vital capacity (VC) as before.

4 Clean Up: Clean up as directed by your instructor

5 **Data Analysis:** Calculate average ERV and VC and record in Table 33.4. Calculate IRC, IC, FRC, and TLC and record values in Table 33.5.

Calculate average ERV and VC

- Add the individual ERVs and divide by 3. Record the average ERV in Table 33.4.
- Add the individual VCs and divide by 3. Record the average VC in Table 33.4.

Calculate IRV and IC using average TV, ERV, and VC Values

- Use the equations for VC and IC in Table 33.3 to calculate IRV and IC. Rearrange the VC equation to formulate an equation to calculate IRV.
- Calculate the values and record them in Table 33.5.

Calculate RV, FRC, and TLC

- Use the appropriate equation in Table 33.6 to calculate your RV based on gender and age, and record the RV in Table 33.5.
- Calculate your FRC and TLC using the equations in Table 33.3 and record the value in Table 33.5.

6 **Graphs:** If graph paper is available, construct a bar graph that compares resting and exercising lung volumes and capacities.

7 Complete the Experimental Report with your lab group.

EXPERIMENTAL REPORT: EFFECT OF EXERCISE ON LUNG VOLUMES AND CAPACITIES

Results: State how the various lung volumes and capacities changed with exercise.

Discussion: Discuss why the various lung volumes and capacities either changed or did not change with exercise.

Conclusion: In one or two sentences, state how exercise affects lung volumes and capacities.

TABLE 33.5 Calculated Lung Volumes and Capacities

VOLUME OR CAPACITY	EQUATION	VALUE
IRV		
IC		
RV		
FRC		
TLC		

TABLE 33.6 Equations for Calculating Residual Volume

GENDER	AGE	EQUATION TO PREDICT RV (LITERS)
Male or Female[1]	<19	RV = (0.029 × height in inches) − 0.9192
Female[2]	19–99	RV = (0.0813 × height in inches) + (0.009 × age in years) − 3.9
Male[2]	19–99	RV = (0.0686 × height in inches) + (0.017 × age in years) − 3.45

1. Polgar G, Promadhat V: Pulmonary Function Testing in Children. Techniques and Standards. Philadelphia: W.B. Saunders Co., 1971.
2. Goldman HI, Becklake MR: Respiratory function tests—normal values at median altitudes and the prediction of normal results. Am Rev Tuberc 79:457–467, 1959.

Timed Forced Expiratory Volumes

A timed forced expiratory volume is used to indicate whether someone has a condition in which air flow is obstructed or restricted. The subject is asked to inhale as much air as possible, then forcibly blow as much air as fast as possible into the spirometer. The volume of air forcibly expelled in the first second (**forced expiratory volume in 1 second or FEV$_1$**) and the total volume of forcibly expelled air [**forced vital capacity (FVC)**] are used to determine if a person has an obstructed air flow. The FEV$_1$/FVC ratio is used to indicate if a person has an obstructive disease such as chronic obstructive pulmonary disease (COPD). In COPD the airway is obstructed and it takes longer to move air out of the lungs, therefore the FEV$_1$ is reduced. During pregnancy, the thoracic volume is restricted so that FVC is reduced, but the ratio of FEV$_1$/FVC is normal (i.e., assumes no obstructive disease).

ACTIVITY 4 **Measuring FEV$_1$ and FVC**

1 The graph in Figure 33.4 illustrates the change in volume over time. The x-axis indicates time and the y-axis indicates volume.

2 Use the values on the y-axis to determine the maximum volume indicated by the curved line, which is the FVC. Record this value in Figure 33.4.

3 Find 1 second on the x-axis. Using a ruler, draw a vertical line that intersects the curve at 1 second. Find the corresponding volume on the y-axis, the FEV$_1$, and record this value in Figure 33.4.

4 Calculate the FEV$_1$/FVC ratio by dividing the FEV$_1$ by the FVC and record this value in Figure 33.4.

FVC = ____ L FEV$_1$ = ____ L FEV$_1$/FVC = ____

FIGURE 33.4 Timed forced expiratory volume.

C. Control of Pulmonary Ventilation Rate: The Role of Carbon Dioxide

Carbon dioxide is formed during aerobic cellular respiration when glucose, fatty acids, and amino acids are broken down to make ATP. Carbon dioxide then diffuses out of cells, into the interstitial fluid and into the bloodstream. Blood travels to the lungs where carbon dioxide diffuses across pulmonary capillaries and into the alveoli, leaving the lungs when we exhale. Exhaled air has a greater percentage of carbon dioxide than atmospheric air.

Carbon dioxide is transported in the bloodstream as dissolved carbon dioxide, bound to hemoglobin, or as **bicarbonate ions.** The majority of carbon dioxide is transported as bicarbonate ions. Bicarbonate ions (HCO_3^-) are formed from carbon dioxide as follows:

1. Carbon dioxide combines with water to form **carbonic acid.**

$$CO_2 + H_2O \rightleftharpoons H_2CO_3$$

2. Carbonic acid dissociates in water to form hydrogen ions and bicarbonate ions.

$$H_2CO_3 \rightleftharpoons H^+ + HCO_3^-$$

The chemical equation for the formation of bicarbonate and hydrogen ions from carbon dioxide and water is:

$$CO_2 + H_2O \rightleftharpoons H_2CO_3 \rightleftharpoons H^+ + HCO_3^-$$

The ratio of carbon dioxide to carbonic acid to bicarbonate ions is constant. If blood carbon dioxide levels increase (as occurs after exercise), then more carbonic acid and bicarbonate and hydrogen ions are formed. If carbon dioxide levels are reduced (as occurs in lungs after exhalation), then more bicarbonate ions and hydrogen ions will form carbonic acid, which will dissociate back into carbon dioxide and water.

The pH of blood is determined by the concentration of hydrogen ions in blood. Blood carbon dioxide levels increase during exercise (carbon dioxide is a by-product of aerobic respiration). The chemical equation shifts to the right and more hydrogen and bicarbonate ions are formed. This results in a decrease in blood pH due to the increased levels of hydrogen ions. The increased levels of hydrogen ions stimulate chemoreceptors in the brain, aortic arch, and carotid bodies, resulting in an increased rate of pulmonary ventilation or breathing. This increased breathing rate allows the lungs to exhale more carbon dioxide. As blood carbon dioxide levels decrease, the chemical equation shifts to the left and more hydrogen ions and bicarbonate ions form carbonic acid. The blood hydrogen ion levels decrease and blood pH returns to normal.

ACTIVITY 5 Formation of Bicarbonate and Hydrogen Ions

1 **Prediction:** With your lab group, predict how exhaling into water after exercise will change the pH of the water. Circle your answer (in italics) below.

- Exhaling into water after exercise will *increase, decrease,* or *not change* the pH.

2 **Materials:** Obtain the materials for Formation of Bicarbonate and Hydrogen Ions given in the Materials list.

3 **Data Collection:** Measure change in pH after exhaling into water while at rest and after exercise. Record your results in Table 33.7.

- Decide who will be the subject, who will set up the experiment, who will be the timer, and who will measure and record.

TABLE 33.7 Exhaled CO_2 Levels and Breathing Rate Before and After Exercise

PROCEDURE: BLOWING AIR INTO BEAKER	BREATHING RATE (BREATHS/MIN)	CONTROL pH	EXPERIMENTAL pH	% pH CHANGE	CLASS AVERAGE OF % pH CHANGE
At rest					
After exercise					

Change of pH at rest

- Measure 20 mL of deionized water using a graduated cylinder and pour into a 50-mL beaker.
- Measure pH of the deionized water and record the value in Table 33.7 as control pH.
- Count the number of breaths the subject takes over a 30-second period, multiply by 2, and record the number of breaths/min in Table 33.7.
- Have the subject take a deep breath and, using a straw, blow air into the water for 40 seconds.
- Measure the pH of the water and record the value in Table 33.7 as experimental pH.
- Discard water and rinse the beaker with deionized water.

Change in pH after exercise

- Measure 20 mL of deionized water using the graduated cylinder and pour into the 50-mL beaker.
- Measure pH of the deionized water and record the value in Table 33.7 as control pH.
- Have the subject run in place or climb steps for at least 3 minutes. During the last 30 seconds, count the number of breaths over this 30-second period, multiply by 2, and record the number of breaths/min in Table 33.7.
- As soon as the subject stops exercising, have him/her take a deep breath and, using a straw, blow air into the water for 40 seconds. The subject might need to inhale once or twice to blow air for 40 seconds.
- Immediately measure the pH of the water and record the value in Table 33.7 as experimental pH.
- Discard the water.

4 Data Analysis: Calculate the percentage change in pH for both procedures and record your values in Table 33.7.

- % change in pH = (control pH − experimental pH)/ control pH × 100
- Pool the class data and calculate average values for percent change in pH and breathing rate for both procedures and record in Table 33.7.

5 Graphs: If graph paper is available, construct a bar graph that compares resting and exercising pH values.

6 Complete the Experimental Report with your lab group.

EXPERIMENTAL REPORT: BICARBONATE AND HYDROGEN ION FORMATION

Results: State the pH values at rest and after exercise and state the change in pH values after exercise.

Discussion:

- Explain how blowing air into the water caused the change in pH.

- Describe how hydrogen ion concentration and bicarbonate ion concentration change as pH changes.

- Describe why exercise caused the observed change in pH.

- Describe why exercise caused the observed change in respiratory rate.

Conclusion: In one or two sentences, state how exercise causes pH changes which, in turn, lead to respiratory changes.

Name _____ Date _____ Section _____

Reviewing Your Knowledge

A. Anatomy Drill and Practice

Match the respiratory muscles to the appropriate muscle function.

_____ **1.** elevates 3rd, 4th, and 5th ribs

_____ **2.** elevates the sternum

_____ **3.** depresses ribs

_____ **4.** compresses abdominal contents and increases abdominal pressure

_____ **5.** elevates first and second ribs

_____ **6.** elevates ribs

_____ **7.** main inspiratory muscle

_____ **8.**
_____ **9.** } muscles used in forced expiration

a. abdominal muscles
b. diaphragm
c. external intercostals
d. internal intercostals
e. pectoralis minor
f. scalenes
g. sternocleidomastoids

B. Volume and Pressure Changes During Pulmonary Ventilation

Indicate whether the volume or pressure increases or decreases during inspiration and expiration in Table 33.8.

TABLE 33.8 Volume and Pressure Changes During Pulmonary Ventilation

VOLUME OR PRESSURE	INSPIRATION	EXPIRATION
Thoracic volume	1.	2.
Intrapleural cavity volume	3.	4.
Intrapleural pressure	5.	6.
Lung volume	7.	8.
Alveolar (intrapulmonic) pressure	9.	10.

C. Lung Volumes and Capacities

Match lung volumes and capacities with the appropriate definition or equation.

_____ **1.** equal to TV + IRV + ERV + RV

_____ **2.** equal to IC − TV

_____ **3.** equal to FRC − ERV

_____ **4.** equal to TV + IRV

_____ **5.** volume of air remaining in lungs after normal expiration

_____ **6.** maximum amount of air that can be exhaled after a normal expiration

_____ **7.** maximum amount of air that can be exhaled after a maximal inspiration

_____ **8.** equal to IC − IRV

a. expiratory reserve volume
b. functional residual capacity
c. inspiratory capacity
d. inspiratory reserve volume
e. residual volume
f. tidal volume
g. total lung capacity
h. vital capacity

D. Control of Pulmonary Ventilation: The Role of Carbon Dioxide

Circle True or False for the following questions. If false underline and change the word(s) that are incorrect to make the statement true.

1. If blood carbon dioxide levels increase, then blood hydrogen ion levels will decrease. True or False

2. If blood carbon dioxide levels increase, then blood pH will decrease. True or False

3. Increasing blood carbon dioxide levels decreases breathing rate. True or False

Using Your Knowledge

A. Physiology of the Respiratory System

Write a short answer for the following questions.

1. A 32-year-old man presented to the emergency room with a pneumothorax. The right lung was collapsed, but the left lung was still inflated. Explain.

2. How does a pneumothorax affect lung volume and alveolar pressure when inspiratory muscles contract? How does this affect air flow?

3. The Valsalva maneuver is forced expiration against a closed glottis. How would you demonstrate the Valsalva maneuver using the model lung?

A 42-year-old woman is breathing rapidly and deeply after exercise. Indicate whether each of the following volumes increases, decreases, or stays the same when compared to resting volumes.

Example: TLC: stays the same _____

4. RV: _____

5. TV: _____

6. IRV: _____

7. ERV: _____

8. A 5-year old child announces to her parents that she is going to hold her breath until she is allowed to watch more television. The parents are not worried. (a) Explain why the parents are not worried. (b) Explain how the child's blood carbon dioxide levels and blood hydrogen ion and bicarbonate ion levels change while she is holding her breath.

9. Explain how it is possible for a person to drink liquid and then have the liquid come out through the nose when the person laughs.

10. When someone is hyperventilating, they are told to breathe into a paper bag or into their cupped hands. (a) Explain why this increases blood carbon dioxide levels. (b) What will this do to the breathing rate?

Digestive System Structure

**After completing this exercise,
you should be able to:**

- Identify the major gastrointestinal (GI) tract organs and accessory digestive organs on a torso model or chart
- Describe the layers of the GI tract walls and the peritoneum
- Describe the functions of each GI tract and accessory digestive organ
- Identify the major features of digestive organs microscopically
- Identify the main digestive organs of a cat or fetal pig

Materials

- human torso
- sagittal head section model
- skull with teeth
- compound microscope, lens paper, prepared slides: esophagus, stomach, small intestine, large intestine, pancreas, and liver
- preserved cat or fetal pig, disposable gloves, dissection equipment, and safety glasses, accompanying cat or fetal pig dissection manual

A. Overview of the Digestive System

The **digestive system** contains the organs of the gastrointestinal (GI) tract and the accessory digestive organs. The **GI tract** or **alimentary** (*alimentary* = nourishment) **canal** is a tube that extends from the mouth to the anus. The lumen of the GI tract opens to the external environment at either end, and anything inside the lumen is therefore considered to be external to the body. The **GI tract organs** include the mouth, most of the pharynx, esophagus, stomach, small intestine (duodenum, jejunum, ileum), large intestine (colon), and anus. The **accessory digestive organs**—the teeth, tongue, salivary glands, liver, gallbladder, and pancreas—assist in digestion. Functions of the digestive system include **ingestion** of food; **secretion** of digestive juices; **mixing** and **propulsion** of food; **digestion** of food into small, absorbable molecules; **absorption** of digestive products across the GI tract wall and into the body; and **defecation.** Digestion includes both **mechanical digestion,** breaking food into smaller pieces, and **chemical digestion,** enzymatic breakdown of large molecules into smaller ones.

ACTIVITY 1 **Organs of the Digestive System**

1 Label and review the structures in Figure 34.1. Refer to Figure 2.1.

2 Identify the structures on a torso or chart.

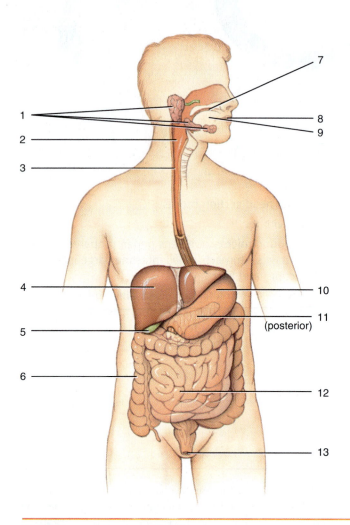

- anus
- esophagus
- gallbladder
- large intestine
- liver
- mouth (oral cavity)
- pancreas
- pharynx
- salivary glands
- small intestine
- stomach
- teeth
- tongue

1 _____

2 _____

3 _____

4 _____

5 _____

6 _____

7 _____

8 _____

9 _____

10 _____

11 _____

12 _____

13 _____

FIGURE 34.1 Organs of the digestive system.

B. Layers of the Gastrointestinal Tract and the Peritoneum

The esophagus, stomach, small intestine, and large intestine have a similar four-layered arrangement to their walls. From the interior (deep layer) to the exterior (superficial layer), the four layers are: the mucosa, submucosa, muscularis, and serosa. The **mucosa** is a moist layer that includes the surface epithelium, connective tissue (lamina propria), and smooth muscle (muscularis mucosae). Inspect the inside of your cheek with your tongue. The wet, slick feeling is the mucosa that lines all the organs of the GI tract. The **submucosa** is areolar connective tissue, and the **muscularis** in all GI tract organs except the stomach is comprised of two layers: an inner layer of circular smooth muscle and an outer layer of longitudinal smooth muscle. The stomach muscularis has three layers: an inner oblique, middle circular, and an outer longitudinal smooth muscle layer. The superficial **serosa** is the visceral layer of the peritoneum.

The largest serous membrane in the body, the **peritoneum,** is a combination of simple squamous mesothelium (epithelium) with underlying connective tissue. As with all serous membranes, the **parietal peritoneum** lines the inner surface of the abdominopelvic wall, and the **visceral peritoneum** (serosa) covers organs within the abdominopelvic cavity. **Peritoneal** (serous) **fluid** is found between the two peritoneal layers in the **peritoneal cavity.** Organs that lie outside the peritoneal cavity are called **retroperitoneal** (*retro-* = behind) **organs** (i.e., pancreas and kidneys).

The peritoneum has large folds not found in other serous membranes that secure organs together and to the abdomi-nal walls. Some of these peritoneal folds are the mesentery, mesocolon, falciform ligament, greater omentum, and lesser omentum. The **mesentery** holds the small intestine to the posterior abdominal wall, whereas the large intestine is similarly secured to the posterior abdominal wall by the **mesocolon.** The liver is bound to the anterior abdominal wall by the **falciform** (*falc-* = sickle-shaped) **ligament.** The **greater omentum** (*omentum* = fat skin) is a large, fatty-looking apron that folds back on itself and attaches to the transverse colon. It forms a pouch that hangs between the body wall and the anterior surface of the small intestine. Attached to the lateral curve of the stomach and duodenum, the greater omentum loosely covers the transverse colon and small intestine. The **lesser omentum** connects the medial curve of the stomach with the liver.

ACTIVITY 2 **Layers of the GI Tract and the Peritoneum**

1 Label the structures in Figure 34.2 and Figure 34.3 (a)–(d).

2 Identify the structures on a model or chart.

3 Pronounce the name of each term as you point to it.

- mucosa (myu-KOH-sa)
- muscularis
- serosa (sir-OH-sa)
- submucosa

1 _____

2 _____

3 _____

4 _____

Lumen

1

2

3 4

FIGURE 34.2 Layers of the gastrointestinal tract.

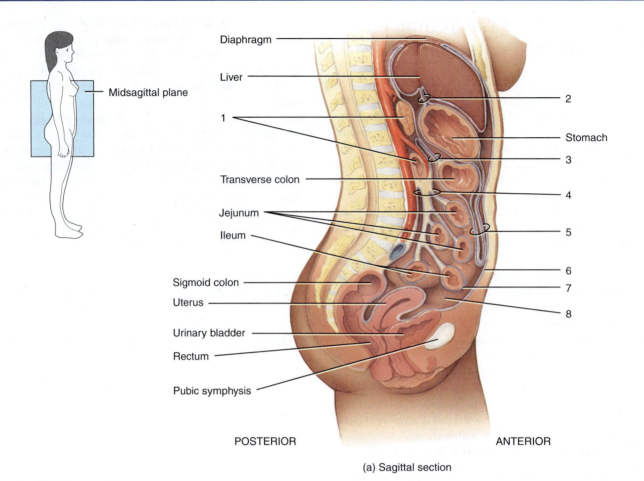

Midsagittal plane

Diaphragm

Liver

1

2

Stomach

3

Transverse colon

4

Jejunum

5

Ileum

6

7

Sigmoid colon

Uterus

8

Urinary bladder

Rectum

Pubic symphysis

POSTERIOR

ANTERIOR

(a) Sagittal section

FIGURE 34.3 Peritoneal folds.

- greater omentum (oh-MEN-tum)
- lesser omentum
- mesentery (MEH-sen-tary)
- mesocolon
- parietal peritoneum (per-ih-toe-NEE-um)
- peritoneal cavity
- retroperitoneal organs
- visceral peritoneum (serosa)

1 _____

2 _____

3 _____

4 _____

5 _____

6 _____

7 _____

8 _____

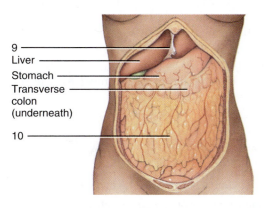

(b) Anterior view

Gallbladder
Liver
Transverse colon
Ascending colon

11
Stomach
Duodenum
Descending colon
Sigmoid colon

(c) Anterior view with liver
and gallbladder lifted

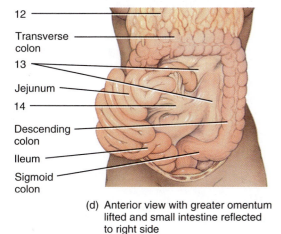

12
Transverse colon
13
Jejunum
14
Descending colon
Ileum
Sigmoid colon

(d) Anterior view with greater omentum
lifted and small intestine reflected
to right side

- falciform ligament
- greater omentum
- greater omentum (reflected)
- lesser omentum
- mesentery
- mesocolon

9 _____

10 _____

11 _____

12 _____

13 _____

14 _____

FIGURE 34.3 Peritoneal folds, *continued.*

C. Gastrointestinal Tract Organs

Mouth

Food first enters the digestive system through the mouth. The **hard** and **soft palates** form the roof of the mouth, the tongue forms the floor, and the **cheeks** form the lateral walls. The anterior border is the **lips,** and the posterior border is the **fauces** (*fauces* = passages), which is the opening to the oropharynx. The mouth is also called the **oral** or **buccal** (*bucca-* = cheeks) **cavity.** The **vestibule** (*vestibule* = courtyard or entrance) is the space between the lips and teeth. The lips are attached to the **gingiva** (*gingiva* = gums) by the **labial frenulum** (*frenulum* = small bridle). The palatine bone and the palatine processes of the maxillae form the **hard palate,** and the soft palate (smooth muscle) is posterior to this. The **soft palate** forms the two arches that border the fauces, and also hangs down to form the oval process called the **uvula.** During swallowing, the soft palate keeps food from entering the nasal cavity. When you look into a mirror with your mouth wide open, you see two arches, the palatoglossal arch (anterior) and palatopharyngeal arch (posterior), with the uvula hanging down in between. The **palatine tonsils** are located between the two arches.

Digestive system functions that occur in the mouth include ingestion, secretion, mechanical and chemical digestion, mixing, and propulsion. The teeth break food into smaller pieces by a mechanical process called **mastication.** The tongue mixes food with saliva that is secreted into the mouth by the salivary glands, initiating chemical digestion. Movements of the tongue and soft palate propel food into the oropharynx during swallowing.

Pharynx

The pharynx is divided into three specific areas, named after structures that are in close proximity. The **nasopharynx,** the superior area, is posterior and inferior to the nose and has respiratory functions. This part of the pharynx begins at the internal naris and ends at the inferior end of the uvula. The middle area, the **oropharynx,** is posterior to the fauces of the oral cavity. This area has both respiratory and digestive functions because air, food, and liquid pass through it. The oropharynx begins at the uvula and extends to the hyoid bone. The inferior area is the **laryngopharynx,** located posterior to the larynx. This area is also a common passageway for air, food, and liquid. The laryngopharynx begins at the hyoid bone and divides into digestive and respiratory pathways just inferior

to the epiglottis. During swallowing, the epiglottis closes off the inferior portion of the laryngopharynx, ensuring food and drink enter the digestive pathway and not the respiratory pathway. The only digestive system function that occurs in the pharynx is propulsion. Contractions of the muscularis layer of the pharynx propel food to the esophagus.

ACTIVITY 3 Mouth and Pharynx

1 Label the structures in Figures 34.4 and 34.5.
2 Identify the structures on a model or chart.
3 Pronounce the name of each term as you point to it.

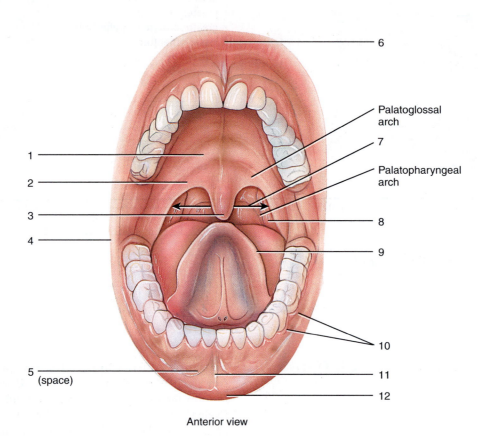

Anterior view

FIGURE 34.4 The mouth.

- cheek
- fauces
- gingivae (gums)
- hard palate
- inferior labial frenulum (FREN-u-lum)
- inferior lip
- palatine tonsil
- soft palate
- superior lip
- tongue
- uvula
- vestibule

1 _____ 7 _____

2 _____ 8 _____

3 _____ 9 _____

4 _____ 10 _____

5 _____ 11 _____

6 _____ 12 _____

Sagittal plane

Sagittal section

FIGURE 34.5 Mouth and pharynx.

- esophagus
- fauces
- hard palate
- laryngopharynx (la-RIN-go-fair-inks)
- nasopharynx (NA-zo-fair-inks)
- oral or buccal cavity
- oropharynx (OR-o-fair-inks)
- soft palate
- tongue
- uvula (YOU-vu-la)

1 _____ 7 _____

2 _____ 8 _____

3 _____ 9 _____

4 _____ 10 _____

5 _____

6 _____

Esophagus

Food and drink move from the mouth into the oropharynx, laryngopharynx, and then the esophagus. The esophagus is a flattened muscular tube about 10 inches long that appears oval in cross-section but distends when food moves through it. The esophagus is posterior to the trachea and is bound with it by connective tissue. The **upper esophageal sphincter** controls the passage of food from the laryngopharynx into the esophagus. Food then continues through the **esophagus** and the **lower esophageal sphincter** into the stomach. The esophageal hiatus is the opening in the diaphragm through which the esophagus passes and is the site of a hiatal hernia.

The only digestive system function that occurs in the esophagus is propulsion, which moves food into the stomach. This is accomplished by **peristalsis,** a series of contractions and relaxations of the muscularis layer.

Stomach

The esophagus leads into the superior part of the stomach, which lies against the diaphragm superiorly. The J-shaped stomach has four areas: the cardia, fundus, body, and pylorus. The **cardia** is the first part of the stomach, inferior to the lower esophageal sphincter. The **fundus** is the elevated area to the left of the cardia that serves as a temporary holding area for food. The **body** is the largest, curved part of the stomach, whereas the **pylorus** continues to the right of the body after the curve. The pylorus is divided into two areas—the **pyloric antrum** (*antrum* = cave), which is next to the body, and the **pyloric canal,** which is nearer the pyloric sphincter. The **pyloric sphincter** is a smooth muscle that allows food to enter the first part of the small intestine, the duodenum. The stomach has large, conspicuous ridges in the mucosa called **rugae** that allow the stomach to be stretched when food and drink are

present. Rugae disappear as the stomach becomes fully distended. The lateral curve of the stomach is the **greater curvature,** and the smaller, medial curve is called the **lesser curvature.**

The digestive system functions that occur in the stomach include secretion of gastric juices, mixing and propulsion, and mechanical and chemical digestion. Contractions of the muscularis layer of the stomach mechanically mix and churn food, breaking it into smaller pieces, and propel food through the stomach and into the small intestine.

Gastric juices begin the chemical digestion of protein and lipids.

ACTIVITY 4 Esophagus and Stomach

1 Label the structures in Figure 34.6.

2 Identify the structures on a model or chart.

3 Pronounce the name of each term as you point to it.

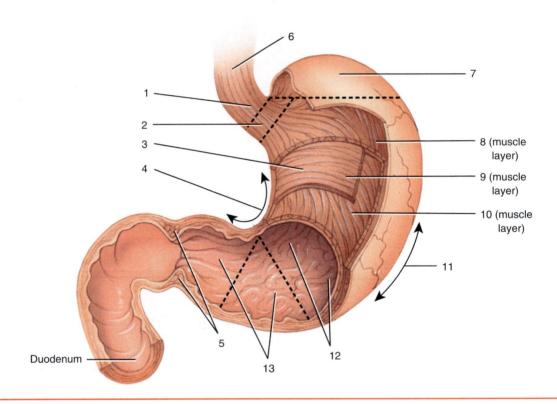

FIGURE 34.6 Esophagus and the stomach.

- body
- cardia
- circular muscle layer
- esophagus
- fundus
- greater curvature
- lesser curvature
- longitudinal muscle layer
- lower esophageal sphincter
- oblique muscle layer
- pyloric sphincter
- pylorus
- rugae

1 _____ 8 _____

2 _____ 9 _____

3 _____ 10 _____

4 _____ 11 _____

5 _____ 12 _____

6 _____ 13 _____

7 _____

Small Intestine

The small intestine is a long tube (about 10 feet long in a living person) divided into three sections: the duodenum, jejunum, and ileum. The shortest section, the **duodenum,** resembles the shape of a "C" and receives food through the pyloric sphincter. The duodenum (*duodenum* = 12 fingers) is about 10 inches long, or approximately the width of 12 fingers. The next section, the **jejunum** (*jejunum* = empty), is about 3 feet long and has a thicker layer of smooth muscle. It is named jejunum because on death it is found to be empty. The **ileum** (*ileum* = intestine), the last section of the small intestine, is about 6 feet long and ends at the **ileocecal sphincter** (valve), where it joins the large intestine. From the exterior, it is difficult to distinguish these three sections of the small intestine.

Digestive system functions of the small intestine include secretion, mixing and propulsion, mechanical and chemical digestion, and absorption. Secretions of the pancreas, liver, and intestinal glands are deposited into the lumen of the small intestine. Local contractions of the muscularis layer of the small intestine mix and churn food by **segmentation** and propel it through the small intestine into the large intestine by peristalsis. Pancreatic and intestinal enzymes digest larger food molecules of carbohydrates, proteins, and lipids into smaller, absorbable molecules.

The small intestine has three special anatomical modifications for increasing the surface area available for absorption. The **circular folds** or **plicae circularis** are permanent folds or ridges of the mucosa that can be observed with the unaided eye. The other two structural specializations, the **villi** and **microvilli,** can only be observed with the microscope and will be studied in the histology of the digestive system.

Large Intestine

The large intestine, also known as the colon (*colon* = food passage) or large bowel, is approximately 5 feet long and is composed of the cecum, colon, rectum, and anal canal. A blind pouch called the **cecum,** which is about 2 to 3 inches long, extends inferiorly from the large intestine at the ileocecal sphincter in the lower right quadrant. The slender **appendix,** which is lymphatic tissue, branches off the cecum and is also about 2 to 3 inches long. From the ileocecal junction, the **ascending colon** extends superiorly to the liver on the right side of the abdomen and makes a 90° turn to the left at the **right colic** (hepatic) **flexure.** The colon continues as the **transverse colon** until it reaches the spleen, where it makes another 90° turn at the **left colic** (splenic) **flexure.** The **descending colon** then continues inferiorly on the left side to an S-shaped curve called the **sigmoid colon** at the level of the iliac crest. **Teniae coli** are three longitudinal bands of smooth muscle that run along the length of the large intestine. Contractions of the teniae coli form pouches called **haustra. Epiploic appendages** are fat-filled pouches of visceral peritoneum that hang from the teniae coli.

Following the sigmoid colon, at about the 3rd sacral vertebra level, is the **rectum.** The last segment of the colon is the **anal canal,** which has longitudinal **anal columns** or folds rich with blood vessels. The anal canal opens to the exterior at an opening called the **anus.** There are two sets of muscle sphincters in the anus: the **internal sphincter** of smooth muscle, which is involuntary, and the **external sphincter,** which is voluntary skeletal muscle.

The large intestine absorbs a large amount of water, concentrating the undigested feces into a solid form. It is the site where bacteria complete digestion and produce vitamins. The large intestine also secretes mucus and eliminates fecal material from the body through the anus.

ACTIVITY 5 **Small and Large Intestines**

1 Label the structures in Figure 34.7(a)–(c).

2 Identify the structures on a model or chart.

3 Pronounce the name of each term as you point to it.

CLINICAL NOTE: *A sigmoidoscopy views the lining of the sigmoid colon, whereas a colonoscopy views the lining of the entire colon.*

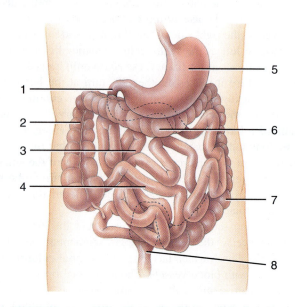

(a) Anterior view

FIGURE 34.7 Small intestine and large intestine.

- ascending colon
- descending colon
- duodenum
- ileum
- jejunum
- rectum
- stomach
- transverse colon

1 _____

2 _____

3 _____

4 _____

5 _____

6 _____

7 _____

8 _____

(b) Large intestine, anterior view

(c) Frontal section of anal canal

FIGURE 34.7 Small intestine and large intestine, *continued.*

- anal canal
- anus
- appendix
- ascending colon
- cecum (SEE-cum)
- descending colon
- epiploic appendages
 (ep-ih-PLOH-ic)
- haustra (HAW-struh)
- ileocecal sphincter
 (ileo-SEE-cal)
- ileum
- left colic (splenic) flexure
- rectum
- right colic (hepatic) flexure
- sigmoid colon
- teniae coli
 (TEE-nee-ee CO-lye)
- transverse colon

9 _____

10 _____

11 _____

12 _____

13 _____

14 _____

15 _____

16 _____

17 _____

18 _____

19 _____

20 _____

21 _____

22 _____

23 _____

24 _____

- anal canal
- anal column
- anus
- external anal sphincter
- internal anal sphincter
- rectum

25 _____

26 _____

27 _____

28 _____

29 _____

30 _____

D. Accessory Digestive Organs

Salivary Glands

The mucous membranes of the mouth and tongue have many small salivary glands that secrete saliva. However, most of the saliva secreted into the oral cavity is secreted by the parotid, submandibular, and sublingual salivary glands. Saliva moistens food and contains enzymes that initiate chemical digestion of carbohydrates and lipids.

The large **parotid** (*par-* = near; *ot-* = ear) **glands** are located anterior and inferior to the ears between the skin and the masseter muscle. Their salivary secretions reach the oral cavity via the **parotid ducts** that open near the second molar in the upper jaw. The **submandibular glands** are located in the posterior part of the mouth floor just medial to the mandible, deep to the mylohyoid muscle. The **submandibular ducts** open into the oral cavity lateral to the lingual frenulum. The **sublingual glands** are located under the tongue as the name suggests and are more medial and superior than the submandibular glands. There are several **lesser sublingual ducts** that open into the floor of the mouth; they are very small compared with the other main salivary gland ducts.

Tongue

Extrinsic muscles attach the tongue to bone and hold it in position, but they also allow movement to maneuver food in the mouth. The tongue is composed of skeletal muscle called **intrinsic muscles** that are used for speech and swallowing. The **lingual frenulum** attaches the tongue to the floor of the mouth. The dorsal and lateral surfaces of the tongue have numerous **papillae** (*papilla* = nipple-shape), many of which have taste buds.

Teeth

Teeth mechanically break up the food into small pieces, aided by the tongue which manipulates the food in the oral cavity. Teeth are composed of three basic regions: the crown, neck, and root. The **crown** is the part that can be seen above the gums, whereas the **neck** is the constricted region at the gum line where the crown and root meet. The **root** is embedded in an alveolar bony socket of the maxilla or mandible and is held in place by the **periodontal** (*peri-* = around; *odont-* = tooth) **ligament** lining the socket. Most of the tooth is formed of **dentin,** a hardened, calcified connective tissue. **Enamel** covers and protects the dentin of the crown. Dentin of the root is covered with **cementum,** attaching the tooth to the periodontal ligament. Surrounding each tooth is **gingiva** or gum tissue.

In the crown of the tooth deep to the dentin is a **pulp cavity** filled with connective tissue called **pulp** that contains nerves, blood vessels, and lymphatic vessels. **Root canals** are slender continuations of the pulp cavity that run the entire length of the root. The small opening at the base of each root, the **apical foramen,** is the location for nerves, blood vessels, and lymphatic vessels to enter and exit the tooth.

Adults have 32 permanent teeth, with each jaw containing four incisors, two cuspids (canines), four premolars (bicuspids), and six molars. **Incisors** cut food, **cuspids** tear food, and **premolars** and **molars** crush and grind food. Premolars have two cusps (pointed surfaces), and molars have four cusps. Incisors, cuspids, and the second premolars have one root, the first bicuspids and mandibular (lower) molars have two roots, and the maxillary (upper) molars have three roots.

Children's teeth, **deciduous** (*decidu-* = falling out) **teeth** or primary teeth, are temporary and are lost by the child between the ages of 6 and 12. They are replaced by **permanent** or **secondary teeth.** The deciduous set has 20 teeth, and the permanent set has 32 teeth.

From the midline moving laterally, the dental formula for *deciduous teeth* is:

$$\frac{\textbf{Upper teeth: } 2 \text{ incisors, } 1 \text{ cuspid,} \quad 0 \text{ premolars, } 2 \text{ molars}}{\textbf{Lower teeth: } 2 \text{ incisors, } 1 \text{ cuspid,} \quad 0 \text{ premolars, } 2 \text{ molars}} \times 2 = 20 \text{ teeth}$$

From the midline moving laterally, the dental formula for *permanent teeth* is:

$$\frac{\textbf{Upper teeth: } 2 \text{ incisors, } 1 \text{ cuspid,} \quad 2 \text{ premolars, } 3 \text{ molars}}{\textbf{Lower teeth: } 2 \text{ incisors, } 1 \text{ cuspid,} \quad 2 \text{ premolars, } 3 \text{ molars}} \times 2 = 32 \text{ teeth}$$

ACTIVITY 6 **The Teeth, Tongue, and Salivary Glands**

1 Label and study the structures in Figure 34.8.

2 Identify the structures on a model or chart.

3 Pronounce the name of each term as you point to it.

4 Identify incisors, cuspids, premolars, and molars on a skull with teeth.

(a) Location of salivary glands

FIGURE 34.8 The salivary glands, tongue, and teeth.

(a)
- openings of sublingual ducts
- opening of parotid duct (near secondary molar)
- parotid duct
- parotid gland
- sublingual gland
- submandibular duct
- submandibular gland

1 _____

2 _____

3 _____

4 _____

5 _____

6 _____

7 _____

(b) Tooth, frontal section

(c) Teeth and tongue

FIGURE 34.8 The salivary glands, tongue, and teeth, *continued.*

(b)
- apical foramen
- blood supply
- cementum
- crown
- dentin
- enamel
- gingiva
- neck
- nerve
- periodontal ligament
- pulp cavity with pulp
- root
- root canal

8 _____

9 _____

10 _____

11 _____

12 _____

13 _____

14 _____

15 _____

16 _____

17 _____

18 _____

19 _____

20 _____

(c)
- cuspid
- incisors
- lingual frenulum
- molars
- premolars
- tongue

21 _____

22 _____

23 _____

24 _____

25 _____

26 _____

Pancreas

The **pancreas,** composed of a head, body, and tail, is located posterior to the stomach. The **head** is the expanded part of the pancreas that lies in the C-shape of the duodenum. The pancreatic **body** is in the middle, and the **tail** tapers to the left toward the spleen. The main **pancreatic duct** runs from the tail to the head of the pancreas and joins the common bile duct to form the **hepatopancreatic ampulla.** This ampulla opens into the duodenum at the **major duodenal papilla,** about 4 inches inferior to the pyloric sphincter. In the head of the pancreas, a smaller **accessory pancreatic duct** branches off the pancreatic duct and opens into the duodenum just superior to the hepatopancreatic ampulla. The pancreas secretes digestive enzymes into the duodenum.

Liver and Gallbladder

The liver is a large organ that lies inferior to the diaphragm and touches the diaphragm all along the liver's superior border. The liver has two main lobes, the **right lobe** and **left lobe,** which are separated by the **falciform** (*falc-* = sickle-shaped) **ligament.** Two smaller lobes, the superior **caudate lobe** and the inferior **quadrate lobe,** lie on the same side of the falciform ligament as the right lobe and can be seen from the posterior view. The **right** and **left hepatic ducts** join to form the **common hepatic duct,** which takes bile to the cystic duct of the gallbladder. The main digestive function of the liver is to make bile to be delivered to the gallbladder for concentration and storage. Bile **emulsifies** lipids, which breaks down large spheres of lipids into smaller ones, preparing them for digestion in the small intestine.

The **gallbladder** is a sac made of smooth muscle that is located on the inferior surface of the liver between the right lobe and the quadrate lobe. Bile leaves the gallbladder via the **cystic duct** which joins with the **common hepatic duct** to form the **common bile duct.** The common bile duct joins the pancreatic duct in the pancreas to form the **hepatopancreatic ampulla** which opens into the duodenum.

ACTIVITY 7 **The Pancreas, Liver, and Gallbladder**

1 Label and study the structures in Figure 34.9(a), (b), and (c).

2 Identify the structures on a model or chart.

3 Pronounce the name of each term as you point to it.

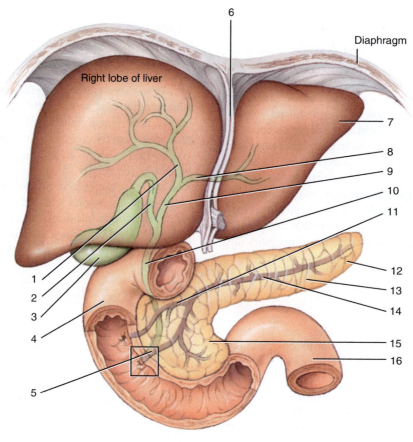

(a) Anterior view

FIGURE 34.9 Pancreas, liver, and gallbladder.

(a)
- accessory pancreatic duct
- body of pancreas
- common bile duct
- common hepatic duct
- cystic duct
- duodenum
- falciform ligament
- gallbladder
- head of pancreas
- hepatopancreatic ampulla
 (heh-PA-toe-pan-cree-A-tic)
- jejunum
- left hepatic duct
- left lobe of liver
- pancreatic duct
- right hepatic duct
- tail of pancreas

1 _____

2 _____

3 _____

4 _____

5 _____

6 _____

7 _____

8 _____

9 _____

10 _____

11 _____

12 _____

13 _____

14 _____

15 _____

16 _____

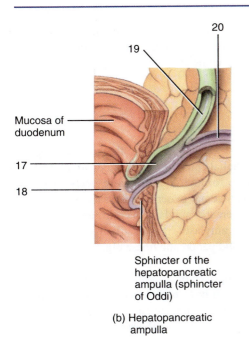

Mucosa of
duodenum

Sphincter of the
hepatopancreatic
ampulla (sphincter
of Oddi)

(b) Hepatopancreatic
ampulla

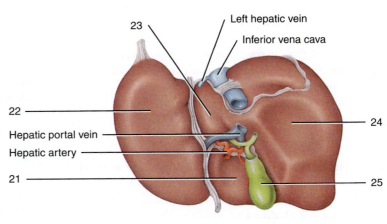

Left hepatic vein

Inferior vena cava

Hepatic portal vein

Hepatic artery

(c) Liver, posterior view

FIGURE 34.9 Pancreas, liver, and gallbladder, *continued.*

(b)
• common bile duct
• duodenal papilla
• hepatopancreatic ampulla
• pancreatic duct

17 _____

18 _____

19 _____

20 _____

(c)
• caudate lobe
• gallbladder
• left lobe
• quadrate lobe
• right lobe

21 _____

22 _____

23 _____

24 _____

25 _____

E. Histology of the Digestive System

Gastrointestinal Tract Organs

The walls of the GI tract organs are composed of four layers: the mucosa (epithelium and lamina propria), submucosa, muscularis, and serosa. The epithelium of the mucosa changes along the digestive tract. Stratified squamous epithelium lines the mouth, oropharynx, laryngopharynx, and esophagus to protect underlying tissues. At the junction of the esophagus and stomach, the epithelium changes to simple columnar epithelium, which lines the stomach, small intestine, and large intestine and reflects the absorptive and secretive functions of these organs. The epithelium changes again to stratified squamous epithelium in the anal canal, providing protection from the external environment.

The simple columnar epithelium of the **stomach** contains **goblet cells** that secrete mucus for stomach lining protection. The gastric epithelium dips down into the underlying **lamina propria** to form narrow channels called **gastric pits.** The gastric pits lead into areas called **gastric glands** that are lined by three types of specialized secretory cells: mucous neck cells, chief cells, and parietal cells. **Mucous neck cells** secrete mucus that protects the stomach lining from being digested. **Chief cells** produce and secrete gastric lipase that digests triglycerides, and pepsinogen which is an inactive form of pepsin. **Parietal cells** produce hydrochloric acid and intrinsic factor. Hydrochloric acid converts pepsinogen into pepsin which begins the digestion of protein. Intrinsic factor facilitates vitamin B_{12} absorption in the small intestine.

The **small intestinal** wall has an unusual feature in that the mucosa forms fingerlike projections called **villi** (*villi* = tufts of hair) that greatly increase the surface area. In the center of each villus is the lamina propria with an arteriole, venule, blood capillary, and a lymphatic capillary called a **lacteal.** Absorptive cells of the **simple columnar epithelium** have **microvilli** on the surface facing the lumen, adding more surface area. When viewed with the compound microscope, the microvilli give the cell surface a brushlike appearance and are called the **brush border.** **Goblet cells** that secrete mucus are also found in the epithelium. Deep fissures in the intestinal mucosa are lined with **intestinal glands,** also called the **crypts of** Lieberkühn. In the submucosa of the duodenum, there are **duodenal (Brunner's) glands,** which secrete mucus to help neutralize the acidic stomach chyme.

Pancreas and the Liver

The **pancreas** has two distinct areas: the *exocrine portion* of glandular epithelial cells in clusters called **acini,** and the *endocrine portion* formed of **pancreatic islets (islets of Langerhans).** The acinar exocrine cells secrete the digestive enzymes pancreatic amylase, trypsin, and pancreatic lipase. Pancreatic amylase digests carbohydrates, trypsin digests proteins, and pancreatic lipase digests triglycerides (lipids).

The **liver** is comprised of structural and functional units called **lobules,** which have **hepatocytes** (*hepato-* = liver) which produce bile. Hepatocytes are arranged in branching, flat plates around a **central vein.** The liver lobules have modified capillaries with large spaces called **sinusoids** where gases, nutrients, and wastes are exchanged with the hepatocytes. Special **fixed macrophages** are in the sinusoids to phagocytize dead or foreign matter from the blood. These fixed macrophages are called **reticuloendothelial (Kupffer's) cells.** The lobules are typically six-sided with portal triads present at each of the corners. A **portal triad** is composed of three structures: a branch of the **hepatic portal vein,** a branch of the **hepatic artery,** and a **bile canaliculus** (*canaliculus* = little canal).

ACTIVITY 8 Histology of the Digestive System

1 Label the structures in Figure 34.10(a)–(e) of the esophagus, stomach, small intestine, pancreas, and liver.

2 Pronounce the terms as you write them.

3 Examine prepared slides of stomach, duodenum, pancreas, and liver using the scanning or low-power objective lens. Identify the structures listed in Figure 34.10(a)–(e).

4 Examine photomicrographs of the digestive system in Appendix-A Histology Atlas Figures HA-47 through HA-54.

Lumen

1
2
3
4
5
6
7
8

LM 17×

(a) Wall of the esophagus

9
10
11
12

Parietal cell

Chief (zymogenic) cell

LM about 250×

(b) Fundic mucosa of the stomach

FIGURE 34.10 Histology of digestive organs.

(a)
- circular layer of smooth muscle fibers
- lamina propria
- longitudinal layer of smooth muscle fibers
- mucosa
- muscularis
- serosa
- stratified squamous epithelium
- submucosa

1 _____
2 _____
3 _____
4 _____
5 _____
6 _____
7 _____
8 _____

(b)
- gastric glands
- gastric pits
- lamina propria
- simple columnar epithelium

9 _____
10 _____
11 _____
12 _____

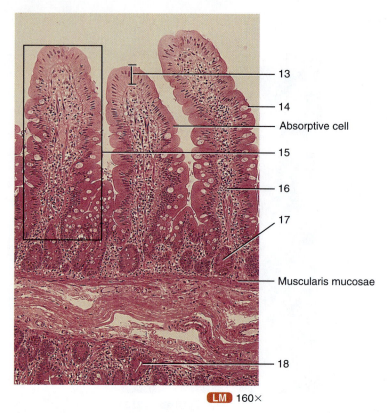

13

14

Absorptive cell

15

16

17

Muscularis mucosae

18

LM 160×

(c) Duodenum of small intestine
• duodenal (Brunner's) glands in submucosa
• goblet cell
• intestinal glands (crypts of Lieberkühn)
• lamina propria
• simple columnar epithelium
• villus

13 _____

14 _____

15 _____

16 _____

17 _____

18 _____

19 20

LM 450×

(d) Histology of the pancreas
• acini (exocrine)
• pancreatic islet (islet of Langerhans)

19 _____

20 _____

21

22

23

LM 180×

(e) Histology of the liver lobule
• central vein
• hepatocytes
• sinusoids

21 _____

22 _____

23 _____

FIGURE 34.10 Histology of digestive organs, *continued.*

Blood flows through each liver lobule from branches of the hepatic artery and the portal vein to the **sinusoids.** The **hepatic artery** carries oxygen-rich blood, and the **hepatic portal vein** carries oxygen-poor blood plus nutrients from the digestive system. The sinusoids empty blood into the **central vein,** and blood then flows from the **central vein** to the **hepatic vein,** and finally into the **inferior vena cava.** **Hepatocytes** produce and secrete **bile** first into narrow **bile canaliculi,** and then into the **bile ducts,** part of the lobule triad. During this time, bile travels in an opposite direction from the blood in the triad and sinusoids. Bile eventually empties into the **hepatic ducts** and on to the **cystic duct** to be stored and concentrated in the **gallbladder.**

> **ACTIVITY 9** **The Flow of Blood and Bile Through the Liver**
>
> **1** Label the structures in Figure 34.11(a) and (b).
> **2** Trace blood from the hepatic portal vein through the liver to the inferior vena cava, listing the structures *in order,* using Figure 34.11(a) and (b).

F. Animal Dissection of Digestive System

If you are dissecting a cat or fetal pig to observe digestive system structures, refer to the accompanying dissection manual.

To inferior vena cava

Hepatic portal vein
Hepatic artery
To hepatic duct

- bile canaliculus
- bile duct
- branch of hepatic portal vein
- branches of hepatic artery
- central vein
- hepatic vein
- hepatocytes
- portal triad
- sinusoids

1 _____
2 _____
3 _____
4 _____
5 _____
6 _____
7 _____
8 _____
9 _____

FIGURE 34.11 Flow of blood and bile through a liver lobule.

Reviewing Your Knowledge

A. Anatomy Drill and Practice

Go to www.wiley.com/college/allen. Click on the picture of your lab manual, then click on Student Companion Site, Anatomy Drill and Practice, and Exercise 34.

B. Layers of the Gastrointestinal Tract

Name the layers of the GI tract that are described.

___Peristalsis___ 1. The layer that contracts to churn food or move food along

___jejunum___ 2. Layer consisting of a membrane that lines a body cavity that opens to the exterior; overlays smooth muscle

___Submucosa___ 3. Areolar connective tissue layer located deep to the mucosa

___Parietal peritoneum___ 4. A serous membrane that is the external layer of a GI tract organ

C. Gastrointestinal Tract Organs

Write the name of the term that is described.

___Peristalsis___ 1. Its only function is propulsion

___alimentary canal___ 2. A continuous digestive tube from the mouth to the anus

___oropharynx___ 3. Conducts both air and food

___Small intestine___ 4. Primary site of nutrient absorption; is composed of 3 sections

___Common bile duct___ 5. Section of small intestine that receives bile, pancreatic secretions, and food from the stomach

___Gastric juices___ 6. Churns food and begins protein digestion

___Pyloric canal___ 7. Has two sphincters that control elimination of feces from the body

___Stomach___ 8. Has regions called the cardia, fundus, body, and pylorus

___Large intestine___ 9. Has regions called the cecum, colon, rectum, and anal canal

___parotid glands___ 10. Receives secretions from salivary glands; mastication occurs here

D. Digestive Organ Structures

Write the name of the organ that is described.

Microvilli **1.** Fingerlike extensions increasing surface area in the small intestine

Circular folds **2.** Folds in the gastric mucosa

rugae **3.** Permanent deep ridges in the small intestine mucosa

Pyloric sphincter **4.** Sphincter valve between the stomach and duodenum

Parietal peritoneum **5.** Serous membrane that lines the abdominal wall

Visceral peritoneum **6.** Serous membrane that covers the abdominal organs

Ileocecal **7.** Sphincter that connects the small and large intestine

Amylase **8.** Fluid that begins digestion of carbohydrates

Hard palate **9.** Bony plate between the mouth and nose

Brush border **10.** Forms a brush border; extension of epithelial cells

laryngopharynx **11.** Keeps food and fluids from going up into the nasopharynx

gingiva **12.** Area between the lips and teeth

Cardia **13.** Region of the stomach where the lower esophageal sphincter meets the stomach

Large intestine **14.** Has regions called ascending, transverse, descending, and sigmoid

Esophagus **15.** Narrowed region of the stomach before the small intestine

Cecum **16.** A series of gathered pouches in the large intestine

E. Accessory Digestive Organs and the Peritoneum

Write the name of the organ that is described.

Mechanical digestion **1.** Mechanically breaks up food during mastication

Pancreatic amylase **2.** Secretes enzymes that digest carbohydrates, proteins, lipids, and nucleic acids

Sublingual gland **3.** Salivary glands located under the tongue with ducts that open in the floor of mouth

liver **4.** Produces and secretes bile into ducts

Serous membrane **5.** Peritoneal membrane that holds the small intestine to the posterior abdominal wall

common bile duct **6.** The secretions of this gland join with bile to enter the duodenum

Mesentary **7.** Peritoneal fold that holds the liver to the anterior abdominal wall

mechanical digestion **8.** Manipulates food in mastication

greater omentum **9.** Fatty, large fold of peritoneum covering the transverse colon and small intestine

submandibular gland **10.** Largest salivary glands whose ducts open by the upper second molars

Gall bladder **11.** Stores and secretes bile into the duodenum

Lesser omentum **12.** Peritoneal membrane that attaches the stomach and duodenum to the liver

Submandibular **13.** Salivary glands whose ducts open lateral to the lingual frenulum

14. Trace bile from its secretion to the gallbladder for storage and concentration, and then to the duodenum, listing the structures *in order,* using Figure 34.9.

a. hepatocytes

b. _bile canulii_

c. _bile ducts_

d. _right hepatic duct_

e. _left hepatic duct_

f. _common hepatic duct_

g. gallbladder

h. _cystic duct_

i. _common bile duct_

j. _hepatopancreatic duct_

k. duodenum

Using Your Knowledge

A. Digestive System Structure

1. A hiatal hernia or failure of the lower esophageal (cardiac) sphincter to close causes the stomach contents to back up into the esophagus. This causes the esophageal wall to have a burning sensation (heartburn) and in serious cases, bleeding. What causes the burning sensation and bleeding to occur?

 This is Gerd due to the fact that the lower esophageal sphicter does not close adequately so there is a back flow which Hydrochloride acid gets up the lower esophageal and burns the esophagus.

2. How does removal of the gallbladder affect digestion?

 The gallbladder is to store bile and without it you cannot eat fatty foods. There fats will not be digested which will cause the person to have frequent diarrhea.

B. Identification of Digestive System Structures

Identify the structures numbered 3–6 on the X-ray of the human stomach shown in Figure 34.12.

3. _____Cardia_____

4. _____Pylorus_____

5. _____Pyloric Sphicter_____

6. _____Oblige muscle_____

FIGURE 34.12 X-ray of the human stomach.

Identify the structures numbered 7–10 on the X-ray of the lower GI tract, shown in Figure 34.13.

7. _____Transverse colon_____

8. _____descending colon_____ (sacs)

9. _____appendix_____

10. _____rectum_____

FIGURE 34.13 X-ray of lower GI tract.

Chemical Digestion

35

Objectives

**After completing this exercise,
you should be able to:**

- Name the substrate for each of the following
 enzymes and the products produced from the
 enzyme-catalyzed reaction: amylase, protease,
 peptidase, and lipase
- Describe the importance of mechanical digestion
- Explain the importance of pH on digestive enzyme
 function
- Describe the action of bile salts on lipids

Materials

- Starch Digestion (per group): 6 test tubes, test tube
 rack, test tube holder, wax marker, Parafilm, 37°C
 water bath, 1 saltine cracker (starch) and 1 dropper
 bottle each of: salivary amylase, 0.5% HCl
 (hydrochloric acid), pH 7.4 buffer, distilled water
- Starch Test (per group): Lugol's iodine in dropper
 bottle, 1 clean spot plate with 6 depressions, wax
 marker
- Sugar Test (per group): Benedict's reagent in
 dropper bottle, 250-mL beaker, hot plate
- Bile Emulsification (per group): bile salts, vegetable
 oil, medicine dropper, distilled water in dropper
 bottle, 2 test tubes, test tube rack, test tube holder,
 wax marker

Digestion includes both mechanical digestion,
mechanically breaking food into smaller pieces,
and chemical digestion, the enzymatic break-
down of large molecules into smaller ones that can be ab-
sorbed into the bloodstream. **Mechanical digestion** begins
in the mouth and continues in the stomach and small intes-
tine. Chewing or **mastication** breaks down large pieces of
food into smaller ones that are swallowed. In the stomach,
peristaltic waves mechanically churn and mix the food
with gastric juices to enhance digestion. Churning contin-
ues in the small intestine by smooth muscle contractions
called **segmentations,** or back and forth movements of
portions of the small intestine, that mix the chyme with di-
gestive enzymes prior to peristaltic contractions moving
the contents along. Mechanical digestion enhances the

speed of chemical digestion by creating a greater surface
area for digestive enzymes to do their work.

A. Chemical Digestion

Chemical digestion involves digestive enzymes breaking
the bonds of the food macromolecules. The food we eat
contains macromolecules of carbohydrates, lipids, and pro-
teins. Because these large molecules cannot be absorbed by
the gastrointestinal tract, enzymes secreted by various parts
of the digestive system **catabolize** or break down large
molecules into small ones that can be absorbed. **Starch,** a
carbohydrate, is digested into **disaccharides** (*di-* = two;
saccharide = sugar) and **monosaccharides** (*mono-* = one)

by the action of specific enzymes called **amylases. Lipids** are macromolecules that are catabolized or digested to **glycerol** and **fatty acids** by the enzyme lipase. **Proteases** digest **proteins** into **peptides** and **amino acids. Peptidases** digest **peptides** into amino acids. Monosaccharides, glycerol and fatty acids, and amino acids are small enough to be absorbed across the wall of the gastrointestinal tract and are used by cells for building new macromolecules or to provide energy (ATP).

In the mouth, salivary amylase begins carbohydrate digestion, and **lingual lipase** begins triglyceride digestion. The stomach has chief cells that secrete **gastric lipase,** continuing triglyceride digestion. Chief cells also secrete **pepsinogen,** the inactive form of pepsin. Parietal cells of the stomach secrete hydrochloric acid that converts pepsinogen into active **pepsin** that digests proteins. The small intestine has epithelial cells that secrete amylases and peptidases that digest carbohydrates and proteins, respectively. The pancreas secretes **pancreatic amylase** for digestion of carbohydrates, **trypsin** that digests protein, and **pancreatic lipase** for the digestion of lipids.

Enzymes are typically large protein molecules produced by cells of the body that are special biological catalysts acting on specific **substrates** to produce **products.** Enzymes speed up the rate of chemical reactions between substrates without becoming chemically changed in the reaction. Some enzymes **anabolize** (build) small molecules into macromolecules, whereas others **catabolize** macromolecules like carbohydrates, lipids, and proteins into small molecules. Enzymes work best on food substrates that are chewed and broken down mechanically into smaller particles with a greater surface area.

Each enzyme has environmental conditions, typically temperature and pH, that allow it to be the most active. Enzymes in our body function well at body temperature (37°C), but will denature and become inactive if temperature becomes too high. The pH of the digestive system changes from the mouth to the small intestine; therefore, the enzymes in these different areas have different optimal pHs.

In this exercise, we are using one specific enzyme, salivary amylase, to demonstrate digestion of carbohydrates into glucose. Whole pieces of cracker (starch) are compared with crumbled crackers to observe the importance of mechanical digestion. The activity of amylase will be observed at a pH similar to that of the oral cavity and a pH similar to the stomach to demonstrate the importance of pH on enzyme activity. Carbohydrate digestion begins in the mouth with a pH of about 7.4, but salivary amylase is denatured by the low pH in the stomach due to the presence of hydrochloric acid (HCl). Pancreatic buffers bring the pH back to 7.4, a favorable pH for pancreatic amylase, allowing carbohydrate digestion to continue in the small intestine.

> **SAFETY PRECAUTIONS**
> - *Wear safety glasses when heating test tubes and working with acids.*
> - *Use a test tube clamp for handling heated test tubes.*

ACTIVITY 1 Experiment: Digestion of Starch

1 **Prediction:** Circle the correct tube choice for each statement.
- The greatest amount of starch digestion will occur using: *whole cracker, crumbled cracker*
- The best pH for amylase to digest starch is: *7.0 (water), 7.4 buffer, acidic pH (HCl)*
- The tube with the most starch-converted sugar will be tube: *1, 2, 3, 4, 5, 6*

2 **Materials and Data Collection: Digestion of starch by amylase**

- Read the procedure and decide who will do each step.
- Materials: Obtain materials for starch digestion (see Materials list).

TABLE 35.1 Digestion of Starch By Salivary Amylase

| | TUBE NO. | | | | | |
	1	2	3	4	5	6
Contents added to test tubes	Whole cracker Buffer Water	Crumbled cracker Buffer Water	Whole cracker Buffer Amylase Water	Crumbled cracker Buffer Amylase Water	Whole cracker Buffer Amylase 0.5% HCl	Crumbled cracker Buffer Amylase 0.5% HCl
IKI test results for starch						
Benedict's test results for sugar						

- Mark the tubes 1–6 with the wax pencil and put your initials on the upper part of the test tube for identification in the water bath. Place tubes in the test tube rack. Test tubes 1 and 2 are your controls to be used as comparisons with your results in tubes 3–6.
- To tubes 1, 3, and 5, place 1/8 of a piece of saltine cracker, uncrumbled. Add 12 drops of buffer to each tube and swirl it around.
- To tubes 2, 4, and 6, crumble 1/8 of a saltine cracker into very small pieces; crush with a mortar and pestle, if available. Add 12 drops of buffer to each tube and swirl it around.
- Swirl to mix the amylase solution before using. To tubes 3–6, add 3 drops of *amylase.*
- To tubes 5 and 6, add 3 drops of *0.5% hydrochloric acid (HCl).*
- Add 6 drops of water to tubes 1 and 2, and 3 drops of water to tubes 3 and 4 to equal the liquid in tubes 5 and 6.
- Cover the test tubes with Parafilm and shake the tubes well to mix the contents.
- Remove the Parafilm and place the test tube holder with all 6 filled tubes in a 37°C water bath. Incubate for at least 30 minutes.
- Appoint one member of your group to be the timer for timing the tubes, and have this person gently shake the test tube rack periodically for best results.
- Proceed to the bile salts preparation setup in Activity 2 while these tubes incubate.
- Start water boiling in the 250-mL beaker for the sugar test.

Determine the amount of starch in each tube

- Read the procedure and decide who will do each step.
- Materials: Obtain materials for starch test (see Materials list).
- Mark the depressions on the spot plates 1–6, matching the test tube identification.
- Obtain test tubes from the water bath and swirl to mix.
- Place 2 drops of each test tube solution in the appropriately marked depression on the spot plate.
- Place one drop of **Lugol's iodine (IKI)** solution (brown) on top of the sample drops.
- Observe the color and record this on Table 35.1. IKI turns *blue* or *black* depending on the amount of starch present and indicates a **positive starch test.** If

there is **no color change,** this indicates a **negative starch test.**
brown or other color (−)
light blue (+)
medium blue (++)
dark blue (+++)
black (++++)

Determine the amount of sugar formed when starch is digested by amylase

- Read the procedure and decide who will do each step.
- Materials: Obtain materials for sugar test (see Materials list).
- Place 3 drops of **Benedict's solution** (light blue) in the remaining solutions in all 6 tubes and swirl.
- Put all 6 tubes into the beaker when the water is boiling and time it for 3 minutes.
- Observe the color of the solution and record this on Table 35.1. A **positive sugar test** is green, yellow, orange, or red, depending on the amount of sugar present. If the color remains blue, you have a **negative sugar test.**
blue (−)
green (+)
yellow (++)
orange (+++)
red (++++)
- Compare the results of the whole pieces of crackers in test tubes 1, 3, and 5 to the crumbled crackers in test tubes 2, 4, and 6.

3 **Clean Up:** Clean up your lab table area and return your supplies as directed by your instructor.

4 Complete the Experimental Report with your lab group.

 EXPERIMENTAL REPORT

Results

- List the test tubes that tested positive for sugar from highest amount of sugar to lowest.
- Identify any test tubes that tested negative for sugar.
- List the test tubes that tested positive for starch from lowest amount of starch to highest.
- Identify any test tubes that tested negative for starch.
- State the pH that was most effective for starch digestion.

Discussion

- Compare and explain the results of the amounts of sugar and starch in test tubes 1 and 3.

- Compare and explain the results of the amounts of sugar and starch in test tubes 2 and 4.

- Compare and explain the results of the amounts of sugar and starch in test tubes 3 and 5.

- Compare and explain the results of the amounts of sugar and starch in test tubes 4 and 6.

- Compare and explain the results of the amounts of sugar and starch in test tubes 3 and 4.

- Compare and explain the results of the amounts of sugar and starch in test tubes 5 and 6.

- Discuss how the results would be different if the incubation time were longer.

- Identify which test tube mimicked the conditions in the stomach.

Conclusion

- State how the presence of an enzyme influences the digestion of starch to sugar.

- State how substrate size and pH influences enzyme activity.

B. Bile Emulsification

Bile is not a digestive enzyme. **Bile** is composed of a mixture of substances, including **bile salts,** that aid in digestion by physically breaking up large molecules of lipids into smaller molecules with a greater surface area, a process called **emulsification.**

ACTIVITY 2 **Bile Emulsification**

1 Read the procedure and decide who will do each step.

2 Obtain materials for Bile Emulsification (see Materials list).

3 Observe the ability of bile salts to emulsify lipids.
 - Mark the tubes 1 and 2 with wax marker.
 - To each test tube, add 12 drops of water and 2 drops of vegetable oil (a lipid).
 - To test tube 2, add a pinch of bile salts.
 - Cover the tubes with Parafilm and shake them to mix well.
 - Let the tube contents stand for 15 minutes.
 - Describe the contents of the 2 tubes:

 Tube 1 _____

 Tube 2 _____

4 Answer the discussion questions with your group.

? **DISCUSSION QUESTIONS:**
BILE EMULSIFICATION

1 Which tube had the oil emulsified? _____

2 Is bile a digestive enzyme? _____
Explain.

3 Explain how emulsification increases chemical digestion of lipids by lipases.

Reviewing Your Knowledge

A. General Terminology

Write the term that matches the description.

absorption macromolecule
amylase product
enzyme protease
lipase substrate

_____ **1.** Large molecules resulting from anabolism

_____ **2.** A biological catalyst

_____ **3.** The molecule an enzyme acts on

_____ **4.** Process that happens to nutrients after digestion

_____ **5.** Enzyme that hydrolyzes lipid

_____ **6.** Enzyme that hydrolyzes protein

_____ **7.** Enzyme that hydrolyzes carbohydrates

_____ **8.** The molecule produced as a result of an enzymatic process

B. Substrates and Products

Fill in the blank with the appropriate term.

1. Proteins are hydrolyzed into and absorbed as _____.

2. Lipids are hydrolyzed into and absorbed as _____ and _____.

3. Carbohydrates are hydrolyzed into and absorbed as _____.

4. Carbohydrate digestion begins in the _____.

5. Bile salts prepare lipids for digestion by a process called _____.

6. Lipid digestion begins in the _____.

7. Protein digestion begins in the _____.

C. Carbohydrate Digestion Exercise

Fill in the blank with the appropriate term.

1. The reagent used to test for the presence of starch was _____.

2. The reagent used to test for the presence of sugar was _____.

3. The three food macromolecules digested in our GI tract are _____, _____,

and _____.

4. Why were there 2 tubes in Activity 1 that had no enzyme added to them? _____

5. The enzymes that digest carbohydrates work best at a(n) _____ pH.

6. The reason why we chew food (and crumbled the cracker in Activity 1) is to _____.

7. A positive test for the presence of sugar is indicated by the colors:

8. A positive test for the presence of starch is indicated by the colors: _____.

Using Your Knowledge

A. Digestion

1–3. Enzymes that digest carbohydrates are produced by what organs?

1. _____

2. _____

3. _____

4–6. Name three organs where lipid digestion occurs.

4. _____

5. _____

6. _____

7–9. Enzymes that digest proteins are produced by what organs?

7. _____

8. _____

9. _____

10. People who practice purging themselves after eating (bulimics) ruin the dentin of their teeth. What causes this?

Urinary System Structures

Objectives

**After completing this exercise,
you should be able to:**

- Locate and identify the organs of the urinary system on models or charts
- Describe the structure of urinary system organs
- Describe the structure of the nephron and trace the path of filtrate through the nephron
- Trace the path of urine from the collecting ducts to the exterior of the body
- Describe the blood supply to the kidneys and trace the pathway of blood from the renal artery to the renal vein
- Identify main urinary organs in the cat or fetal pig

Materials

- male and female urogenital models or charts
- human kidney and nephron models
- disposable gloves, dissection equipment, safety glasses, fresh or preserved sheep or pig kidney
- preserved cat or fetal pig, accompanying cat or fetal pig dissection manual

Urinary system organs include the kidneys, ureters, urinary bladder, and urethra. Urine is formed in the kidneys and flows through the ureters to the urinary bladder, which stores urine until it is eliminated from the body through the urethra.

A. Location and Gross Anatomy of Urinary System Organs

Kidneys

Kidneys are bean-shaped structures that are **retroperitoneal** (*retro* = behind), located between the abdominal wall and the peritoneum. They are found at waist level between the 12th thoracic vertebra and the 3rd lumbar vertebra. The concave surface of each kidney faces the vertebral column and contains a vertical fissure, the **renal hilus.** The ureter, renal arteries and veins, nerves, and lymphatics pass through the renal hilus.

The kidney is attached to the abdominal wall by the **renal fascia,** an outer layer of dense irregular connective tissue. The **renal capsule,** a thin fibrous membrane, covers the outer surface of the kidney. Between the renal fascia and the renal capsule is adipose tissue which forms the **adipose capsule** for protection and padding.

Within the kidney are three main regions: the cortex, medulla, and sinus. The **renal cortex,** a smooth area, is the most superficial region, whereas the **renal medulla** is deeper. The medulla contains cone-shaped **renal pyramids**

that have extensions of the cortex, **renal columns,** that are between each pyramid. The base of each pyramid faces the cortex, and the **renal papilla** is the apex that is pointed toward the renal sinus. The cortex and medulla contain **nephrons,** the structural and functional units of the kidney that form urine. Urine drains into **papillary ducts** that exit through the openings in the renal papilla. The **renal sinus** is a space or cavity that is adjacent to the medulla and extends to the renal hilus. Within the renal sinus are minor and major calyces and the renal pelvis which collect urine from the papillary ducts and deliver it to the ureters. The **minor calyces** are cuplike structures adjacent to the renal papillae which receive urine from the papillary ducts. Several minor calyces drain into a major calyx. Each kidney has 8 to 18 minor calyces and 2 to 3 major calyces. The **major calyces** drain into the **renal pelvis,** which is continuous with the ureter. Blood vessels, lymphatics, and nerves also travel through the renal sinus. Adipose tissue fills the remaining space within the renal sinus.

ACTIVITY 1 **Location and Structure of Kidneys**

1 Label the structures in Figures 36.1, 36.2, and 36.3.
2 Locate the kidneys on a human torso.
3 Identify the structures from Figure 36.3 on a dissectable kidney model.
4 Pronounce the terms as you point to them.
5 Answer the discussion questions with your laboratory partners.

? DISCUSSION QUESTIONS: LOCATION AND GROSS ANATOMY OF THE KIDNEY

1 Use an articulated skeleton to identify which ribs help protect the kidneys.
2 Observe the kidneys in a dissectable human torso model.
 a. Which kidney is lower?
 b. Identify the organ that forces this kidney into a lower position.

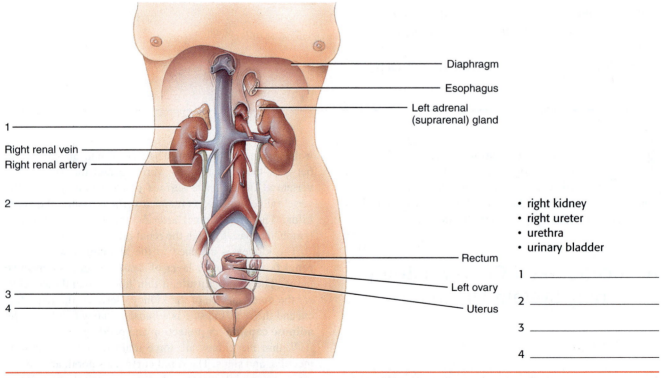

Diaphragm
Esophagus
Left adrenal (suprarenal) gland

1 _____
Right renal vein _____
Right renal artery _____

2 _____

Rectum
Left ovary
Uterus

• right kidney
• right ureter
• urethra
• urinary bladder

1 _____
2 _____
3 _____
4 _____

3 _____
4 _____

FIGURE 36.1 Organs of the female urinary system.

- adipose capsule
- left kidney
- peritoneum
- renal capsule
- renal fascia
- renal hilus (HIGH-lus)
- right kidney

1 _____

2 _____

3 _____

4 _____

5 _____

6 _____

7 _____

Transverse section through
the body (lying down and
viewed from the feet)

FIGURE 36.2 Location and coverings of the kidneys.

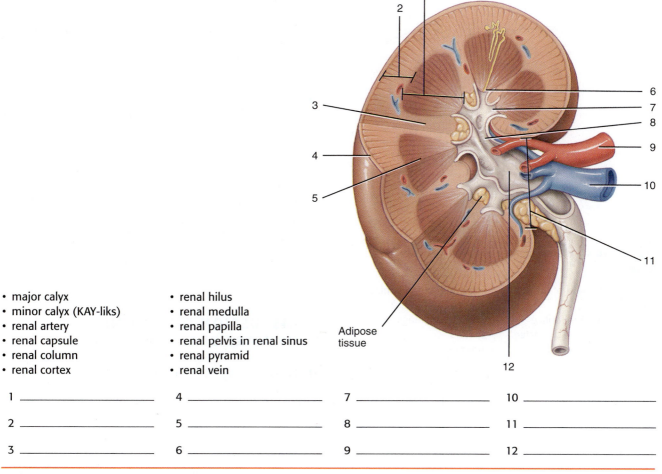

- major calyx
- minor calyx (KAY-liks)
- renal artery
- renal capsule
- renal column
- renal cortex

- renal hilus
- renal medulla
- renal papilla
- renal pelvis in renal sinus
- renal pyramid
- renal vein

Adipose
tissue

1 _____ 4 _____ 7 _____ 10 _____

2 _____ 5 _____ 8 _____ 11 _____

3 _____ 6 _____ 9 _____ 12 _____

FIGURE 36.3 Internal structure of the kidney.

Ureters, Urinary Bladder, and Urethra

The **ureters** are narrow, 25 to 30 cm long muscular tubes located behind the peritoneum (retroperitoneal). The diameter of the ureters varies from 1 to 10 mm. They descend toward the urinary bladder, curving medially as they approach the inferior portion of the bladder and enter the posterior wall of the bladder at an oblique angle. Urine is propelled through the ureters by peristalsis, hydrostatic pressure, and gravity.

The **urinary bladder** is a hollow, muscular organ that distends to store urine. It is covered by visceral peritoneum and is secured to the pelvic wall by parietal peritoneal folds. The smooth muscle within the wall of the urinary bladder is called the **detrusor muscle,** and the epithelial lining of the bladder forms folds or **rugae.** The inferior surface of the urinary bladder contains three openings that form a triangle called the **trigone.** The two posterior openings are the **ureteral openings,** whereas the anterior opening is the opening into the urethra, the **internal urethral orifice.** In males, the urinary bladder is anterior to the rectum and posterior to the pubic symphysis. In females, the urinary bladder is anterior to the vagina, inferior to the uterus, and posterior to the pubic symphysis.

The tubelike **urethra** carries urine from the internal urethral orifice to the **external urethral orifice,** the opening through which urine exits the body. The **internal urethral sphincter** is a layer of circular, involuntary smooth muscle that controls passage of urine into the urethra from the urinary bladder. Voluntary skeletal muscle within the urogenital diaphragm (deep muscles of perineum) forms the **external urethral sphincter** that permits the passage of urine to the external urethral orifice. During **micturition,** the detrusor muscle contracts and forces urine out of the urinary bladder, and the internal and external urethral sphincters relax to allow the passage of urine. In females, the urethra is short (4 cm), whereas the urethra in males is longer (15–20 cm). The male urethra has three regions: the **prostatic urethra** that passes through the prostate gland, the **membranous urethra** that passes through the urogenital diaphragm, and the **spongy (penile) urethra** that passes through the penis.

FIGURE 36.4 Ureters, urinary bladder, and urethra in a female.

- detrusor muscle of urinary bladder
- external urethral orifice
- external urethral sphincter
- internal urethral orifice
- internal urethral sphincter
- trigone (TRY-goan)
- ureteral openings
- ureters
- urethra

1 _____ 6 _____

2 _____ 7 _____

3 _____ 8 _____

4 _____ 9 _____

5 _____

> **CLINICAL NOTE:** *BPH, or benign prostatic hypertrophy, is a noncancerous enlargement of the prostate gland that restricts the prostatic urethra and inhibits urine flow, causing urinary retention.*

ACTIVITY 2 **Location and Gross Anatomy of the Ureters, Urinary Bladder, and Urethra**

1 Label the urinary system structures on Figure 36.4 and Figure 36.5.

2 Identify the structures from Figures 36.4 and 36.5 on urogenital models or charts.

3 Pronounce the names as you point to them.

4 Answer discussion questions with your lab partners.

? DISCUSSION QUESTION: URINARY BLADDER

1 Is the urinary bladder retroperitoneal? Explain your answer.

- anus
- external urethral orifice
- internal urethral orifice
- membranous urethra
- prostatic urethra
- rectum
- spongy urethra
- ureter
- ureteral opening
- urinary bladder
- urogenital diaphragm

1 _____

2 _____

3 _____

4 _____

5 _____

6 _____

7 _____

8 _____

9 _____

10 _____

11 _____

(a) Sagittal section through male pelvis

- anus
- external urethral orifice
- internal urethral orifice
- rectum
- urethra
- urinary bladder
- urogenital diaphragm
- uterus
- vagina

12 _____

13 _____

14 _____

15 _____

16 _____

17 _____

18 _____

19 _____

20 _____

(b) Sagittal section through female pelvis

FIGURE 36.5 Comparison of the urinary bladder and urethra in males and females.

B. Structure of the Nephron

The 1 million nephrons in each kidney filter the blood and form urine. Each **nephron** is composed of a renal corpuscle that filters blood, and a renal tubule that modifies the filtrate to form urine. The **renal corpuscle,** which is located in the renal cortex, consists of a **glomerulus** (capillary network) and a **glomerular (Bowman's) capsule,** a cup-shaped epithelial membrane surrounding the glomerulus. Blood is filtered across a filtration membrane into the **capsular space** (glomerular cavity), and the liquid, now called filtrate, drains into the renal tubule.

Each renal tubule is subdivided into three structural and functional sections. Starting at the glomerular capsule, these sections are: proximal convoluted tubule, loop of Henle, and distal convoluted tubule. The **proximal convoluted tubule** is located in the renal cortex and connects to the **loop of Henle,** which dips down into the renal medulla. The third portion is the **distal convoluted tubule,** which is located in the cortex.

Distal convoluted tubules from several nephrons drain the liquid (now called urine) into **collecting ducts** that descend through the renal pyramids and merge to form larger **papillary ducts.** These ducts connect to renal papillae, which immediately drain urine into minor calyces, and then into major calyces, the renal pelvis, ureters, and finally into the urinary bladder for storage.

The vasculature of the nephron is important to the formation of urine. The glomerulus is located between two arterioles—an **afferent arteriole** that delivers systemic blood to the glomerulus and an **efferent arteriole** that receives blood from the glomerulus. The efferent arteriole then delivers blood to a second capillary bed, the **peritubular capillary network** that entwines the renal tubule.

There are two types of nephrons: cortical nephrons and juxtamedullary nephrons. **Cortical nephrons** have short loops of Henle that only descend slightly into the medulla, whereas **juxtamedullary nephrons** have long loops of Henle that extend deep into the medulla. In addition to peritubular capillaries, the loops of Henle of juxtamedullary nephrons are supplied by **vasa recta,** long capillary loops that extend from the efferent arteriole.

ACTIVITY 3 Structure and Function of the Nephron

1 Label the renal tubule in Figure 36.6(a), and the blood vessels and renal tubule in Figure 36.6(b).
2 Identify the structures in Figure 36.6(a) and (b) on a nephron model or chart.
3 Pronounce the terms as you point to them.

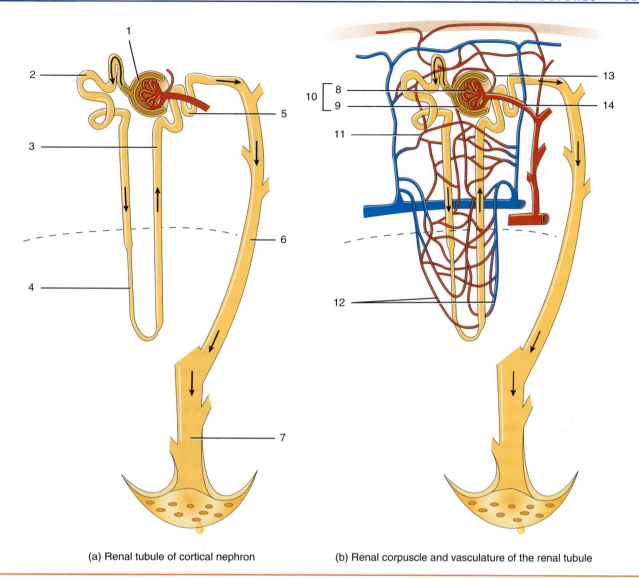

(a) Renal tubule of cortical nephron

(b) Renal corpuscle and vasculature of the renal tubule

FIGURE 36.6 Structure and vascular supply of a cortical nephron.

(a)
- ascending limb of loop of Henle
- collecting duct
- descending limb of loop of Henle
- distal convoluted tubule
- glomerular capsule (glow-MER-u-lar)
- papillary duct
- proximal convoluted tubule

(b)
- afferent arteriole
- efferent arteriole
- glomerular capsule
- glomerulus
- peritubular capillary
- renal corpuscle
- vasa recta

1 _____

2 _____

3 _____

4 _____

5 _____

6 _____

7 _____

8 _____

9 _____

10 _____

11 _____

12 _____

13 _____

14 _____

C. Blood Flow Through the Kidneys

During rest, the renal arteries carry 20 to 25% of the cardiac output to the kidneys and the renal veins return blood to the inferior vena cava. The pathway of blood through the kidneys is outlined below.

Renal artery
↓
Segmental arteries
↓
Interlobar arteries
↓
Arcuate arteries
↓
Interlobular arteries renal vein
↓ ↑
Afferent arteriole interlobar veins
↓ ↑
Glomerular capillaries arcuate veins
↓ ↑
Efferent arterioles → peritubular → interlobular veins
 capillaries
 and/or vasa
 recta

- afferent arteriole
- arcuate artery
- arcuate vein
- efferent arteriole
- interlobar artery
- interlobar vein
- interlobular artery

- interlobular vein
- peritubular capillary
- renal artery
- renal vein
- segmental artery
- vasa recta

1 _____
2 _____
3 _____
4 _____
5 _____
6 _____
7 _____

8 _____
9 _____
10 _____
11 _____
12 _____
13 _____

ACTIVITY 4 Blood Flow Through the Kidneys

1 Label the blood vessels in Figure 36.7. Find the renal artery and label it first to follow the blood flow.

2 Identify the blood vessels on a model or chart.

3 Pronounce the names as you point to them.

4 Answer the discussion questions with your lab partners.

Blood supply of the nephron

FIGURE 36.7 Blood supply of the kidney.

? DISCUSSION QUESTIONS: BLOOD FLOW THROUGH THE KIDNEY

Use Figure 36.7 to answer the following questions.

1 Identify the arteries and veins that travel through the renal sinus.

2 Identify the artery and vein that travel through the renal columns.

3 Identify the artery and vein that arch over the base of the renal pyramid between the cortex and medulla.

4 Identify the blood vessels located in the renal cortex.

D. Dissection of a Pig (or Sheep) Kidney

Pig and sheep kidneys are very similar to human kidneys. This dissection will illustrate the location of the calyces and pelvis within the renal sinus.

ACTIVITY 5 Kidney Dissection

1 Identify the bolded structures listed below on a pig or sheep kidney, using Figure 36.8 as a reference.

2 Protect your hands with disposable gloves.

3 Obtain a fresh or preserved kidney, dissecting tray, scalpel, or knife.

4 If you have a fresh kidney, carefully remove any fat on its exterior. If you are lucky, the adrenal glands may still be located on the superior surface. Adrenal glands are usually not present on preserved kidneys.

5 Locate the **renal hilus** on the concave surface. Identify the **ureter, renal artery,** and **renal vein** that travel through the hilus. The renal artery will have a thicker wall than the renal vein.

6 Using a scalpel or knife, carefully cut along the frontal plane to separate the kidney into anterior and posterior sections.

7 Observe the cut surface. Using a needle probe, pull the thin **renal capsule** away from the surface of the kidney. The outer region of the kidney, the **cortex,** will be a slightly lighter color than the deeper **medulla.**

FIGURE 36.8 Frontal section of preserved pig kidney.

8 **Glomeruli** may be observed in preserved kidneys that are injected with red and blue latex. The glomeruli will appear as red and blue dots within the cortex.

9 Within the medulla, extensions of light-colored cortical tissue, **renal columns,** are found between the darker, cone-shaped **renal pyramids.** Identify the **renal papillae** located at the apices of the renal pyramids.

10 Between the medulla and the renal hilus is a space called the **renal sinus.** Within the renal sinus are the **minor** and **major calyces,** and the **renal pelvis.** Adipose tissue, blood vessels, lymphatics, and nerves fill the remaining space within the renal sinus.

E. Animal Dissection of the Urinary System

The urinary system organs of the cat and fetal pig are similar to those of the human. This dissection will illustrate the retroperitoneal location of the urinary system organs, the relationship of urinary system organs to other organs, and how well these organs are secured by connective tissues. Refer to the appropriate accompanying cat or fetal pig dissection manual for dissection instructions and figures.

F. Histology of the Urinary System

Kidneys

The **glomerular** (Bowman's) **capsule** is composed of a **visceral** and **parietal epithelial layer.** The **visceral layer** is adjacent to the glomerulus and the **parietal layer** forms a funnel-like structure that collects the filtrate within the capsular space (glomerular cavity). Both layers are simple squamous epithelium; however, the simple squamous epithelial cells of the visceral layer have footlike projections that wrap around glomerular capillary walls and these cells are named **podocytes** (*podo* = foot).

The epithelium lining the **proximal convoluted tubule** is **simple cuboidal epithelium.** The **microvilli** forming the **brush border** on the apical surface of these epithelial cells give a brushlike appearance.

The **descending limb of the loop of Henle** is lined by **simple squamous epithelium.** The initial part of the ascending limb or **thin section of the ascending limb** is also lined by simple squamous epithelium, whereas the **thick section of the ascending limb of the loop of Henle** is lined by **simple cuboidal** to low **columnar cells.**

The **distal convoluted tubule** is lined with **simple cuboidal epithelium.** These cells do not have a brush border and appear shorter than the simple cuboidal cells lining the proximal convoluted tubule. The final part of the distal convoluted tubule and the **collecting ducts** are lined with **simple cuboidal epithelium.**

ACTIVITY 6 **Histology of the Kidney**

1 Label the tissues in Figure 36.9. Notice that with this stain, the distal convoluted tubules stain a lighter pink than the proximal convoluted tubules.

2 Pronounce the terms as you write them.

3 Observe a prepared slide of the kidney.
- Using a low-power objective lens, identify **renal corpuscles** and **proximal and distal renal tubules** in the cortex and the **loops of Henle** and **collecting ducts** that extend into the renal medulla.
- Place a renal corpuscle in the center of the field of view, and using a high-power objective lens, identify the following structures: **simple squamous epithelium of the parietal layer of the glomerular capsule, capsular space,** and **glomerulus.**
- Return to a low-power objective lens, and place a section of the cortex with proximal and distal convoluted tubules in the center of the field of view. Switch to a high-power objective lens and identify the **simple cuboidal epithelium lining a proximal convoluted tubule** and **a distal convoluted tubule.**
- Return to a low-power objective lens and place a section of the renal medulla with collecting tubules in the center of the field of view. Switch to a high-power objective lens and identify the **epithelial lining of the collecting tubules.**

4 Refer to Figure HA-55 in Appendix-A Histology Atlas.

5 Answer discussion questions with your laboratory partners.

DISCUSSION QUESTIONS: HISTOLOGY OF THE KIDNEY

1 Describe the function of the epithelium of the visceral layer of the glomerular capsule.

2 The proximal convoluted tubules and distal convoluted tubules are lined with simple cuboidal epithelium. The epithelium lining of the proximal convoluted tubule has a brush border, but that lining the distal convoluted tubule does not. Explain how this difference complements the functions of these two areas of the renal tubule.

(a) Survey photomicrograph

Lumen of proximal convoluted tubule

Lumen of distal convoluted tubule

4 5 6

LM 1225×

(b) Renal corpuscle and proximal and distal convoluted tubules

FIGURE 36.9 Sectional view of the kidney.

(a)
• loop of Henle and collecting ducts
• proximal and distal convoluted tubules (outer cortex)
• renal corpuscle

1 _____

2 _____

3 _____

(b)
• parietal layer of glomerular capsule (simple squamous epithelium)
• capsular space
• glomerulus

4 _____

5 _____

6 _____

Ureters, Urinary Bladder, and Urethra

The ureters are composed of three layers: a mucosa that lines the lumen, a muscular layer called the muscularis, and an outer adventitial layer. The **mucosa** contains transitional epithelium that can distend to accommodate urine, and an underlying lamina propria of areolar connective tissue. The **muscularis** layer contains smooth muscle. **Peristalsis,** the function of the muscularis layer, moves urine from the renal pelvis to the urinary bladder. The **adventitia** secures the ureters in place.

The urinary bladder contains the same layers as the ureter. The epithelium of the **mucosa** is transitional epithelium, and the **muscularis** contains smooth muscle. Contraction of the muscularis layer forces urine out of the urinary bladder, and the **adventitia** secures the bladder in place.

The urethra consists of a **mucosa** surrounded by a **muscularis** layer of circular smooth muscle. In both males and females, the epithelium of the mucosa changes along the course of the urethra. The epithelium is transitional near the urinary bladder, becomes stratified columnar or pseudostratified columnar, distally, and changes to stratified squamous epithelium near the external urethral orifice.

ACTIVITY 7 Histology of the Ureters, Urinary Bladder, and Urethra

1 Label the structures in Figure 36.10.

2 Pronounce the names as you write them.

3 Obtain a prepared microscopic slide of a transverse section through the ureter or urinary bladder.
 • Using a low-power objective lens, identify the **mucosa,** the **muscularis,** and the **adventitia.**
 • Using the high-power objective lens, identify the **transitional epithelium** lining the lumen (refer to the photomicrograph in Activity 6).

4 Refer to HA-56 and HA-57 in Appendix-A Histology Atlas.

5 Answer discussion questions with your lab partners.

? DISCUSSION QUESTION: HISTOLOGY OF URETERS, URINARY BLADDER, AND URETHRA

1 Explain why the epithelial layer changes to stratified squamous epithelium near the external urethral orifice.

• adventitia
• lamina propria
• lumen
• mucosa
• muscularis
• transitional epithelium

1 _____
2 _____
3 _____
4 _____
5 _____
6 _____

LM 40×

FIGURE 36.10 Sectional view of the ureter.

Reviewing Your Knowledge

A. Anatomy Drill and Practice

Go to www.wiley.com/college/allen. Click on the picture of your lab manual, then on Student Companion Site, Anatomy Drill and Practice, and Exercise 36.

B. Renal Structures

Write the renal structure that the phrase describes.

Urinary system **1.** Urine-forming structure of the kidney.

nephrons **2.** Region of the kidney deep to cortex; contains collecting duct.

renal medulla **3.** Extensions of renal cortex found in between renal pyramids.

renal sinus **4.** Urine flowing through this structure drains into a minor calyx.

renal hilus **5.** Located between renal fascia and renal capsule.

renal papilla **6.** Apex of renal pyramid.

major calyces **7.** Urine flowing through this structure drains into the renal pelvis.

Urinary bladder **8.** Vertical fissure in concave surface of kidney through which blood vessels and ureters pass.

renal capsule **9.** Covers outer surface of kidney.

renal fascia **10.** Dense irregular connective tissue that covers the adipose capsule and attaches the kidney to the abdominal wall.

renal pelvis **11.** Receives urine from the major calyces.

renal sinus **12.** Space within kidney that is adjacent to renal medulla, contains calyces and renal pelvis.

minor calyces **13.** Cuplike structure that is located in renal sinus that receives urine from openings of papillary ducts.

renal pyramids **14.** Cone-shaped structures located within the renal medulla.

nephron **15.** Outermost region of the kidney, contains renal corpuscles.

C. Ureters, Urinary Bladder, and Urethra

Write the name of the structure that the phrase describes.

<u>urinary bladder</u> **1.** The detrusor muscle is the main muscle for this structure.

<u>Prostatic urethra</u> **2.** Region of male urethra that passes through the prostate gland.

<u>Uretal openings</u> **3.** Two openings in posterior urinary bladder wall.

<u>Spongy urethra</u> **4.** Region of male urethra that passes through penis.

<u>Anterior openings</u> **5.** Area bounded by ureteral openings and internal urethral orifice.

<u>external urethral sphincter</u> **6.** Voluntary skeletal muscle in urogenital diaphragm that allows passage of urine to exterior of body.

<u>Ureters</u> **7.** Carries urine from renal pelvis to urinary bladder.

<u>external urethral orifice</u> **8.** Urine is excreted through this opening.

<u>internal urethral sphincter</u> **9.** Circular smooth muscle that involuntarily controls passage of urine from the urinary bladder to the urethra.

<u>Ureteral openings</u> **10.** Anterior opening in urinary bladder that leads into urethra.

<u>membranous urethra</u> **11.** Region of male urethra that passes through urogenital diaphragm.

D. The Nephron

Write the name of the structure the phrase describes.

<u>afferent arteriel</u> **1.** Blood vessel that delivers blood to glomerulus.

<u>Peritubular capillaries</u> **2.** Blood from the efferent arteriole flows into this capillary bed.

<u>Renal tubule</u> **3.** Structure that surrounds glomerulus and collects filtrate.

<u>loop of henle</u> **4.** Section of renal tubule that descends into medulla.

<u>Glomerullus</u> **5.** Capillary network within the renal corpuscle.

<u>Renal corpuscle</u> **6.** Structure composed of glomerulus and glomerular capsule.

<u>efferent arterial</u> **7.** Blood vessel that drains blood from glomerulus.

<u>Vasa recta</u> **8.** Capillary loops that extend from the efferent arteriole and run along loop of Henle of juxtamedullary nephrons.

E. Urine Formation and Flow

Trace the flow of filtrate and urine through the urinary system. Write the structures in order, starting with the glomerulus.

1. gomerulus
2. efferent arteriole
3. Proximal convoluted tubue
4. Peritubular capillaries
5. Cortical nephron
6. Ascending loop of henle
7. Descending loop of henle
8. Distal convoluted tubule
9. Collecting duct
10. minor calyx
11. Major calyx
12. Renal Pelvis
13. Ureter
14. external urethral orifice

F. Blood Flow Through the Kidneys

Trace blood flow through the kidney. Start at the renal artery and number the blood vessels in sequence.

6 **1.** afferent arteriole

4 **2.** arcuate artery

11 **3.** arcuate vein

8 **4.** efferent arteriole

7 **5.** glomerulus

12 **6.** interlobar artery

3 **7.** interlobar vein

5 **8.** interlobular artery

10 **9.** interlobular vein

9 **10.** peritubular capillary

1 **11.** renal artery

13 **12.** renal vein

2 **13.** segmental artery

Using Your Knowledge

Observe the X-ray in Figure 36.11 and identify the structures.

1. _Major Calyx_
2. _Renal Pelvis_
3. _Kidney_
4. _Ureter_
5. _Bladder_

FIGURE 36.11 Urinary system structures in the cat.

6. Is the urinary bladder full or partially voided (refer to Figure 2.7(d) for another view of the urinary bladder).

Partially voided

7. Explain why pregnancy increases the frequency of urination.

The Baby sits on the bladder.

8. Epithelial cells that reabsorb solutes are cuboidal but epithelial cells involved in filtration are thin, squamous cells. The cuboidal cells contain more cytoplasm and organelles, especially mitochondria and rough endoplasmic reticulum (RER), than the squamous cells. Why do the cuboidal cells of the kidney tubules need more mitochondria and RER? *Because the mitochondria is the powerhouse.*

9. Define *ptosis* of the kidney.

Drooping, as of the eyelid or the kidney.

10. Define incontinence and describe how this situation would be different in a child under 2 years old versus in an adult. *due to the external urethral sphincter the child has not have brain the internal sphincter. So the child can not understand when to pee.*

Urine Formation and Urinalysis

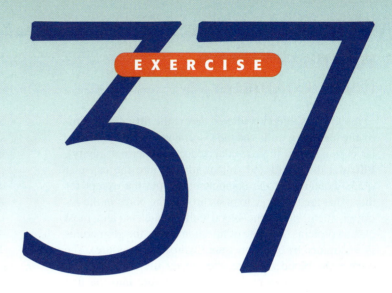

Objectives

After completing this exercise, you should be able to:

- Explain the function of the nephron
- Explicate the process of urine formation
- Describe the major features of a urinalysis
- Describe the composition and properties of normal urine
- Compare and contrast the characteristics of normal and abnormal urine

Materials

- Filtration: filter paper, funnel, ring stand with ring, 250-mL beaker, 5% powdered charcoal solution, and 1% copper sulfate solution
- Urinalysis: 100-mL beakers (4 per group), 50-mL beakers (4 per group), wax pencils, urine test strips, 1 normal and 3 abnormal artificial urines (obtained from a supply house or made with instructions on the Instructor Companion Site of the Wiley Web site: www.wiley.com/college/allen)

The formation of urine by the 1 million nephrons in each kidney enables the body to remove metabolic wastes from blood and to maintain homeostasis by regulating water volume, concentration of ions in blood, and blood pH. A urinalysis is a simple test that can determine if urine formation is normal or abnormal. This test analyzes the volume of urine, and its physical, chemical, and microscopic composition of urine. In addition, the level of some metabolic wastes in the blood is assessed.

A. Function of the Nephron

Urine is produced by the nephon and collecting ducts through three different processes—glomerular filtration, tubular reabsorption, and tubular secretion.

- **Filtration**—The first step in urine production. Water and most solutes in blood pass through the filtration membrane (glomerular capillary walls and visceral wall of glomerular capsule) into the glomerular cavity.
- **Tubular reabsorption**—Water and solutes cross the wall of the renal tubule, diffuse through the interstitial fluid, and return to blood by entering the peritubular capillaries or vasa recta.
- **Tubular secretion**—Solutes leave the peritubular capillaries or vasa recta, diffuse through the interstitial fluid, cross the wall of the renal tubule, and are eliminated in urine.

Function of the Renal Corpuscle

In the kidneys, blood is filtered across the **filtration membrane** (capillary walls and the visceral wall of the glomerular capsule) into the capsular space (glomerular cavity). **Filtrate,** the liquid and solutes filtered into the capsular space, contains all the components of plasma except for those that are too large to pass through the pores in the filtration membrane—the blood cells, platelets, and most plasma proteins. The rate of filtration across the glomerulus is determined by blood pressure (hydrostatic pressure). The greater the blood pressure, the greater the filtration rate. Also, filtrate moves by hydrostatic pressure into the first section of the renal tubule, the proximal convoluted tubule.

ACTIVITY 1 Experiment: Filtration

In this activity, you will observe the **selectivity** of the filtration membrane and the effect of hydrostatic pressure on **filtration rate.**

1 **Prediction:** With your lab partners, predict which substances will pass through the filter: water, copper sulfate (blue), or charcoal.

_____ will pass through the filter.

_____ will not pass through the filter.

2 Read the procedure and decide who will do each step.

3 **Materials:** Obtain materials for Filtration: (see Material list).

4 **Data Collection:** Observe and record the substances that pass through a filter paper in the kidney filtration simulation. Record your data in Table 37.1.

- Fold the circular filter paper in half twice, and open it to form a cone. Place the cone in the funnel and place the funnel in a ring on the ring stand. Put the beaker under the funnel (Fig. 37.1).
- Stir or mix the solution comprised of water, copper sulfate (blue), and powdered charcoal, and pour it into the funnel to just below the top of the filter paper. Watch so the edges of the filter paper do not capsize or that the solution doesn't go over the top.
- As soon as the liquid slows down enough to count individual drops, count number of drops in 15, 30, 45, 60, and 90 seconds. Record your information in Table 37.1.

5 **Clean Up:** Clean up as directed by your instructor.

6 Complete the Experimental Report with your lab group.

Funnel

Filter paper

Filtrate

FIGURE 37.1 Filtration setup.

TABLE 37.1 Filtration Results

TIME (SEC)	#DROPS
15	
30	
45	
60	
90	

EXPERIMENTAL REPORT: FILTRATION

Results:

- State how filtration rate changed over time.

- Name the substance(s) that passed through the filter paper.

- Name the substance(s) that did not pass through the filter paper.

Discussion:

- Discuss why you saw a difference in the filtration rate during the activity.

- Identify the force that drives the filtration rate.

- Identify the cell structure that the filter paper represents.

- Compare the substances that were not filtered in this experiment to the substances that are not filtered in the kidney.

- Compare your results of this experiment with your predictions.

Conclusion:

Write a statement that postulates what substances can pass through the glomerulus of the kidney and how blood pressure affects the filtration rate.

Function of the Proximal Convoluted Tubule

The **proximal convoluted tubule** is the major site of **tubular reabsorption of water and solutes** from the filtrate. Proximal convoluted tubules reabsorb 100% of most organic solutes and nutrients (glucose, amino acids, lactic acid, water soluble vitamins); 65% of water, sodium ions, and potassium ions; 50% of chloride ions; and 80 to 90% of the bicarbonate ions. Water follows the solutes via osmosis.

The proximal convoluted tubule also **secretes hydrogen ions and urea** (nitrogenous waste) into the tubular fluid. The hydrogen ions secreted into the tubular fluid assist in the reabsorption of bicarbonate ions.

Because the amount of water reabsorbed is approximately equal to the amount of solutes reabsorbed (minus the solutes secreted into the tubule), the osmolarity (solute concentration) of the filtrate at the end of the convoluted tubule is the same as at the beginning of the convoluted tubule.

Function of the Loop of Henle

The remaining filtrate travels on through the renal tubule to the **descending limb of the loop of Henle** which is permeable to water, but not to solutes. As the filtrate travels through the descending limb of the loop of Henle toward the medulla of the kidney, the high solute content (higher osmolarity) of the interstitial fluid in the renal medulla causes 15% of the original filtered **water to be reabsorbed by osmosis.** This causes the osmolarity of the filtrate to increase.

The **ascending limb of the loop of Henle** is impermeable to water, but permeable to solutes. Therefore, the opposite occurs. As the filtrate travels toward the cortex, the decrease in solute content (lower osmolarity) of the interstitial fluid causes **reabsorption of Na^+, Cl^-, and K^+.** The osmolarity of the filtrate decreases and becomes more dilute than the proximal convoluted tubule filtrate as it heads toward the distal convoluted tubule.

Function of the Distal Convoluted Tubule

As the filtrate enters the **distal convoluted tubule,** which is located in the cortex of the kidney, a total of 80% of water has been reabsorbed back into the blood. In this segment of the tubule, further **Na$^+$ and Cl$^-$ are reabsorbed** into the blood and an additional 10 to 15% of the original filtered **water follows the solute by osmosis.**

Function of the Late Distal Tubules and Collecting Ducts

When the filtrate reaches the collecting ducts, 90 to 95% of the water and solutes present in the original filtrate have been removed. The filtrate is now more dilute than in the proximal convoluted tubule. In the late distal tubules and collecting ducts, there is **additional reabsorption of Na$^+$ and HCO$_3^-$ ions and secretion of K$^+$ and H$^+$ ions.**

Hormones affect reabsorption and secretion in the late distal tubules and collecting ducts. **ADH** determines whether a concentrated (high osmolarity) or dilute (low osmolarity) urine is produced. High levels of ADH increase water reabsorption producing concentrated urine. When ADH levels are low, dilute urine is produced. **Aldosterone** acts on the collecting ducts only, increasing the reabsorption of Na$^+$ and Cl$^-$ and the secretion of K$^+$.

ACTIVITY 2 **Function of the Nephron**

1 Identify the function of each part of the renal tubule in Figure 37.2.

2 Complete Table 37.2 by comparing the osmolarity of the filtrate in each portion of the renal tubule or collecting duct with the osmolarity of the filtrate in the glomerular capsule by writing in one of the following choices: *no change, increased, decreased, change dependent on presence of ADH and/or aldosterone.*

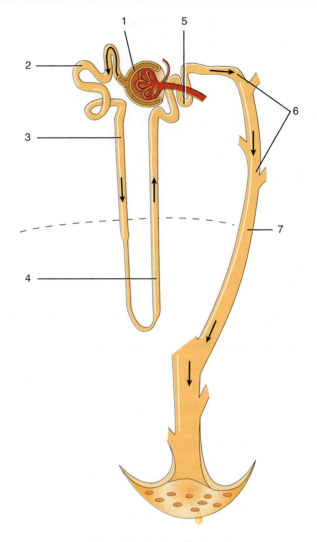

Renal tubule of cortical nephron

FIGURE 37.2 Structure and vascular supply of a cortical nephron.

- additional reabsorption of solutes and water
- ADH increases reabsorption of water, additional reabsorption and secretion of solutes
- aldosterone increases reabsorption of Na$^+$ and Cl$^-$ and secretion of K$^+$
- filtration
- major site of tubular reabsorption of water and solutes; secretion of solutes
- Na$^+$ and Cl$^-$ reabsorbed but not water
- water reabsorbed, but not solutes

1 _____

2 _____

3 _____

4 _____

5 _____

6 _____

7 _____

TABLE 37.2	Comparison of Filtrate Osmolarity in Renal Tubule and Glomerular Capsule	
OSMOLARITY OF FILTRATE IN RENAL TUBULE		**COMPARED TO OSMOLARITY OF FILTRATE IN GLOMERULAR CAPSULE**
End of proximal convoluted tubule		
End of ascending limb of the loop of Henle		
End of descending limb of the loop of Henle		
Distal convoluted tubule		
Collecting duct		

B. Urinalysis

A **urinalysis** is an analysis of the physical, chemical, and microscopic characteristics of urine and a measure of urine volume. The normal characteristics of urine are listed in Table 37.3. **Urine volume** varies depending on the water content of the body, and decreases when body fluid volume is low. ADH secreted by the posterior pituitary gland acts on the collecting ducts to stimulate reabsorption of water from the filtrate. If body fluid volume is high, ADH secretion is inhibited, the collecting ducts do not reabsorb water from the filtrate, and dilute urine is excreted. The **specific gravity of urine** is the weight of a volume of urine divided by the weight of the same volume of distilled water. Urine weight per volume is higher than distilled water because of the presence of solutes in urine. The more solutes present in urine, the higher the specific gravity. Normal urine contains 95% water and 5% solutes. The solutes found in normal urine include **electrolytes** (sodium, potassium, chloride, and other ions), **urea** (formed from breakdown of amino acids), **creatinine** (formed from breakdown of creatine phosphate), **uric acid** (formed from breakdown of nucleic acids), and **metabolic end products** of hormones and other substances. Although drugs are not a normal solute in urine, they will be excreted in urine if present in the bloodstream.

Selected abnormal components of urine are listed in Table 37.4. Your textbook may include a more extensive list. The presence of these substances in urine indicates an abnormality in metabolism or kidney function.

TABLE 37.3	Normal Characteristics of Urine
CHARACTERISTIC	**DESCRIPTION**
Volume	One to two liters per 24 hours; considerable variation in normal volume
Color	Yellow or amber color; urine color is darker in concentrated urine
Turbidity	Transparent in freshly voided urine; will turn cloudy after standing; microbes, pus, epithelial cells, or crystals may cause cloudiness in fresh urine
Odor	Aromatic when fresh; ammonia-like after standing because of breakdown of urea to ammonia by bacteria
pH	Normal range is 4.6–8.0, with an average of 6.0; high protein diets produce an acidic urine; vegetarian diets produce an alkaline urine
Specific gravity	Normal range is 1.001–1.035; low specific gravity represents dilute urine, higher values represent a concentrated urine

TABLE 37.4 Selected Abnormal Constituents of Urine

ABNORMAL CONSTITUENTS	DISORDER AND CAUSE
Glucose	**Glucosuria** (glucose in urine) is usually caused by diabetes mellitus, but sometimes may be caused by stress (epinephrine stimulates glycogen breakdown); diabetes mellitus and stress result in high blood glucose levels and therefore high levels of glucose in the filtrate; glucose transporters cannot work fast enough to reabsorb all glucose from filtrate.
Red blood cells	**Hematuria** (erythrocytes in urine) may be caused by inflammation of urinary system organs, irritation by kidney stones, kidney disease, trauma to urinary system organs, or polyps or tumors within the urinary system.
White blood cells	**Pyuria** (white blood cells in urine) is caused by an infection in the kidney, ureters, urinary bladder, or urethra.
Albumin	**Albuminuria** (excess albumin in urine) is from an increase in filtration membrane permeability caused by high blood pressure, kidney trauma, disease, or inflammation; trace amounts of albumin in urine is normal.
Ketone bodies	**Ketonuria** (ketone bodies in urine) is caused by a metabolic condition called ketosis; ketosis occurs when cells do not have enough glucose to completely break down fatty acids; ketosis can occur from starvation, low carbohydrate diets, or untreated diabetes mellitus.
Casts	**Casts** are hardened clumps of material formed by protein precipitation and/or cell agglutination within renal tubules; the following conditions may result in cast formation: albuminuria, abnormally acidic urine, and highly concentrated urine.
Calculi	**Calculi** are insoluble salt crystals that can form anywhere within the kidney tubules, ureters, urinary bladder, or urethra; commonly called kidney stones, calculi can cause considerable pain as they pass through the lumens or urinary system organs.
Microbes	Presence of bacteria or other microorganisms indicate the presence of a **urinary tract infection (UTI);** normal urine is sterile (no microorganisms present), however microbes present on the skin surrounding the external urethral orifice may contaminate the urine sample if the urine sample is not carefully obtained.

Experiment: Urinalysis

1 **Prediction:** Predict the normal urine values for each of the following characteristics by circling your choice (in italics).
 - Color: *yellow* or *red*
 - Turbidity: *clear* or *cloudy*
 - pH range: *3.0–4.0* or *4.6–8.0* or *8.5–9.5*
 - Presence of glucose: *negative* or *positive*
 - Presence of protein: *negative, trace,* or *positive*
 - Presence of ketone bodies: *negative, trace,* or *positive*
 - Presence of hemoglobin: *negative, trace,* or *positive*
 - Presence of leukocytes: *negative, trace,* or *positive*

2 **Materials:** Obtain materials for Urinalysis (from Materials list).

3 **Data Collection:** Determine color, turbidity, specific gravity, pH, and presence or absence of glucose, protein, ketone bodies, hemoglobin, or leukocytes and record the results in Table 37.5.
 - Label each of four 100-mL beakers as Urine Sample #1, 2, 3, or 4.
 - Put 50 mL from each of the 4 stock artificial urine samples into each appropriately labeled 100-mL beaker. You will have 1 normal (control) and 3 abnormal urine samples.
 - Look at the urine samples to determine the color and turbidity. Record the color and turbidity in Table 37.5.
 - Using a urinometer, measure the specific gravity for each urine sample using instructions provided by your instructor. Do not discard urine; use it for the next test.

 - Urine test strips contain small squares of different reagent paper that change color when they contact specific reagents (chemicals). Use a urine test strip to measure the pH and presence or absence of glucose, albumin (protein), hemoglobin (blood), leukocytes, and ketone bodies. Follow the instructions on the test strip container and record the results in Table 37.5.

4 **Clean Up:** Clean up as instructed.

5 Complete the Experimental Report with your lab group.

 EXPERIMENTAL REPORT

Results:

For each urine sample, state all abnormal results.

Urine Sample #1:

Urine Sample #2:

Urine Sample #3:

Urine Sample #4:

TABLE 37.5 **Urinalysis Results**

TEST	SAMPLE 1	SAMPLE 2	SAMPLE 3	SAMPLE 4
Color				
Turbidity				
Specific gravity				
pH				
Glucose				
Albumin				
Blood				
Ketone bodies				

Discussion:

For each urine sample, state whether the urine sample was normal or abnormal. If abnormal, state which component of the urinalysis was abnormal, what could have caused abnormality, and what condition the patient might have.

Urine Sample #1:
 Possible condition:

Urine Sample #2:
 Possible condition:

Urine Sample #3:
 Possible condition:

Urine Sample #4:
 Possible condition:

Conclusion:

Give an overall statement regarding the characteristics of normal urine.

Reviewing Your Knowledge

A. Nephron Functions

Write the term the phrase describes.

_____ **1.** Hormone that increases sodium and chloride ion reabsorption and potassium secretion.

_____ **2.** Major secretion region of renal tubule.

_____ **3.** Movement of substances from renal tubule into peritubular capillary.

_____ **4.** ADH acts mainly at this portion of the nephron.

_____ **5.** Movement of substances from the peritubular capillary into renal tubule.

_____ **6.** Major reabsorptive region of renal tubule.

_____ **7.** Process occurring across wall of glomerular capillary and visceral layer of glomerular capsule that forms filtrate.

_____ **8.** The act of voiding (emptying) the urinary bladder.

_____ **9.** Hormone that increases water reabsorption in the collecting duct.

B. Urinalysis

Answer the following questions with a short answer.

1. What is the normal volume of urine output per day in an adult?

2. What is the normal color of urine?

3. What could make urine be cloudy?

4. What is the normal specific gravity of urine?

5. What makes urine have a higher specific gravity than distilled water?

6. Name five normal solutes found in urine.

7. Normal urine contains _____% water and _____% solutes.

8. Is normal daytime urine output dilute or concentrated?

9. What is the normal pH of urine?

10. What is pyuria?

11. What is ketoturia?

12. What is hematuria?

13. What is glucosuria?

14. True or false. Normal urine has some bacteria in it even if the urine sample was taken properly.

EXERCISE

37

Using Your Knowledge

A. Urine Formation

1. Explain how a kidney stone blocking urine flow through the ureter eventually may decrease filtrate formation in the glomerular capsule. (*Hint:* hydrostatic pressure.)

2. Explain how inflammation of the glomerular capsule causes hematuria.

3. Compare the contents of the filtrate in the glomerular capsule and the contents of urine in the renal pelvis. (Yes, you can answer this question without research.)

B. Urinalysis

For each of the urine samples in Table 37.6, indicate whether the urine is normal or abnormal. If abnormal indicate the abnormal characteristics and/or constituents. State the probable cause of the abnormal urine.

4. Urine A

5. Urine B

6. Urine C

7. Urine D

8. Which urine has the
 (a) lowest concentration of solutes?
 (b) highest concentration of solutes?

9. Which urines are:
 (a) acidic?
 (b) basic?

10. Which specific gravity value is the most concentrated? Urine _____.

TABLE 37.6 **Urinalysis Results**

MEASUREMENT	URINE A	URINE B	URINE C	URINE D
Color	Yellow	Slightly yellow	Amber	Red
Turbidity	Clear	Clear	Clear	Turbid
pH	6.0	6.5	4.0	7.5
Specific gravity	1.01	1.001	1.04	1.02
Glucose	Negative	Positive	Negative	Negative
Red blood cells	Negative	Negative	Negative	Positive
White blood cells	Negative	Negative	Negative	Positive
Albumin	Negative	Negative	Negative	Negative
Ketone bodies	Negative	Positive	Negative	Negative
Casts	Negative	Negative	Positive	Negative
Calculi	Negative	Negative	Positive	Negative
Microbes	Negative	Negative	Negative	Positive

Male Reproductive System Structure and Function

Objectives

After you complete this exercise, you should be able to:

- Describe the structure and function of the male reproductive organs
- Identify the location of male reproductive organs on a urogenital model or chart
- Describe spermatogenesis
- Trace a spermatozoan from its production to ejaculation
- Name the cells that produce male hormones and describe the functions of the hormones
- Identify the microscopic structure of the testes on microscope slides
- Identify the main male reproductive structures in the cat or fetal pig

Materials

- dissectible models or charts of the male reproductive system
- compound microscopes, lens paper, and prepared slides of sperm and a cross-section of testis
- preserved cats or fetal pigs, disposable gloves, dissection equipment and trays, accompanying cat or fetal dissection manual

A. Male Reproductive Organs

The male reproductive system has **testes** that produce **gametes** called **sperm (spermatozoa),** ducts that store and transport sperm, and accessory sex glands that secrete fluid that, along with sperm, form **semen.** Supporting reproductive structures are the **scrotum** that houses the testes and the **penis** that transfers sperm into the female vagina.

Scrotum and Testes

The **scrotum** (scrotal sac) is a pouch of loose skin that is suspended from the root of the penis outside the abdominopelvic cavity. The scrotum is internally divided into two portions by a central **scrotal septum** (partition) with one testis in each section. The **dartos muscle** (*dartos* = skinned) is within the scrotal septum and scrotal wall. This smooth muscle wrinkles and raises the scrotum when

contracted. The **cremaster muscles** (*cremaster* = hanging) are continuations of the internal oblique muscles, and are located in the spermatic cords superior to the testes, one on each side of the groin. The contraction or relaxation of the cremaster muscles occurs when it is cold or hot, respectively, to adjust the distance of the testes from the body wall. Being on the exterior of the body, the scrotum is provided with a cooler temperature, 2 to 3°C lower than body temperature for normal sperm development.

The **testes** (**testicles** or **male gonads**) are oval-shaped glands that are covered with a white fibrous capsule called the **tunica albuginea** (*tunic* = coat or covering; *albus* = white). The tunica albuginea is dense fibrous connective tissue that extends into the testis to form **septa,** partitions that divide the testes into small compartments or **lobules.** External to this coat is an extension of the peritoneum called the **tunica vaginalis,** and **spermatic fascia** composed of loose connective tissue.

The testes produce and release testosterone and sperm which will travel through a series of tubular structures before being ejaculated from the body. **Sperm** are produced by a process called **spermatogenesis** in tiny **seminiferous tubules** that are located within the **lobules** of the testes. Seminiferous tubules form **straight tubules** that carry the sperm into a network of tubules called the **rete testis** (*rete* = network). **Efferent ducts** connect the rete testis to the **epididymis** (*epi-* = on or upon; *didymus* = pair), the first part of the male duct system.

ACTIVITY 1 Testes and Scrotum

1 Label the structures on Figure 38.1 and Figure 38.2.

2 Locate the structures on a model or chart.

3 Pronounce the terms as you point to them.

- cremaster muscle (kree-MAS-ter)
- dartos muscle (DAR-tohs)
- scrotum (SCROH-tum)
- tunica albuginea of the testis
 (TUNE-ih-kah
 ahl-byoo-GIN-ee-uh)

1 _____

2 _____

3 _____

4 _____

FIGURE 38.1 Anterior view of the testes and scrotum.

- efferent duct
- lobule
- rete testis (reet)
- seminiferous tubules (sem-ih-NIF-er-us)
- septum
- straight tubules
- tunica albuginea

1 _____

2 _____

3 _____

4 _____

5 _____

6 _____

7 _____

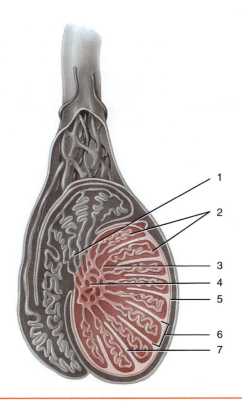

FIGURE 38.2 Sagittal section of the testis.

The Male Duct System

The male duct system includes the ducts of the testes, epididymis, ductus (vas) deferens, ejaculatory duct, and urethra. Each **epididymis,** the site of sperm maturation and storage, curves over the posterior of the corresponding testis. Peristaltic contractions of smooth muscle in the wall of the epididymis move sperm into the **ductus deferens** (*duct* = to lead; *deferens* = carrying away) or **vas deferens** (*vas* = vessel). The initial portion of the ductus deferens leaves the scrotum and enters the spermatic cord that travels between the testes and the inguinal canal.

Each **spermatic cord** contains an artery, vein, lymphatic vessels, nerves, ductus (vas) deferens, and a cremaster muscle. Spermatic cord structures enter the pelvic cavity through passageways in the anterior abdominal wall called **inguinal** (*inguinal* = groin) **canals.** The exterior and interior openings of the inguinal canals, respectively, are called **superficial inguinal** and **deep inguinal rings.** Because the testes descended through the inguinal canal before the male was born, this is a weaker area in the abdominal wall where inguinal hernias can occur.

Within the pelvic cavity, each ductus (vas) deferens travels posteriorly, passing over a ureter as it turns medially and inferiorly to the posterior side of the urinary bladder. The ductus (vas) deferens broadens into an **ampulla** (*ampulla* = flasklike bottle) that unites with the **seminal vesicle duct** to form the short **ejaculatory duct.** Two ejaculatory ducts, one on either side, enter the **prostate gland** (*prostate* = standing before) located inferior to the urinary bladder. Peristaltic smooth muscle contractions in the wall of the ductus deferens and the ejaculatory ducts propel the **semen** (*semen* = seed) into the urethra. The urethra has three regions, the **prostatic urethra** that passes through the prostate gland, the **membranous urethra** within the urogenital diaphragm, and the **spongy (penile) urethra** within the penis. Semen leaves the urethra through the **external urethral orifice.**

ACTIVITY 2) Male Duct System

1 Label the structures in Figure 38.3 and Figure 38.4.

2 Identify the structures on a model or chart of the male duct system.

3 Pronounce the words as you point to them.

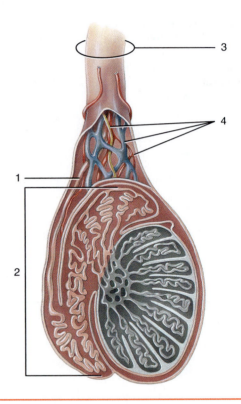

FIGURE 38.3 The epididymis and ductus deferens, sagittal section.

- blood vessels and nerves 1 _____
- ductus (vas) deferens
 (DUK-tus DEAF-er-ens) 2 _____
- epididymis
 (ep-ih-DID-ih-miss) 3 _____
- spermatic cord
 (sper-MAT-ic) 4 _____

FIGURE 38.4 Male reproductive organs and duct system, sagittal section.

- ampulla of ductus deferens (am-POOL-a)
- ductus (vas) deferens in inguinal canal
- ejaculatory duct
- epididymis
- external urethral orifice
- membranous urethra
- penis
- prostate gland (PRAH-state)
- prostatic urethra (prah-STA-tic)
- scrotum
- spongy (penile) urethra
- testis

1 _____

2 _____

3 _____

4 _____

5 _____

6 _____

7 _____

8 _____

9 _____

10 _____

11 _____

12 _____

Accessory Sex Glands and Semen

Three types of accessory glands—the seminal vesicles, prostate gland, and bulbourethral (Cowper's) glands—make and secrete most of the fluid portion of the semen. To assist sperm movement, the testes secrete a small amount of fluid before the addition of fluid from these three glands.

The **seminal vesicles** (*seminal* = pertaining to semen) are saclike glands posterior to the bladder, flanking either side. When sperm cells reach the end of the **ampulla** (expanded end) of the ductus deferens, additional fluid is added from the seminal vesicle through the **seminal vesicle duct.** This alkaline fluid comprises about 60% of the semen volume and neutralizes the acidic male urethra and female vagina, because sperm motility is affected by an acidic environment. It also contains fructose, an energy source for sperm, and prostaglandins that stimulate sperm motility and muscular contractions in the female reproductive tract.

The chestnut-size **prostate gland** surrounds the prostatic urethra inferior to the urinary bladder. Its milky secretion has several proteolytic (*proteo-* = protein; *lytic* = to break apart) enzymes, including prostate-specific antigen (PSA). This slightly acidic secretion comprises approximately 25% of the semen volume and also increases sperm viability and motility. The prostate gland increases in size during male development and is known to enlarge after age 45, a condition known as BPH or benign prostatic hypertrophy. BPH can constrict the prostatic urethra causing difficulty in urination and ejaculation.

The smallest accessory glands are the paired **bulbourethral (Cowper's) glands** (*bulbus* = swollen root) that are located on either side of the membranous urethra within the urogenital diaphragm, a skeletal muscle. These peasized glands secrete an alkaline secretion that neutralizes the acidic male urethra and secrete mucus that lubricates the urethra and glans penis.

ACTIVITY 3 Accessory Sex Glands

1 Label the structures on Figure 38.5 and Figure 38.6.
2 Identify the structures on a model or chart.
3 Pronounce the terms as you point to them.

- ampulla of ductus deferens
- bulbourethral gland
 (buhl-boh-your-REETH-ruhl)
- ductus (vas) deferens
- ejaculatory duct
- membranous urethra
- prostate gland
- prostatic urethra
- seminal vesicle (SEM-ih-nul
 VES-sih-kul)
- spongy (penile) urethra

1 _____

2 _____

3 _____

4 _____

5 _____

6 _____

7 _____

8 _____

9 _____

FIGURE 38.5 Male reproductive organs, posterior view.

FIGURE 38.6 Accessory male sex glands, sagittal section.

• bulbourethral gland
• prostate gland
• seminal vesicle

1 _____

2 _____

3 _____

Penis

The **penis** is a cylindrical shaft which contains the distal portion of the urethra, a passageway for urine and semen and the ejaculation of semen. The penis is located medial and anterior to the scrotum and testes and contains the spongy (penile) urethra. The penis is composed of three main parts: a body, a root, and the glans penis. The **body of the penis** contains three separate cylinders of erectile tissue, two dorsolateral **corpora cavernosa penis** (*corpor-* = body; *cavernosa* = hollow) and one midventral **corpus spongiosum penis,** all containing blood sinuses. Fibrous connective tissue, fascia, and skin surround all three cylinders. The spongy (penile) urethra is located posteriorly in the corpus spongiosum penis.

The **glans penis** (*glans* = acorn) is the distal, expanded portion of the corpus spongiosum that is mostly covered with **prepuce** (foreskin) in an uncircumcised male. **Circumcision** removes the prepuce, exposing the glans penis. The **external urethral orifice** is the external opening of the urethra in the glans penis.

ACTIVITY 4 **The Penis**

1 Label the structures in Figure 38.7.
2 Identify the structures on a model or chart.
3 Pronounce the terms as you point to them.

- bulbourethral gland
- corpora cavernosa penis (core-PORE-ah cav-er-NOH-suh)
- corpus spongiosum penis (CORE-pus spun-gee-OH-sum)
- external urethral orifice
- glans penis
- membranous urethra
- prepuce (PREE-pyoos)
- prostate gland
- prostatic urethra
- spongy (penile) urethra

1 _____

2 _____

3 _____

4 _____

5 _____

6 _____

7 _____

8 _____

9 _____

10 _____

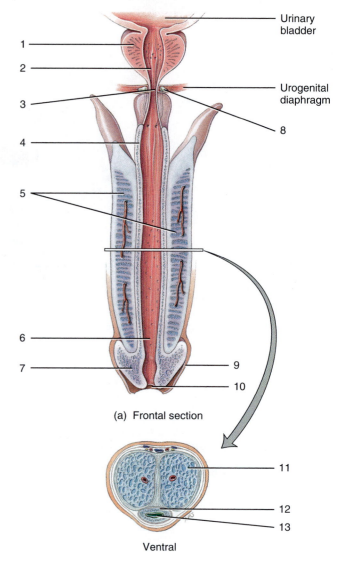

(a) Frontal section

(b) Transverse section

- corpora cavernosa
- corpus spongiosum penis
- spongy (penile) urethra

11 _____

12 _____

13 _____

FIGURE 38.7 Penis.

ACTIVITY 5 **Function of Male
Reproductive Organs**

1 Identify the structure (by number) in Figure 38.8 that
each phrase describes.

2 Write the number in the blank next to function.

FIGURE 38.8 Accessory male sex glands, sagittal section.

_____ carries sperm from epididymis to ejaculatory duct

_____ copulatory organ; passageway for urine and sperm

_____ organ that produces sperm

_____ secretes 60% of semen volume; alkaline pH; prostaglandins

_____ secretes 25% of semen volume; increases viability and motility of sperm

_____ secretes mucus for lubrication of glans; alkaline pH

_____ site of storage and maturation of sperm

B. Microscopic Anatomy of the Testes

Histology of the Testis and Spermatogenesis

Spermatogenesis (*spermato-* = sperm; *genesis* = origin) or the formation of haploid sperm with half the number of chromosomes, occurs in the walls of numerous seminiferous tubules of each testis. This process takes about 2 to 2½ months for each sperm. Between the seminiferous tubules of the testes are triangular-shaped spaces called **interstitial** (*interstitial* = stand between) **spaces** that contain **Leydig cells** or **interstitial endocrinocytes** (endocrine cells). Leydig cells, when stimulated by luteinizing hormone from the anterior pituitary, produce and secrete **testosterone.**

Spermatogenesis begins with **spermatogonia,** diploid (43 or 2n chromosomes) stem cells located in the outer portion of the seminiferous tubule wall just beneath the basement membrane. Spermatogonia divide to form primary spermatocytes (diploid) and more spermatogonia stem cells. Each division moves the cells closer to the lumen of the tubule. **Primary spermatocytes** with 46 chromosomes enlarge and their nucleus divides in meiosis I to form two haploid (n) secondary spermatocytes, each with 23 *double* chromosomes. The two **secondary spermatocytes** divide by a second nuclear division, meiosis II, to form four haploid (n) **spermatids,** each with 23 *single* chromosomes. Spermatids go through a drastic change in shape and composition in a process called **spermiogenesis,** the making of sperm. **Sertoli cells** surround the cells undergoing spermatogenesis, providing the chemical environment for this process and assisting the cells to move from the basement membrane to the lumen.

Sperm

Sperm are also called **sperm cells** or **spermatozoa.** Mature sperm have lost most of their cytoplasm and cellular organelles, and do not survive in the female reproductive tract more than a few days after ejaculation. Sperm are composed of a head, midpiece, and a tail. The **head** consists mostly of the nucleus with an acrosome (*acro-* = a top) covering it. The **acrosome** is a vesicle filled with hydrolytic enzymes that allow sperm to penetrate the female oocyte for fertilization. The **midpiece** contains many mitochondria that are needed to produce energy (ATP) for the tail once the sperm cells are ejaculated. The **tail** is a flagellum that performs a whiplike motion, propelling the sperm.

ACTIVITY 6 **Histology of the Testis, Spermatogenesis, and Sperm**

1 Label the structures in Figure 38.9, Figure 38.10, and Figure 38.11.

2 Observe a prepared slide of the testis.
- Using low power, identify the walls of the **seminiferous tubules, tails of sperm** in the lumen, and **interstitial spaces.**
- Using high power, identify the **basement membrane, spermatogonia, spermatocytes, spermatids,** and **sperm.**

3 Observe a prepared slide of sperm.
- Using a lower power, locate the sperm.
- Using high power, identify the **head, midpiece,** and **flagellum.**

4 Answer the discussion question with your lab partners.

? **DISCUSSION QUESTION**

1 How many sperm are produced in meiosis from one primary spermatocyte? _____

C. Animal Dissection

If you are dissecting a cat or fetal pig, consult the appropriate dissection manual.

(a) Events of spermatogenesis

- primary spermatocyte (spur-MAH-toe-site)
- secondary spermatocyte
- spermatid (SPERM-uh-tid)
- spermatogonium (spur-ma-toe-GOHN-ee-um)
- spermatazoon (spur-ma-toe-ZOH-un)

1 _____

2 _____

3 _____

4 _____

5 _____

- basement membrane
- early spermatid
- late spermatid
- Leydig (interstitial) cells
- primary spermatocyte
- secondary spermatocyte
- Sertoli cells
- sperm cell
- spermatogonium

6 _____

7 _____

8 _____

9 _____

10 _____

11 _____

12 _____

13 _____

14 _____

(b) Transverse section through a seminiferous tubule

FIGURE 38.9 Spermatogenesis.

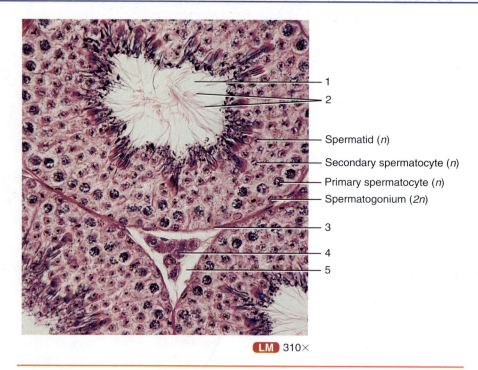

LM 310×

FIGURE 38.10 Photomicrograph of transverse section of seminiferous tubules.

- basement membrane
- interstitial space (inter-STIH-shee-ul)
- Leydig (interstitial) cell
- lumen
- tails of sperm

1 _____

2 _____

3 _____

4 _____

5 _____

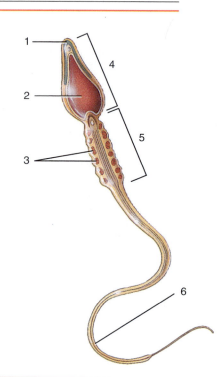

- acrosome (AK-row-sohm)
- head
- midpiece
- mitochondria
- nucleus
- tail (flagellum)

1 _____

2 _____

3 _____

4 _____

5 _____

6 _____

FIGURE 38.11 Sperm cell.

Reviewing Your Knowledge

A. Anatomy Drill and Practice

Go to www.wiley.com/college/allen. Click on the picture of your lab manual, then on Student Companion Site, Anatomy Drill and Practice, and Exercise 38.

B. Male Reproductive Organ Structure and Function

Write the name of the structure that the phrase describes.

Seminal vesicles **1.** Secretes 60% of semen volume; alkaline pH; prostaglandins

testis **2.** Organ that produces sperm

Membranous urethra **3.** Part of urethra within the urogenital diaphragm

Epididymis **4.** Site of storage and maturation of sperm

Prostate gland **5.** Secretes 25% of semen volume; increases viability and motility of sperm

Ejaculatory duct **6.** Located in prostate; formed by union of ducts of the ductus deferens and seminal vesicle

Scrotum **7.** Covers and protects testes

Cremaster muscle **8.** Contracts to pull testes close to body; located in spermatic cord

external urethra orifice **9.** Last section of urethra located in penis

Vas deferens **10.** Tube located in both the spermatic cord and in pelvic cavity for transporting sperm

dartos muscle **11.** Contracts to wrinkle scrotum; located in scrotal wall

Prostate urethra **12.** Ejaculatory duct opens into this part of urethra

urethra **13.** Copulatory organ; passageway for urine and semen

Bulbourethra **14.** Secretes mucus for lubrication of glans; alkaline pH

C. Male Duct System

List the following ducts *in order* (1–10), tracing sperm from their production to ejaculation.

_____6_____ ductus (vas) deferens

_____5_____ epididymis

_____3_____ efferent ducts

_____7_____ ejaculatory duct

_____9_____ membranous urethra

_____8_____ prostatic urethra

_____4_____ rete testis

_____1_____ seminiferous tubules

_____10_____ spongy (penile) urethra

_____2_____ straight tubules

D. Spermatogenesis and Sperm Cells

Write the name of the structure that the phrase describes.

_____Head Piece_____ **1.** Part of sperm that contains chromosomes

_____Spermatids_____ **2.** Haploid cells formed when secondary spermatocytes undergo meiosis II

_____Leydig cells_____ **3.** Cells that secrete testosterone

_____Spermatogonia_____ **4.** Diploid stem cells

_____Meiosis cells_____ **5.** Process that produces haploid sperm from diploid stem cells

_____Sertoli cells_____ **6.** Cells that surround spermatocytes and spermatids and provide chemical environment for spermatogenesis

_____Interstitial cells_____ **7.** Space within testes where Leydig cells are located

_____Primary spermatocyte_____ **8.** Diploid cells that undergo meiosis I to form two haploid secondary spermatocytes

_____Spermiogenesis_____ **9.** Process that transforms spermatids into sperm

_____Mid Piece_____ **10.** Part of sperm that produces ATP

_____Tail_____ **11.** Propels the sperm

Using Your Knowledge

A. Male Reproductive System

Write a short answer for the following questions.

1. Can sperm move in reverse as well as forward? Explain.
 No this kan not go in reverse due to they can only perform forward

2. Which men would have a higher testicular temperature, those who wear briefs or those who wear boxer shorts? Which would be more likely to cause sterility? Explain.
 Briefs because their scrotum will be relaxed.

3. The male reproductive system has both exocrine and endocrine functions. Name one endocrine secretion.
 Hormonal secretions.

Name four exocrine secretions

4. _____ *Sweat* _____

5. _____ *mucus* _____

6. _____ *saliva* _____

7. _____ *Semen* _____

8. To perform a vasectomy, does the doctor have to surgically enter the abdominal cavity? Explain.
 No, because they put a catheter thru and corterize the ductus deferens.

9. A vasectomy involves cutting the ductus deferens. Does a vasectomy impede testosterone levels, having a penile erection, or having an ejaculation? Explain each situation.
 No, because it does not disrupt any other functions.

10. Many body builders who take high doses of anabolic steroids have undesirable side effects. Name four side effects.
 sterility, low level of kstosterone, premature ejaculation,

Female Reproductive System Structure and Function

Objectives

After completing this exercise, you should be able to:

- Describe the structure and function of the female reproductive organs
- Identify the location of the female reproductive organs on a model or chart
- Describe oogenesis
- Trace an oocyte from its production until implantation
- Identify the selected structures on the microscopic slides of the ovary and uterus
- Identify the main male reproductive structures in the cat or fetal pig

Materials

- dissectible models or charts of the female reproductive system
- compound microscopes, lens paper, prepared slides of the ovary and uterus
- preserved animals (cats or fetal pigs), dissection trays and tools, disposable gloves, accompanying cat or fetal pig dissection manual

A. Female Reproductive Organs

The female reproductive system has ovaries that produce gametes (oocytes), tubes to transport the oocytes, and a uterus to protect and nourish the developing embryo and fetus. In addition, the female has accessory structures: the vagina, external genitalia, and mammary glands.

Ovaries and Uterine Tubes

The paired, oval-shaped **gonads** or **ovaries** (*ovary* = egg receptacle) are the size of unshelled almonds (about an inch long) and are located on either side of the uterus. **Oocytes** (eggs or gametes) are produced in the ovary and are released during a process called **ovulation.** The

hormones **estrogen** and **progesterone** are produced by the ovaries. The ovaries are anchored in place by three paired ligaments: the broad ligaments, the ovarian ligaments, and the suspensory ligaments. The sheetlike **broad ligament** of the uterus is a double layer of peritoneum that attaches to each ovary and also secures the uterine tube in place. The **ovarian ligament** on the superior medial side of each ovary attaches the ovary to the uterus. The **suspensory ligament** on the superior lateral side of each ovary attaches the ovary to the pelvic wall in the lumbar region and surrounds the ovarian artery and ovarian vein.

Paired **uterine (fallopian) tubes** or **oviducts** curve around the ovaries to receive and transport the oocyte to the superior part of the uterus. There are three main regions of the uterine tube: the infundibulum (with fimbriae), ampulla, and isthmus. The funnel-shaped **infundibulum** composes the distal portion of the uterine tube and has fingerlike structures projecting from its expanded end called **fimbriae.** During ovulation, the oocyte is projected toward the infundibulum, and the beating of the fimbriae forms a current that sweeps the oocyte into the infundibulum. The oocyte then enters the longest region of the uterine tube, the **ampulla,** that comprises about two-thirds of the length

of the uterine tube. Fertilization of the oocyte by a sperm cell usually occurs in the ampulla. The third region, the **isthmus,** is proximal to and opens into the superior lateral part of the uterus. The shorter isthmus region has a narrower lumen and thicker walls than the rest of the uterine tube.

The lumen of the uterine tubes is lined with **ciliated columnar epithelia.** The fertilized ovum (**zygote**) is moved through the uterine tubes by the beating of the cilia and by peristaltic, smooth muscle contractions of the uterine tube walls. The zygote reaches the uterus about 1 week after ovulation.

ACTIVITY 1 Ovaries and Uterine Tubes

1 Label the structures on Figure 39.1.

2 Identify the structures on a urogenital model or chart.

3 Pronounce the terms as you point to them.

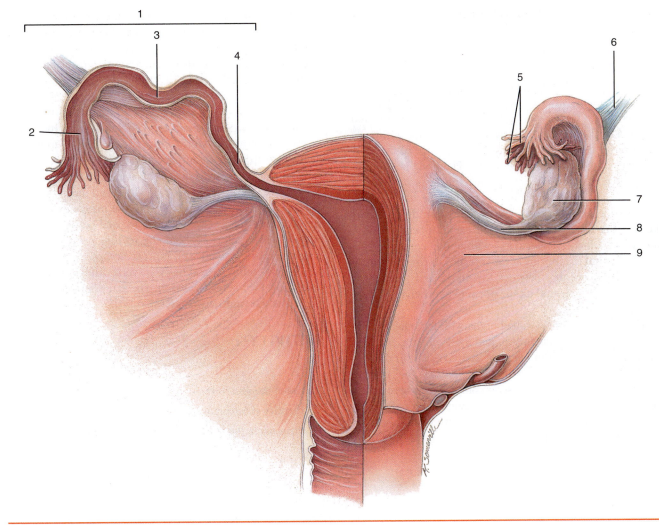

FIGURE 39.1 Posterior view of the ovaries, uterine tubes, and uterus.

- ampulla
- broad ligament
- fimbriae (FIHM-brih-ee)
- infundibulum (in-fun-DIB-you-lum)
- isthmus
- ovarian ligament
- ovary
- suspensory ligament
- uterine tube

1 _____

2 _____

3 _____

4 _____

5 _____

6 _____

7 _____

8 _____

9 _____

Uterus and Vagina

The pear-shaped **uterus** is posterior to the urinary bladder and anterior to the rectum. It is a little smaller than the size of your fist, but obviously larger in a recently pregnant female and smaller after menopause. Externally, three main regions of the uterus can be seen: the fundus, body, and cervix. The **fundus** is the dome-shaped superior portion of the uterus that is wider and meets the entrance of the uterine tubes on both sides of the uterus. The **body** starts inferior to the openings of the uterine tubes, is the largest region of the uterus, and decreases in size to meet the narrower cervix inferiorly. The space within the uterus is the **uterine cavity.** The narrowed **cervix** opens into the vagina inferiorly. The cavity within the cervix is the **cervical canal** that begins at the **internal os** (*os* = mouth), a constricted opening between the uterus and the cervix, and ends at the **external os,** the opening between the cervix and the vagina.

The uterus is a thick-walled organ with three layers: the perimetrium, the myometrium, and the endometrium. The outer, thin **perimetrium** (*metr-* = uterus) is part of the visceral peritoneum, also called the serosa. The middle **myometrium** is the thickest layer composed of three layers of smooth muscle that have fibers going in three different directions: oblique, circular, and longitudinal. The thin **endometrium** is the inner, vascular layer lining the uterine cavity.

Although the uterus is secured in its position by four paired ligaments, only two of these ligaments are easily found on models: the broad ligaments and the round ligaments. The large, flat **broad ligaments** attach the uterus to the lateral pelvic walls. The **round ligaments,** located between the two layers of the broad ligaments, begin below the uterine tubes and attach the uterus to the labia majora of the external genitalia. The ligaments of the uterus keep it in an **anteflexion** (*ante-* = forward) position as it leans forward over the urinary bladder.

The **vagina** (*vagina* = sheath) is a cylindrical passageway for sexual intercourse, childbirth, and menstrual flow. Inferior to the cervix of the uterus, the muscular vagina is angled posteriorly. The vagina is located between the urinary bladder anteriorly and the rectum posteriorly. The **vaginal canal** is lined with mucosa composed of stratified squamous epithelia, and has transverse folds called **rugae** that allow the vagina to stretch. The exterior opening of the vagina is the **vaginal orifice.**

ACTIVITY 2 The Uterus and the Vagina

1 Label the structures on Figure 39.2 and Figure 39.3.

2 Identify the structures on a urogenital model or chart.

3 Pronounce the terms as you point to them.

- cervix
- myometrium of uterus
- ovary
- round ligament
- urinary bladder
- uterine cavity
- vaginal canal

1 _____

2 _____

3 _____

4 _____

5 _____

6 _____

7 _____

FIGURE 39.2 Female reproductive organs, sagittal section.

FIGURE 39.3 Posterior view of the uterus and vagina.

- body of uterus
- broad ligament
- cervical canal
- cervix (SIR-vex)
- endometrium (endo-MEE-tree-um)
- external os
- fundus
- internal os
- myometrium (my-oh-MEE-tree-um)
- perimetrium (peri-MEE-tree-um)
- round ligament
- rugae of vagina
- uterine cavity
- vagina

1 _____

2 _____

3 _____

4 _____

5 _____

6 _____

7 _____

8 _____

9 _____

10 _____

11 _____

12 _____

13 _____

14 _____

External Genitalia

The external female genitalia are referred to as the **vulva** (*vulva* = to wrap around) or **pudendum** (pyoo-DEN-duhm) (*pudens* = modest). The vulva consists of the mons pubis, labia majora, labia minora, clitoris, and vestibule. The **mons** (*mons* = mountain) **pubis** is an anterior pad of adipose tissue beneath the skin that covers the pubic symphysis. Posterior to this are two longitudinal folds of adipose tissue covered with skin and pubic hair called the **labia majora** (sing. = labium majus). Medial to these larger folds are smaller paired longitudinal folds of skin without pubic hair called the **labia minora** (sing. = labium minus). The **clitoris,** a cylindrical erectile tissue, is posterior to the mons pubis and anterior to the external urethral orifice. The **prepuce,** an anterior extension of the labia minora, covers the body of the clitoris like a hood, but leaves part of the clitoris exposed. The **vestibule** is the area medial to the paired labia minora and contains the external urethral orifice, vaginal orifice, hymen (if present), and openings of glands that secrete mucus for lubrication. The **hymen** (*hymen* = membrane) is a thin membrane around the perimeter of the vaginal orifice that partially blocks the orifice.

> ### ACTIVITY 3 External Genitalia
>
> **1** Label the structures on Figure 39.4 and Figure 39.5.
> **2** Identify the structures on a urogenital model or chart.
> **3** Pronounce the terms as you point to them.

FIGURE 39.4 Female reproductive structures, sagittal section.

- anus
- clitoris (KLIH-tor-us)
- external urethral orifice
- labium majus (LAY-bee-um MAY-jus)
- labium minus
- mons pubis (mohnz PYOO-bis)
- pubic symphysis (PYOO-bik SYM-fih-sis)
- vaginal orifice

1 _____

2 _____

3 _____

4 _____

5 _____

6 _____

7 _____

8 _____

FIGURE 39.5 Female genitalia.

- anus
- clitoris
- external urethral orifice
- hymen (HIGH-men)
- labia majora
- labia minora
- mons pubis
- prepuce (PREE-pyoos)
- vaginal orifice

1 _____

2 _____

3 _____

4 _____

5 _____

6 _____

7 _____

8 _____

9 _____

Mammary Glands

Although **mammary** (*mamma-* = breast) **glands** are derivatives of the skin and are categorized as modified sweat glands, they belong to the reproductive system because in females they secrete milk for the offspring. The mammary glands in both males and females lie anterior to the pectoralis major and serratus anterior muscles. Female breasts are enlarged compared with those of the male due to a greater amount of adipose tissue that surrounds the mammary glands. The main external structures of the breasts are the darkened **areola** and the raised **nipple** with **lactiferous duct openings** (*lact-* = milk). The internal structures are the alveoli, lobules, lobes, mammary ducts, lactiferous sinuses, lactiferous ducts, and adipose tissue. **Alveoli** (mammary glands) are milk-producing glands that are arranged in clusters called **lobules.** Several lobules make up a **lobe.** Milk is secreted from alveoli into **mammary ducts** that drain one lobe, and then into dilated **lactiferous sinuses** that merge to form lactiferous ducts. Milk is ejected through **lactiferous ducts** in the nipple.

ACTIVITY 4 Mammary Glands

1 Label the structures on Figure 39.6.

2 Identify the structures on a model or chart.

3 Pronounce the terms as you point to them.

FIGURE 39.6 Mammary glands.

- adipose tissue
- areola (ah-REE-oh-la)
- lactiferous duct (lak-TIF-er-us)
- lactiferous sinus
- lobule with alveoli
- mammary duct
- nipple

1 _____

2 _____

3 _____

4 _____

5 _____

6 _____

7 _____

ACTIVITY 5 **Function of Female Reproductive Organs**

1 Write the number of the organ shown in Figure 39.7 next to the bulleted phrase that describes its function. Numbers may be used more than once.

2 Discuss the pathway of a secondary oocyte from its production until implantation in the uterus.

FIGURE 39.7 Posterior view of the ovaries, uterine tubes, and uterus.

____ Layer of uterus that sheds during menstruation

____ Produces secondary oocytes

____ Transports secondary oocytes toward uterus

____ Usual site of fertilization

____ Beat together to bring ovulated oocyte into uterine tube

____ Implantation of zygote and development of fetus occurs here

____ Produces estrogen and progesterone

____ Narrow region of uterine tube that opens into uterus

B. Oogenesis and Microscopic Anatomy of the Ovary and Uterus

Oogenesis

Oogenesis (*oo-* = egg) is the formation of haploid ova with half the number of chromosomes that occurs in the ovary. Early in fetal development, diploid (2n) stem cells called **oogonia** are produced that divide to create millions of germ cells, most of which degenerate.

Meiosis I

Some of the **oogonia** develop into **primary oocytes** (2n) that begin meiosis I before birth, but stop in prophase I. At puberty, several primary oocytes complete meiosis I each month, forming two haploid cells of unequal sizes. The much larger cell with most of the cytoplasm and organelles is called the **secondary oocyte.** The smaller one is called the **first polar body** that may or may not survive to undergo the meiosis II division. Polar bodies are mostly DNA or genetic material and will eventually degenerate.

Meiosis II

The secondary oocyte begins meiosis II, but stops in metaphase II prior to ovulation. If the ovulated secondary oocyte is fertilized, it will complete meiosis II and form two haploid cells, a large **ovum** and a **second polar body.** When the nucleus from the sperm (n) unites with the nucleus of the ovum (n), a diploid single-celled **zygote** (2n) is formed.

ACTIVITY 6 Oogenesis

1 Label the structures and terms in Figure 39.8.
2 Pronounce the terms as you write them.
3 Answer the discussion questions with your lab partners.

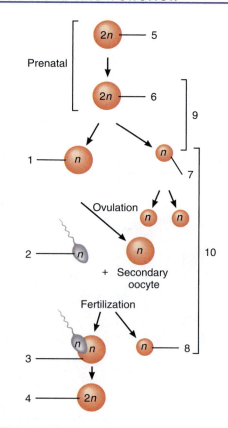

FIGURE 39.8 Oogenesis.

- first polar body
- meiosis I
- meiosis II
- oogonium (oh-oh-GOHN-ee-um)
- ovum
- primary oocyte (OH-oh-site)
- second polar body
- secondary oocyte
- sperm cell
- zygote

1 _____

2 _____

3 _____

4 _____

5 _____

6 _____

7 _____

8 _____

9 _____

10 _____

? DISCUSSION QUESTIONS

1 Compare the size of the secondary oocyte with the polar bodies.

2 Compare the number of gametes produced from one stem cell by oogenesis and spermatogenesis.

- Using low power, identify the **medulla, follicles** of various sizes, and the **hilus.**
- Search for **primordial follicles,** a **primary follicle, secondary follicle,** and a mature **(Graafian) follicle.**
- Center the mature follicle, and observe the **secondary oocyte** with its **granulosa cells** surrounding it.

4 Refer to Figures HA-59 and HA-60 in Appendix A: Histology Atlas.

5 Answer the discussion questions with your laboratory partners.

Histology of the Ovary

The regions of the ovary from superficial to deep are the **germinal epithelium** (simple epithelium), cortex, and medulla. The stringy looking **medulla** contains connective tissue, blood vessels, and nerves. The **cortex** contains follicles (oocytes surrounded by cells) in various stages of development. **Primordial follicles,** the smallest follicles, each contain a **primary oocyte** surrounded by a single layer of cells called follicular cells. These follicles mature into **primary follicles** that have **granulosa cells** surrounding the follicular cells. As the primary follicle matures, the granulosa cells secrete follicular fluid and become secondary follicles. **Secondary follicles** contain an antrum (space within granulosa cell layers) that is filled with follicular fluid. Each month one secondary follicle develops into a **mature** or **graafian follicle.** This follicle, which contains a secondary oocyte, is the largest follicle and has the greatest amount of fluid in its antrum. When a graafian follicle expels the secondary oocyte during ovulation, the remaining follicular tissue becomes a **corpus luteum** (*corpus* = body; *luteum* = yellow). When this structure is no longer functional, it disintegrates and a scarlike **corpus albicans** (*alba* = white) is formed. The corpus albicans doesn't remain in the ovary very long and also disintegrates. Because of this, neither structure may be seen in your slides of the ovary.

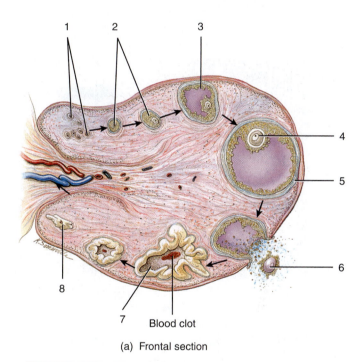

(a) Frontal section

FIGURE 39.9 Histology of the ovary, frontal section.

- corpus albicans (KOR-pus AL-bih-kanz) 1 _____
- corpus luteum (LOO-tee-um) 2 _____
- graafian (mature) follicle (GRAF-ee-uhn) 3 _____
- ovulated secondary oocyte 4 _____
- primary follicles
- primordial follicles (pry-MORE-dee-uhl) 5 _____
- secondary follicle 6 _____
- secondary oocyte 7 _____
 8 _____

ACTIVITY 7 Histology of the Ovary

1 Label the structures in Figure 39.9(a–c).

2 Pronounce the terms as you write them.

3 Examine a prepared slide of the ovary.

- Hold the ovary slide up to the light and determine whether or not you have a section of a complete ovary. If you do not have a complete ovary, some of the structures may not be present.

? DISCUSSION QUESTIONS

1 Which follicle types contain fluid?

2 Describe the difference in appearance between a corpus luteum and a corpus albicans.

(b) Primordial and primary follicles

Germinal epithelium

9 10 11 12 **LM** about 200×

(c) Secondary follicle

13 14 15 **LM** 60×

FIGURE 39.9 Histology of the ovary, frontal section, *continued*.

- granulosa cells
- ovarian cortex
- primary oocyte
- primordial follicles

9 _____

10 _____

11 _____

12 _____

- follicular fluid in antrum
- granulosa cells
- primary oocyte

13 _____

14 _____

15 _____

Histology of the Uterus

The endometrium of the uterus has epithelium facing the uterine cavity with **endometrial (uterine) glands** lying deep to the epithelium. The endometrium is further divided into the stratum functionalis and the stratum basalis. The epithelial layer and the endometrial glands together form the **stratum functionalis** (functional layer). Each month estrogen and progesterone stimulate the stratum functionalis to greatly increase in size in preparation to nourish a zygote. The stratum functionalis is the layer that is sloughed off in menstruation if implantation does not occur. The **stratum basalis,** the layer that remains after menstruation is completed for that month, gives rise to the new stratum functionalis the following month.

ACTIVITY 8 **Histology of the Uterus**

1 Label the structures in Figure 39.10.

2 Pronounce the terms as you write them.

3 Observe a prepared slide of the uterus.
 • Using scanning power or low power, observe a prepared slide of the uterus to identify the **endometrium** and the **myometrium.** You will have to move the slide to see both layers. If your slide is taken from a uterus that is in the secretory phase, it will have a thick endometrium.
 • Using high power, identify the **simple columnar epithelium** lining the lumen, **endometrial glands, stratum functionalis,** and **stratum basalis.**

LM 25×

FIGURE 39.10 Histology of the uterus.

• endometrial glands
• endometrium
• myometrium
• stratum basalis (bay-SAHL-is)
• stratum functionalis

1 _____

2 _____

3 _____

4 _____

5 _____

Reviewing Your Knowledge

A. Anatomy Drill and Practice

Go to www.wiley.com/college/allen. Click on the picture of your lab manual, then on Student Companion Site, Anatomy Drill and Practice, and Exercise 39.

B. The Ovaries, Uterine Tubes, and Uterus

Write the name of the structure that the phrase describes.

fundus _____ **1.** Domed portion of uterus

uterine tubes _____ **2.** Transports secondary oocyte toward uterus

ampulla _____ **3.** Usual site of fertilization

fimbriae _____ **4.** Found on end of infundibulum of uterine tubes

internal os _____ **5.** Opening between uterus and cervix

body _____ **6.** Largest portion of uterus

uterus _____ **7.** Implantation of zygote and development of fetus occurs in this organ

cervix _____ **8.** Narrower portion of uterus; connects with vagina

ovaries _____ **9.** Produces secondary oocytes and hormones

myometrium _____ **10.** Middle layer of uterus

broad ligament _____**11.** Sheetlike ligament attaches uterus to lateral body wall

endometrium _____**12.** Layer of uterus that sheds during menstruation

round ligament _____**13.** Ligament attaches uterus to labia majora

C. Vagina, External Genitalia, and Mammary Gland

Write the name of the structure that the phrase describes.

Vestibute **1.** Area between the labia minora with openings of urethra and vagina

rugae **2.** Folds in vagina

Clitoris **3.** Erectile tissue

adipose tissue **4.** Tissue that pads mons pubis

mammary glands **5.** Milk-producing glands

lactiferous duct openings. **6.** Openings in the nipple

Vagina **7.** Organ for intercourse; canal for childbirth and menstruation

Vulva **8.** Another name for external genitalia

D. Histology of Ovary

Number in correct order (1–6) from the earliest to the latest stage of development.

6 corpus albicans

5 corpus luteum

4 mature (graafian) follicle

1 primary follicle

3 primordial follicle

2 secondary follicle

E. Pathway of a Secondary Oocyte

Number the structures (1–7) for the pathway of a secondary oocyte from the ovary to the uterus.

4 ampulla of uterine tube

7 body of uterus

2 fimbriae of uterine tube

6 fundus of uterus

3 infundibulum of uterine tube

5 isthmus of uterine tube

1 ovary

Using Your Knowledge

A. Female Reproductive System

Write a short answer for the following questions.

1. An ovariectomy is a surgical procedure that removes one or both ovaries, and a hysterectomy is a procedure that

removes the uterus. State whether ovulation and menstruation will occur after:

a. A bilateral (both ovaries) ovariectomy _Menstruation will occur but not ovulation._

b. A hysterectomy _Nothing will occur._

2. A hysterectomy can cause a protrusion of the urinary bladder through the wall of the vagina. Based on your

knowledge of anatomy, why would a hysterectomy cause this problem? _there is an infection_
of the urinary bladder which will cause a protrusion.

3. Vasectomies and female tubal ligations both involve severing and tying a tube to prevent a pregnancy. Which

procedure is less invasive surgically? Explain. _Vasectomies are less invasive_
because it is easier to get in there.

4. An ectopic pregnancy is a pregnancy somewhere other than in the uterus. Explain how a pregnancy could occur in

the abdominal cavity. _When it grows in the ampulla._

5. Some women notice a small blood spot shortly after ovulation. Explain why ovulation can cause this. _____
blood vessels bursting.

6. Endometriosis is a condition in which the endometrial tissue grows on the outside of the uterus and in other loca-

tions in the pelvic cavity. Explain how the endometrial tissue travels to these locations. _travels thru_
under the influence of estrogen + progesterone,

7. A female who has contracted gonorrhea (bacterial infection) through intercourse can develop scarring of the uterine tubes. Explain how an ectopic pregnancy may then result. _____ The scarring of the uterine tubes will effect the growth of the zygote & will later result in an ectopic pregnancy.

8. a. Compare the amount of cytoplasm and number of organelles in ova to that in sperm. _____
Ova = 1 Sperm = millions.

b. Explain why this difference is important to their function. _____
Women are born with ovaries but the men make sperm every day.

9. A homologous (*homo-* = same; *logos* = relation) structure is an organ that has the same embryonic origin. Using your textbook, look up which male structure is homologous to the clitoris.
The ejaculatory duct.

10. Explain why it would be advisable for a patient to have an empty bladder when having a pelvic exam.
The pressure inside makes you to pee.

Human Development

Objectives

**After completing this exercise,
you should be able to:**

- Describe the major events that occur during embryonic development and give the time period during which they occur
- Describe the process of fertilization and early cleavage to the blastocyst stage
- Describe the process of implantation and placenta formation
- Name the three primary germ layers and the body systems that are formed by each
- Name the embryonic membranes and give the importance of each
- Describe the structure and function of the placenta and umbilical cord
- Describe the major developmental events that occur during fetal development and give the time period during which they occur

Materials

- Fertilization and Blastocyst Formation: sea urchin egg suspension, sea urchin sperm suspension, medicine droppers, depression microscope slides, coverslip, Vaseline, toothpicks, compound microscopes, lens paper
- prepared slides of sea urchin embryonic development (early and late cleavage), compound microscopes, lens paper
- models or charts of human embryo development
- models or charts of human fetal development
- model or chart of pregnant human torso
- fresh or preserved placenta or placenta model
- preserved pregnant cat or other animal, dissecting trays and equipment, rubber gloves

Fertilization is the process by which the nuclei of an egg cell and a sperm cell, each with 23 chromosomes, fuse to form a single cell with 46 chromosomes. The fertilized ovum is a single cell called a **zygote** (*zygote* = a pair or yoked). After fertilization, the zygote undergoes a phenomenal series of rapid mitotic divisions that result in an organized ball of cells called an **embryo** (*en-* = in; *-bryein* = to grow). The developing human embryo continues to grow, and its cells differentiate (specialize), forming organs. By 9 weeks, the developing human is called a **fetus** (*fetus* = fruitful, unborn offspring). During this stage, all organs continue to develop and grow, taking on the appearance of adult organs. The study of early human development by embryologists offers us definite information that life truly begins at conception.

A. Embryonic Period

The embryonic period is the time from fertilization through the 8th week of development. The major events during the embryonic period are: fertilization, cleavage, blastocyst formation, implantation, formation of the three primary germ layers, and formation of embryonic membranes.

Fertilization

After ovulation, the secondary oocyte is swept into the uterine tube by the motion of fimbriae in the distal part of the uterine tube, the infundibulum. The secondary oocyte is surrounded by two layers, the corona radiata and the zona pellucida. The **corona radiata** (*corona* = crown) is the outer layer of cells, and the **zona pellucida** (*pellucida* = allowing light to pass through) is the translucent layer of glycoprotein just beneath the corona radiata and exterior to the plasma membrane. Sperm must break through both of these layers to reach the oocyte. After one sperm cell penetrates the plasma membrane, the zona pellucida is enzymatically hardened so other sperm may not enter. The secondary oocyte now completes the second meiotic division and is called an **ovum.** The 26-chromosome sperm cell then fertilizes the 26-chromosome ovum and the two fuse to form a one-celled **zygote** with 46 chromosomes. Fertilization usually occurs within the ampulla of the uterine tube.

Cleavage

The rapid series of early mitotic divisions of the zygote are referred to as **cleavage** (*cleavage* = to split into two or more). The one-celled zygote divides into two cells that are called **blastomeres** (*blasto-* = to sprout, germ; *mere-* = part) between 24–30 hours after fertilization. The two blastomeres further divide into 4-, 8-, 16-, and 32-cells. Although there is an increase in the number of cells because of rapid mitosis, each cell becomes increasingly smaller in size so that the entire ball of cells remains approximately the same size as the original zygote. By the 4th day, the 32 cell stage is a solid ball of cells called a **morula** (*morula* = raspberry) that passes through the isthmus of the uterine tube and into the uterine cavity.

Blastocyst Formation

By the 5th day, the solid morula develops a hollow cavity called the **blastocyst cavity (blastocele),** and the developing zygote is now called a **blastocyst** (*blasto-* = to sprout, germ; *cyst* = bag). Blastocyst cells begin the process of differentiation by forming two different cell regions, the inner cell mass and the trophoblast. The **inner cell mass** is a group of cells in the interior of the blastocyst which will develop into the embryo and form the developing human body. The **trophoblast** (*troph-* = to nourish) is the superficial layer of cells that form the wall of the blastocyst and which will burrow into the endometrium and subsequently form the fetal side of the placenta.

ACTIVITY 1 Fertilization and Blastocyst Formation

1 Label the structures in Figure 40.1 and pronounce the terms as you write them.

2 Examine the models of fertilization, cleavage, and early embryologic development.

3 Fertilization and Blastocyst Formation.
Observe Fertilization
- Obtain materials for Fertilization (see Materials list).
- Using a medicine dropper, transfer one drop of sea urchin egg-cell suspension to the well of a depression slide.
- Examine the drop under a low-power objective lens and focus on one egg cell. Now go to high-power and observe the egg cell.
- Add a drop of sea urchin sperm-cell suspension to the egg-cell suspension on your slide.
- Observe the sperm cells with high-power lens to see them cluster around the egg cells. The eggs secrete chemicals that attract the sperm.
- Watch the egg cell for the rapid appearance of a **fertilization membrane.** This membrane appears as a clear halo and is formed when a sperm cell has penetrated the egg cell. Because of its small size, you will not actually see the nucleus of the sperm.
Observe Blastocyst Formation
- Using a toothpick, place a thin line of Vaseline around the edge of the depression and gently press a clean coverslip over the Vaseline and depression. This will prevent water evaporation.
- The first cleavage should occur in about 45 to 60 minutes. **It is important to keep the slide at 72°F or 22°C.** You may need to use an incubator set at that temperature.

- Use low power to view the slide approximately every 30 minutes. After the first mitotic division, examine the slide about every 15 to 20 minutes.
- Find 2-, 4-, 8-, and 16-cell stages depending on lab time available.
- **Optional note to the instructor:** You may want to fertilize several eggs an hour or more before class and have the students examine those zygotes for cleavage while waiting for their slides.

4 Examine the prepared slides of sea urchin embryonic development, both early and late cleavage. Find 2-, 4-, 8-, and 16-cell stages.

5 Describe the processes of fertilization and cleavage to the blastocyst stage to your lab partners.

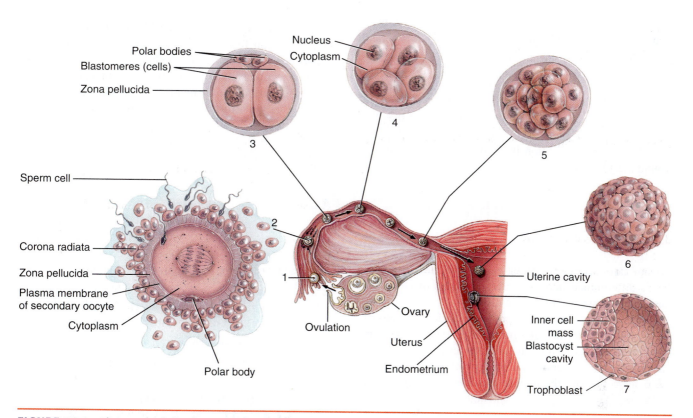

FIGURE 40.1 First week of development—fertilization to blastocyst.

- 2-cell stage of zygote
- 4-cell stage of zygote
- blastocyst, external view
- fertilization of secondary oocyte
- implantation of blastocyst, internal view
- morula
- secondary oocyte

1 _____

2 _____

3 _____

4 _____

5 _____

6 _____

7 _____

Implantation

At approximately day 7, the blastocyst sheds the zona pellucida and usually begins **implantation,** or insertion of the blastocyst into the wall of the uterus. This process starts when the blastocyst adheres to the wall of the uterus at approximately the 6th day after fertilization. Implantation typically occurs in the posterior wall of the fundus or body of the uterus with the inner cell mass facing the endometrium. When this occurs, the endometrium of the uterus is in its secretory phase with a rich blood supply. The trophoblast secretes enzymes that allow the blastocyst to penetrate and bury itself in the endometrium. A hormone, **human chorionic gonadotropin (hCG),** is secreted by the trophoblast to maintain secretion of estrogen and progesterone by the corpus luteum of the ovary. The endometrium is now called the **decidua** (*decida* = falling off), and the basal portion (next to the embryo) is the **deciduas basalis** which will form the maternal side of the placenta. Although, the process of implantation started in the first week after fertilization it is not completed until the around the 14th day.

Formation of the Three Primary Germ Layers

About 8 days after fertilization, cells of the inner mass begin to differentiate into two different layers of cells that form the **bilaminar embryonic disc** (*bi-* = two; *lamina* = flat layer or membrane). At approximately day 15, cells of the bilaminar embryonic disc further differentiate and form three layers of cells. This structure, now called the **trilaminar embryonic disc,** consists of three **primary germ layers,** the outer ectoderm, the middle mesoderm, and the inner endoderm. All tissues, organs, and body systems will arise from these three primary germ layers. The **ectoderm** gives rise to the epidermis, glands, nervous tissue, and parts of the eye and ear; the **mesoderm** gives rise to cartilage, bones, all three muscle types, and blood; and the **endoderm** gives rise to the epithelial lining of the digestive and respiratory systems, and other internal organs. The tissues and organs are all growing rapidly during the embryonic period. Around day 22 the heart starts beating and by the end of the 8th week, the embryo takes on a human appearance and brain waves can be detected.

Formation of Embryonic (Extraembryonic) Membranes

The thin, protective membrane that surrounds and protects the embryo is called the **amnion.** This transparent sac is formed around the 8th day after fertilization. The cavity located between the amnion and the embryo is called the amniotic cavity. **Amniotic fluid** fills the amniotic cavity, cushioning and protecting the embryo from temperature extremes and trauma. The amnion is also known as the "bag of waters."

The **yolk sac,** formerly called the blastocyst cavity, is another embryonic membrane that lies outside the embryo and also begins to develop around the 8th day. The bilaminar disc is now located between the yolk sac and the amniotic cavity. During the 2nd and 3rd week while the placenta is being established, the yolk sac, although small, gives nourishment to the embryo.

The **chorion** (*chorion* = a skin or membrane), a membrane that surrounds all the other embryonic membranes, develops into the embryonic portion of the placenta. Cells of the chorion form the **connecting (body) stalk** which connects the trophoblast and the bilaminar embryonic disc. The connection stalk will become the umbilical cord.

The fourth embryonic membrane, the **allantois** (*allas* = sausage; *-eidos* = form), is an extension of the yolk sac and extends into the connecting stalk of the embryo. It contains blood vessels that will become the umbilical arteries and the umbilical vein.

ACTIVITY 2 Primary Germ Layers and Embryonic Membrane Development

1 Label the structures in Figure 40.2(a) and (b).
2 Observe models or charts of the 4 embryonic membranes and locate the **amnion, yolk sac, chorion,** and **allantois.** State the function of each embryonic membrane.
3 Answer discussion questions with your laboratory partners.

? DISCUSSION QUESTIONS: IMPLANTATION, PRIMARY GERM LAYERS, AND EMBRYONIC MEMBRANES

1 How does the blastocyst implant into the endometrium?

2 What body structures are produced from the three primary germ layers?

3 Discuss the difference between the four embryonic membranes and the importance of each.

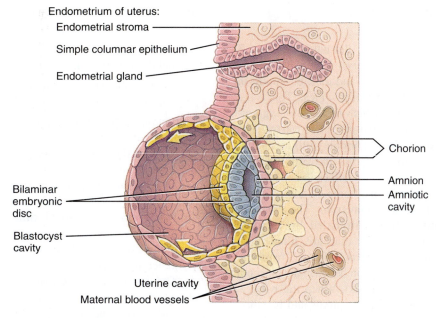

Endometrium of uterus:
 Endometrial stroma
 Simple columnar epithelium
 Endometrial gland

Chorion

Amnion
Amniotic cavity

Bilaminar embryonic disc

Blastocyst cavity

Uterine cavity
Maternal blood vessels

(a) Implantation of blastocyst, about 8 days after fertilization

Endometrium of uterus:
 Simple columnar epithelium
 Endometrial stroma

Sinusoid

Lacunae

1

Chorion

2
3

4

Lacunar network

Uterine cavity

Endometrial gland (right) and sinusoid (left) emptying into lacunar network

(b) Development of embryonic membranes, about 12 days after fertilization

- amnion
- amnionic cavity
- bilaminar embryonic disc
- yolk sac

1 _____

2 _____

3 _____

4 _____

FIGURE 40.2 Second week of development—embryonic membrane development.

Placenta and Umbilical Cord

At birth the placenta (*placenta* = flat cake) is about 8 inches in diameter and approximately 1 inch to 1⅜ inches thick. The **placenta** is disc shaped and dark burgundy in color. It almost has the appearance of raw meat. This is the structure in the uterus through which the developing embryo and fetus obtain nutrients and get rid of wastes. The placenta has an embryonic side (chorionic villi) and a maternal side (decidua basalis). During the 3rd week of growth, the chorion forms **chorionic villi** (*villus* = hairlike projection) with blood vessels that lengthen and protrude into the **decidua basalis** of the endometrium. The chorionic villi lie in **intervillus spaces** surrounded by **lacunae** (*lacuna* = a pit, hollow space) that are filled with maternal blood. The **umbilical cord** is formed from the connecting stalk and is the connection between the placenta and the embryo (and later the fetus). When the baby is born, the placenta separates from the uterus and is called the **afterbirth.** The area where the umbilical cord was attached to the baby is called the **umbilicus** (navel).

ACTIVITY 3 **Placenta and Umbilical Cord**

1 Label the structures in Figures 40.3 and 40.4.

2 Pronounce the terms as you look at a model or chart of the placenta.

3 Observe the demonstration of a fresh or preserved placenta, if available.
 • Observe the maternal surface which is spongy with a cobblestone appearance. This side of the placenta is textured due to the fetal chorionic villi which extensively branched into the endometrium.
 • Now observe the smooth, shiny **fetal surface** of the placenta which is somewhat translucent. The umbilical cord is attached to this side of the placenta, generally in the center.
 • The umbilical cord contains 2 arteries and 1 vein which are surrounded by mucous connective tissue called Wharton's jelly. The shiny layer on the exterior of the umbilical cord is a layer of the amnion.

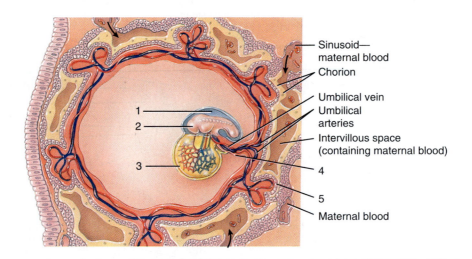

Sinusoid—maternal blood
Chorion
Umbilical vein
Umbilical arteries
Intervillous space (containing maternal blood)
4
5
Maternal blood

FIGURE 40.3 Development of chorionic villi in a 21-day embryo.

• amniotic cavity
• chorionic villus
• connecting stalk
• embryo
• yolk sac

1 _____

2 _____

3 _____

4 _____

5 _____

4 Observe the demonstration of a pregnant cat or other animal, if available. Identify the **chorion, amnion, embryo, placenta,** and **umbilical cord.**

5 Answer discussion questions with your laboratory partners.

2 How is the placenta formed? What are the embryonic and maternal parts of the placenta?

3 What is the function of the umbilical arteries? The umbilical vein?

② DISCUSSION QUESTIONS: PLACENTA AND UMBILICAL CORD

1 Create a sentence including the following three terms: fetus, placenta, umbilical cord.

FIGURE 40.4 Placenta and umbilical cord.

- chorion
- chorionic villi
- decidua basalis
- fetal blood vessels
- umbilical arteries
- umbilical vein

1 _____

2 _____

3 _____

4 _____

5 _____

6 _____

B. Fetal Period

The fetal period is from week 9 to week 38. By the 9th week all organ systems are present in the fetus, and the fetal head is about half of the size of the body. The fetus doubles in size between the 12th and 20th fetal weeks, doubles again between the 16th and the 24th weeks, and more than doubles between the 20th and 38th weeks. The uterus expands as the fetus grows, and the mother's abdominal organs are shifted from their normal location. Although at 8 weeks toward the end of the embryonic period the developing embryo looks like a human, the 16-week fetus takes on even more of a human appearance.

ACTIVITY 4 Fetal Development

1 Examine the drawings in Figure 40.5 and the photographs in Figure 40.6.
2 Examine the fetal development models and describe the position of the fetus in the uterus.

? DISCUSSION QUESTIONS:

1 At what stage can you first identify the:
 • upper and lower limbs
 • future auditory meatus
 • eye tear
 • fingers and toes

2 Does the position of the fetus in the uterus change very much toward the end of the 38th week?

| 4 | 8 | 12 | 16 | 20 | 24 | 28 | 32 | 36 | (weeks) |

FIGURE 40.5 Fetal development.

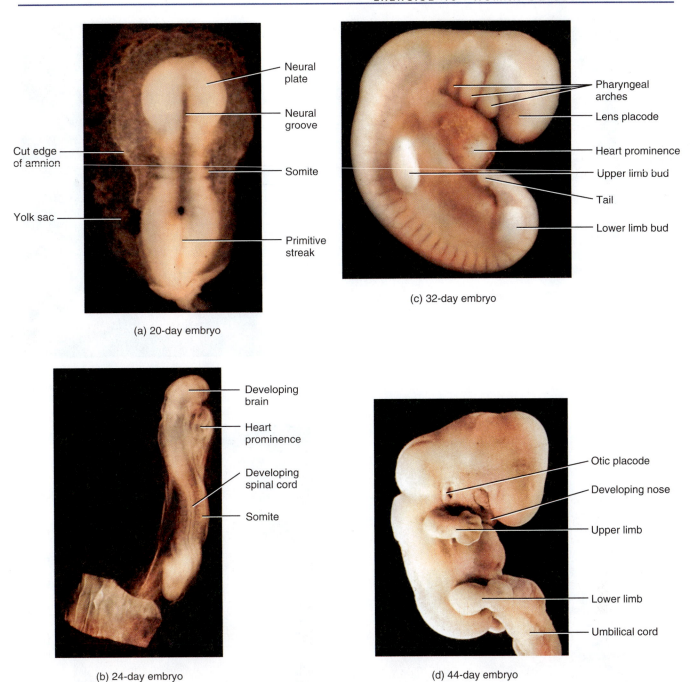

(a) 20-day embryo

Neural plate
Neural groove
Cut edge of amnion
Somite
Yolk sac
Primitive streak

(c) 32-day embryo

Pharyngeal arches
Lens placode
Heart prominence
Upper limb bud
Tail
Lower limb bud

(b) 24-day embryo

Developing brain
Heart prominence
Developing spinal cord
Somite

(d) 44-day embryo

Otic placode
Developing nose
Upper limb
Lower limb
Umbilical cord

FIGURE 40.6 Fetal development.

(e) 52-day embryo

(g) Thirteen-week fetus

(f) Ten-week fetus inside amniotic sac in uterus

(h) Twenty-six-week fetus

FIGURE 40.6 Fetal development, *continued*.

EXERCISE

40

Reviewing Your Knowledge

A. Embryonic Development

Give the name of the structure that the phrase describes.

_____ **1.** The name of the ovulated structure prior to fertilization.

_____ **2.** The blastocyst structure that becomes the embryo.

_____ **3.** The anatomical structure where fertilization normally takes place.

_____ **4.** The stage of the embryo that implants in the uterine wall.

_____ **5.** After fertilization, implantation normally occurs on this day.

_____ **6.** The membrane that forms the embryonic side of the placenta.

_____ **7.** The structure that is formed after the morula stage.

_____ **8.** The structure that the blastocyst sheds before implantation.

_____ **9.** The structure that forms the maternal side of the placenta.

_____ **10.** The name of the thin, protective membrane that encompasses the embryo.

_____ **11.** The side of the placenta that is smooth and shiny.

_____ **12.** The number of umbilical arteries in the umbilical cord.

_____ **13.** The number of umbilical veins in the umbilical cord.

_____ **14.** This placental structure has villi.

_____ **15.** The number of primary germ layers that form all body organs.

_____ **16.** Sustains the embryo with nutrients and removes wastes.

_____ **17.** The name of the female gamete.

_____ **18.** The name of the male gamete.

_____ **19.** The inner layer of the uterus where implantation occurs.

_____ **20.** The name of the fertilized egg.

_____ **21.** Primary germ layer that becomes the outer skin and nervous system.

_____ **22.** Primary germ layer that gives rise to epithelial lining of digestive and respiratory organs.

_____ **23.** Primary germ layer that produces bone, muscle, cartilage, and blood.

B. Fetal Development

Give the name of the structure that the phrase describes.

_____ **1.** The "bag of waters" that breaks at birth.

_____ **2.** The fetal period lasts from week _____ to week _____.

_____ **3.** By week _____ all organ systems are present.

_____ **4.** Toward the end of the embryonic period, week _____, the developing embryo looks like a human.

_____ **5.** The fetus first doubles in size between weeks _____ and _____.

_____ **6.** The fetus doubles in size again between weeks _____ and _____.

_____ **7.** The fetus more than doubles between weeks _____ and _____.

Using Your Knowledge

A. Human Development

Write a short answer for each of the following questions.

1. The acrosome of a sperm contains hydrolytic enzymes. Using your knowledge, what is the purpose of these enzymes.

2. Ectopic pregnancies occur in the uterine tubes or pelvic cavity. Explain how this can occur.

3. Amniocentesis is a test conducted on fetal cells found in amnionic fluid to determine whether the fetal cells have genetic abnormalities. Amniocentesis is usually conducted at 12 weeks. Amnionic fluid is collected by inserting a needle into the mother's abdomen. List the structures the needle must go through to collect the amniotic fluid.

4. Are the cells obtained for culture from amniocenteses most likely derived from ectoderm, mesoderm, or endoderm? Explain your answer.

5. Chorionic villi sampling (CVS) is a test to determine whether fetal cells have genetic abnormalities and can be performed as early as 8 weeks. Chorionic villi cells are removed using a small tube that is passed through the cervix. These cells are cultured and undergo chromosomal analysis. Explain why chorionic villi cells will determine whether the fetus has genetic abnormalities.

6. Drinking alcohol during pregnancy can lead to fetal alcohol syndrome, a condition that can cause heart defects, damage to the central nervous system, deformed limbs, slow growth, and abnormal facial features. Explain how the ingested alcohol is transferred from mother to fetus.

7. Early pregnancy tests used antibodies to detect the presence of human chorionic gondaotropin (hCG) in maternal urine. Explain how hCG, produced by the chorion, gets in maternal urine.

8. Human embryonic stem cell research uses stem cells cultured from the blastula. These cells are pluripotent, meaning that these stem cells can develop many different types of human cells. Identify the part of the blastula that they are using and explain your choice.

9. During an ultrasonography, the doctor observes that all four chambers of the fetal heart are developed normally and that blood can be observed passing from the right atrium to the left atrium. Explain why it is normal for blood to pass from the right to left atrium in the fetal heart.

10. Placenta previa (*previa* = before or in front of) is a condition in which either part or the entire placenta implants in the inferior part of the uterus near the cervix. Explain why this can cause spontaneous abortions or premature birth.

Heredity

Heredity is the transfer of genetic traits, such as hair or eye color, from parents to offspring, generation after generation. **Genetics** is the science that studies the principles and patterns governing heredity. Geneticists can study the complete set of an organism's genes (located on chromosomes) called a **genome.** The complete set of genes and alleles for all humans that is studied in genetics is called the **human gene pool.** Geneticists have discovered that environmental influences like nutrition play a very critical part in our appearances (height, weight, etc.) and whether or not we will inherit a genetic disease.

A. Basic Genetic Language

Each chromosome contains one **DNA** (deoxyribonucleic acid) molecule. **Genes** are short segments of DNA that are the functional or heritable units of DNA. Each gene is situated at a specific location on a chromosome and provides a template for the cell to make one or more proteins.

Each human cell contains the same 46 chromosomes, 22 pairs of **homologous** (*homo-* = same; *logos* = proportion) chromosomes called **autosomes** and 2 **sex chromosomes.** The homologous chromosomes look identical in size and shape under the microscope and have the same genes at identical locations. There are two types of human sex chromosomes, an X chromosome and a Y chromosome. The X and Y chromosomes not only look different in size and shape, but they also carry genes for different characteristics.

Each gene has two or more alternate forms, called **alleles,** which occupy the same **loci** (location) on homologous chromosomes. The different alleles code for different proteins, and therefore will result in the expression of different traits. For example, there are two alleles pertaining to dimples. One allele of this gene codes for dimples (D) and the other allele is for no dimples (d). If the pair of alleles of a particular gene are identical (i.e. DD or dd), the

individual is **homozygous** (*homo-* = same; *zygous* = yoked) for that trait. If the alleles are different (Dd), the person is **heterozygous** (*hetero-* = different). If alleles are heterozygous, one allele is usually completely expressed and is called **dominant,** while the other allele is not expressed and is called **recessive.** Note that a capital letter is used to designate a dominant allele which is written first, and a lower case letter is used for a recessive allele. The particular genetic composition of a person is the **genotype** (DD, Dd, or dd), and the expressed physical traits that are observable are called the **phenotype** (dimples or no dimples).

ACTIVITY 1 Genotypes and Phenotypes

1 For each genotype, indicate whether it is homozygous (hom) or heterozygous (het).

_____ Zz _____ bb _____ MM

_____ Tt _____ cc _____ AA

2 For each genotype, indicate the phenotype.

The freckles allele (F) is dominant to the no freckles allele (f). The attached earlobes allele (E) is dominant to unattached earlobes allele (e).

ff _____ Ee _____

FF _____ EE _____

Ff _____ ee _____

3 For each phenotype, list the genotype(s).

Widow's peak (W) is dominant to a straight hairline (w).

_____ widow's peak (homozygous)

_____ widow's peak (heterozygous)

_____ straight hairline

Mid-digital finger hair (M) is dominant to no finger hair (m).

_____ mid-digital hair (homozygous)

_____ mid-digital hair (heterozygous)

_____ no mid-digital hair

B. Monohybrid Crosses and the Punnett Square

Gregor Mendel was the first person to systematically study, record, analyze, and publish information regarding the patterns of inheritance. Mendel's first rule of inheritance, developed while studying inheritance in garden peas, was the **principle of segregation.** He showed that when gametes are produced, the 2 units (genes) for a characteristic separate from each other and the sperm and ovum get only one unit each. For instance, if the genotype for one parent was Dd, as the sperm or ovum is formed, each sperm or ovum gets only a 'D' or a 'd'.

An uncomplicated way to view patterns of inheritance for a single pair of alleles is by using a diagram called the **Punnett square.** This diagram is used to predict possible genotypes and therefore phenotypes of the offspring from two parents. In a **monohybrid cross,** the inheritance of only one trait is studied using a single pair of alleles. In a **dihybrid** or **trihybrid cross,** there are two and three traits, respectively, that are studied. All of the genetic activities in this exercise are monohybrid crosses. The presence or absence of dimples is used as the trait in the next activity. Let D = dimples and d = no dimples. The genotype of the female is homozygous dominant (DD) and the male is homozygous recessive (dd).

ACTIVITY 2 Dimples or No Dimples

1 In Figure 41.1, look at the genotype of the parents that displays the **diploid** state (2n). There are 2 sets of chromosomes, one inherited from the father and one from the mother. According to the first principle of inheritance, observe that the two alleles for any trait will separate during meiosis. Therefore, each "D" or each "d" is now **haploid** (n) with half the number of chromosomes (alleles) present in the ova or sperm.

2 Put the alleles of the sperm on the side of the square, and the alleles of the ova on top. In order to have a pattern, always put the same parent on the same axis.

3 Combine the letters of the sperm and ova in the squares that now represent the diploid genotype(s) of their offspring.

4 In Figure 41.1, indicate the number of offspring with each **genotype** and the % chance offspring has the genotype.

5 On Figure 41.1, indicate the number of offspring with each **phenotype** and the % chance offspring has the phenotype.

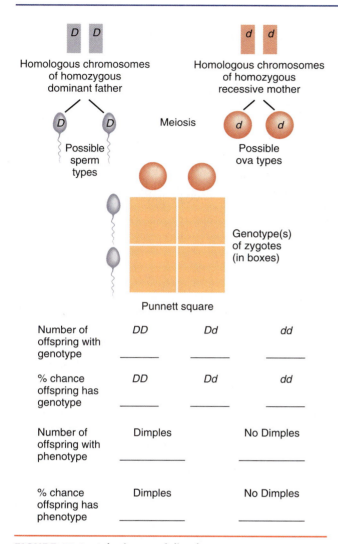

Number of offspring with genotype

	DD	Dd	dd
	_____	_____	_____

% chance offspring has genotype

	DD	Dd	dd
	_____	_____	_____

Number of offspring with phenotype

	Dimples	No Dimples
	_____	_____

% chance offspring has phenotype

	Dimples	No Dimples
	_____	_____

FIGURE 41.1 Inheritance of dimples.

FIGURE 41.2(a) Tongue roller.

FIGURE 41.2(b) Non-tongue roller

ACTIVITY 3 Tongue Rolling or No Tongue Rolling

1 Some people can roll the tongue into a U-shape (Figure 41.2a), while others do not have the ability to roll the tongue (Figure 41.2b). Let R = tongue roller, r = non-tongue roller. Cross a heterozygous tongue-rolling mother and father.

 • In Figure 41.3, choose the dominant and recessive alleles for the sperm and the ova and write the alleles on the top and left side of the Punnett square.

 • Combine the letters of the sperm and ova in the squares. These letters represent all of the possible diploid genotypes of the offspring.

2 In Figure 41.3, indicate the number of offspring with each **genotype** and the % chance offspring has the genotype.

3 On Figure 41.3, indicate the number of offspring with each **phenotype** and the % chance offspring has the phenotype.

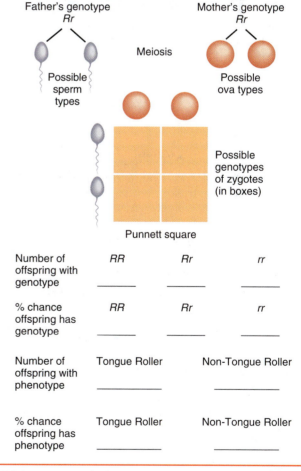

Number of offspring with genotype

	RR	Rr	rr
	_____	_____	_____

% chance offspring has genotype

	RR	Rr	rr
	_____	_____	_____

Number of offspring with phenotype

	Tongue Roller	Non-Tongue Roller
	_____	_____

% chance offspring has phenotype

	Tongue Roller	Non-Tongue Roller
	_____	_____

FIGURE 41.3 Inheritance of tongue rolling.

ACTIVITY 4 Mid-Digital Hair or No Mid-Digital Hair

1 Some people have mid-digital hair on the medial digits (Figure 41.4a), while others have not inherited this trait (Figure 41.4b). Let M = mid-digital hair and m = no mid-digital hair. Cross a heterozygous mid-digital hair father with a homozygous no mid-digital hair mother.
 • In Figure 41.5, choose the alleles for the sperm and the ova and write them by the square.
 • Combine the letters of the sperm and ova in the squares that now represent all of the possible diploid genotypes of their offspring.

2 Calculate the percentages of each genotype and phenotype of the offspring from the Punnett square.

FIGURE 41.4(a) Mid-digital hair.

FIGURE 41.4(b) No mid-digital hair.

Punnett square

% chance offspring has genotype	*MM*	*Mm*	*mm*
	_____	_____	_____

% chance offspring has phenotype	Mid-digital hair	No mid-digital hair
	_____	_____

FIGURE 41.5 Inheritance of mid-digital hair.

C. Incomplete Dominance

A few human traits are inherited because neither allele for a gene is dominant over the other and the offspring show an intermediate phenotype or blending of the two characteristics. This type of inheritance is called **incomplete dominance.** An example of this is curly, wavy, or straight hair. Curly hair (CC) is incompletely dominant over straight hair (cc), therefore the heterozygous genotype (Cc) produces wavy hair (Cc), a phenotype intermediate between curly and straight hair.

ACTIVITY 5 Curly, Wavy, and Straight Hair

1 Cross a mother with curly hair with a father with straight hair in Figure 41.6.

2 Calculate the percentages of each genotype and phenotype of the offspring from the Punnett square.

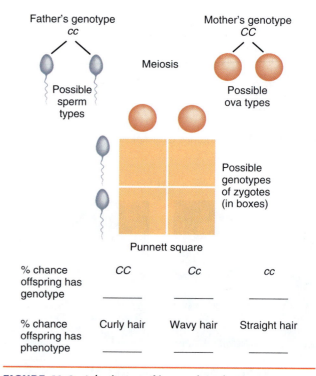

Punnett square

% chance offspring has genotype	*CC*	*Cc*	*cc*
	_____	_____	_____

% chance offspring has phenotype	Curly hair	Wavy hair	Straight hair
	_____	_____	_____

FIGURE 41.6 Inheritance of incomplete dominance.

D. Codominance in Monohybrid Crosses

Codominance is a pattern of inheritance in which two different alleles of a gene are equally expressed in a heterozygote and no blending of the characteristic occurs. There are three alleles for the ABO human blood type system: A, B, and O. The A and B alleles are codominant and both are dominant over the O allele. The A allele causes an A antigen to appear on the surface of red blood cells, the B allele causes a B antigen to appear on the surface of red blood cells, and the O allele results in no antigen on the surface of the red blood cell. Because each of us inherits only two of these alleles, there are six possible genotypes: AA, AB, AO, BB, BO, and OO. Both A and B alleles are dominant over O, and alleles A and B are codominant. This means that there are only four phenotypes for blood type: A, B, AB, and O.

ACTIVITY 6 ABO Blood Type

1 Cross a father with blood type AO and a mother with BO blood type in Figure 41.7.

2 Calculate the percentages of each genotype and phenotype of the offspring from the Punnett square.

3 Cross a father with AB blood type with a mother with O blood type in Figure 41.8.

4 Calculate the percentages of each genotype and phenotype of the offspring from the Punnett square.

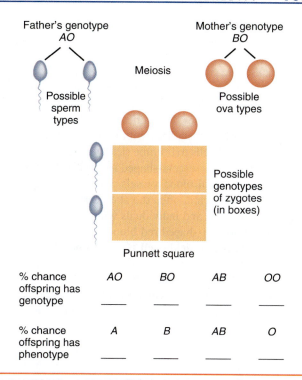

FIGURE 41.7 Codominance with ABO blood types.

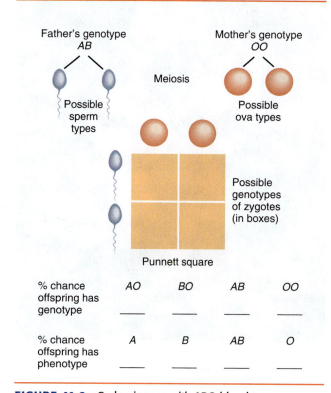

FIGURE 41.8 Codominance with ABO blood types.

Sickle-cell anemia is another example of codominance. Two alleles are present that are involved in the production of hemoglobin (Hb), a protein that carries oxygen in red blood cells. The sickle-cell allele is designated S and produces an abnormal Hb. Individuals who are SS produce only the abnormal Hb. When oxygen levels in the red blood cells are low (examples: high altitudes or during exercise), the sickle-cell Hb crystallizes within red blood cells causing the red blood cells to become sickle or crescent-shaped (Figure 41.9). The crescent-shaped sickle cell clumps and forms clots in small blood vessels. Untreated individuals with sickle cell anemia often die at an early age. The normal Hb allele is A, and individuals who are AA have normal Hb and normal-shaped red blood cells. Heterozygous individuals have the **sickle-cell trait** (but not the disease) with half of their Hb normal and half is sickle-cell Hb. These individuals are usually healthy, and only a small number experience some symptoms of sickle-cell disease. Heterozygous individuals are also resistant to malaria.

Therefore, in regions of the world where malaria is a problem, heterozygous individuals have an advantage and more individuals with the SA genotype survive compared to individuals with the AA genotype.

ACTIVITY 7 Sickle-Cell Anemia

1 The abbreviation for hemoglobin is Hb. Let $Hb^S Hb^S$ = sickle-cell anemia,
$Hb^A Hb^A$ = normal Hb (unaffected), and $Hb^S Hb^A$ = sickle-cell trait.

2 Using Figure 41.10, cross a mother with sickle cell anemia and father with the sickle-cell trait.

3 Calculate the percentages of each genotype and phenotype of the offspring from the Punnett square.

FIGURE 41.9 Sickle-cell anemia.

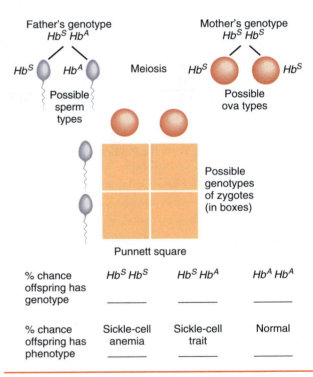

FIGURE 41.10 Codominance with sickle-cell anemia.

E. Sex-Linked Inheritance

The majority of genes are carried on **autosomes,** chromosomes which code for traits not directly involved in determining the sex of the individual. There are 22 pairs of autosomes and 2 **sex chromosomes** (X and Y) for each person (Figure 41.11). Females have two X chromosomes (XX) while males have one X and one Y (XY). The X chromosome is as large as many of the autosomes, but the Y chromosome is about one-third of the size of the X chromosome and is missing genes present on the X chromosome. The genes carried *only* on the X chromosome are called **sex-liked genes** and do not produce female characteristics. Sex-linked genes have recessive alleles that cause baldness, hemophilia, and color blindness. For a woman to express these phenotypes, she must have two recessive alleles, one each X chromosome. However, because the male only has one X chromosome and these sex-linked genes are not present on the Y chromosome, the presence of one recessive allele will give a male the recessive phenotype. In sex-linked inheritance, the terms *homozygous* and *heterozygous* in the male do not apply. It is interesting to note that with sex-linked inheritance, the father can not pass the condition to his son. The son can only receive a Y chromosome from his father and an X chromosome from his mother.

FIGURE 41.11 Autosomes and sex chromosomes.

ACTIVITY 8 Sex-Linked Inheritance

1 Let X^C = dominant allele for **normal** vision and X^c = recessive allele for **color blindness.** Indicate the phenotype of the following alleles for recessive sex-linked color blindness.

$X^C X^C$_____ $X^C X^c$_____ $X^c X^c$_____

$X^C Y$_____ $X^c Y$_____

2 Using Figure 41.12, cross a heterozygous female carrying the allele for color blindness (a carrier) with a male with normal vision.

3 Calculate the percentages of each genotype and phenotype of the offspring from the Punnett square.

4 Answer the discussion questions.

? DISCUSSION QUESTIONS

1 What percentage of girls, if any, would be a carrier for the color-blind allele?

2 What percentage of boys, if any, would be a carrier for the color-blind allele?

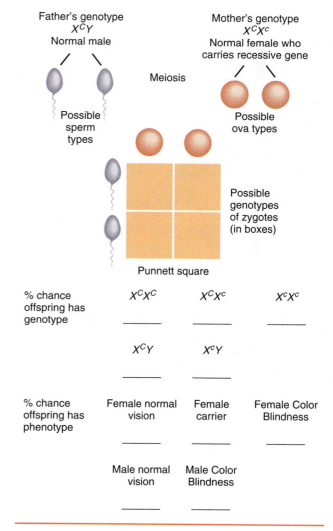

FIGURE 41.12 Sex-linked inheritance for color blindness.

Reviewing Your Knowledge

A. Basic Genetic Language

Write the name of the term that describes the phrase.

_____ 1. Two different alleles of a gene are both expressed with no blending in a heterozygous individual

_____ 2. Non-identical chromosomes that determine male or female

_____ 3. The functional units of DNA; short segments of chromosomes

_____ 4. The allele that is expressed in a heterozygous individual

_____ 5. Two or more alternate forms or variations of a gene

_____ 6. The genetic composition for a trait

_____ 7. Individual having identical alleles for a gene

_____ 8. 22 pairs of chromosomes that determine most of the body traits

_____ 9. The allele that is not expressed in heterozygous alleles

_____ 10. The physical or functional expression of a genetic trait

_____ 11. Possessing all of the normal number of chromosomes

_____ 12. Possessing half of the normal number of chromosomes

13. Explain what a "carrier" for an allele of a gene means.

14. Distinguish between a phenotype and a genotype.

15. How do we obtain two copies of each gene?

B. Monohybrid Dominant–Recessive Inheritance

1. Unattached earlobes (E) are dominant over attached earlobes (e). (Figures 41.13a and b). If a woman with unattached earlobes and a man with attached earlobes have children, what percentage of their children have the possibility of having unattached earlobes if the mother is homozygous for the trait?

FIGURE 41.13(a)
Unattached earlobes.

FIGURE 41.13(b)
Attached earlobes.

2. What percent of the offspring in question 1 above have the possibility of having unattached earlobes if the mother is heterozygous for the trait?

3. For questions 1 and 2, could the phenotype of the children born allow you to determine whether or not the mother is homozygous or heterozygous for the trait? Explain.

C. Incomplete Dominance Inheritance

1. If a woman with wavy hair and a man with curly hair have children calculate the percentages of each genotype and phenotype of the offspring from the Punnett square.

2. If a woman with straight hair and a man with wavy hair have children calculate the percentages of each genotype and phenotype of the offspring from the Punnett square.

D. Monohybrid Codominant Inheritance

1. If a father with type O blood and a mother with AB blood have children calculate the percentages of each genotype and phenotype of the offspring from the Punnett square.

2. If a mother with sickle-cell anemia and a father with the sickle-cell trait have children calculate the percentages of each genotype and phenotype of the offspring from the Punnett square.

E. Sex-Linked Inheritance

1. Cystic fibrosis is due to a defective inherited gene. The recessive allele produces a defective version of a protein. Do a cross with a heterozygous (carrier) mother and a father who has a normal gene for the trait. Let C = the normal allele and c = cystic fibrosis allele. Calculate the percentages of each genotype and phenotype of the offspring from the Punnett square.

Using Your Knowledge

Answer the following questions with a short answer.

1. Why are marriages between blood relatives prohibited in the United States and other countries?

2. If a phenotypic ratio for a single trait is 3:1 (75%:25%) in the offspring, what genotypes would you propose the parents might be? Draw a Punnett square to explain.

3. Can any sons obtain the allele for baldness from the father with the allele? Draw a Punnett square to explain.

4. Why are men more likely than women to be color blind? Explain.

5. Can a son be a carrier for an X-linked trait? Explain.

6. Could a father who has blood type A and mother who has blood type B possibly have a child who has O blood type? Draw a Punnett square to explain.

7. Huntington's disease is due to inheriting a lethal dominant allele that affects the nervous system and causes early dementia around age 40. Cross a heterozygous mother who has the trait with a normal father. Draw a Punnett square to explain.
 a. What are the percentages and types of genotypes?
 b. What are the percentages and types of phenotype?

8. Why does the lethal dominant allele of Huntington's disease stay in the gene pool?

9. If there is a possibility of four different genotypes in the offspring that result from a dominant-recessive inheritance, and if the couple has four children, would they definitely have children with one expressing each genotype? Explain.

10. List the 10 possible genotypes of the four blood types. Identify the parental genotypes that could produce children with blood type AO.

Answer Key

Exercise 1

Figure 1.1

1. armpit
2. forearm
3. wrist
4. thigh
5. ankle
6. chest
7. arm
8. abdomen
9. groin
10. pelvis
11. shoulder
12. back
13. buttock
14. leg

Figure 1.2

1. cranial
2. facial
3. cervical
4. acromial
5. axillary
6. brachial
7. antecubital
8. antebrachial
9. carpal
10. digital
11. palmar
12. femoral
13. patellar
14. crural
15. tarsal
16. digital
17. frontal
18. orbital
19. otic
20. buccal
21. nasal
22. oral
23. mental
24. sternal

25. pectoral
26. umbilical
27. coxal
28. inguinal
29. manual
30. pubic
31. pedal
32. occipital
33. scapular
34. vertebral
35. olecranal
36. gluteal
37. popliteal
38. sural
39. fibular (peroneal)
40. plantar
41. lumbar
42. calcaneal

Activity 4 Directional Terms: Complete the sentence.

1. a. anterior
 b. superior
 c. medial
 d. distal
 e. ipsilateral
 f. superficial

Figure 1.3

1. superior or cephalic
2. posterior or distal
3. anterior or ventral
4. proximal
5. distal
6. inferior or caudal

Figure 1.4

1. frontal
2. oblique
3. transverse
4. parasagittal
5. midsagittal

Exercise 2

Figure 2.1

1. integumentary system
2. skeletal system
3. muscular systems
4. cardiovascular system
5. lymphatic and immune system
6. nervous system
7. endocrine system
8. respiratory system
9. digestive system
10. urinary system
11. reproductive system

Activity 2 Identification of Organs

1. a, 2. b, 3. d 4. c, 5. d, 6. b, 7. b, 8. a,
9. b, 10. b

Figure 2.2

1. cranial cavity
2. vertebral canal
3. dorsal cavity
4. thoracic cavity
5. diaphragm
6. abdominal cavity
7. pelvic cavity
8. abdominopelvic cavity
9. ventral cavity
10. thoracic cavity
11. diaphragm
12. abdominal cavity
13. brim of pelvis
14. pelvic cavity
15. abdominopelvic cavity

Table 2.2

1. brain
2. spinal cord
3. lungs
4. heart

5–8. esophagus, trachea, bronchi, thymus

9–17. stomach, pancreas, liver, gall bladder, small intestine, large intestine, appendix, part of kidneys, spleen, adrenal glands

18–24. ovaries, uterine tubes, uterus, vagina, prostate gland, ductus deferens, urinary bladder, part of large intestine, rectum

Figure 2.3

1. parietal pericardium
2. pericardial cavity
3. visceral pericardium
4. visceral pleura
5. pleural cavity
6. parietal pleura
7. visceral pericardium
8. pericardial cavity
9. parietal pericardium
10. visceral pleura
11. pleural cavity
12. parietal pleura

Figure 2.4

1. lesser omentum
2. mesentery
3. greater omentum
4. parietal peritoneum
5. visceral peritoneum
6. peritoneal cavity
7. visceral peritoneum
8. peritoneal cavity
9. parietal peritoneum

Activity 4 Discussion Questions: Serous Membranes

1. It was easier to push a fist into the bag with water. The water served as a lubricant, decreasing friction.
2. visceral layer
3. parietal layer
4. serous fluid

Activity 5 Abdominopelvic Quadrants and Regions

Quadrants

3. a. mainly RLQ
 b. all four quadrants
 c. mainly RUQ
 d. LLQ and RLQ
 e. LUQ
 f. all four quadrants
 g. LUQ
 h. mainly LUQ

Regions

3. a. right inguinal (iliac)
 b. right hypochondriac
 c. right inguinal (iliac)
 d. hypogastric
 e. left hypochondriac
 f. epigastric

Exercise 3

Activity 2 Using the Microscope

11. When you move the slide to the left, through the ocular it moves to the right.
 When you move the slide to the right, through the ocular it moves to the left.
 When you move the slide forward (away from you), through the ocular it moves backward (toward you).
 When you move the slide backward (toward you), through the ocular it moves forward (away from you).

Activity 5 Discussion Questions: Cheek Smear

1. To increase contrast of cell structure.
2. cell membrane, cytoplasm, and nucleus

Exercise 4

Figure 4.1

1. mitochondrion
2. peroxisome
3. smooth endoplasmic reticulum
4. lysosome
5. plasma membrane
6. centrioles
7. microvillus
8. microtubules (cytoskeleton)
9. cilia
10. secretory vesicle
11. chromatin
12. nuclear envelope
13. nucleolus
14. nucleus
15. cytoplasm
16. rough endoplasmic reticulum
17. ribosome
18. golgi complex
19. microfilament (cytoskeleton)

Activity 1 Discussion Questions: Cell Organelles

1. microvilli, cilia, flagella

Activity 2 Cell Specialization

2. a. spindle shaped; nucleus
 b. columnar; cilia, nucleus
 c. irregular shape with processes that extend from cell body; nucleus
 d. oval with a flagellum and is tapered for movement; nucleus
 e. biconcave; none
 f. round; nucleus

Activity 2 Discussion Questions: Cell Specialization

1. size, shape, and number of nuclei
2. A cell's size and shape depends on it's specific function

Figure 4.3

a. interphase
b. prophase
c. metaphase
d. anaphase
e. telephase

Exercise 5

Activity 2 Experiment: Osmosis in Living RBCs

1. 0.9% saline solution—isotonic
 5% saline solution—hypertonic
 distilled water—hypotonic

Figure 5.3

1. hypotonic solution
2. hypertonic solution

Activity 5 Discussion Question: Phagocytosis

It is Irregular with cytoplasmic processes that change as it moves.

Exercise 6

Figure 6.3

1. nucleus of connective tissue
2. connective tissue
3. simple squamous epithelium
4. nucleus of simple squamous epithelial cell

Figure 6.4

1. simple cuboidal epithelium
2. apical surface of simple cuboidal epithelial cell
3. lumen of kidney tubule
4. nucleus of simple cuboidal epithelial cell

Figure 6.5

1. microvilli on apical surface of simple columnar epithelial cell
2. nucleus of simple columnar epithelial cell
3. simple columnar epithelium
4. connective tissue

Figure 6.6

1. nucleus of squamous epithelial cell
2. stratified squamous epithelium
3. nucleus of epithelial cell in basal layer of epithelium
4. connective tissue

Figure 6.7

1. nucleus of transitional epithelial cell in apical layer
2. nucleus of transitional cell in basal layer
3. transitional epithelium
4. connective tissue

Figure 6.8

1. nucleus of ciliated columnar epithelial cell
2. cilia on apical surface of columnar epithelial cell
3. pseudostratified ciliated columnar epithelium
4. connective tissue

Figure 6.10

1. collagen fiber
2. elastic fiber
3. connective tissue cells

Figure 6.11

1. reticular fiber
2. reticular cell

Figure 6.12

1. nucleus of adipocyte
2. lipid storage area

Figure 6.13

1. fibroblast
2. collagen fiber bundle

Figure 6.14

1. fibroblast
2. collagen fiber bundles running in different directions
3. parallel collagen fiber bundles

Figure 6.15

1. fibroblast
2. individual elastic fiber
3. bundle of elastic fibers

Figure 6.16

1. extracellular matrix
2. lacuna
3. nucleus of chondrocyte

Figure 6.17

1. lacuna
2. nucleus of chondrocyte
3. elastic fibers

Figure 6.18

1. collagen fibers
2. lacuna
3. chondrocytes

Figure 6.19

1. lamella
2. canaliculus
3. lacuna
4. central canal

Figure 6.20

1. trabecula
2. osteocyte

Figure 6.21

1. red blood cells
2. nucleus of white blood cell
3. platelet

Figure 6.23

1. width of individual muscle fiber
2. nucleus
3. striation

Figure 6.24

1. width of cardiac muscle fiber
2. nucleus
3. branches of cardiac muscle fiber
4. intercalated discs

Figure 6.25

1. nucleus of smooth muscle fiber in cross-section
2. nucleus of smooth muscle fiber in longitudinal section

Figure 6.27

1. process
2. cell body of multipolar neuron
3. nucleus

Exercise 7

Figure 7.1

1. dermal papillae
2. epidermis
3. papillary layer
4. reticular layer
5. hypodermis

Figure 7.2

1. stratum corneum
2. stratum lucidum
3. stratum granulosum
4. stratum spinosum
5. stratum basale

Figure 7.3

1. hair shaft
2. hair root
3. sebaceous gland
4. arrector pili muscle
5. hair follicle
6. hair bulb
7. eccrine sweat gland
8. papillae of hair
9. apocrine sweat gland

Figure 7.4

1. sebaceous gland
2. hair follicle
3. hair root
4. hair bulbs
5. papilla of hair

Figure 7.5

1. free edge
2. nail body
3. lunula
4. eponychium
5. nail matrix
6. nail root
7. eponychium
8. lunula
9. nail body
10. free edge
11. hyponychium

Exercise 8

Figure 8.1

a. short
b. flat
c. irregular
d. sesamoid
e. long

Figure 8.2

1. proximal epiphysis
2. diaphysis
3. distal epiphysis
4. articular cartilage
5. epiphyseal line
6. compact bone
7. medullary cavity
8. yellow marrow
9. periosteum
10. nutrient artery
11. endosteum
12. spongy bone

Figure 8.3

1. spongy bone
2. trabeculae
3. concentric lamellae
4. blood vessels
5. canaliculi
6. lacuna
7. osteocyte
8. periosteum
9. central canal
10. perforating canal
11. compact bone
12. trabeculae
13. osteocyte in lacuna
14. interstitial lamellae

Figure 8.4

1. concentric lamella
2. lacuna
3. central canal
4. canaliculi
5. trabecula
6. osteocyte

Activity 5 Discussion Questions: Collagen and Mineral Salts

1. The baked bone is the least flexible and the acid-treated bone is the most flexible.
2. The baked bone is the most brittle.
3. Collagen has been damaged in the brittle bone. Collagen provides tensile strength and flexibility so bones do not break easily under normal stress.
4. The acid-treated bone is the softest.
5. Calcium phosphate and mineral salts have been leached from this bone. Osteomalacia.

Exercise 9

Figure 9.2

1. parietal bone
2. coronal suture
3. squamous suture
4. temporal bone
5. lambdoid suture
6. occipital bone
7. frontal bone
8. sphenoid bone
9. ethmoid bone
10. lacrimal bone
11. nasal bone
12. zygomatic bone
13. maxilla
14. mandible

Figure 9.3

1. frontal bone
2. coronal suture
3. parietal bone
4. sagittal suture
5. lambdoid suture
6. occipital bone

Figure 9.4

1. maxilla
2. vomer
3. sphenoid bone
4. occipital bone
5. zygomatic bone
6. palatine bone
7. temporal bone

Figure 9.5

1. parietal bone
2. occipital bone
3. frontal bone
4. ethmoid bone
5. sphenoid bone
6. temporal bone
7. lambdoid suture

Figure 9.6

1. parietal bone
2. sphenoid bone
3. ethmoid bone
4. lacrimal bone
5. inferior nasal concha
6. vomer
7. frontal bone
8. squamous suture
9. temporal bone
10. nasal bone
11. zygomatic bone
12. maxilla
13. mandible

Figure 9.7

1. frontal bone
2. sphenoid bone
3. zygomatic bone
4. ethmoid bone
5. lacrimal bone
6. maxilla

Figure 9.8

1. supraorbital foramen
2. orbit of eye
3. inferior orbital fissure
4. perpendicular plate of ethmoid
5. mental foramen of mandible
6. supraorbital margin
7. superior orbital fissure
8. middle nasal concha

Figure 9.9

1. external auditory meatus
2. mastoid process
3. lacrimal fossa
4. zygomatic process of temporal bone
5. condylar process
6. coronoid process
7. ramus of mandible
8. body of mandible

Figure 9.10

1. palatine process of maxilla
2. palatine bone
3. hard palate
4. pterygoid process
5. foramen ovale
6. jugular foramen
7. occipital condyle
8. hypoglossal foramen
9. mandibular fossa
10. foramen lacerum
11. carotid foramen (canal)
12. stylomastoid foramen
13. mastoid process
14. foramen magnum

Figure 9.11

1. crista galli
2. olfactory foramina
3. cribriform plate
4. sella turcica
5. foramen ovale
6. internal auditory (acoustic) meatus
7. foramen magnum
8. lesser wing of sphenoid
9. optic foramen
10. greater wing of sphenoid
11. foramen rotundum
12. foramen lacerum
13. jugular foramen

Figure 9.12

1. frontal sinus
2. ethmoidal sinus

3. sphenoidal sinus
4. maxillary sinus

Figure 9.13

1. perpendicular plate of ethmoid
2. septal cartilage
3. vomer

Figure 9.14

1. anterior (frontal) fontanel
2. posterior (occipital) fontanel
3. posterior (occipital) fontanel
4. posterolateral (mastoid) fontanel
5. anterior (frontal) fontanel
6. anterolateral (sphenoidal) fontanel

Figure 9.15

1. cervical curve
2. thoracic curve
3. lumbar curve
4. sacral curve
5. cervical vertebrae
6. thoracic vertebrae
7. lumbar vertebrae
8. sacrum
9. coccyx

Figure 9.17

1. superior articular facet
2. transverse foramen
3. transverse process
4. lamina
5. dens (odontoid process)
6. spinous process
7. body
8. transverse process
9. bifurcated spinous process
10. pedicle

Figure 9.18

1. transverse process
2. facet for articular part of tubercle
3. superior articular facet
4. superior demifacet
5. facet for articular part of tubercle
6. slanted spinous process
7. superior demifacet
8. inferior demifacet

Figure 9.19

1. pedicle
2. superior articular process
3. transverse process
4. vertebral foramen
5. body
6. hatchet-shaped spinous process
7. inferior articular facet

Figure 9.20

1. intervertebral foramen
2. nucleus pulposus
3. annulus fibrosus
4. intervertebral disc
5. herniation
6. nucleus pulposus
7. annulus fibrosus

Figure 9.21

1. sacral ala
2. base of sacrum
3. sacral promontory
4. sacral foramen
5. coccyx
6. sacral canal
7. superior articular facet
8. auricular surface (for sacroiliac joint)

Figure 9.22

a. scoliosis
b. kyphosis
c. lordosis

Figure 9.23

1. suprasternal(jugular) notch
2. manubrium
3. body of sternum
4. xiphoid process
5. sternum
6. costal cartilage
7. sternal angle
8. true ribs
9. floating ribs
10. false ribs

Figure 9.24

1. facet for articular part of tubercle
2. tubercle (nonarticular part)
3. articular part of tubercle
4. head of rib
5. tubercle (nonarticular part)
6. intervertebral foramen
7. inferior demifacet of vertebra
8. head of rib
9. superior demifacet of vertebra
10. body of rib

Exercise 10

Figure 10.1

1. acromial end
2. sternal end
3. acromion (acromial process)
4. coracoid process

5. glenoid cavity or fossa
6. lateral (axillary) border
7. subscapular fossa
8. (medial) vertebral border
9. supraspinous fossa
10. infraspinous fossa
11. acromion (acromial process)
12. spine of scapula
13. glenoid cavity

Figure 10.2

1. greater tubercle
2. intertubercular groove
3. lesser tubercle
4. lateral epicondyle
5. capitulum
6. head
7. anatomical neck
8. deltoid tuberosity
9. coronoid fossa
10. medial epicondyle
11. trochlea
12. olecranon fossa
13. medial epicondyle
14. lateral epicondyle

Figure 10.3

1. head of radius
2. styloid process of radius
3. olecranon (process)
4. trochlear notch (semilunar)
5. coronoid process
6. radial notch
7. styloid process of ulna
8. ulna
9. radius

Figure 10.4

1. carpals
2. metacarpals
3. 5th proximal phalanx
4. 5th middle phalanx
5. 5th distal phalanx

Figure 10.5

1. ilium
2. posterior superior iliac spine
3. posterior inferior iliac spine
4. greater sciatic notch
5. ischium
6. ischial spine
7. lesser sciatic notch
8. obturator foramen
9. ischial tuberosity
10. iliac crest
11. anterior superior iliac spine
12. anterior inferior iliac spine

13. acetabulum
14. pubis

Figure 10.6

1. iliac crest
2. ilium
3. ischial spine
4. pelvic brim
5. pubic symphysis
6. false pelvis
7. true pelvis
8. pubis
9. ischial spine
10. false pelvis
11. sacroiliac joint
12. sacrum
13. coccyx
14. true pelvis
15. pubis

Figure 10.7

1. head of femur
2. greater trochanter
3. neck
4. lesser trochanter
5. medial epicondyle
6. medial condyle
7. linea aspera
8. lateral epicondyle
9. lateral condyle

Figure 10.8

1. lateral condyle
2. head of fibula
3. fibula
4. lateral malleolus
5. medial condyle
6. tibial tuberosity
7. tibia
8. medial malleolus

Figure 10.9

1. calcaneus
2. talus
3. 2nd proximal phalanx
4. 2nd middle phalanx
5. 2nd distal phalanx
6. tarsals
7. metatarsals
8. phalanges
9. calcaneus
10. tibia
11. lateral malleolus of fibula
12. talus
13. tarsals
14. metatarsals
15. phalanges

Exercise 11

Figure 11.1

1. synovial
2. synovial
3. cartilaginous
4. synovial
5. fibrous
6. fibrous
7. cartilaginous
8. cartilaginous
9. synovial
10. synovial
11. fibrous

Figure 11.2

1. chondrocyte
2. lacuna
3. extracellular matrix
4. lacuna
5. chondrocyte
6. collagen fiber

Figure 11.3

1. fibrocartilage
2. fibrocartilage
3. hyaline cartilage
4. hyaline cartilage
5. hyaline cartilage
6. fibrocartilage

Figure 11.4

1. synarthrosis
2. amphiarthrosis
3. diarthrosis
4. amphiarthrosis
5. diarthrosis
6. diarthrosis
7. diarthrosis
8. diarthrosis
9. amphiarthrosis
10. synarthrosis

Figure 11.5

1. articular bone
2. synovial fluid
3. synovial cavity
4. articular cartilage
5. fibrous capsule
6. synovial membrane
7. articular capsule

Figure 11.6

1. articular cartilage of femur
2. lateral (fibular) collateral ligament
3. lateral meniscus

4. posterior cruciate ligament
5. anterior cruciate ligament
6. medial meniscus
7. medial (tibial) collateral ligament
8. patellar ligament
9. articular capsule
10. articular cartilage
11. synovial cavity with synovial fluid
12. bursae
13. infrapatellar fat pad
14. patellar ligament

Figure 11.7

1. flexion
2. extension
3. hyperextension
4. hyperextension
5. extension
6. flexion
7. flexion
8. extension
9. flexion
10. extension
11. hyperextension
12. hyperextension
13. extension
14. flexion
15. flexion
16. extension

Figure 11.8

1. abduction
2. adduction
3. circumduction
4. abduction
5. adduction
6. circumduction
7. adduction
8. abduction
9. abduction
10. adduction

Figure 11.9

1. rotation
2. lateral rotation
3. medial rotation

Figure 11.10

1. elevation
2. depression
3. protraction
4. retraction
5. pronation
6. supination

Figure 11.11

1. inversion
2. eversion

3. dorsiflexion
4. plantar flexion

Exercise 12

Figure 12.1

1. epimysium
2. perimysium
3. fascicle
4. endomysium
5. muscle fiber

Figure 12.2

1. endomysium
2. skeletal muscle fiber in cross-section
3. fascicle
4. perimysium
5. width of skeletal muscle fiber
6. nucleus
7. striation

Figure 12.3

1. triad
2. sarcolemma
3. myofibril
4. t-tubule
5. terminal cisernae
6. sarcoplasmic reticulum
7. sarcomere
8. thin filament
9. thick filament
10. actin
11. tropomyosin
12. troponin
13. myosin cross bridges
14. myosin tails

Figure 12.5

1. Z disc
2. M line
3. H zone
4. I bands
5. A band

Activity 3 Discussion Questions: Sarcomere Structure

1. No. Filament length does not change, but there is greater overlap of thin and thick filaments when the muscle fiber contracts (shortens).
2. No. By definition, the A band is the length of the thick filaments and since filament length does not change, A band length does not change.
3. Yes. The I band contains the portion of the thin filaments that does not overlap with thick filaments. When a muscle fiber

contracts, thin filaments are pulled over the thick filaments, so there is more overlap of thin and thick filaments and a shorter I band.
4. Yes. The H zone is the portion of the thick filaments that does not overlap with thin filaments. As thin filaments are pulled over thick filaments, this zone becomes smaller and eventually disappears when muscle is fully contracted.
5. Fully relaxed because the H zone is still present.

Figure 12.6

1. axon terminal
2. synaptic end bulb
3. synaptic vesicle with acetylcholine
4. synaptic cleft
5. motor end plate
6. receptors

Figure 12.7

1. skeletal muscle fibers
2. axon terminal with synaptic end bulbs
3. motor neuron axon

Exercise 13

Figure 13.1

1. slow oxidative fiber
2. fast glycolytic fiber
3. fast oxidative-glycolytic fiber

Activity 2 Discussion Questions: Influence of Muscle Fiber Type on Muscle Contraction

1. Slow oxidative fibers obtain most of their ATP by aerobic cellular respiration. This enables them to produce ATP for prolonged periods and this method of

ATP production does not produce lactic acid which contributes to fatigue.
2. Slow oxidative fibers have more capillaries to provide enough oxygen to produce ATP by cellular respiration in mitochondria.
3. Fast glycolytic fibers produce most of their ATP for contraction by glycolysis. Therefore they have more glycogen stores. Slow oxidative fibers produce most of their ATP by cellular respiration and don't need large glycogen stores.
4. The larger diameter fibers have more myofibrils, and therefore more cross-bridges. The amount of force a muscle can develop is dependent on the number of cross-bridges in the muscle. Therefore the more cross-bridges, the more force a muscle can develop.
5. Walking—fast oxidative-glycolytic
 Standing—slow oxidative
 Lifting a heavy object and immediately putting it down—fast glycolytic

Activity 3 Discussion Questions: Contraction of Skeletal Muscle Fibers

1. Incomplete tetanus and asynchronous contractions allow you to maintain a sustained contraction over time.
2. Motor unit recruitment. More motor units are recruited, therefore more cross-bridges are attached and more force is generated.
3. More motor units are contracting at a given time. As motor units fatigue, there are less motor units to take over. Therefore, it takes less time to fatigue all motor units.
4. Isotonic contraction occurs when flexing the knee. Isometric contraction occurs when maintaining the knee in a flexed position.

TABLE 13.2

	SLOW OXIDATIVE	FAST OXIDATIVE-GLYCOLYTIC	FAST GLYCOLYTIC
Fiber diameter	smallest	intermediate	largest
Force	lowest	intermediate	greatest
Myosin ATPase	Slow	Fast	Fast
Methods of ATP generation	Aerobic respiration	Aerobic respiration and Glycolysis	glycolysis
Glycogen stores	Low	Intermediate	high
Capillaries	Many	many	few
Myoglobin content	high	high	low
Color	White	Red-pink	Red

5. Concentric. Extend the leg slowly to original position.

Activity 4 Recording Muscle Contractions

1. Force = 1.5 g; latent period = 30 msec; contraction period = 50 msec; relaxation period = 80 msec (160 msec if count time to baseline)

Exercise 14

Figure 14.1

1. frontalis
2. orbicularis oculi
3. zygomaticus major
4. zygomaticus minor
5. orbicularis oris
6. platysma
7. frontalis
8. occipitalis
9. orbicularis oculi
10. masseter
11. zygomaticus minor
12. zygomaticus major
13. buccinator
14. temporalis
15. lateral pterygoids
16. medial pterygoids
17. orbicularis oris
18. buccinator

Figure 14.2

1. splenius capitis
2. trapezius
3. sternocleidomastoid
4. levator scapulae
5. platysma
6. middle scalene
7. digastric, anterior belly
8. mylohyoid
9. stylohyoid
10. digastric, posterior belly
11. omohyoid
12. sternocleidomastoid
13. sternohyoid
14. mylohyoid
15. scalenes
16. levator scapulae

Figure 14.3

1. pectoralis major
2. serratus anterior
3. external oblique
4. pectoralis minor
5. internal intercostal
6. external intercostal

7. internal oblique
8. transverse abdominis
9. rectus abdominis

Figure 14.4

1. deltoid
2. trapezius
3. teres minor
4. teres major
5. infraspinatus
6. latissimus dorsi
7. rhomboid minor
8. supraspinatus
9. teres minor
10. infraspinatus
11. teres major
12. rhomboid major

Figure 14.5

1. biceps brachii
2. brachialis
3. triceps brachii, lateral head
4. triceps brachii, long head
5. triceps brachii, medial head

Figure 14.6

1. pronator teres
2. flexor carpi radialis
3. brachioradialis
4. palmaris longus
5. flexor carpi ulnaris
6. supinator
7. flexor digitorum profundus
8. flexor pollicis longus
9. extensor carpi radialis longus
10. flexor carpi ulnaris
11. extensor carpi radialis brevis
12. extensor carpi ulnaris
13. extensor digiti minimi
14. extensor digitorum
15. supinator
16. extensor pollicis longus

Figure 14.7

1. sartorius
2. rectus femoris
3. tensor fasciae latae
4. vastus intermedius
5. vastus medialis
6. vastus lateralis
7. iliacus
8. psoas major
9. pectineus
10. gracilis
11. adductor longus
12. adductor magnus
13. adductor brevis
14. adductor magnus
15. gracilis

16. gluteus maximus
17. gluteus medius
18. semitendinosus
19. semimembranosus
20. biceps femoris
21. gluteus minimus
22. piriformis

Figure 14.8

1. extensor digitorum longus
2. tibialis anterior
3. fibularis (peroneus) longus
4. gastrocnemius
5. soleus
6. flexor digitorum longus
7. tibialis anterior
8. extensor digitorum longus
9. fibularis (peroneus) longus
10. soleus
11. gastrocnemius
12. gastrocnemius
13. soleus
14. flexor hallucis longus
15. flexor digitorum longus

Exercise 15

Figure 15.1(a & b)

1. frontalis muscle
2. zygomaticus major muscle
3. mental protuberance
4. supraorbital margin
5. nasal bone
6. body of mandible
7. temporalis muscle
8. occipitalis muscle
9. external occipital protuberance
10. mastoid process
11. ramus of mandible
12. temporomandibular joint
13. masseter muscle

Figure 15.1(c & d)

1. common carotid artery
2. sternocleidomastoid muscle
3. hyoid bone
4. thyroid cartilage
5. cricoid cartilage
6. trapezius muscle
7. suprasternal (jugular) notch

Figure 15.2(a)

1. manubrium of stenum
2. sternal angle
3. pectoralis major muscle
4. serratus anterior muscle
5. body of sternum

6. xiphisternal joint
7. xiphoid process of sternum
8. costal margin

Figure 15.2(b)

1. trapezius muscle
2. infraspinatus muscle
3. teres major muscle
4. erector spinae muscle
5. spinous process of vertebrae
6. vertebral border of scapula
7. triangle of auscultation
8. latissimus dorsi muscle

Figure 15.2(c)

1. rectus abdominis muscle
2. external oblique muscle
3. iliac crest
4. anterior superior iliac spine
5. serratus anterior muscle
6. linea alba
7. tendinous intersection
8. linea semilunaris
9. umbilibus
10. McBurney's point

Figure 15.3(a)

1. acromion of scapula
2. spine of scapula
3. greater tubercle of humerus
4. clavicle
5. deltoid muscle

Figure 15.3(b)

1. biceps brachii muscle
2. tendon of biceps brachii muscle
3. groove for brachial artery
4. olecranon of ulna
5. medial epicondyle of humerus
6. triceps brachii muscle

Figure 15.3(c)

1. cephalic vein
2. cubital fossa
3. brachioradialis muscle
4. cephalic vein
5. basilic vein
6. median cubital vein

Figure 15.4(a)

1. sacrum
2. greater trochanter
3. coccyx
4. ischial tuberosity
5. iliac crest
6. gluteus medius muscle
7. gluteus maximus muscle
8. gluteal (natal) cleft

Figure 15.4(b)

1. femoral triangle
2. sartorius muscle
3. adductor longus muscle
4. rectus femoris muscle
5. gracilis muscle
6. vastus lateralis muscle
7. vastus medialis muscle

Figure 15.4(c)

1. semitendinosus and semimembranosus muscles
2. biceps femoris muscle
3. biceps femoris tendon
4. popliteal fossa
5. tendon of semitendinosus

Figure 15.4(d)

1. lateral condyle of femur
2. patella
3. lateral condyle of tibia
4. medial condyle of femur
5. medial condyle of tibia
6. patellar ligament
7. tibial tuberosity

Figure 15.4(e)

1. fibularis (peroneus) longus muscle
2. anterior border of tibia
3. tibialis anterior muscle
4. lateral malleolus of fibula
5. medial malleolus of tibia

Figure 15.4(f)

1. gastrocnemius muscle (medial and lateral heads)
2. soleus muscle
3. calcaneal (Achilles) tendon
4. calcaneus

Exercise 16

Table 16.1

1. Schwann cells - PNS
2. ependymal cells - CNS
3. microglia - CNS
4. astrocytes - CNS
5. satellite cells - PNS
6. oligodendrocytes - CNS

Figure 16.1

1. oligodendrocytes
2. microglia
3. astrocytes
4. ependymal cells
5. satellite cells
6. Schwann cell

Figure 16.2

1. processes
2. cell body
3. dendrites
4. axon
5. axon hillock
6. cell body
7. nucleus

Figure 16.3

1. axon terminals
2. Schwann cell
3. node of Ranvier
4. axon collateral
5. axon
6. dendrites
7. cell body
8. axon hillock
9. trigger zone
10. myelin sheath of Schwann cell

Figure 16.4

1. process
2. satellite cells
3. nucleus
4. neuron cell body
5. axon
6. dendrites
7. neuron cell body

Figure 16.5

1. unipolar
2. multipolar
3. bipolar
4. sensory neuron
5. interneuron
6. motor neuron

Figure 16.6

(a) myelinated
1. axon
2. myelin sheath
3. node of Ranvier
4. Schwann cell cytoplasm
(b) unmyelinated
5. axons
6. Schwann cell cytoplasm

Figure 16.7

1. neurolemma
2. myelin sheath
3. node of Ranvier
4. axon

Figure 16.8

1. white matter
2. gray matter

Exercise 17

Figure 17.1

1. pia mater
2. dura mater
3. epidural space
4. subarachnoid space
5. arachnoid mater, web-like projection

Figure 17.2

1. cervical enlargement
2. lumbar enlargement
3. conus medullaris
4. cauda equina
5. filum terminale

Figure 17.3

1. spinal nerve
2. posterior (dorsal) root ganglion
3. posterior (dorsal) root
4. posterior median sulcus
5. central canal
6. anterior (ventral) root
7. anterior median fissure

Figure 17.4

1. anterior white column
2. anterior gray horn
3. lateral white column
4. lateral gray horn
5. posterior gray horn
6. posterior white column
7. gray commissure

Figure 17.5

1. posterior (dorsal) root
2. lateral white column
3. central canal
4. posterior median sulcus
5. posterior white column
6. posterior gray horn
7. posterior (dorsal) root ganglion
8. anterior (ventral) root
9. anterior gray horn
10. anterior white column
11. gray commissure
12. anterior median fissure

Exercise 18

Figure 18.1

1. spinal nerve
2. epineurium
3. perineurium
4. fascicle
5. endoneurium
6. axon

Figure 18.2

1. perineurium
2. axon
3. myelin sheath
4. endoneurium

Figure 18.3

1. spinal nerve
2. anterior (ventral) ramus
3. posterior (dorsal) ramus
4. rami communicantes
5. sympathetic ganglion
6. spinal nerve
7. intervertebral foramen
8. anterior (ventral) ramus
9. posterior (dorsal) ramus
10. rami communicantes

Figure 18.4

1. sacral plexus
2. lumbar plexus
3. brachial plexus
4. cervical plexus
5. cervical nerves
6. thoracic nerves
7. lumbar nerves
8. sacral nerves
9. coccygeal nerve

Figure 18.5

1. axillary n.
2. musculocutaneous n.
3. median n.
4. ulnar n.
5. radial n.
6. phrenic n.

Figure 18.6

1. obturator n.
2. pudendal n.
3. femoral n.
4. sciatic n.
5. tibial n.
6. common fibular (peroneal) n.

Exercise 19

Table 19.1

1. somatic, spinal nerve reflex
2. autonomic, cranial nerve reflex
3. autonomic, cranial nerve reflex
4. somatic, spinal nerve reflex

Figure 19.1

1. integrating center
2. sensory neuron axon

3. sensory neuron cell body
4. sensory receptor
5. effector
6. motor neuron axon
7. motor neuron cell body

Activity 1 Discussion Questions: Reflex Arc

1. polysynaptic
2. ipsilateral

Figure 19.2

1. receptor
2. sensory neuron
3. integrating center for patellar reflex arc
4. motor neuron for patellar reflex arc
5. effector for patellar reflex arc
6. integrating center for reciprocal innervation
7. motor neuron for reciprocal innervation
8. effector for reciprocal innervation

Activity 2 Discussion Question: Patellar Reflex

1. ipsilateral
2. inhibited
3. reinforces and intensifies the response

Activity 3 Discussion Question: Biceps Reflex

1. musculocutaneous nerve
2. triceps brachii

Activity 4 Discussion Questions: Triceps Reflex

1. radial nerve
2. biceps brachii

Activity 5 Discussion Questions: Achilles Reflex

1. tibial nerve
2. tibialis anterior

Activity 6: Plantar Flexion

3. tibial nerve

Exercise 20

Figure 20.1

1. cerebrum
2. cerebellum
3. diencephalon
4. brain stem

Figure 20.2

1. superior colliculus
2. inferior colliculus
3. corpora quadrigemina
4. cerebral peduncle
5. pons
6. medulla oblongata
7. brain stem
8. pons
9. medulla oblongata
10. midbrain

Figure 20.3

1. cerebellar hemispheres
2. vermis
3. folia
4. superior colliculus
5. inferior colliculus
6. cerebellar cortex (gray matter)
7. arbor vitae (white matter)
8. folia
9. cerebellar peduncle
10. pons
11. medulla oblongata

Figure 20.4

1. pineal gland
2. thalamus
3. hypothalamus
4. diencephalon
5. mammillary body
6. pineal gland
7. thalamus
8. intermediate mass of thalamus
9. hypothalamus
10. infundibulum
11. pituitary gland

Figure 20.5

1. internal capsule
2. insula
3. cerebral cortex (gray matter)
4. white matter
5. corpus callosum
6. fornix
7. basal nuclei (gray matter)
8. projection fibers
9. association fibers
10. commissural fibers (corpus callosum)
11. commissural fibers (fornix)

Figure 20.6

1. occipital lobe
2. temporal lobe
3. transverse fissure
4. parieto-occipital sulcus
5. parietal lobe
6. postcentral gyrus
7. central sulcus
8. precentral gyrus
9. lateral cerebral sulcus
10. frontal lobe
11. frontal lobe
12. longitudinal fissure
13. precentral gyrus
14. central sulcus
15. postcentral gyrus
16. parietal lobe
17. gyrus
18. sulcus
19. gray matter
20. white matter
21. longitudinal fissure

Figure 20.7

1. Broca's speech area
2. primary gustatory cortex
3. primary motor cortex
4. central sulcus
5. primary somatosensory cortex
6. primary visual cortex
7. Wernicke's area
8. primary auditory cortex

Figure 20.8

1. meningeal layer of dura mater
2. superior sagittal sinus
3. periosteal layer of dura mater
4. arachnoid villus
5. parietal bone
6. arachnoid mater
7. subarachnoid space
8. pia mater
9. cerebral cortex
10. white matter
11. falx cerebri

Figure 20.9

1. lateral ventricles
2. interventricular foramen
3. third ventricle
4. cerebral aqueduct
5. fourth ventricle
6. central canal of spinal cord
7. superior sagittal sinus
8. arachnoid villus
9. subarachnoid space
10. lateral ventricle
11. choroid plexus
12. third ventricle
13. cerebral aqueduct
14. fourth ventricle
15. central canal

Activity 10 Sheep Brain Dissection

4. have two smaller mammillary bodies pons is smaller; medulla is larger
6. similar

Exercise 21

Figure 21.1

1. abducens
2. facial
3. vestibulocochlear
4. glossopharyngeal
5. vagus
6. olfactory tract
7. optic
8. oculomotor
9. trochlear
10. trigeminal
11. accessory
12. hypoglossal

Exercise 22

Figure 22.1

1. type I cutaneous mechanoreceptor
2. corpuscle of touch
3. type II cutaneous mechanoreceptor
4. hair root plexus
5. laminated (Pacinian) corpuscle
6. muscle spindle
7. tendon organ

Figure 22.2

1. third order neuron
2. second order neuron
3. first order neuron
4. third order neuron
5. second order neuron
6. first order neuron

Discussion Questions: Sensory Pathways

1. unipolar
2. posterior (dorsal) root ganglion, posterior (dorsal) horn of spinal cord
3. posterior (dorsal) horn of spinal cord, thalamus
4. thalamus, primary somatosensory cortex
5. medulla oblongata
6. spinal cord

Activity 4 Discussion Questions: Adaptation of Sensory Receptors

1. less than 1 minute
2. Receives decreased number of signals
3. (a) some rapidly and some slowly
 (b) some rapidly and some slowly
 (c) not at all (d) not at all

Activity 5 Discussion Questions: Referred pain

1. elbow, forearm, hand
2. ulnar

Exercise 23

Figure 23.1

1. conjunctival fold
2. palpebral conjunctiva
3. bulbar conjunctiva
4. lacrimal gland
5. lacrimal canals
6. lacrimal sac
7. nasolacrimal duct

Figure 23.2

1. superior oblique
2. superior rectus
3. lateral rectus
4. medial rectus
5. inferior oblique
6. inferior rectus

Table 23.2

inferior oblique—moves eye superiorly and laterally
inferior rectus—moves eye inferiorly
lateral rectus—moves eye laterally
medial rectus—moves eye medially
superior oblique—moves eye inferiorly and laterally
superior rectus—moves eye superiorly

Figure 23.3

1. scleral venous sinus
2. ciliary muscle
3. ciliary process
4. ciliary body
5. choroid
6. sclera
7. retina
8. cornea
9. pupil
10. iris
11. suspensory ligaments (zonular fibers)
12. ora serrata

Figure 23.4

1. anterior chamber
2. posterior chamber
3. anterior cavity
4. scleral venous sinus
5. vitreous chamber
6. lens

Figure 23.6

1. optic disc
2. central retinal vein
3. optic nerve
4. central retinal artery
5. central fovea
6. optic nerve fibers
7. ganglion cell layer
8. bipolar cell layer
9. photoreceptor layer
10. neural portion of retina
11. pigmented epithelium of retina

Figure 23.7

1. macula lutea
2. central fovea
3. blood vessel
4. optic disc

Figure 23.8

1. cornea
2. pupil
3. iris
4. ciliary body
5. lens
6. choroid
7. sclera
8. retina
9. optic disc
10. optic nerve
11. ganglion cell layer
12. bipolar cell layer
13. photoreceptor layer
14. neural portion of retina
15. pigmented epithelium of retina

Activity 7 Discussion Questions: Visual Tests

1. To accommodate the lens of the eye for distance vision.
2. To accommodate the lens of the eye for near vision.

Figure 23.12

1. helix
2. external ear
3. malleus
4. incus
5. stapes attached to oval window
6. middle ear
7. internal ear
8. auditory tube
9. tympanic membrane
10. external auditory canal
11. lobule
12. auricle

Figure 23.13

1. anterior semicircular canal
2. posterior semicircular canal
3. lateral semicircular canal
4. ampulla of semicircular canal & duct
5. utricle
6. saccule
7. oval window
8. membranous semicircular duct
9. vestibule
10. round window
11. cochlea
12. cochlear duct

Figure 23.14

1. scala vestibuli
2. scala tympani
3. vestibular membrane
4. basilar membrane
5. cochlear duct
6. spiral organ of Corti
7. vestibular membrane
8. tectorial membrane
9. basilar membrane
10. hair cells
11. cochlear duct

Figure 23.15

1. otoliths
2. otolithic membrane
3. hair bundle
4. hair cell
5. vestibular branches of vestibulocochlear nerve
6. cupula
7. hair bundle
8. hair cell

Figure 23.16

1. olfactory bulb
2. cribiform plate
3. olfactory tract
4. olfactory nerve
5. olfactory receptor cell
6. olfactory hairs

Figure 23.17

1. vallate (circumvallate) papilla
2. foliate
3. papilla filiform
4. fungiform papilla
5. vallate papillae
6. filiform papillae
7. fungiform papillae
8. taste bud
9. taste pore
10. gustatory hairs
11. gustatory receptor cell
12. sensory axons

Activity 13 Discussion Questions: Olfactory Adaptation

1. Similar
2. The receptors on that side have not adapted.

Exercise 24

Figure 24.1

1. right sympathetic trunk ganglia
2. splanchnic nerves
3. left sympathetic trunk ganglia
4. prevertebral ganglia

Figure 24.2

1. otic ganglion
2. ciliary ganglion
3. pterygopalatine ganglion
4. submandibular ganglion

Figure 24.3

1. sensory receptor
2. sensory neuron (afferent)
3. interneuron
4. preganglionic motor neuron
5. autonomic ganglion
6. postganglionic motor neuron
7. visceral effector

Activity 4 Discussion Questions: Pupillary Light Reflex

1. autonomic
2. optic nerve
3. occulomotor nerve
4. polysynaptic
5. both

Activity 5 Discussion Questions: Lie Detector Test and the ANS

1. The type of test done in class will give erroneous results because data collection is inaccurate.
2. Some people have learned to lie without feeling emotion or lie without showing ANS responses.

Exercise 25

Figure 25.1

1. pineal gland
2. parathyroid glands
3. adrenal glands
4. pancreas
5. hypothalamus
6. pituitary gland

TABLE 26.4

BLOOD TYPE	TYPE A	TYPE B	AB TYPE	O TYPE
Antigen on RBCs	A	B	A, B	neither A or B
Antibody in plasma	anti-B	anti-A	none	anti-A; anti-B
Compatible donor	A; O	B; O	A; B; AB; O	O
Incompatible donor	B; AB	A; AB	none	A; B; AB

7. thyroid gland
8. thymus
9. ovaries
10. testes

Figure 25.2

1. hypothalamus
2. infundibulum
3. anterior pituitary
4. posterior pituitary
5. hypothalamus
6. infundibulum
7. anterior pituitary
8. posterior pituitary
9. posterior pituitary
10. anterior pituitary

Figure 25.3

1. right lobe of thyroid gland
2. thyroid cartilage of larynx
3. left lobe of thyroid gland
4. isthmus of thyroid gland
5. trachea
6. left parathyroid glands
7. thyroid gland
8. right parathyroid glands
9. trachea

Figure 25.4

1. parathyroid gland
2. thyroid gland
3. colloid-filled follicle
4. follicular cell
5. parafollicular cell

Activity 3 Discussion Question: Thyroid and Parathyroid Glands

1. Storage space for thyroglobulin is much greater due to the large follicles. Calcitonin and PTH are stored only within the small parafollicular cells and principal cells respectively.

Figure 25.5

1. right adrenal gland
2. left adrenal gland
3. kidney
4. capsule
5. adrenal cortex

6. adrenal medulla
7. capsule
8. zona glomerulosa
9. zona fasciculata
10. zona reticularis
11. adrenal medulla
12. adrenal cortex

Figure 25.6

1. tail of pancreas
2. body of pancreas
3. head of pancreas
4. exocrine or acinar cells
5. pancreatic islet

Exercise 26

Figure 26.1

1. plasma
2. RBCs
3. WBCs and platelets

Activity 5 Discussion Questions: Hematocrit

1. Tube (a)—45% = normal
 Tube (b)—22% = anemia

Activity 6 Discussion Questions: Hemoglobin Content of Blood

2. Lower
3. RBCs could have different amounts of hemoglobin

Activity 7 Discussion Questions: Coagulation Time

1. Longer than normal coagulation time
2. Coagulation time would be longer

Exercise 27

Figure 27.2

1. base of heart
2. right lung
3. left lung
4. apex of heart
5. fibrous pericardium

6. diaphragm
7. fibrous pericardium
8. parietal layer of serous pericardium
9. pericardial cavity
10. visceral layer of serous pericardium
11. endocardium
12. myocardium

Figure 27.3

1. auricle of right atrium
2. coronary sulcus
3. right ventricle
4. auricle of left atrium
5. anterior interventricular sulcus
6. left ventricle
7. left auricle
8. coronary sulcus
9. left ventricle
10. adipose tissue
11. posterior interventricular sulcus
12. right auricle
13. right ventricle

Figure 27.4

1. superior vena cava
2. right pulmonary artery
3. ascending aorta
4. pulmonary trunk
5. right pulmonary veins
6. inferior vena cava
7. aortic arch
8. ligamentum arteriosum
9. left pulmonary artery
10. descending aorta
11. left pulmonary veins
12. aortic arch
13. ligamentum arteriosum
14. left pulmonary artery
15. left pulmonary veins
16. coronary sinus
17. superior vena cava
18. ascending aorta
19. right pulmonary artery
20. right pulmonary veins
21. inferior vena cava

Figure 27.5

1. superior vena cava opening
2. pulmonary (semilunar) valve
3. right atrium
4. coronary sinus opening
5. inferior vena cava opening
6. tricuspid valve
7. right ventricle
8. trabeculae carneae
9. pulmonary vein openings
10. left atrium
11. aortic (semilunar) valve
12. bicuspid valve (mitral)
13. chordae tendineae

14. interventricular septum
15. papillary muscle
16. left ventricule

Activity 6

Blood Vessels with oxygen-poor blood

1. pulmonary arteries
2. pulmonary trunk
3. venae cavae

Blood Vessels with oxygen-rich blood

4. aorta
5. pulmonary veins

Trace blood flow

1. right atrium
2. tricuspid valve
3. right ventricle
4. pulmonary (semilunar) valve
5. pulmonary trunk
6. right and left pulmonary arteries
7. pulmonary capillaries
8. pulmonary veins
9. left atrium
10. bicuspid valve
11. left ventricle
12. aortic (semilunar) valve
13. aorta
14. systemic arteries
15. systemic capillaries
16. systemic veins
17. venae cavae

Figure 27.7

1. posterior interventricular branch
2. marginal branch
3. right coronary artery
4. left coronary artery
5. circumflex branch
6. anterior interventricular branch (LAD)
7. small cardiac vein
8. middle cardiac vein
9. coronary sinus
10. great cardiac vein

Activity 8: Dissection of a Sheep Heart

8. The three cusps are half moon shaped and have no chordae tendineae.
10. Two main cusps that are counted; yes; yes; lesser number; left ventricle is thicker to pump blood through the systemic circulation

Figure 27.9

1. cardiac muscle fiber
2. intercalated discs
3. branching fiber
4. nucleus

Exercise 28

Activity 1 Discussion Questions: Heart Sounds

1. Lubb occurs when ventricles contracting.

Figure 28.2

1. Purkinje fibers in right ventricle
2. right bundle branch
3. AV bundle (of His)
4. AV node
5. SA node
6. left bundle branch
7. Purkinje fibers in left ventricle

Figure 28.4

1. P wave
2. P-Q interval
3. QRS complex
4. S-T segment
5. T wave
6. Q-T interval

Figure 28.5

(a) bradycardia 50 squares = 2 sec/beat = 30 beats/min
(b) NSR 25 squares = 1 sec/beat = 60 beats/min
(c) tachycardia 13 squares = 0.52 sec/beat = 115 beats/min

Activity 6: Calculate Lengths of the ECG Segments

PQ interval = 4.5 squares = 0.18 sec
QRS complex = 3 squares = 0.12 sec
QT interval = 14 squares = 0.56 sec
Relaxation period = 11 squares = 0.44 sec
Heart rate = 53 bpm

Activity 6 Discussion Questions: Calculate Lengths of the ECG Segments

1. QT is abnormal; QRS is long but not over 0.12 sec
2. The relaxation period decreases when heart rate increases and increases when heart rate decreases.

Exercise 29

Figure 29.1

(b) is vasoconstricted

Figure 29.2

1. endothelium
2. basement membrane

3. internal elastic lamina
4. tunica interna
5. smooth muscle
6. external elastic lamina
7. tunica media
8. tunica externa

Figure 29.3

1. endothelium
2. basement membrane
3. tunica interna
4. valve
5. smooth muscle
6. tunica media
7. tunica externa

Figure 29.4

1. lumen of artery
2. tunica interna
3. tunica media
4. tunica externa
5. tunica externa
6. tunica media
7. tunica interna
8. lumen of vein

Figure 29.5

1. capillary
2. arteriole

Figure 29.6

1. intercellular cleft
2. vesicles involved in transcytosis
3. fenestration

Activity 4 Discussion Questions: Blood Pressure Measurements

1. Blood pressure is lower in a supine position and increases upon standing.
2. Heart rate may or may not increase.
3. When in the supine position, the head is not higher than the heart. It does not take as much pressure to drive blood to the head. Pressure increases when standing to provide enough pressure to overcome gravity and deliver blood to brain.

Exercise 30

Figure 30.1

1. right common carotid artery
2. right subclavian artery
3. brachiocephalic trunk
4. ascending aorta
5. right coronary artery
6. left common carotid

7. left subclavian artery
8. aortic arch
9. left coronary artery

Figure 30.2

1. basilar artery
2. internal carotid artery
3. vertebral artery
4. middle cerebral artery
5. anterior cerebral artery
6. posterior communicating artery
7. posterior cerebral artery
8. external carotid artery
9. common carotid artery
10. brachiocephalic trunk
11. anterior cerebral artery
12. middle cerebral artery
13. internal carotid artery
14. posterior communicating artery
15. posterior cerebral artery
16. basilar artery
17. vertebral artery
18. anterior communicating artery

Figure 30.3

1. subclavian artery
2. axillary artery
3. brachial artery
4. radial artery
5. ulnar artery

Figure 30.4

1. celiac trunk
2. suprarenal
3. gonadal
4. common iliac
5. thoracic aorta
6. superior mesenteric
7. renal
8. abdominal aorta
9. inferior mesenteric

Figure 30.5

1. left gastric artery
2. celiac trunk
3. common hepatic artery
4. splenic artery

Figure 30.6

1. common iliac
2. external iliac
3. internal iliac
4. femoral
5. popliteal
6. anterior tibial
7. posterior tibial
8. fibular (peroneal)
9. dorsal artery of the foot (dorsalis pedis)

Figure 30.7

1. superior vena cava
2. coronary sinus
3. inferior vena cava

Figure 30.8

1. vertebral vein
2. internal jugular vein
3. external jugular vein
4. subclavian vein
5. brachiocephalic vein

Figure 30.9

1. subclavian vein
2. axillary vein
3. brachial veins
4. basilic vein
5. cephalic vein
6. radial veins
7. median cubital vein
8. ulnar veins

Figure 30.10

1. right brachiocephalic vein
2. left brachiocephalic vein
3. superior vena cava
4. azygous
5. hepatic veins
6. right suprarenal
7. right renal
8. right gonadal
9. inferior vena cava
10. right common iliac
11. right internal iliac
12. right external iliac
13. accesory hemiazygous
14. hemiazygous
15. left suprarenal vein
16. left renal vein
17. left gonadal vein
18. left common iliac vein
19. left internal iliac vein
20. left external iliac vein

Figure 30.11

1. hepatic vein
2. hepatic portal vein
3. superior mesenteric vein
4. splenic vein
5. inferior mesenteric vein

Figure 30.12

1. external iliac vein
2. femoral vein
3. great saphenous vein
4. anterior tibial veins
5. small saphenous vein
6. posterior tibial veins

7. femoral vein
8. popliteal vein
9. great saphenous vein
10. posterior tibial veins
11. small saphenous veins
12. fibular (peroneal) veins

Figure 30.13

1. right pulmonary artery
2. right pulmonary vein
3. left pulmonary artery
4. pulmonary trunk
5. left pulmonary vein

Figure 30.14

1. foramen ovale
2. ductus venosus
3. umbilical vein
4. umbilical arteries
5. placenta
6. ductus arteriosus
7. ligamentum arteriosum
8. fossa ovalis
9. ligamentum venosum
10. ligamentum teres
11. medial umbilical ligaments

Exercise 31

Figure 31.1

1. submandibular node
2. cervical node
3. thymus
4. intestinal node
5. appendix
6. red bone marrow
7. axillary node
8. mammary node
9. thoracic node
10. spleen
11. aggregated lymphatic follicle
 (Peyer's patches)
12. iliac node
13. inguinal node
14. pharyngeal tonsil
15. palatine tonsil
16. lingual tonsil

Figure 31.2

1. tissue cell
2. interstitial fluid
3. blood capillary (with plasma)
4. lymphatic capillary (with lymph)
5. interstitial fluid
6. tissue cell
7. lymph in lymphatic capillary

Figure 31.3

1. right lymphatic duct
2. lymphatic vessel in arm
3. thoracic duct in thorax
4. cisterna chyli
5. lymphatic vessel in leg
6. thoracic duct draining into subclavian vein
7. area drained by right lymphatic duct
8. area drained by thoracic duct
9. right jugular trunk
10. right lymphatic duct
11. right subclavian trunk
12. cisterna chyli
13. thoracic duct
14. thoracic (left lymphatic) duct in thorax

Figure 31.4

1. trabecular sinus
2. afferent vessels
3. valve
4. subcapsular sinus
5. cortex
6. medulla
7. medullary sinus
8. efferent vessels
9. hilus
10. capsule
11. trabecula
12. cortex
13. medulla
14. capsule
15. germinal center in lymphatic nodule
16. reticulocytes
17. reticular fiber

Figure 31.5

1. splenic artery
2. splecnic vein
3. hilus
4. capsule
5. trabecula
6. white pulp

Figure 31.6

1. e
2. e
3. e
4. c
5. e
6. e
7. a, b, d
8. e
9. e
10. e
11. e
12. e

Exercise 32

Figure 32.1

1. pharynx
2. right primary bronchus
3. lungs
4. nose
5. nasal cavity
6. larynx
7. trachea

Figure 32.2

1. internal naris
2. superior nasal concha
3. middle nasal concha
4. inferior nasal concha
5. superior nasal meatus
6. middle nasal meatus
7. inferior nasal meatus
8. external naris
9. hard palate
10. middle meatus
11. nasal septum
12. inferior meatus
13. middle concha
14. inferior concha

Figure 32.3

1. nasopharynx
2. oropharynx
3. laryngopharynx
4. pharyngeal tonsil
5. nosopharynx
6. orifice of the auditory tubes
7. uvula
8. oropharynx
9. laryngopharynx
10. internal naris
11. soft palate
12. palatine tonsil
13. lingual tonsil

Figure 32.4

1. epiglottis
2. thyroid cartilage
3. cricoid cartilage
4. epiglottis
5. thyroid cartilage
6. cricoid cartilage
7. arytenoid cartilage
8. cricoid cartilage
9. epiglottis
10. ventricular folds
11. vocal folds
12. thyroid cartilage
13. ventricular folds
14. vocal folds
15. glottis

Figure 32.5

1. trachea
2. right primary bronchus
3. carina
4. right secondary bronchus
5. diaphragm
6. larynx
7. tracheal cartilage
8. left primary bronchus
9. left tertiary bronchus
10. bronchiole
11. primary bronchus
12. secondary bronchus
13. tertiary bronchus
14. terminal bronchiole
15. respiratory bronchiole
16. alveolar sac

Figure 32.6

1. respiratory bronchiole
2. alveolar ducts
3. alveolar sac
4. alveoli

Figure 32.7

1. apex
2. superior lobe
3. middle lobe
4. inferior lobe
5. base
6. apex
7. superior lobe
8. cardiac notch
9. inferior lobe
10. base
11. superior lobe
12. hilus
13. middle lobe
14. inferior lobe
15. superior lobe
16. inferior lobe
17. hilus
18. cardiac notch

Figure 32.8

1. visceral pleura
2. parietal pleura
3. pleural cavity
4. diaphragm
5. right lung
6. left lung
7. visceral pleura
8. pleural cavity
9. parietal pleura
10. pericardium

Figure 32.9

1. lumen of esophagus
2. trachealis muscle

3. lumen of trachea
4. epithelial lining of trachea
5. tracheal cartilage
6. respiratory bronchiole
7. alveolar ducts
8. simple squamous epithelium
9. alveoli
10. alveolar sac

Activity 7 Discussion Questions: Microscopic Anatomy of Respiratory System

1. elastic tissue
2. hyaline cartilage
3. The type of epithelia present reflects function. Pseudostratified columnar epithelium traps and removes particles from air in the nasal cavity, nasopharynx, and large airways. Stratified squamous epithelium provide protection in the passageways where food and liquids also pass. Simple squamous epithelium is present where gas exchange occurs.

Exercise 33

Figure 33.1

1. sternocleidomastoid
2. scalenes
3. pectoralis minor
4. rectus abdominis
5. external oblique
6. internal oblique
7. transverse abdominis
8. internal intercostals
9. external intercostals
10. diaphragm

Table 33.1

external intercostals—increase diameter
internal intercostals—decrease diameter
sternocleidomastoids—increase diameter
scalenes—increase diameter
pectoralis minors—increase diameter
diaphragm—increases length
abdominal muscles—decrease length

Figure 33.2

a. expiration
b. inspiration

Figure 33.3

1. parietal pleura and thoracic wall
2. trachea
3. bronchus
4. lungs covered with visceral pleura

5. pleural cavity
6. diaphragm

Table 33.2

1. decreased
2. increased
3. deflated
4. out of balloons
5. increased
6. decreased
7. inflated
8. into balloons

Activity 2 Discussion Questions: Bell Jar Demonstration

1. serous fluid; larger than pleural cavity
2. (a) subatmospheric (If rubber membrane is not sealed to bell jar, then pressure inside bell jar might be atmospheric.) (b) subatmospheric
3. (a) It would increase and become atmospheric. (b) Lung size decreases. (c) Pneumothorax.
4. two separate pleural cavities
5. The trachea and bronchi provide a patent, semirigid airway. The bronchial tree has many branches.
6. The thorax increases in size with inspiration and decreases with expiration.

Figure 33.4

$FVC = 3.1\ L$
$FEV_1 = 1.9\ L$
$FEV_1/FVC\ ratio = 0.61$

Exercise 34

Figure 34.1

1. salivary glands
2. pharynx
3. esophagus
4. liver
5. gallbladder
6. large intestine
7. mouth
8. teeth
9. tongue
10. stomach
11. pancreas
12. small intestine
13. anus

Figure 34.2

1. mucosa
2. submucosa
3. serosa
4. muscularis

Figure 34.3

1. retroperitoneal organs
2. lesser omentum
3. mesocolon
4. mesentery
5. greater omentum
6. parietal peritoneum
7. visceral peritoneum
8. peritoneal cavity
9. falciform ligament
10. greater omentum
11. lesser omentum
12. greater omentum reflected
13. mesocolon
14. mesentery

Figure 34.4

1. hard palate
2. soft palate
3. uvula
4. cheek
5. vestibule
6. superior lip
7. fauces
8. palatine tonsil
9. tongue
10. gingiva (gums)
11. inferior labial frenulum
12. inferior lip

Figure 34.5

1. nasopharynx
2. soft palate
3. uvula
4. fauces
5. oropharynx
6. laryngopharynx
7. esophagus
8. hard palate
9. oral cavity
10. tongue

Figure 34.6

1. lower esophageal sphincter
2. cardia
3. body
4. lesser curvature
5. pyloric sphincter
6. esophagus
7. fundus
8. longitudinal muscle layer
9. circular muscle layer
10. oblique muscle layer
11. greater curvature
12. rugae
13. pylorus

Figure 34.7

1. duodenum
2. ascending colon
3. jejunum
4. ileum
5. stomach
6. transverse colon
7. descending colon
8. rectum
9. right colic flexure
10. ascending colon
11. ileum
12. ileocecal sphincter
13. cecum
14. appendix
15. rectum
16. transverse colon
17. left colic flexure
18. descending colon
19. teniae coli
20. epiploic appendages
21. haustra
22. sigmoid colon
23. anal canal
24. anus
25. rectum
26. anal canal
27. internal anal sphincter
28. external anal sphincter
29. anal column
30. anus

Figure 34.8

1. parotid gland
2. parotid duct
3. opening of parotid duct
4. opening of sublinguinal duct
5. sublinguinal gland
6. submandibular duct
7. submandibular gland
8. crown
9. neck
10. root
11. enamel
12. dentin
13. gingiva
14. blood supply
15. pulp cavity with pulp
16. cementum
17. root canal
18. peridontal ligament
19. nerve
20. apical foramen
21. molars
22. premolar
23. cuspids
24. inscisors
25. tongue
26. lingual frenulum

Figure 34.9

1. right hepatic duct
2. cystic duct
3. gall bladder
4. duodenum
5. hepatopancreatic ampulla
6. falciform ligament
7. left lobe of liver
8. left hepatic duct
9. common hepatic duct
10. common bile duct
11. accessory pancreatic duct
12. tail of pancreas
13. body of pancreas
14. pancreatic duct
15. head of pancreas
16. jejunum
17. hepatopancreatic ampulla
18. duodenal papilla
19. common bile duct
20. pancreatic duct
21. quadrate lobe
22. left lobe of liver
23. caudate lobe
24. right lobe
25. gall bladder

Figure 34.10

1. stratified squamous epithelium
2. lamina propria
3. circular layer of smooth muscle fibers
4. longitudinal layer of smooth muscle fibers
5. mucosa
6. submucosa
7. muscularis
8. serosa
9. gastric pits
10. simple columnar epithelium
11. lamina propria
12. gastric glands
13. simple columnar epithelium
14. goblet cell
15. villus
16. lamina propria
17. intestinal glands
18. duodenal glands
19. acini
20. pancreatic islet
21. heptocytes
22. sinusoids
23. central vein

Figure 34.11

1. bile canaliculus
2. hepatic vein
3. central vein
4. sinusoids

5. portal triad
6. bile duct
7. hepatocytes
8. branch of hepatic portal vein
9. branches of hepatic artery

Exercise 36

Figure 36.1

1. right kidney
2. right ureter
3. urinary bladder
4. urethra

Figure 36.2

1. peritoneum
2. right kidney
3. renal capsule
4. adipose capsule
5. renal fascia
6. left kidney
7. renal hilus

Figure 36.3

1. renal medulla
2. renal cortex
3. renal column
4. renal capsule
5. renal pyramid
6. renal papilla
7. minor calyx
8. major calyx
9. renal artery
10. renal vein
11. renal hilus
12. renal pelvis in renal sinus

Activity 1 Discussion Questions: Location and Gross Anatomy of the Kidney

1. Ribs 11 and 12
2. Right kidney. Liver forces the right kidney lower.

Figure 36.4

1. internal urethral sphincter
2. ureters
3. detrusor muscle of urinary bladder
4. ureteral openings
5. trigone
6. internal urethral orifice
7. urethra
8. external urethral sphincter
9. external urethral orifice

Figure 36.5

1. rectum
2. anus

3. ureter
4. urinary bladder
5. ureteral opening
6. internal urethral orifice
7. prostatic urethra
8. membranous urethra
9. spongy urethra
10. external urethral opening
11. urogenital diaphragm
12. rectum
13. anus
14. uterus
15. urinary bladder
16. internal urethral orifice
17. urethra
18. urogenital diaphragm
19. external urethral orifice
20. vagina

Activity 2 Discussion Questions: Ureters, Urinary Bladder, and Urethra

1. No, it is covered by visceral peritoneum.

Figure 36.6

1. glomerular capsule
2. proximal convoluted tubule
3. ascending limb of loop of Henle
4. descending limb of loop of Henle
5. distal convoluted tubule
6. collecting duct
7. papillary duct
8. glomerulus
9. glomerular capsule
10. renal corpuscle
11. peritubular capillaries
12. vasa recta
13. efferent arteriole
14. afferent arteriole

Figure 36.7

1. afferent arteriole
2. efferent arteriole
3. vasa recta
4. peritubular capillaries
5. interlobular artery
6. arcuate artery
7. interlobar artery
8. segmental artery
9. renal artery
10. renal vein
11. interlobar vein
12. arcuate vein
13. interlobular vein

Activity 4 Discussion Questions: Blood Flow through the Kidney

1. segmental and interlobar arteries and interlobar veins

2. interlobar arteries and veins
3. arcuate arteries and veins
4. interlobular arteries and veins, afferent and efferent arterioles, glomerulus, peritubular capillaries

Figure 36.9

1. proximal and distal convoluted tubules (outer cortex)
2. renal corpuscle
3. loops of Henle and collecting ducts
4. glomerulus
5. parietal layer of glomerular capsule
6. capsular space

Activity 6 Discussion Questions: Histology of the Kidney

1. Filtration
2. PCT needs extra area of brush border for reabsorption

Figure 36.10

1. lamina propria
2. transitional epithelium
3. lumen
4. mucosa
5. muscularis
6. adventitia

Activity 7 Discussion Question: Histology of Ureters, Urinary Bladder, and Urethra

1. Stratified squamous epithelium offers more protection from external environment.

Exercise 38

Figure 38.1

1. cremaster muscle
2. dartos muscle
3. scrotum
4. tunica albuginea of testis

Figure 38.2

1. efferent duct
2. seminiferous tubules
3. straight tubules
4. rete testis
5. tunica albuginea
6. lobule
7. septum

Figure 38.3

1. ductus (vas) deferens
2. epididymis

3. spermatic cord
4. blood vessels and nerves

Figure 38.4

1. ampulla of ductus (vas) deferens
2. ejaculatory duct
3. prostate gland
4. membranous urethra
5. epididymis
6. testis
7. scrotum
8. ductus (vas) deferens (in inguinal canal)
9. prostatic urethra
10. spongy (penile) urethra
11. penis
12. external urethral orifice

Figure 38.5

1. seminal vesicle
2. prostate gland
3. prostatic urethra
4. membranous urethra
5. ductus (vas) deferens
6. ampulla of ductus deferens
7. ejaculatory duct
8. bulbourethral gland
9. spongy (penile) urethra

Figure 38.6

1. seminal vesicle
2. prostate gland
3. bulbourethral gland

Figure 38.7

1. prostate gland
2. prostatic urethra
3. membranous urethra
4. corpus spongiosum penis
5. corpora cavernosa penis
6. spongy (penile) urethra
7. glans penis
8. bulbourethral gland
9. prepuce
10. external urethral orifice
11. corpora cavernosa penis
12. corpus spongiosum penis
13. spongy (penile) urethra

Figure 38.8

1. secretes 60% of semen volume; alkaline pH; prostaglandins
2. secretes 25% of semen volume; increases viability and motility of sperm
3. secretes mucus for lubrication of glans; alkaline pH
4. site of storage and maturation of sperm
5. organ that produces sperm

6. copulatory organ; passageway for urine and sperm
7. carries sperm from epididymis to ejaculatory duct

Figure 38.9

1. spermatogonium
2. spermatozoa (n)
3. spermatid (n)
4. secondary spermatocyte (n)
5. primary spermatocyte (2n)
6. spermatozoa
7. late spermatid
8. early spermatid
9. secondary spermatocyte
10. Sertoli cells
11. primary spermatocyte
12. spermatogonium
13. basement membrane
14. Leydig (interstitial) cells

Activity 5 Discussion Question

1. 4

Figure 38.10

1. lumen
2. tails of spermatozoa
3. basement membrane
4. Leydig (interstitial) cells
5. interstitial space

Figure 38.11

1. acrosome
2. nucleus
3. mitochondria
4. head
5. midpiece
6. tail (flagellum)

Exercise 39

Figure 39.1

1. uterine tube
2. infundibulum
3. ampulla
4. isthmus
5. fimbriae
6. suspensory ligament
7. ovary
8. ovarian ligament
9. broad ligament

Figure 39.2

1. uterine cavity
2. myometrium of uterus
3. vaginal canal
4. ovary
5. round ligament

6. cervix
7. urinary bladder

Figure 39.3

1. uterine cavity
2. endometrium
3. myometrium
4. perimetrium
5. round ligament
6. internal os
7. cervical canal
8. cervix
9. rugae of vagina
10. fundus of uterus
11. broad ligament
12. body of uterus
13. external os
14. vagina

Figure 39.4

1. anus
2. vaginal orifice
3. pubic symphysis
4. mons pubis
5. clitoris
6. external urethral orifice
7. labium minus
8. labium majus

Figure 39.5

1. labia majora
2. labia minora
3. hymen
4. anus
5. mons pubis
6. prepuce
7. clitoris
8. external urethral orifice
9. vaginal orifice

Figure 39.6

1. lobule with alveoli
2. mammary duct
3. lactiferous sinus
4. lactiferous duct
5. adipose tissue
6. areola
7. nipple

Figure 39.7

1. beat together to bring ovulated oocyte into uterine tube
2. transports secondary oocytes toward uterus
3. usual site of fertilization
4. narrow area of uterine tube that opens into uterus
5. produces secondary oocytes produces estrogen and progesterone

6. layer of uterus that sheds during menstruation
implantation of zygote and development of fetus occurs here

Figure 39.8

1. secondary oocyte
2. sperm cell
3. ovum
4. zygote
5. oogonium
6. primary oocyte
7. 1st polar body
8. 2nd polar body
9. meiosis I
10. meiosis II

Figure 39.9

1. primordial follicles
2. primary follicles
3. secondary follicle
4. secondary oocyte
5. Graafian (mature) follicle
6. ovulated secondary oocyte
7. corpus luteum
8. corpus albicans
9. primordial follicles
10. ovarian cortex
11. primary oocyte
12. granulosa cells
13. follicular fluid in antrum
14. granulosa cells
15. primary oocyte

Figure 39.10

1. endometrium
2. endometrial gland
3. stratum functionalis
4. stratum basalis
5. myometrium

Exercise 40

Figure 40.1

1. secondary oocyte
2. fertilization of secondary oocyte
3. 2-cell stage
4. 4-cell stage
5. morula
6. blastocyst, external view
7. implantation of blastocyst, internal view

Figure 40.2

1. yolk sac
2. amnion
3. amnionic cavity
4. bilaminar embryonic disc

Activity 2 Discussion Questions: Implantation, Primary Germ Layers, and Embryonic Membranes

1. The blastocyst adheres to the wall of the uterus at approximately 6 days after fertilization. The trophoblast of the blastocyst secretes enzymes that allow the blastocyst to bury itself in the endometrium.
2. ectoderm: epidermis, glands, nervous tissue, and parts of the eye and ear
mesoderm: cartilage, bones, 3 muscle types, and blood
endoderm: epithelial lining of digestive system organs, respiratory system organs, and other organs
3. amnion: surrounds and protects the embryo
yolk sac: membrane surrounding the former blastocyst cavity, contains nutrients for embryo while pacenta is developing
chorion: surrounds amnion, yolk sac, and allantois and forms the embryonic portion of placenta; chorion also forms connecting stalk which becomes the umbilical cord.
allantois: extension of the yolk sac into connecting stalk. Surrounds blood vessels that will become umbilical arteries and veins

Figure 40.3

1. amniotic cavity
2. embryo
3. yolk sac
4. connecting stalk
5. chorionic villus

Figure 40.4

1. chorionic villi
2. fetal blood vessels
3. decidua basalis
4. chorion
5. umbilical arteries
6. umbilical vein

Activity 3 Discussion Questions: Placenta and Umbilical Cord

1. Example: The placenta surrounds the fetus, supplying nutrients and removing wastes via blood vessels in the umbilical cord.
2. The umbilical cord is formed from the connecting stalk. The chorionic villi form the embryonic portion of the placenta and the decidua basalis forms the maternal side of the placenta.
3. The umbilical vein supplies oxygen and nutrient rich blood to the fetus while the umbilical arteries carry oxygen poor and waste laden blood back to the maternal side of the placenta.

Activity 4 Discussion Questions: Fetal Development

1. upper and lower limbs: upper and lower limb buds are observed in the 32-day embryo and actual limbs are observed in the 44-day embryo
2. future auditory meatus: otic placode observed in 44-day embryo
3. eye: 52-day embryo
4. fingers and toes: 52-day embryo

Exercise 41

Activity 1

1. het = Zz, Tt; hom = bb, MM, cc, AA
2. freckles = FF, Ff; no freckles = ff; attached earlobes = EE, Ee; unattached earlobes = ee
3. homozygous widow's peak = WW; heterozygous widow's peak = Ww; straight hairline = ee
4. homozygous mid-digital hair = MM; heterozygous mid-digital hair = Mm; no mid-digital hair = mm

Figure 41.1

DD = 0; Dd = 4; dd = 0
DD = 0%; Dd = 100%; dd = 0%
dimples = 4; no dimples = 0
dimples = 100%; no dimples = 0%

Figure 41.3

RR = 1; Rr = 2; rr = 1
RR = 25%; Rr = 50%; rr = 25%
Tongue-roller = 3; non-tongue roller = 1
Tongue-roller = 75%; non-tongue roller = 25%

Figure 41.5

MM = 0%; Mm = 50%; mm = 50%
Mid-digital hair = 50%; no mid-digital hair = 50%

Figure 41.6

CC = 0%; Cc = 100%, cc = 0%
Curly hair = 0%; wavy hair = 100%; straight hair = 0%

Figure 41.7

AO = 25%; BO = 25%; AB = 25%; OO = 25%
A = 25%; B = 25%; AB = 25%; O = 25%

Figure 41.8

AO = 50%; BO = 50%; AB = 0%;
OO = 0%
A = 50%; B = 50%; AB = 0%; O = 0%

Figure 41.10

Hb^SHb^S = 50%; Hb^SHb^A = 50%;
Hb^AHb^A = 0%
Sickle-cell anemia = 50%; sickle-cell
trait = 50%; Normal = 0%

Activity 8

1. X^CX^C = normal; X^CX^c = normal;
 X^cX^c = color blind X^CY = normal;
 X^c = color blind

Activity 8 Discussion Questions

1. 50% of the girls in Figure 41.12 would
 be carriers.
2. No boys would be a carrier. There are
 no male carriers for sex-linked
 inheritance.

Figure 41.12

X^CX^C = 25%; X^CX^c = 25%; X^cX^c = 0%;
X^CY = 25%; X^cY = 25%
Female normal = 25%; Female carrier =
25%; female color blind = 0%; male
normal = 25%; male color blind = 25%

Appendix A: Histology Atlas

Plasma membrane

Cytoplasm

Nucleus

FIGURE HA-1 Surface view, Simple Squamous Epithelium (Peritoneum) **LM** 243×

Renal Capsule

Glomerulus

Nucleus of Squamous Cell

FIGURE HA-2 Side view, Simple Squamous Epithelium (Renal Corpuscle) **LM** 117×

Simple Cuboidal cell

Lumen of Tubule

Nucleus

FIGURE HA-3 Simple Cuboidal Epithelium (Kidney Tubules) **LM** 117×

Goblet Cell

Simple Columnar Cell

Nucleus

FIGURE HA-4 Simple Columnar Epithelium (Villus of Small Intestine) **LM** 117×

Goblet
Cell

Nucleus of
Columnar Cell

Cilia

Pseudostratified
Columnar Cell

FIGURE HA-5 Pseudostratified Ciliated Columnar
Epithelium (Trachea) **LM** 341×

Stratified
Nuclei

Duct
Lumen

FIGURE HA-7 Stratified Cuboidal Epithelium (Sweat
Gland) **LM** 403×

Nucleus of Squamous Cell

FIGURE HA-6 Stratified Squamous Epithelium
Nonkeratinized (Vagina) **LM** 50×

Stratified columnar
epithelium

Lumen of duct

Nucleus of stratified
columnar cell

Connective tissue

FIGURE HA-8 Stratified Columnar Epithelium (Salivary
Gland Duct) **LM** 420×

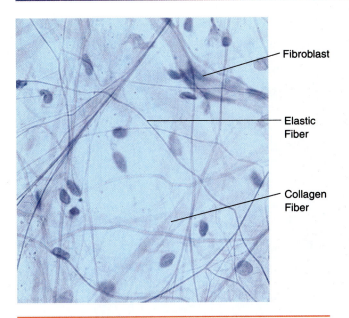

Fibroblast

Elastic Fiber

Collagen Fiber

FIGURE HA-9 Areolar Connective Tissue (Mesentery)
LM 537×

Reticular Fiber

Reticular Cell

FIGURE HA-11 Reticular Connective Tissue (Lymph Node) LM 422×

Nucleus of Adipocyte

Blood Vessel

Lipid Storage Vacuole

Plasma Membrane

FIGURE HA-10 Adipose Tissue (White fat) LM 386×

Nucleus of Fibroblast

Collagen Fiber

FIGURE HA-12 Dense Regular Connective Tissue (Tendon) LM 453×

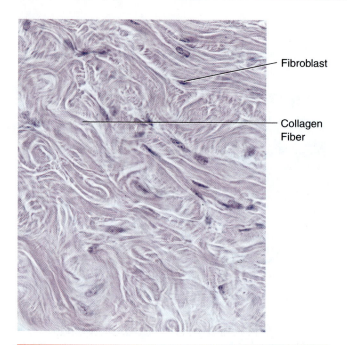

FIGURE HA-13 Dense Irregular Connective Tissue (Dermis) `LM` 294×

FIGURE HA-15 Hyaline Cartilage (Trachea) `LM` 295×

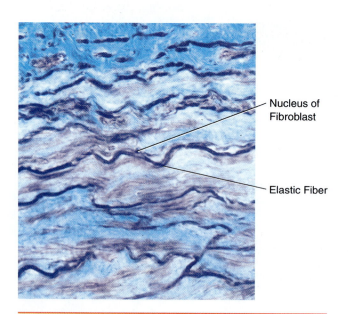

FIGURE HA-14 Elastic Connective Tissue (Aorta) `LM` 448×

FIGURE HA-16 Fibrocartilage (Intervertebral Disc) `LM` 434×

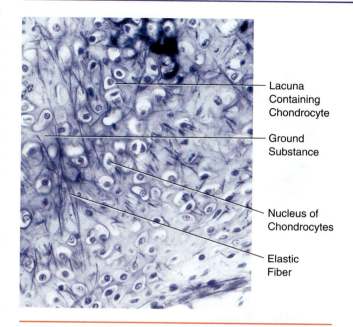

FIGURE HA-17 Elastic Cartilage (Epiglottis)
LM 272×

Lacuna Containing Chondrocyte

Ground Substance

Nucleus of Chondrocytes

Elastic Fiber

FIGURE HA-19 Human Blood (with Neutrophil and Lymphocyte) **LM** 1017×

Erythrocytes

Neutrophil

Lymphocyte

Platelet

FIGURE HA-18 Compact Bone **LM** 92×

Perforating Canal

Central Canal

Lacunae

Lamella

Canaliculi

FIGURE HA-20
Eosinophil **LM** 1500×

FIGURE HA-21 Basophil
LM 1500×

FIGURE HA-22 Lympho-cyte **LM** 1500×

FIGURE HA-23 Mono-cyte **LM** 1500×

Fiber (Cell)

Nucleus

FIGURE HA-24 Visceral Smooth Muscle (LM) 840×

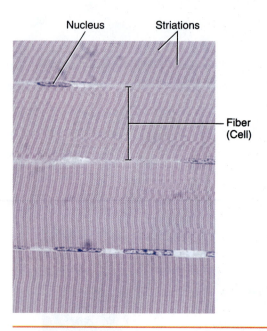

Nucleus Striations

Fiber (Cell)

FIGURE HA-25 Skeletal Muscle (LM) 504×

Nucleus

Fiber (Cell)

Intercalated Disc

Striations

FIGURE HA-26 Cardiac Muscle (LM) 516×

Dendrite

Body

Nucleus

FIGURE HA-27 Multipolar Neuron (Spinal Cord) (LM) 521×

Posterior median sulcus

Posterior white column

Posterior gray horn

Lateral white column

Gray commissure

Anterior gray horn

Anterior white column

Anterior median fissure

FIGURE HA-28 Transverse Section of Spinal Cord (LM) 3.8×

Node of Ranvier

FIGURE HA-29 Nodes of Ranvier (Nerve, Long. Sec.)
(LM) 679×

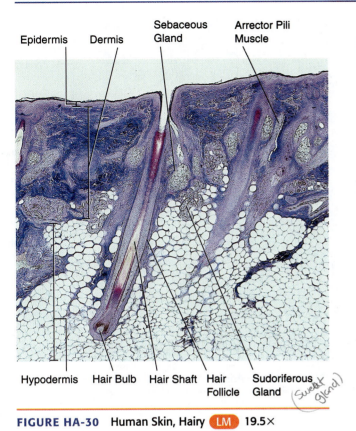

Epidermis Dermis Sebaceous Gland Arrector Pili Muscle

Hypodermis Hair Bulb Hair Shaft Hair Follicle Sudoriferous Gland *(sweat gland)*

FIGURE HA-30 Human Skin, Hairy LM 19.5×

Stratum Corneum
Stratum Lucidum
Stratum Granulosum
Stratum Spinosum
Stratum Basale

FIGURE HA-32 Human Epidermis LM 442×

Epidermis
Dermal Papillae
Dermis
Sudoriferous Gland
Hypodermis

FIGURE HA-31 Human Skin, Hairless LM 30×

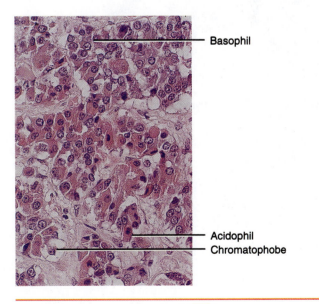

Basophil
Acidophil
Chromatophobe

FIGURE HA-33 Pituitary Gland LM 1000×

Basement membrane

Follicular cell

Thyroid follicle

Thyroglobulin (TGB)

Parafollicular (C) cell

FIGURE HA-34 Thyroid Gland LM 500×

Capsule

Adrenal corte

Zona glome

Zona fascic

Zona reticul

Adrenal med
chromaffin ce

FIGURE HA-36 Adrenal Gland LM 50×

Chief (principal) cell

Oxyphil cell

Blood vessel

FIGURE HA-35 Parathyroid Gland LM 340×

Exocrine acin

Pancreatic islet

Beta cell

Alpha cell

FIGURE HA-37 Pancreas LM 340×

Tunica Externa

Lumen

External Elastic Laminum

Tunica Media

Internal Elastic Laminum

Tunica Interna

FIGURE HA-38 Artery and Vein, cross sections
LM 40×

Capsule

Red pulp

White pulp

Central artery

Trabecula

FIGURE HA-40 Spleen LM 66×

Blood vessels

Capsule

Lobule:
— Cortex

Medulla

Trabecula

FIGURE HA-39 Thymus Gland LM 30×

Capsule
Subcapsular sinus
Trabecula
Trabecular sinus
Outer cortex
Germinal center in secondary lymphatic nodule
Inner cortex

Medullary sinus

Medulla

FIGURE HA-41 Lymph Node LM 55×

Lumen of esophagus

POSTERIOR

Trachealis muscle

Lumen of trachea

Tracheal cartilage
(hyaline)

ANTERIOR

FIGURE HA-42 Transverse Section of the Trachea and
Esophagus LM 2.6×

Lumen of
bronchiole

Smooth muscle

Epithelium
(ciliated simple
columnar)

FIGURE HA-44 Transverse Section, Bronchiole
LM 140×

Lumen of trachea

Mucosa

Epithelium (pseudostratified
ciliated columnar epithelium)

Lamina propria

Submucosa (areolar
connective tissue)

Seromucous gland

Tracheal cartilage
(hyaline)

Perichondrium

Adventitia (areolar
connective tissue)

FIGURE HA-43 Transverse Section, Tracheal Wall LM 80×

Terminal
bronchiole

Blood
vessel

Respiratory
bronchiole

Alveolar
ducts

Alveoli

Alveolar
sacs

Visceral
pleura

FIGURE HA-45 Lung Lobule LM about 24×

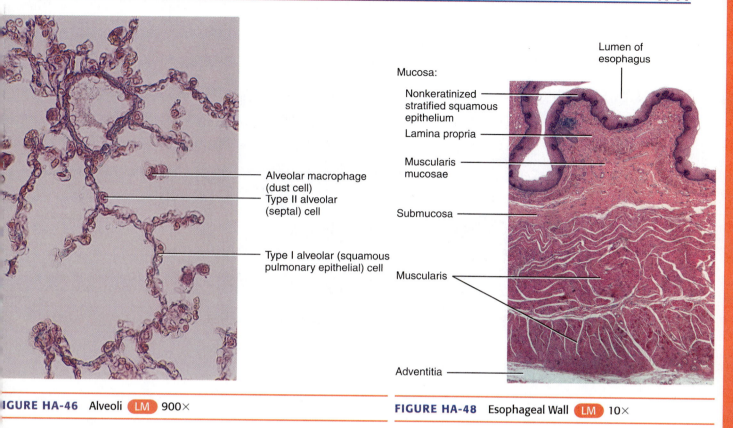

Alveolar macrophage (dust cell)

Type II alveolar (septal) cell

Type I alveolar (squamous pulmonary epithelial) cell

Lumen of esophagus

Mucosa:
Nonkeratinized stratified squamous epithelium
Lamina propria
Muscularis mucosae
Submucosa
Muscularis
Adventitia

FIGURE HA-46 Alveoli LM 900×

FIGURE HA-48 Esophageal Wall LM 10×

Gastric pit

Lumen of stomach
Surface mucous cell
Lamina propria
Parietal (oxyntic) cell
Chief (zymogenic) cell

Mucous acini

Serous acini

FIGURE HA-47 Submandibular Gland LM 350×

FIGURE HA-49 Fundic Mucosa of Stomach LM 250×

Hepatocytes
(liver cells)

Central vein
of liver lobule

Sinusoid

FIGURE HA-50 Liver Lobule LM 150×

Villi

Lumen of duodenum

Brush border

Simple columnar
epithelium

Goblet cell

Absorptive cell

Lamina propria

Intestinal glands (crypts
of Lieberkühn)

Muscularis mucosae

Duodenal (Brunner's)
gland in submucosa

FIGURE HA-52 Villi from the Duodenum LM 160×

Lumen of duodenum

Villi

Mucosa

Intestinal gland (crypt
of Lieberkühn)

Muscularis mucosae

Duodenal (Brunner's)
gland

Submucosa

Muscularis

FIGURE HA-51 Duodenum LM 27×

Lumen of large intestine

Lamina propria

Intestinal gland

Mucosa

Muscularis mucosae

Lymphatic nodule

Submucosa

Muscularis

Serosa

FIGURE HA-53 Large Intestine LM 252×

Opening of intestinal gland

Lumen

Goblet cell

Intestinal gland

Lamina propria

FIGURE HA-54 Mucosa of the Large Intestine
LM 300×

Adventitia

Lumen of ureter

Mucosa

Blood vessel

Muscularis

Adipose tissue

FIGURE HA-56 Transverse Section of Ureter LM 7.5×

Lumen of urinary bladder

Mucosa:
Transitional epithelium

Lamina propria

Muscularis:
Inner longitudinal

Middle circular

Outer longitudinal

FIGURE HA-57 Urinary Bladder LM 65×

Glomerulus

Podocytes of visceral layer of glomerular capsule

Capsular space

Simple squamous epithelial cells

Glomerular capsule:
Parietal layer
Visceral layer

Afferent arteriole
Juxtaglomerular cell

Ascending limb of loop of Henle

Macula densa cell

Efferent arteriole
Proximal convoluted tubule

FIGURE HA-55 Renal Corpuscle LM 690×

Leydig cell (interstitial endocrinocyte)

Basement membrane

Spermatogonium (*2n*) (stem cell)

Primary spermatocyte (*2n*)

Secondary spermatocyte (*n*)

Spermatid (*n*)

Sperm cell or spermatozoan (*n*)

Lumen of seminiferous tubule

FIGURE HA-58 Seminiferous Tubules LM 310×

Antrum filled with follicular fluid

Corona radiata

Secondary (antral) follicle granulosa

Theca folliculi

Zona pellucida

Primary oocyte

FIGURE HA-60 Secondary Follicle LM 60×

Primordial follicles

Germinal epithelium

Tunica albuginea

Ovarian cortex

Theca folliculi

Zona pellucida

Primary oocyte

Primary (preantral) follicle granulosa cells

Primary follicle

FIGURE HA-59 Ovarian Cortex LM about 200×

Lumen of uterus

Simple columnar epithelium

Endometrial stroma

Endometrial gland

Endometrium:
Stratum functionalis

Stratum basalis

Myometrium

FIGURE HA-61 Uterus LM 24×

Appendix B: Word Roots

WORD ROOT	DEFINITION	NOUNS AND/OR ADJECTIVES FORMED FROM WORD ROOT AND A SUFFIX	WORDS FORMED BY COMBINING WORD ROOT WITH ANOTHER WORD ROOT, PREFIX AND/OR SUFFIX
abdomin-	abdomen	**Noun:** abdominis **Adjective:** abdominal	**abdomin**opelvic
acromi-	high point of shoulder	**Noun:** acromion (sing.) **Adjective:** acromial	**acromi**oscapular
axill-	armpit	**Noun:** axilla (sing.) and axillae (pl.) **Adjective:** axillary	**axill**ofemoral
brachi-	arm	**Noun:** brachium (sing.) and brachia (pl.) **Adjective:** brachial	**brachi**ocephalic, ante**brachi**um
bucc-	cheek	**Noun:** bucca (sing.), buccae (pl.) **Adjective:** buccal	**bucc**inator
calcane-	heel	**Noun:** calcaneus (sing.) and calcanei (pl.) **Adjective:** calcaneal	**calcane**ofibular
carp-	wrist bones	**Noun:** carpus (sing.) and carpi (pl.) **Adjective:** carpal	**carp**ectomy
cervic-	neck or neck-like structure	**Noun:** cervix (sing.) and cervices (plural) **Adjective:** cervical	**cervic**othorax, **cervic**ectomy, **cervic**itis
cox-	hip	**Noun:** coxa (sing.) and coxae (pl.) **Adjective:** coxal	**cox**algia
crani-	cranium	**Noun:** cranium (sing.), crania (pl.) **Adjective:** cranial	**crani**osacral, **crani**ofacial
cubit-	elbow or ulna	**Noun:** cubitus (sing.), cubiti (pl.) **Adjective:** cubital	ante**cubit**al
digit-	finger or toe	**Noun:** digit and digitus (sing.) and digiti (pl.) **Adjective:** digital	**digit**ation
dors-	back	**Noun:** dorsum (sing.) and dorsa (pl.) **Adjective:** dorsal	**dors**iflexion
faci-	face	**Noun:** facies (sing.) **Adjective:** facial	**faci**oplasty
fibul-	lateral leg	**Noun:** fibula **Adjective:** fibular	**fibul**ocalcaneal
front-	forehead	**Noun:** frons (sing.) **Adjective:** frontal	**front**onasal
glute-	buttocks	**Noun:** gluteus **Adjective:** gluteal	**glute**ofemoral
lumb-	lower back or loin	**Noun:** lumbus (sing.) and lumbi (pl.) **Adjective:** lumbar	**lumb**ago, **lumb**osacral
ment-	mind or chin	**Noun:** mentum (sing.) **Adjective:** mental	**ment**alis muscle, **ment**al foramen

WORD ROOT	DEFINITION	NOUNS AND/OR ADJECTIVES FORMED FROM WORD ROOT AND A SUFFIX	WORDS FORMED BY COMBINING WORD ROOT WITH ANOTHER WORD ROOT, PREFIX AND/OR SUFFIX
nas-	nose	**Noun:** nasus (sing.) **Adjective:** nasal	**nas**opharynx, **nas**osinusitis
occipit-	back of head	**Noun:** occipitis, occiput (sing.) **Adjective:** occipital	**occipit**ofrontal, **occipit**omental
olecran-	proximal projection of ulna	**Noun:** olecranon (sing.) and olecrana (pl.) **Adjective:** olecranal	**olecran**arthropathy
or-	mouth	**Noun:** oris (sing.) **Adjective:** oral	**or**opharynx
orbit-	eye	**Noun:** orbit (sing), orbitae (pl.) **Adjective:** orbital	**orbit**onasal
ot-	ear	**Adjective:** otic	**ot**itis, **ot**olaryngology
palm -	flexor surface of hand	**Noun:** palm and palma (sing.) and palmae (pl.) **Adjective:** palmar	**palm**ature
patell-	kneecap	**Noun:** patella (sing.) and patellae (pl.) **Adjective:** patellar	**patell**ectomy
pector-	breast or chest	**Noun:** pectus (sing.) and pectora (pl.) **Adjective:** pectoral	**pect**oralgia
ped-	foot or child	**Noun:** pes (sing.) and pedes (pl.) **Adjective:** pedal	**ped**icle, **ped**ometer
pelvi-	pelvis, basin	**Noun:** pelvis (sing.) and pelves (pl.) **Adjective:** pelvic	**pelvi**operitonitis, abdomino**pelvic**
plant-	sole of foot	**Noun:** planta (sing.) and plantae (pl.) **Adjective:** plantar	**plant**arflexion, plantaris
pub-	pubis or genital area	**Noun:** pubis (sing.) **Adjective:** pubic	**pub**erty
scapul-	scapula (shoulder bone)	**Noun:** scapula (sing.) and scapulae (pl.) **Adjective:** scapular	**scapul**ectomy
stern-	chest	**Noun:** sternum (sing.) **Adjective:** sternal	**stern**ocleidomastoid
tars-	ankle bones or edge of eyelid	**Noun:** tarsus (sing.) and tarsi (pl.) **Adjective:** tarsal	**tars**ometatarsal, tibio**tarsal**
thorac-	chest	**Noun:** thorax (sing.) and thoraces (pl.) **Adjective:** thoracic	**thorac**olumbar
umbilic-	navel	**Noun:** umbilicus (sing.) and umbilici (pl.) **Adjective:** umbilical	**umbilic**ation
vertebr-	vertebra, spinal column	**Noun:** vertebra (sing.) and vertebrae (pl.) **Adjective:** vertebral	**vertebr**oarterial

Appendix C: Skeletal Muscle Origins and Insertions

TABLE C1 Muscles of the Head and Neck

MUSCLE	ORIGIN	INSERTION
MUSCLES OF FACIAL EXPRESSION		
Occipitofrontalis muscle		
Frontalis	Galea (epicranial) aponeurosis	Skin superior to supraorbital margin
Occipitalis	Occipital bone and mastoid process of temporal bone	Galea (epicranial) aponeurosis
Orbicularis oculi	Medial wall of orbit	Skin around margin of orbit
Zygomaticus minor	Zygomatic bone	Upper lip
Zygomaticus major	Zygomatic bone	Skin at angle of mouth and orbicularis oris
Orbicularis oris	Muscle fibers surrounding mouth	Skin at corner of mouth
Platysma	Fascia covering deltoid and pectoralis major	Mandible, muscles around angle of mouth, and skin of lower face
MUSCLES OF MASTICATION		
Superficial Muscles		
Temporalis	Temporal bone	Coronoid process and ramus of mandible
Masseter	Maxilla and zygomatic arch	Ramus and angle of mandible
Deep Muscles		
Buccinator	Alveolar processes of maxilla and mandible	Orbicularis oris
Lateral pterygoid	Greater wing of sphenoid and lateral part of pterygoid process of sphenoid bone	Mandibular condyle; temporomandibular joint (TMJ)
Medial pterygoid	Medial part of pterygoid process of sphenoid bone and maxilla bone	Ramus and angle of mandible
MUSCLES THAT MOVE THE HEAD AND NECK		
Anterior and Lateral Muscles		
Sternocleidomastoid	Sternum and clavicle	Mastoid process of temporal bone
Scalenes	Transverse processes of vertebrae C_3 through C_7	Ribs 1 and 2
Posterior Muscles		
Splenius capititis	Ligamentum nuchae and spinous processes of vertebrae C_7 through T_4	Occipital bone and mastoid process of temporal bone
Trapezius (superior portion)	Occipital bone, ligamentum nuchae, and spines of vertebrae C_7 through T_{12}	Clavicle, acromion (process) of scapula, and spine of scapula

TABLE C1 (Continued)

MUSCLE	ORIGIN	INSERTION
MUSCLE THAT ONLY MOVES THE SCAPULA		
Levator scapulae	Vertebrae C_1–C_5	Superior vertebral border of scapula
MUSCLES THAT MOVE HYOID BONE		
Suprahyoid muscles		
Digastric		
Anterior belly	Inner surface of inferior border of mandible	Body of hyoid bone
Posterior belly	Mastoid process of temporal bone	Body of hyoid bone
Stylohyoid	Styloid process of temporal bone	Body of hyoid bone
Mylohyoid	Inner surface of mandible	Body of hyoid bone
Infrahyoid Muscles		
Omohyoid	Superior border of scapula	Body of hyoid bone
Sternohyoid	Manubrium and medial end of clavicle	Body of hyoid bone

TABLE C2 Muscles of the Anterior Trunk

MUSCLE	ORIGIN	INSERTION
MUSCLES THAT MOVE THE ARM AT THE SHOULDER JOINT		
Superficial muscles		
Deltoid	**Anterior fibers:** acromial end of clavicle **Lateral fibers:** acromion of scapula **Posterior fibers:** spine of scapula	Deltoid tuberosity of humerus
Pectoralis major	Clavicle (clavicular head), sternum, and costal cartilages of 2nd–6th ribs (sternocostal head)	Greater tubercle and intertubecular sulcus of humerus
Deep muscle		
Subscapularis	Subscapular fossa of scapula	Lesser tubercle of humerus
MUSCLES THAT MOVE THE SCAPULA		
Serratus anterior	Ribs 1–9	Vertebral border of scapula and inferior angle of scapula
Pectoralis minor	Ribs 2–5, 3–5, or 2–4	Coracoid process of scapula
MUSCLES THAT MOVE THE ABDOMINAL WALL		
Superficial muscles		
Rectus abdominis	Pubic crest and pubic symphysis	Cartilage of ribs 5–7 and the xiphoid process
External oblique	Ribs 5–12	Iliac crest and linea alba
Deep muscles		
Internal oblique	Iliac crest, inguinal ligament, and thoracolumbar fascia	Cartilage of ribs 9–12 and linea alba
Transversus abdominis	Iliac crest, inguinal ligament, lumbar fascia, and cartilage of ribs 7–12	Xiphoid process, linea alba, and pubis
MUSCLES USED IN BREATHING		
Deep muscles		
External intercostals	Inferior border of rib superior to muscle	Superior border of rib inferior to muscle
Internal intercostals	Superior border of rib inferior to muscle	Inferior border of rib superior to muscle
Diaphragm	Xiphoid process of sternum, costal cartilages of inferior six ribs, and lumbar vertebrae	Central tendon

TABLE C3 Muscles of the Posterior Trunk

MUSCLE	ORIGIN	INSERTION
MUSCLES THAT MOVE THE ARM AT THE SHOULDER JOINT		
Superficial muscles		
Deltoid	**Posterior fibers:** spine of scapula	Deltoid tuberosity of humerus
Infraspinatus	Infraspinous fossa of scapula	Greater tubercle of humerus
Teres minor	Inferior lateral border of scapula	Greater tubercle of humerus
Teres major	Inferior angle of scapula	Intertubercular sulcus of humerus
Latissimus dorsi	Spines of T_7–T_{12} vertebrae, lumbar vertebrae, crests of sacrum and ilium, ribs 9–12	Intertubercular sulcus of humerus
Deep muscles		
Supraspinatus	Supraspinous fossa of scapula	Greater tubercle of humerus
MUSCLES THAT MOVE THE SCAPULA		
Superficial muscle		
Trapezius	Occipital bone, ligamentum nuchae, and spines of vertebrae C_7–T_{12}	Clavicle, acromion of scapula, and spine of scapula
Deep muscles		
Levator scapulae	Vertebrae C_1–C_5	Superior vertebral border of scapula
Rhomboideus minor	Spines of vertebrae C_7 and T_1	Vertebral border of scapula superior to spine
Rhomboideus major	Spines of vertebrae T_2–T_5	Vertebral border of scapula inferior to spine
MUSCLES THAT MOVE THE VERTEBRAL COLUMN		
Erector spinae consists of iliocostalis, longissiumus, and spinalis groups	Large group of muscles with origins on ribs, iliac crest, and cervical, thoracic, and lumbar vertebrae	Skull, ribs, and cervical, thoracic, and lumbar vertebrae

TABLE C4 Muscles of the Arm

MUSCLE	ORIGIN	INSERTION
ANTERIOR SURFACE		
Biceps brachii	**Long head:** tubercle superior to glenoid cavity of scapula	Radial tuberosity of radius
	Short head: coracoid process of scapula	
Brachialis	Anterior surface of distal humerus	Ulnar tuberosity and coronoid process of ulna
POSTERIOR SURFACE		
Triceps brachii	**Long head:** inferior to glenoid cavity of scapula	All three heads insert on the olecranon of ulna
	Medial head: posterior humerus inferior to groove for radial nerve	
	Lateral head: posterior surface of lateral humerus superior to radial groove	

TABLE C5 Muscles of the Forearm

MUSCLE	ORIGIN	INSERTION
ANTERIOR SURFACE		
Superficial muscles (lateral to medial)		
Brachioradialis	Distal lateral border of humerus	Superior to styloid process of radius
Pronator teres	Medial epicondyle of humerus and coronoid process of ulna	Midlateral surface of radius
Flexor carpi radialis	Medial epicondyle of humerus	2nd and 3rd metacarpals
Palmaris longus	Medial epicondyle of humerus	Flexor retinaculum and palmar aponeurosis
Flexor carpi ulnaris	Medial epicondyle of humerus and superior posterior border of humerus	Pisiform, hamate, and base of 5th metacarpal
Flexor digitorum superficialis	Medial epicondyle of humerus, coronoid process of ulna, and anterior surface of radius	Middle phalanx of each finger
Deep muscles		
Supinator	Lateral epicondyle of humerus and ridge near radial notch of ulna (supinator crest)	Lateral surface of proximal one-third of radius
Flexor pollicis longus	Anterior surface of radius and interosseous membrane	Base of distal phalanx of thumb
POSTERIOR SURFACE		
Superficial muscles (medial to lateral)		
Extensor carpi ulnaris	Lateral epicondyle of humerus and posterior ulna	5th metacarpal
Extensor digitorum	Lateral epicondyle of humerus	distal and middle phalanges of each finger
Extensor carpi radialis longus and brevis	**Brevis:** lateral epicondyle of humerus	3rd metacarpal
	Longus: lateral supracondylar ridge of humerus	2nd metacarpal
POSTERIOR SURFACE		
Deep muscles		
Supinator	Lateral epicondyle of humerus and ridge near radial notch of ulna	Lateral surface of proximal one-third of radius
Extensor pollicis longus	Posterior surface of middle of ulna, and interosseous membrane	Distal phalanx of thumb

TABLE C6 **Muscles of the Thigh**

MUSCLE	ORIGIN	INSERTION
ANTERIOR SURFACE		
(*Lateral to medial*)		
Tensor fasciae latae	Iliac crest	Iliotibial tract (fascia)
Quadriceps femoris		Patella (quadriceps tendon) and tibial tuberosity (patellar ligament)
Rectus femoris	Anterior inferior iliac spine	
Vastus lateralis	Greater trochanter and linea aspera of femur	
Vastus medialis	Linea aspera of femur	
Vastus intermedius	Anterior and lateral surface of body of femur	
Sartorius	Anterior superior iliac spine	Medial surface of body of tibia
Iliopsoas		
Psoas major	Psoas major: Transverse processes and bodies of lumbar vertebrae	Psoas major and iliacus insert together into lesser trochanter of femur
Iliacus	Iliacus: Iliac fossa	
Pectineus	Superior ramus of pubis	Pectineal line of femur between lesser trochanter and linea aspera
Adductor longus	Pubic crest and pubic symphysis	Linea aspera of femur
Adductor magnus	Inferior ramus of pubis and ischium to ischial tuberosity	Linea aspera of femur
Adductor brevis	Inferior ramus of pubis	Proximal linea aspera of femur
Gracilis	Pubic symphysis and pubic arch	Medial surface of body of tibia
POSTERIOR SURFACE		
Gluteus maximus	Iliac crest, sacrum, and coccyx	Greater trocanter of femur and iliotibial tract
Gluteus medius	Ilium	Greater trochanter of femur
Gluteus minimus	Ilium	Greater trochanter of femur
Piriformis	Anterior sacrum	Superior border of greater trochanter of femur
Hamstrings		
Biceps femoris	**Long head:** ischial tuberosity **Short head:** linea aspera of femur	Head of fibula and lateral condyle of tibia
Semitendinosus	Ischial tuberosity	Medial surface of proximal part of tibia
Semimembranosus	Ischial tuberosity	Medial condyle of tibia

TABLE C7 Muscles of the Leg and Foot

MUSCLE	ORIGIN	INSERTION
ANTERIOR AND LATERAL SURFACE		
Superficial muscles		
Tibialis anterior	Lateral condyle and body of tibia and interosseous membrane	1^{st} metatarsal and 1^{st} (medial) cuneiform (tarsal bone)
Extensor digitorum longus	Lateral condyle of tibia, anterior fibula, and interosseous membrane	Middle and distal phalanges 2–5
Fibularis (peroneus) longus	Lateral condyle of tibia and head and body of fibula	1^{st} metatarsal and 1^{st} (medial) cuneiform
POSTERIOR SURFACE		
Superficial muscles		
Gastrocnemius	Lateral and medial condyles of femur	Calcaneus via calcaneal tendon
Soleus	Medial tibia and head of fibula	Calcaneus via calcaneal tendon
POSTERIOR SURFACE		
Deep muscles		
Flexor digitorum longus	Posterior tibia	Distal phalanges 2–5
Flexor hallucis longus	Inferior fibula	Distal phalanx of great toe

Photo Credits

Exercise 1 Figure 1.6: Mark Nielsen. Figure 1.8a: Royal Victoria Infirmary, Newcastle Upon Tyne/Photo Researchers. Figure 1.8b: Alfred Pasieka/Photo Researchers. Figure 1.8c: Mark Nielsen.

Exercise 2 Figure 2.7a,d: CNRI/Phototake. Figure 2.7b: Benedet/Phototake. Figure 2.7c: Geoff Tompkinson/Photo Researchers.

Exercise 3 Figure 3.1: Courtesy Olympus America, Inc. Figure 3.2a: Bruce Iverson/Photo Researchers.

Exercise 4 Figure 4.2a,b: Courtesy Michael Ross, University of Florida. Figure 4.2c: Carolina Biological/Visuals Unlimited. Figure 4.2d,e: M. Abbey/Photo Researchers. Figure 4.3: Courtesy Michael Ross, University of Florida.

Exercise 5 Figure 5.2: David Phillips/Photo Researchers.

Exercise 6 Figure 6.3 to 6.25: Courtesy Michael Ross, University of Florida. Figure 6.27: Jan Leesma/Custom Medical Stock Photo, Inc. Figure 6.28 #12: Biophoto Associates/Photo Researchers. Figure 6.28 #13 to #15: Ed Reschke. Figure 6.29 #1.3: Biophoto Associates/Photo Researchers. Figure 6.29 #4, 5: Ed Reschke. Figure 6.29 #6: Courtesy Michael Ross, University of Florida. Figure 6.29 #7: Courtesy Andrew J. Kuntzman. Figure 6.29 #8: Ed Reschke. Figure 6.29 #9: Biophoto Associates/Photo Researchers. Figure 6.29 #10: Ed Reschke. Figure 6.29 #11: Biophoto Associates/Photo Researchers. Figure 6.29 #12: Ed Reschke. Figure 6.20b: Courtesy Michael Ross, University of Florida. Figure 6.29 #13 to #16: Biophoto Associates/Photo Researchers.

Figure 6.29 #17: Courtesy Michael Ross, University of Florida. Figure 6.29 #18: Biophoto Associates/Photo Researchers. Figure 6.30 #1, 2: Courtesy Michael Ross, University of Florida. Figure 6.30 #3: Ed Reshke.

Exercise 7 All Photos: Courtesy Michael Ross, University of Florida.

Exercise 8 Figure 8.3b: Ed Reschke. Figure 8.5: Don W. Fawcett/Visuals Unlimited. Figure 8.6: L. Bassett/Visuals Unlimited. Figure 8.7a: Mehau Kulyk/Photo Researchers. Figure 8.7b: Scott Camazine & Sue Trainor/Photo Researchers. Figure 8.8: P. Motta/Photo Researchers.

Exercise 9 Figure 9.2 to 9.24: Mark Nielsen. Figure 9.29a: Dept. of Clinical Radiology, Salisbury District Hospital/Photo Researchers. Figure 9.29b: AFIP/Science Source/Photo Researchers. Figure 9.30: Courtesy Chaim J. Margolin.

Exercise 10 Figure 10.1 to 10.6: Mark Nielsen. Figure 10.7: John Wilson White. Figure 10.8 and 10.9a: Mark Nielsen. Figure 10.10a: Larry Mulvehill/Photo Researchers. Figure 10.10b: SIU/Visuals Unlimited. Figure 10.10c: Mehau Kulyk/Photo Researchers. Figure 10.11a: Jim Erickson/Corbis Stock Market. Figure 10.11b: Dick Luria/Photo Researchers.

Exercise 11 Figure 11.2: Biophoto Associates/Photo Researchers. Figure 11.7 to 11.11: John Wilson White. Figure 11.12: Mark Nielsen.

Exercise 12 Figure 12.2: Courtesy Michael Ross, University of Florida. Figure 12.5: Courtesy Denah Appelt and Clara Franzini-Armstrong. Figure 12.7: Courtesy Michael Ross, University of Florida.

Exercise 13 Figure 13.1: Biophoto Associates/Photo Researchers.

Exercise 16 Figure 16.2a: Courtesy Michael Ross, University of Florida. Figure 16.2b: Jan Leesma/Custom Medical Stock Photo, Inc. Figure 16.4a: Courtesy Michael Ross, University of Florida. Figure 16.4b: Science VU/Visuals Unlimited. Figure 16.7: Courtesy Michael Ross, University of Florida. Figure 16.8a: Glauberman/Photo Researchers. Figure 16.8b: John D. Cunningham/Visuals Unlimited.

Exercise 17 Figure 17.5: CABISCO/Visuals Unlimited. Figure 17.6 and 17.7: Mark Nielsen.

Exercise 18 Figure 18.2: Courtesy Michael Ross, University of Florida.

Exercise 20 Figure 20.1 #1: ISM/Phototake. Figure 20.2 to 20.5a: Mark Nielsen. Figure 20.5b #2: N. Gluhbegovic and T. H. Williams, *The Human Brain: A Photographic Guide,* Harper & Row, Publishers, Inc. Hagerstown, MD, 1980. Figure 20.6a #2: CNRI/Phototake. Figure 20.10 to 20.13: Mark Nielsen. Figure 20.17: From Stephen A. Kieffer and E. Robert Heitzman, *An Atlas of Cross-Sectional Anatomy,* Harper & Row, Publishers, Inc. Figure 20.18 to 20.20: Mark Nielsen.

Exercise 21 Figure 21.1 #2: Ralph Hutchings/©imagingbody.com.

Exercise 23 Figure 23.5: Courtesy William Radke, University of Central Oklahoma. Figure 23.7: Scott Camazine/Photo Researchers. Figure 23.8 to 23.18: Courtesy Michael Ross, University of Florida. Figure 23.19: Science VU-NEINIH/Visuals Unlimited.

Exercise 25 Figure 25.2 to 25.6: Courtesy Michael Ross, University of

Florida. Figure 25.7: From *The Color Atlas of Anatomy,* 3e., by J. W. Rohen and C. Yokochi. Reprinted by permission of Lippincott Wiliams and Wilkins.

Exercise 26 Figure 26.2: From Lennart Nilsson, *Our Body Victorious,* Boehringer Ingelheim International GmbH. Reproduced with permission. Figure 26.3a #1 to 26.3c #3: Courtesy Michael Ross, University of Florida. Figure 26.3c #4: John D. Cunningham/Visuals Unlimited. Figure 26.3d #1 to #3: Courtesy Michael Ross, University of Florida. Figure 26.3d #4: John D. Cunningham/Visuals Unlimited. Figure 26.3e #1, #2: Courtesy Michael Ross, University of Florida. Figure 26.3e #3: John D. Cunningham/Visuals Unlimited. Figure 26.7: James Hayden/ Phototake. Figure 26.6: JC REVY/ ISM/Phototake.

Exercise 27 Figure 27.8a,b: Mark Nielsen. Figure 27.8c: Courtesy William Radke, University of Central Oklahoma. Figure 27.9: Ed Reshke.

Exercise 29 Figure 29.1: From Phelps, P. C., and J. H. Luft. Am. J. Anat. 125:399, 1969. Figure 29.4: Carolina Biological/Visuals Unlimited. Figure 29.5: T. Kuwabara-D. Fawcett/ Visuals Unlimited. Figure 29.6a: Don Fawcett/Visuals Unlimited. Figure 29.6b: Biophoto Associates/Photo Researchers.

Exercise 30 Figure 30.16: CNRI/ Photo Researchers.

Exercise 31 Figure 31.3d: Courtesy Michael Ross, University of Florida. Figure 31.4b: Leroy, Biocosmos/ Photo Researchers. Figure 31.4c: Courtesy Michael Ross, University of Florida. Figure 31.5b: Biophoto Associates/Photo Researchers. Figure 31.6b: Arnold Brody/Photo Researchers.

Exercise 32 Figure 32.9a: John Cunningham/Visuals Unlimited. Figure 32.9b: Biophoto Associates/ Photo Researchers. Figure 32.10: CNRI/Science Photo Library/Photo Researchers. Figure 32.11: CNRI/ Phototake. Figure 32.12a: Ed Reschke/Peter Arnold, Inc. Figure 32.12b: Martin Rotker/Phototake. Figure 32.12c: Ida Wyman/Phototake.

Exercise 33 Figure 33.2: ©SIU/ Visuals Unlimited.

Exercise 34 Figure 34.10a: Courtesy Michael Ross, University of Florida. Figure 34.10b: Ed Reschke. Figure 34.10c: Willis/Biological Photo Service. Figure 34.10d,e: Courtesy Michael Ross, University of Florida. Figure 34.12a: Alfred Pasieka/Photo Researchers. Figure 34.12b: CNRI/Science Photo Library/Photo Researchers.

Exercise 36 Figure 36.8: Courtesy Dr. William Benyak. Figure 36.9 and 36.10: Courtesy Michael Ross, University of Florida. Figure 36.11: Salisbury District Hospital/Photo Researchers.

Exercise 38 Figure 38.10: Ed Reschke.

Exercise 39 Figure 39.9: Biophoto Associates/Photo Researchers. Figure 39.10: Andrew J. Kuntzman.

Exercise 40 Figure 40.6a–c: Courtesy National Museum of Health and Medicine, Armed Forces Institute of Pathology. Figure 40.6d: Photo by Lennart Nilsson/Albert Bonniers Forlag AB, *A Child is Born,* Dell Publishing Company. Reproduced with permission. Figure 40.6e, f: Courtesy Kohei Shiota, Congenital Anomaly Research Center, Kyoto University, Graduate School of Medicine.

Exercise 41 Figure 41.1a: A. Lowrey/Custom Medical Stock Photo, Inc. Figure 41.1b: David Young.Wolff/ PhotoEdit. Figure 41.3a: Richard Wehr/Custom Medical Stock Photo, Inc. Figure 41.3b: Schleichkorn/Custom Medical Stock Photo, Inc. Figure 41.8: Dr. Stanley Flegler/Visuals Unlimited. Figure 41.13a: Custom Medical Stock Photo. Figure 41.13b: Elyse Rieder.

Histology Figure HA.1: Biophoto Associates/Photo Researchers. Figure HA.2 to HA.7: Courtesy Victor Eroschenko, University of Idaho. Figure HA.8: Biophoto Associates/ Photo Researchers. Figure HA.9– HA.19: Courtesy Victor Eroschenko, University of Idaho. Figure HA.20 and HA.21: Biophoto Associates/ Photo Researchers. Figure HA.22: Ed Reschke. Figure HA.23: Lester Bergman/Project Masters, Inc. Figure HA.24: Ed Reschke. Figure HA.25–HA.27: Courtesy Victor Eroschenko, University of Idaho. Figure HA.28: Jean Claude Revy/ Phototake. Figure HA.29: Courtesy Victor Eroschenko, University of Idaho. Figure HA.30: ©CABISCO/ Visuals Unlimited. Figure HA.31: Astrid & Hanns.Frieder Michler/Photo Researchers. Figure HA.32: Lester Bergaman/Project Masters, Inc. HA.33: Dennis Strete. Figure HA.34 to HA.36: Courtesy Michael Ross, University of Florida. Figure HA.37: CNRI/Phototake. Figure HA.38: ©CABISCO/Visuals Unlimited. HA.39–HA.41: Courtesy Michael Ross, University of Florida. Figure HA.42: John D. Cunningham/ Visuals Unlimited. Figure HA.43: Courtesy Andrew J. Kuntzman. Figure HA.44: Courtesy Michael Ross, University of Florida. Figure HA.45 and HA.46: Biological Photo Service/Photo Researchers. Figure HA.47 to HA.48: Courtesy Michael Ross, University of Florida. Figure HA.49: Ed Reschke. Figure HA.50: Courtesy Michael Ross, University of Florida. Figure HA.51: Fred E. Hossler/Visuals Unlimited. Figure HA.52: G. W. Willis/Visuals Unlimited. Figure HA.53: Roland Birke/Peter Arnold, Inc. Figure HA.54: Courtesy Michael Ross, University of Florida. Figure HA.55: Dennis Strete. Figure HA.56 and HA.57: Biophoto Associates/Photo Researchers. Figure HA.58: Ed Reschke. Figure HA.59–HA.60: Biophoto Associates/Photo Researchers. Figure HA.61: John D. Cunningham/Visuals Unlimited.

Index